Lecture Notes in Networks and Systems 837

The series "Lecture Notes in Networks and Systems" publishes the latest developments in Networks and Systems—quickly, informally and with high quality. Original research reported in proceedings and post-proceedings represents the core of LNNS.

Volumes published in LNNS embrace all aspects and subfields of, as well as new challenges in, Networks and Systems.

The series contains proceedings and edited volumes in systems and networks, spanning the areas of Cyber-Physical Systems, Autonomous Systems, Sensor Networks, Control Systems, Energy Systems, Automotive Systems, Biological Systems, Vehicular Networking and Connected Vehicles, Aerospace Systems, Automation, Manufacturing, Smart Grids, Nonlinear Systems, Power Systems, Robotics, Social Systems, Economic Systems and other. Of particular value to both the contributors and the readership are the short publication timeframe and the world-wide distribution and exposure which enable both a wide and rapid dissemination of research output.

The series covers the theory, applications, and perspectives on the state of the art and future developments relevant to systems and networks, decision making, control, complex processes and related areas, as embedded in the fields of interdisciplinary and applied sciences, engineering, computer science, physics, economics, social, and life sciences, as well as the paradigms and methodologies behind them.

Indexed by SCOPUS, INSPEC, WTI Frankfurt eG, zbMATH, SCImago.

All books published in the series are submitted for consideration in Web of Science.

For proposals from Asia please contact Aninda Bose (aninda.bose@springer.com).

Yousef Farhaoui · Amir Hussain · Tanzila Saba ·
Hamed Taherdoost · Anshul Verma
Editors

Artificial Intelligence, Data Science and Applications

ICAISE'2023
Volume 1

Springer

Editors
Yousef Farhaoui
Department of Computer Science
Moulay Ismail University
Errachidia, Morocco

Amir Hussain
Centre of AI and Robotics
Napier University
Edinburgh, UK

Tanzila Saba
Prince Sultan University
Riyadh, Saudi Arabia

Hamed Taherdoost
University Canada West
Vancouver, BC, Canada

Anshul Verma
Institute of Science
Banaras Hindu University
Varanasi, India

ISSN 2367-3370 ISSN 2367-3389 (electronic)
Lecture Notes in Networks and Systems
ISBN 978-3-031-48464-3 ISBN 978-3-031-48465-0 (eBook)
https://doi.org/10.1007/978-3-031-48465-0

This Springer imprint is published by the registered company Springer Nature Switzerland AG
The registered company address is: Gewerbestrasse 11, 6330 Cham, Switzerland

Preface

Introduction In the dynamic landscape of technology, Artificial Intelligence (AI) and Data Science have emerged as pivotal forces reshaping the way we perceive and interact with information. The convergence of these two domains has given rise to a plethora of innovative applications that span industries, academia, and everyday life. As we navigate through the complexities of an interconnected world, the significance of understanding and harnessing the power of AI and Data Science becomes increasingly evident.

The book "Artificial Intelligence, Data Science, and Applications" delves into the multifaceted realm of these transformative technologies, offering a comprehensive exploration of their theoretical foundations, practical applications, and the synergies that arise when they are combined. This book is designed to cater to a diverse audience, ranging from seasoned researchers and practitioners to students eager to embark on a journey into the cutting-edge advancements in AI and Data Science.

Key Themes:

1. **Foundations of Artificial Intelligence:**
   - Unravel the fundamental principles and algorithms that underpin AI, providing readers with a solid understanding of the field's core concepts.
2. **Data Science Techniques and Methodologies:**
   - Explore the methodologies, tools, and best practices in Data Science, addressing the entire data lifecycle from collection and preprocessing to analysis and interpretation.
3. **Integration of AI and Data Science:**
   - Investigate the seamless integration of AI and Data Science, showcasing how the synergy between these domains enhances the development of intelligent systems, predictive models, and decision-making processes.
4. **Real-world Applications:**
   - Showcase a diverse array of practical applications in various domains, including healthcare, finance, cybersecurity, and more, illustrating how AI and Data Science are actively shaping industries and improving societal outcomes.
5. **Ethical and Societal Implications:**
   - Delve into the ethical considerations and societal implications of deploying AI and Data Science solutions, emphasizing the importance of responsible innovation and addressing potential biases.

6. **Future Perspectives:**

- Anticipate and discuss emerging trends, challenges, and future directions in AI and Data Science, offering insights into the evolving landscape of these rapidly advancing fields.

This comprehensive compilation serves as a guide through the intricate web of Artificial Intelligence and Data Science, providing readers with a holistic view of the theories, methodologies, applications, and ethical considerations that define these disciplines. Each chapter is crafted by experts in the respective fields, ensuring a rich and diverse tapestry of knowledge that will inspire and inform both novices and seasoned professionals alike. "Artificial Intelligence, Data Science, and Applications" invites readers to embark on an enlightening journey into the heart of technological innovation and its transformative impact on our world.

Errachidia, Morocco Yousef Farhaoui

Organisation

Chairman of ICAISE'2023

Yousef Farhaoui	Moulay Ismail University of Meknes, Faculty of sciences and Techniques, Errachidia, Morocco

International Organizing Committee

Seyed Ghaffar	Brunel University London, UK
Zouhaier Brahmia	University of Sfax, Tunisia
Amir Hussain	Director of the Centre of AI and Robotics at Edinburgh Napier University, UK
Tanzila Saba	Prince Sultan University, Saudi Arabia
Hamed Taherdoost	University Canada West, Vancouver, Canada
Anshul Verma	Institute of Science, Banaras Hindu University, Varanasi, India
Lara Brunelle Almeida Freitas	University of Mato Grosso do Sul—UEMS/Dourados, Brazil
Youssef Agrebi Zorgani	ISET Sfax, Tunisia
Bharat Bhushan	School of Engineering and Technology (SET), Sharda University, India
Fathia Aboudi	High Institute of Medical Technology of Tunis, Tunisia
Agbotiname Lucky Imoize	University of Lagos, Nigeria
Javier González Argote	President and CEO of Fundación Salud, Ciencia y Tecnología, Argentina

Committee Members

Ahmad El Allaoui	FST-UMI, Errachidia, Morocco
Yousef Qarrai	FST-UMI, Errachidia, Morocco
Fatima Amounas	FST-UMI, Errachidia, Morocco
Mourad Azrour	FST-UMI, Errachidia, Morocco
Imad Zeroual	FST-UMI, Errachidia, Morocco
Said Agoujil	ENCG-UMI, Errachidia, Morocco
Laidi Souinida	FSTE-UMI, Errachidia, Morocco
Youssef El Hassouani	FSTE-UMI, Errachidia, Morocco
Abderahman El Boukili	FSTE-UMI, Errachidia, Morocco
Abdellah Benami	FSTE-UMI, Errachidia, Morocco

Ernst L. Leiss	University of Houston, Texas USA
Mehdi Shadaram	University of Texas at San Antonio, USA
Lennart Johnsson	University of Houston, USA
Nouhad Rizk	Houston University, USA
Jaime Lloret Mauri	University of Politécnica de Valencia, Spain
Janusz Kacprzyk	Systems Research Institute, Polish Academy of Sciences in Warsaw, Poland
Mahmoud Al-Ayyoub	University of Science and Technology, Jordan
Houbing Song	West Virginia University, USA
Mohamed Béchir Dadi	University of Gabès, Tunisia
Amel Ltifi	University of Sfax, Tunisia
Mohamed Slim Kassis	University of Tunis, Tunisia
Sebastian Garcia	Czech Technical University in Prague, Czech Republic
Younes Asimi	University of Iben Zohr, Agadir, Morocco
Samaher Al-Janabi	University of Babylon, Iraq
Safa Brahmia	University of Sfax, Tunisia
Hind Hamrouni	University of Sfax, Tunisia
Anass El Haddadi	University of Abdelmalek Essaadi, Morocco
Abdelkhalek Bahri	University of Abdelmalek Essaadi, Morocco
El Wardani Dadi	University of Abdelmalek Essaadi, Morocco
Ahmed Boujraf	University of Abdelmalek Essaadi, Morocco
Ahmed Lahjouji El Idrissi	University of Abdelmalek Essaadi, Morocco
Aziz Khamjane	University of Abdelmalek Essaadi, Morocco
Nabil Kannouf	University of Abdelmalek Essaadi, Morocco
Youness Abou El Hanoune	University of Abdelmalek Essaadi, Morocco
Mohamed Addam	University of Abdelmalek Essaadi, Morocco
Fouzia Moradi	University of Abdelmalek Essaadi, Morocco
Hayat Routaib	University of Abdelmalek Essaadi, Morocco
Imad Badi	University of Abdelmalek Essaadi, Morocco
Mohammed Merzougui	FSJES, Oujda, Morocco
Asmine Samraj	Quaid-E-Millath Government College, India
Esther Asare	Anhui University of Science and Technology, China
Abdellah Abarda	FEM-Hassan 1 University, Morocco
Yassine El Borji	University of Abdelmalek Essaadi, Morocco
Tarik Boudaa	University of Abdelmalek Essaadi, Morocco
Ragragui Anouar	University of Abdelmalek Essaadi, Morocco
Abdellah Benami	FSTE-UMI, Errachidia, Morocco
Abdelhamid Zouhair	University of Abdelmalek Essaadi, Morocco
Mohamed El Ghmary	FSDM-USMBA, Morocco

Youssef Mejdoub	ESTC-UH2, Casablanca, Morocco
Amine Tilioua	FST-UMI, Errachidia, Morocco
Azidine Geuzzaz	EST Essaouira, University of Cadi Ayyad Marrakech, Morocco
Said Benkirane	EST Essaouira, University of Cadi Ayyad Marrakech, Morocco
Ali Ouacha	FS-Mohammed V University, Morocco
Fayçal Messaoudi	ENCG, USMBA, Morocco
Tarik Chanyour	Ain Chock Faculty of Sciences-UH2, Morocco
Naercio Magaia	School of Engineering and Informatics, University of Sussex, UK
Imane Halkhams	UPE, Fez, Morocco
Elsadig Musa Ahmed	Faculty of Business Multimedia University, Malaysia
Valliappan Raju	Professor, International Islamic University of Malaysia, Malaysia
David Jiménez-Castillo	Faculty of Economics and Business, University of Almeria, Spain

Introduction

Data is becoming an increasingly decisive resource in modern societies, economies, and governmental organizations. Data science, Artificial Intelligence and Smart Environments inspires novel techniques and theories drawn from mathematics, statistics, information theory, computer science, and social science. This book reviews the state of the art of big data analysis, Artificial Intelligence and Smart Environments. It includes issues which pertain to signal processing, probability models, machine learning, data mining, database, data engineering, pattern recognition, visualisation, predictive analytics, data warehousing, data compression, computer programming, smart city, etc. Papers in this book were the outcome of research conducted in this field of study. The latter makes use of applications and techniques related to data analysis in general and big data and smart city in particular.

The book appeals to advanced undergraduate and graduate students, post-doctoral researchers, lecturers and industrial researchers, as well as anyone interested in big data analysis and Artificial Intelligence.

Contents

Sustainability in Internet of Things: Insights and Scope

Swati Sharma(✉)

Jindal Global Business School, O. P. Jindal Global University, Sonipat, India
swati@jgu.edu.in

Abstract. The study explores the sustainability in Internet of Things (IoT) by conducting bibliometric analysis of extant literature on the topic. For bibliometric analysis, we employ Scientific Procedures and Rationales for Systematic Literature Reviews (SPAR-4-SLR) method of systematic literature review. Year-wise, Author-wise, Citation-wise, Country-wise, Source-wise, Affiliation-wise and keywords-wise listing are the parameters to identify the trend and future scope of sustainability in IoT. We use Scopus database to list the extant literature. The study suggests that sustainability in IoT is explored in different fields like smart cities, agriculture, edge computing, block-chain, forecasting and energy utilization. This study provides insights on current trends and future scope of sustainable use of IoT.

Keywords: IoT · Sustainability · Green IoT · Sustainable IoT · SPAR-4-SLR · Bibliometric analysis

1 Introduction

Technological advancement has brought effectiveness and efficiency to almost all operational area of human life. The question that emerges due to such advancement is whether such effectiveness and efficiency is sustainable or not? Internet of Things (IoT) has revolutionized the accessibility of information by exchanging and connecting data with electronic devices and systems over the internet. Though IoT offers numerous benefits to the users, it also generates volumes of solid and toxic waste [1]. To minimize such wastages and look for sustainable options in IoT is an ongoing researched topic. Only few studies have reviewed existing literature on green IoT [2–5]. Hence, the present study reviews extant literature on sustainability in internet of things and present comprehensive view of current trend and future scope.

2 Research Methodology

This study employs widely used method of systematic literature review known as Scientific Procedures and Rationales for Systematic Literature Reviews (SPAR-4-SLR) developed by Paul et al. [6] to examine the extant literature on Sustainability in Internet

Y. Farhaoui et al. (Eds.): ICAISE 2023, LNNS 837, pp. 1–7, 2024.
https://doi.org/10.1007/978-3-031-48465-0_1

of Things. Paul et al. [6] also suggests looking for four basic characteristics of existing studies on underlying topic before employing the SPAR-4-SLR method of literature review as unlike empirical studies, literature review studies need to have certain base. These four basic criteria are:

- When a substantial body of work in the domain exists
- When no systematic literature review in the domain exists in recent years
- When no review of the domain exists in high-quality journals
- When existing systematic literature reviews have gaps or short-changes

The underlying topic i.e., Sustainability in Internet of Things fulfill all four criteria as more than 40 articles exist on topic, no systematic literature review is conducted on exact same topic for given time-period in highly reputed source title. Hence, SPAR-4-SLR method for literature review is suitable for fulfilling research objectives of the present study. Furthermore, the SPAR-4-SLR protocol consists of three stages and each stage has two sub-stages named assembling, arranging, and assessing. Each of these three stages has 2 sub-stages explained in Fig. 1 and as follows:

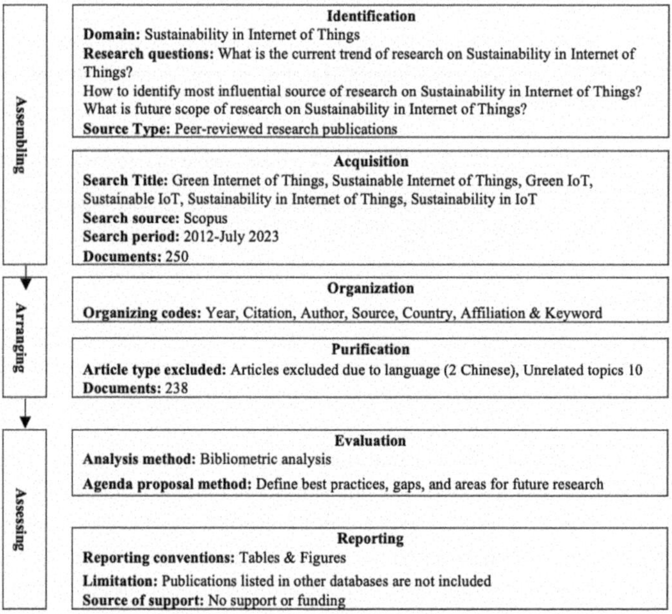

Fig. 1. SPAR-4-SLR methodology

2.1 Assembling

The assembling includes the identification and acquisition of literature on the topic. The sub-stage identification specifies domain, research questions and source type whereas sub-stage acquisition specifies search-title, search source, search period, and filters e.g., keyword, subject etc. Other specifications of assembling stage are depicted in Fig. 1. These search titles have generated 250 articles ranging from 2012 to July 2023.

2.2 Arranging

The second stage of arranging includes organization and purification of articles assembled in stage 1. The first sub-stage of organization of articles includes organizing code and organizing framework. The present study conducts bibliographic analysis of literature; hence, the organizing code includes year, citation, authors, source-titles, country, affiliation, and keywords analysis. As the present study is not analyzing literature as per any framework, this part is not applicable (NA). The second sub-stage of arranging is the purification of literature. We exclude the article which are non-English and unrelated topic. Hence, 250 articles of assembling stage, are further curtailed to 238 articles.

2.3 Assessing

Assessing is the final stage of SPR-4-SLR protocol. It has two sub-stages i.e., Evaluation & Reporting. For evaluation, the present study employs bibliometric analysis and finds best practices, gaps, and areas for future research on topic. The result of evaluation is presented in the form of tables and figure. Figure 1 describes all these six sub-stages in detail.

3 Results and Discussion

We have summarized findings with quantifying the research trends on existing literature as follows:

3.1 RO1. To Quantify the Trends of Publication for Sustainability in IoT

Table 1 summarizes the trends on publication. The publication year ranges from 2012 to July 2023. Average publication per year is found 23 publications. The highest publication is the year 2022 with 75 publications followed by the year 2021 with 55 publications. Year 2023 has 39 publications by July. Hence, there is possibility that 2023 years publication will have more publication recording year for Sustainability in Internet of Things.

Table 1. Descriptive statistics

Statistics	Value	Statistics	Value	Statistics	Value
Mean	22.82	Range	74	Standard deviation	25.24
Standard error	7.61	Minimum	1	Count	11.00
Median	10	Maximum	75	Kurtosis	0.12
Mode	1	Sum	251	Skewness	1.03

Table 2. Influential articles on sustainability in Internet of Things

Authors	Title	Year	Source title	TC
Allam Z.; Dhunny Z.A	On big data, artificial intelligence and smart cities	2019	Cities	461
Mosavi A.; Salimi M.; Ardabili S.F.; Rabczuk T.; Shamshirband S.; Varkonyi-Koczy A.R	State of the art of machine learning models in energy systems, a systematic review	2019	Energies	251

3.2 RO2. To Identify Most Influential Research Source on Sustainability in IoT

Most cited articles are considered to be the most influential study of the subject area. These 238 studies has been cited 2012 times. Considering number of citations, Table 2 lists publications which has more than 200 citations. Two such articles are identified which are accounted for more than 15% of total citations on topic. Seven articles have more than 100 citations. Table 3 lists source titles that have minimum 10 articles. Sustainability Switzerland has highest TP (total publications), highest TC (total citation) and the highest TCP (total cited publication), although IEEE Access highest TC/TP and TC/TCP. Table 4 present data on top contributing author. Akram, S.V., Gehlot, A. and Singh, R., as co-authors, have highest TP, TCP, TC AND TC/TP. Total 159 authors have contributed on the topic.

Table 3. Highest contributing journals on sustainability in Internet of Things

Source title	TP	TCP	TC	TC/TP	TC/TCP	IMF
Sustainability Switzerland	22	16	404	18.36	25.25	2.60
IEEE Access	10	9	378	37.80	42	3.48

Table 4. Top contributing authors on sustainability in Internet of Things

Author name	TP	TCP	TC	TC/TP
Akram, S.V., Gehlot, A., Singh, R	4	4	38	9.5

Table 5 shows the publication trends country-wise and lists countries with minimum 20 publications. Three such countries are identified which are accounted for 50% of total published articles. India has the highest TP, TCP and TC. United States has the highest TC/TP and TC/TCP. Table 6 lists affiliation with minimum 5 articles on topic. King Saud University has the highest TP, TCP, TC, TC/TP and TC/TCP.

Table 5. Top countries on sustainability in Internet of Things

Country	TP	TCP	TC	TC/TP	TC/TCP
India	69	48	942	13.65	19.63
United States	24	21	682	28.42	32.48
China	22	18	406	18.45	22.56

Table 6. Top affiliation on sustainability in Internet of Things

Affiliation	TP	TCP	TC	TC/TP	TC/TCP
King Saud University	6	5	302	50.33	60.4
Instituto de Telecomunicações	5	4	107	21.4	26.75

RO3. To draw inferences about future scope of studies on Sustainability in IoT

This study investigates the keywords frequency to draw inference about current trends and future scope of sustainable IoT. Keywords are analyzed to find under-researched and promising topics, and VOSviewer tool is used for keyword visualization. There are 2534 keywords used in all 238 articles. Table 7 and Fig. 2 represents the keywords which have appeared minimum 10 times (F) and their total strength link (TLS). Total strength link represents number of documents which have co-occurrence of two keywords [7]. Based on Table 7 and Fig. 2, the future scope of sustainable IoT can be identified. Those listed keywords are mostly used research area in green IoT currently and are going to be substantially explored in near future as well.

Table 7. Keyword frequency analysis

Keyword	F	TLS	Keyword	F	TLS
Smart city	70	527	Agriculture	27	225
Big data	35	304	Decision making	24	186
Energy efficiency	66	571	Smart power grid	26	290
Forecasting	30	246	Edge computing	14	127
Block chain	36	373	Electric power transmission network	10	107

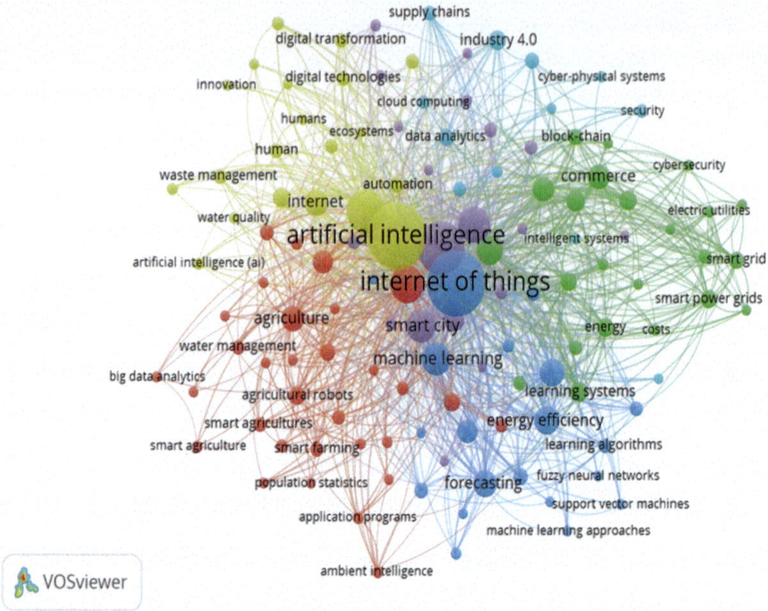

Fig. 2. Network visualization of keyword co-occurrence

4 Scope and Limitation of Study

Though present studies reviews 238 articles for presenting a comprehensive view sustainable IoT there are few limitations that exists with the study. We search only one research engine for finding literature i.e., Scopus, that may limit the view of the present paper. The keyword used for searching the articles of the topic may also include other non-related keywords as there may be studies on IoT business and management in different domain/industries. These limitations can be overcome by using more keyword and exploring other research databases for finding the articles on the topic.

5 Conclusion

Sustainable IoT has been explored in different field of studies including health, defense, logistics, education, governance and trading due to its numerous benefits. Still, it is important to develop energy efficient IoT technology and sustainable solutions to minimize wastage generated with its usage [8]. Many economies have pledged to be using renewable energy. Use of renewable energy brings IoT in its purview as well. The present study explores existing literature on sustainable IoT and present current trends in form of highest cited article, highest contributing source title, highest contributing author, highest contributing country and affiliation. The keyword analysis of underlying studies concludes that employment of sustainable IoT will be explored in agriculture, smart cities, energy utilization, forecasting, decision making, power grid, edge computing and block-chain [9–12].

References

1. Maksimovic, M.: Greening the future: green Internet of Things (G-IoT) as a key technological enabler of sustainable development. In: Internet of Things and Big Data Analytics Toward Next-Generation Intelligence, pp. 283–313 (2018)
2. Albreem, M.A., El-Saleh, A.A., Isa, M., Salah, W., Jusoh, M., Azizan, M.M., Ali, A.: Green internet of things (IoT): an overview. In: 2017 IEEE 4th International Conference on Smart Instrumentation, Measurement and Application (ICSIMA), pp. 1–6. IEEE (2017)
3. Bashar, A.: Review on sustainable green Internet of Things and its application. IRO J. Sustain. Wirel. Syst. **1**(4), 256–264 (2019)
4. Huang, J., Meng, Y., Gong, X., Liu, Y., Duan, Q.: A novel deployment scheme for green internet of things. IEEE Internet Things J. **1**(2), 196–205 (2014)
5. Charef, N., Mnaouer, A.B., Aloqaily, M., Bouachir, O., Guizani, M.: Artificial intelligence implication on energy sustainability in Internet of Things: a survey. Inf. Process. Manage. **60**(2), 103212 (2023)
6. Paul, J., Lim, W.M., O'Cass, A., Hao, A.W., Bresciani, S.: Scientific procedures and rationales for systematic literature reviews (SPAR-4-SLR). Int. J. Consum. Stud. **45**(4), O1-6 (2021)
7. Guo, Y.M., Huang, Z.L., Guo, J., Li, H., Guo, X.R., Nkeli, M.J.: Bibliometric analysis on smart cities research. Sustainability **11**(13), 3606 (2019)
8. Alsharif, M.H., Jahid, A., Kelechi, A.H., Kannadasan, R.: Green IoT: a review and future research directions. Symmetry **15**(3), 757 (2023)
9. Almalki, F.A., et al.: Green IoT for eco-friendly and sustainable smart cities: future directions and opportunities. Mobile Netw. Appl. **17**, 1–25 (2021)
10. Benhamaid, S., Bouabdallah, A., Lakhlef, H.: Recent advances in energy management for Green-IoT: an up-to-date and comprehensive survey. J. Netw. Comput. Appl. **1**(198), 103257 (2022)
11. Tuysuz, M.F., Trestian, R.: From serendipity to sustainable green IoT: technical, industrial and political perspective. Comput. Netw. **9**(182), 107469 (2020)
12. Muzafar, S.: Energy harvesting models and techniques for green IoT: a review. In: Role of IoT in Green Energy Systems, pp. 117–43 (2021)

Machine Learning in Finance Case of Credit Scoring

Driss El Maanaoui[1(✉)], Khalid Jeaab[1(✉)], Hajare Najmi[1], Youness Saoudi[2], and Moulay El Mehdi Falloul[1]

[1] Economics and Management, Laboratory USMS, Beni Mellal, Morocco
{elmaanaoui.driss.feg,jeaab.khalid.feg}@usms.ac.ma

[2] Advanced Systems and Engineering, Laboratory Ibn Tofail University, Kenitra, Morocco

Abstract. In recent years, the financial services industry has prioritized the use of artificial intelligence to surpass consumer expectations; lower operating costs, and make better business decisions.

Because the finance business accumulates a large volume of big data from its consumers, it is perfectly suited to the benefits of data mining. Banks and financial institutions are already using several innovative financial apps based on machine learning algorithms. The objective of this paper is to study the two different credit scoring techniques: classical statistical methods and advanced Machine Learning methods. Their interests are a powerful tool to predict the creditworthiness of borrowers.

Keywords: Credit Scoring · Prediction · Machine Learning · Classification · Accuracy · Recall · F1 score · Precision · Confusion matrix

1 Introduction

A bank's viability might be jeopardized by many factors. Market risk, options risk, credit risk, operational risk, and so on are examples of these hazards. The most common risk is credit risk, often known as counterparty risk.

The first credit scoring models were developed in the 1960s and 1970s, based on failure models used by businesses and financial organizations. Beaver's (1966) univariate classification, Altman's (1968) discriminant multivariate analysis, and Meyer and Pifer's (1970) use a linear probability model.

Thomas (2000) offered the first literature study on scoring models that used Machine Learning methods. According to Baesens et al. (2003), credit scoring has too few non-linearities for ML's predicted performance advantages to be meaningful. Lessmann et al. (2015) conducted a comparative analysis using additional assessment criteria and the most recent machine learning methods.

Machine Learning algorithms provide significant productivity benefits by reducing data management and preprocessing steps before the modeling stage Farhaoui (2018).

Y. Farhaoui et al. (Eds.): ICAISE 2023, LNNS 837, pp. 8–16, 2024.
https://doi.org/10.1007/978-3-031-48465-0_2

2 Objectif

The objective of this study is to apply credit scoring methods to a large database in order to make a comparative approach between the results obtained.

3 Methodology

The purpose is to study the different methods of credit scoring: logistic regression, support vector machine, decision trees, adaboost classifier, random forest classifier, xgboost.

4 Credit Scoring

4.1 Definition

Scoring is a technique used to assign a score to a customer. The score obtained generally reflects the probability that an individual meets a need or belongs to the desired target.

The risk scores that can be found:

- Credit risk or credit scoring: predicting the delay of credit repayment.
- Financial risk: predicting the good or bad health of a company.

We will then focus on the credit risk:

Credit scoring is the statistical assessment of an individual's or a company's risk of credit default.

4.2 Principle

The scoring model is a technique used to forecast a borrower's solvency by combining some ratios into a single indication that distinguishes healthy corporations from failing organizations. It is determined by the borrower's attributes (Baesens et al. 2003; Farhaoui et al. 2023).

5 Statistical Method: Logistic Regression

The link between independent and dependent variables is measured using logistic regression.

In binary (dichotomous) logistic regression, we consider a binary target variable $S = 0$ or 1.

X_j are binary or qualitative explanatory variables.

Under the following assumptions, logistic regression calculates the a posteriori probability:

$$p_i = P(S_i = 1/X_i) = \frac{1}{1 + e^{-\beta - \alpha X_i}}$$

$$1 - p_i = P(S_i = 0/X_i) = \frac{1}{1 + e^{\beta + \alpha X_i}}$$

With $S_i = 1$ if the firm $i \in N$ and $S_i = 0$ if the firm $i \in D$.
The likelihood $\prod_{i=1}^{n} p_i^{S_i}(1 - p_i)^{1-S_i}$ where n is the sample size, $n = n_D + n_N$.
The maximum likelihood approach is used to calculate the parameters α and β.
p_i is the a posteriori probability that the firm is healthy.
As a result, $logitp_i = ln\left(\frac{p_i}{1-p_i}\right) = \beta + \alpha X_i$, and the decision rule can be stated:
"Company i is classified as healthy." $<=> p_i > 1 - p_i <=> logitp_i > 0 <=> \beta + \alpha X_i > 0$.

Another decision rule that might be used is $\beta + \alpha X_i > K$.. The addition of a threshold K allows the choice to be tailored to the bank's objectives, as measured by the costs of misclassification (Hand and Henly (1997)).

6 Machine Learning Methods

6.1 Definition: Machine Learning

Machine learning is a subfield of artificial intelligence that makes predictions using statistical models. Machine learning algorithms are used in finance to detect fraud, automate trading processes, compute credit scores, and give investors financial advice. Without being explicitly programmed, machine learning can examine millions of data sets in a short period to improve outcomes.

6.2 Support Vector Machine (SVMs)

SVMs are supervised learning methods used to solve classification and regression problems. They construct a maximum-margin hyperplane in a modified input space and split example classes while maximizing distance to closest examples (Fig. 1).

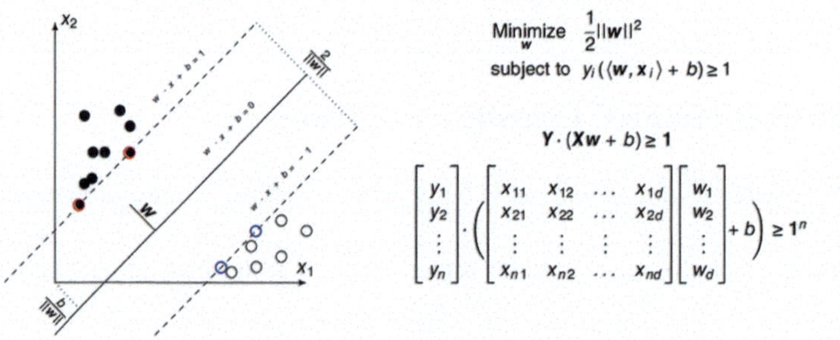

Fig. 1. Illustration of the SVM process

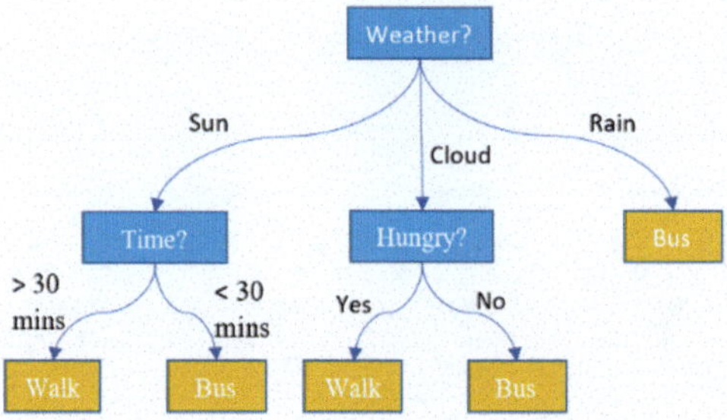

Fig. 2. The graphic up depicts a decision tree that forecasts how to go to work.

6.3 Decision Trees

Decision Trees (DTs) are non-parametric supervised learning approaches that are used to create models that predict the value of a target variable based on decision rules generated from data characteristics (Fig. 2).

6.4 AdaBoost Classifier

AdaBoost classifiers are meta-estimators that adjust the weights of incorrectly classified instances to focus more on difficult cases (Fig. 3).

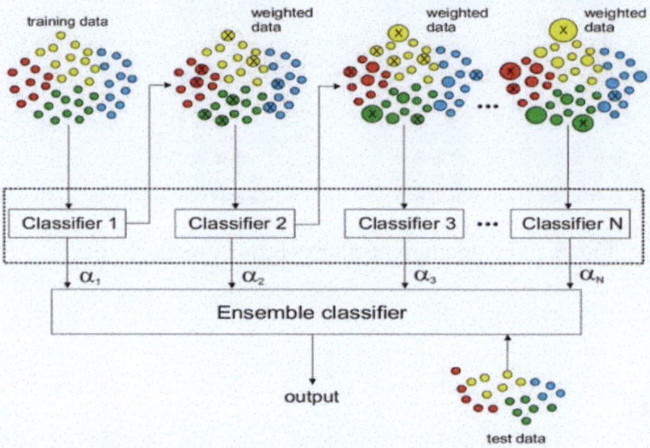

Fig. 3. Diagram of the AdaBoost algorithm exemplified for multi-class classification problems.

6.5 Random Forest Classifier

A random forest is a meta estimator that employs averaging to improve projected accuracy and control over-fitting by training decision tree classifiers on distinct sub-samples of a dataset (Fig. 4).

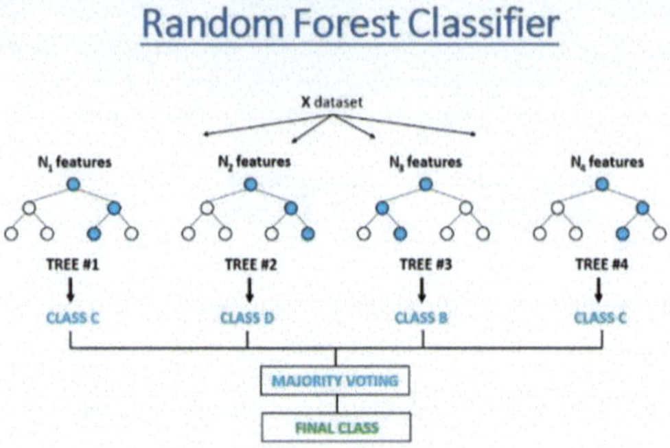

Fig. 4. A graphic representation of the Random Forest.

6.6 XGBoost

XGBoost is a parallel tree boosting toolkit that delivers optimal distributed gradient boosting to address data science challenges rapidly and reliably (Fig. 5).

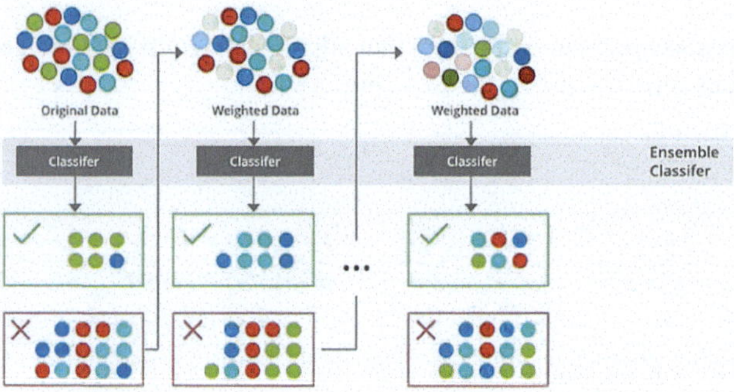

Fig. 5. The diagram up shows how to work XGBoost.

7 Case Study

7.1 Presentation of the Database

Mortgage-level data from 2016 provided by International Financial Research.

The initial database contains 50,000 with 12 explanatory variables. In our study we were interested in 5690 observations with 10 explicative variables (Ratios).

7.2 Ratios Used

- LOAN: Amount of the loan application mortgage
- MORTGAGE: Amount payable on existing
- VALUE: Current property value
- YOJ: Years in present position
- DEROG: Number of significant derogatory reports
- DELINQ: Number of outstanding credit lines
- CLAGE: Age of oldest credit line in months
- NINQ: Number of recent credit applications
- CLNO: Number of credit lines
- DEBT: The debt-to-income ratio

7.3 Definition of the Parameters

Model accuracy (Accuracy Score) is a machine learning model performance statistic that assesses how frequently a machine learning model predicts an outcome correctly.

Recall assesses the model's ability to properly forecast positives, whereas precision measures how many of the models' positive forecasts are correct.

F1 model score is a machine learning model performance statistic that evaluates accuracy by providing equal weight to precision and recall, making it an alternative to precision measurements. It gives high-level information about the model's output quality.

The Precision score, which represents the ratio of genuine positives to the sum of real positives and false positives, is a useful predictability statistic.

The Confusion Matrix is an overview of classification issue prediction findings that emphasizes and categorizes right and incorrect predictions and compares them to real values to uncover inaccuracies.

Understanding the Confusion Matrix Terminologies

To completely comprehend how a confusion matrix works, it is necessary to first comprehend the four primary terminologies: TP, TN, FP, and FN. Each of these words is defined precisely below:

- True Positives (TP) are instances in which the forecast is positive and the true value is also positive.
- True Negatives (TN): When the prediction is negative and the actual value is negative.
- FP (False Positive): When a positive forecast is made but the actual number is negative.
- FN (False Negative): When a negative forecast is made but the actual number is positive.

$$Accuracy = \frac{TP + TN}{TP + TN + FP + FN}$$

$$Recall = \frac{TP}{TP + FN}$$

$$Precision = \frac{TP}{TP + FP}$$

$$F1Score = \frac{2 * Recall * Precision}{Recall + Precision}$$

$$ErrorRate = \frac{FP + FN}{TP + TN + FP + FN}$$

7.4 Discussion

The following table presents the results obtained in our case study. Using the python programming language and its integrated libraries and the Sckikit-Learn machine learning family (Table 1).

Table 1. Comparison between the methods.

	Logistic regression (%)	Support vector machines (%)	Decision tree (%)	AdaBoost (%)	Random forest (%)	XGBoost (%)
Accuracy Score	80.62	80.62	86.99	89.09	92.19	92.28
F1 Score	42.28	45.59	79.22	81.35	87.01	87.08
Precision Score	68.95	90.30	79.27	84.03	88.79	89.12
Recall Score	50.70	50.21	79.17	79.33	85.51	85.40

Machine learning approaches outperform statistical methods, with the exception of the support vector machine method, which has the same accuracy as logistic regression but is preferred in terms of score correctness. SVM is, in fact, a regression approach. Finally, machine learning approaches are more powerful, with increased accuracy, precision, and recall power.

Conclusion

Banks strive to maintain efficient scoring models to identify potential hazards on each credit line. These models require historical information on the borrower to predict future solvency and identify potential hazards in a timely and effective manner.

We looked at two credit risk prediction techniques, statistical models and machine learning models, and used Python as an application programming language for all models, which is free and open source.

The results suggest that the machine learning strategy beats the statistical approach in terms of credit risk prediction.

References

Altman, E.I.: Financial ratios, discriminant analysis and the predictions of corporate bankruptcy. J. Finance **23** (1968)

Altman, E.I., Haldeman, R.G., Narayanan, P.: Zeta analysis a new model to identify bankruptcy risk of corporations. J. Bank. Finance **1** (1977)

Avery, R.B, Hanweck, G.A.: A dynamic analysis of bank failures. In: Research Papers in Banking and Financial Economics 74, Board of Governors of the Federal Reserve System (U.S.) (1984)

Baesens, B., Van Gestel, T., Viaene, S., Stepanova, M., Suykens, J., Vanthienen, J.: Benchmarking state-of-the-art classification algorithms for credit scoring. J. Oper. Res. Soc. **54**(6) (2003)

Barr, R.S., Siems, T.F.: Predicting bank failure using DEA to quantify management quality. In: Financial Industry Studies Working Paper, Federal Reserve Bank of Dallas (1994)

Beaver, W.: Financial ratios as predictors of failure, empirical research in accounting: selected studies. J. Acc. Res. supplément au **4** (1966)

Desai, V.S., Crook, J.N., Overstreet, G.A.: A comparison of neural networks and linear scoring models in the credit environment. Eur. J. Oper. Res. **95** (1996)

Lessmann, S., Baesens, B., Seow, H.V., Thomas, L.C.: Benchmarking state-of-the-art classification algorithms for credit scoring: an update of research. Eur. J. Oper. Res. **247**(1) (2015)

Martin, D.: Early warning of bank failure: a logit regression approach. J. Banking Finance **1** (1977)

Meyer, P.A., Pifer, H.W.: (1970) Prediction of bank failure. J Finance **2** (1970)

Pantalone, C.C., Platt, M.B.: Predicting commercial bank failures since deregulation. N. Engl. Econ. Rev. (1987)

Thomas, L.C.: A survey of credit and behavioural scoring: forecasting financial risk of lending to customers. Int. J. Forecast. **16** (2000)

Gotway, C.A., Stroup, W.W.: A generalized linear model approach to spatial data analysis and prediction. J Agric. Biol., Environ. Stat. **2**(2), 157–78 (1997). JSTOR, https://doi.org/10.2307/1400401

Pearce, J., Ferrier, S.: Evaluating the predictive performance of habitat models developed using logistic regression. Ecol. Model. **133**(3), 225–245 (2000). https://doi.org/10.1016/s0304-3800(00)00322-7

Hassanipour, S., Ghaem, H., Arab-Zozani, M., Seif, M., Fararouei, M., Abdzadeh, E., …, Paydar, S.: Comparison of artificial neural network and logistic regression models for prediction of outcomes in trauma patients: a systematic review and meta-analysis. Injury (2019). https://doi.org/10.1016/j.injury.2019.01.007

Kabra, R.R., Bichkar, R.S.: Performance prediction of engineering students using decision trees. Int. J. Comput. Appl. (0975–8887) **36**(11) (2011)

Farhaoui, Y.: Big data analytics applied for control systems. In: Lecture Notes in Networks and Systems, vol. 25, pp. 408–415 (2018). https://doi.org/10.1007/978-3-319-69137-4_36

Hindman, M.: Building better models: prediction, replication, and machine learning in the social sciences. Ann. Am. Acad. Polit. Soc. Sci. **659**, 48–62 (2015). Toward Computational Social Science: Big Data in Digital Environments

Lemmens, A., Croux, C.: Bagging and boosting classification trees to predict churn. J. Mark. Res. **43**(2), 276–286 (2006), Published by: Sage Publications, Inc. on behalf of American Marketing Association Stable URL: https://www.jstor.org/stable/30163394

Hand , D.J., Henley, W.E.: Statistical classification methods in consumer credit scoring: a review. J Royal Stat. Soc. Ser. A (Statistics in Society) **160**(3), 523–541 (1997), Published by: Wiley for the Royal Statistical Society Stable URL: https://www.jstor.org/stable/2983268

Amemiya, T., Powell, J.: A comparison of Logit model and normal discriminant analysis when the independent variables are binary. In: Karlin, A. (ed.) Studies in Econometrics, Time Series and Multivariate Statistics. Academic Press, New York

Baesens, Van Gestel, T., Viaene, S., Stepanova, M., Suykens, J, Vanthienen, J.: Benchmarking state-of-art classification algorithms for credit scoring. J. Oper. Res. Soc.

Farhaoui, Y., et al.: Big data mining and analytics 6(3), I–II. https://doi.org/10.26599/BDMA.2022.9020045

Farhaoui, Y., et al.: Big data mining and analytics **5**(4), I–II (2022). https://doi.org/10.26599/BDMA.2022.9020004

Prediction of Coefficient of Friction and Wear Rate of Stellite 6 Coatings Manufactured by LMD Using Machine Learning

Ricardo-Antonio Cázares-Vázquez[1], Viridiana Humarán-Sarmiento[1,2], and Ángel-Iván García-Moreno[1](✉)

[1] Center for Engineering and Industrial Development (CIDESI), Querétaro, Qro., Mexico
angel.garcia@cidesi.edu.mx

[2] Instituto Tecnológico Superior de Guasave, Guasave, Sinaloa, México

Abstract. Laser Metal Deposition (LMD) is a Direct Energy Deposition (DED) technique, which uses a laser source to melt the input material layer by layer, creating the desired geometry with high deposition volume.s Due to its ability to produce exceptional surface properties, LMD is widely used in coatings. The present work presents a comparative study of different Machine Learning (ML) architectures derived from the information of the monitoring process of Stellite-6 coatings on AISI 304 substrates, for the prediction of friction coefficient and wear rate. Random Forest (RF), Support Vector Regressor (SVR), and Artificial Neural Networks (ANN) were compared, where RF obtained a score performance of 0.93 for the wear rate and 0.82 for the friction coefficient. The results show that the geometry of the melt pool, i.e. width, length, and eccentricity, has the greatest influence on the forecast.

Keywords: Coefficient of Friction · Random Forest · Laser Metal Deposition · Additive Manufacturing

1 Introduction

In recent years, Additive Manufacturing (AM) is being adopted in a wide variety of industries due to its versatility of applications. Within this wide range of techniques, Direct Energy Deposition (DED) processes have gained wide recognition for applications in the production of large, high-precision components. This technique allows the construction of three-dimensional objects by selectively melting the feed material layer by layer with high deposition volumes, making it ideal for coatings. Within DED is Laser Metal Deposition (LMD), which has applications in complex and innovative structures [1]. LMD processes offer a number of unique advantages over other additive manufacturing techniques particularly when working with large volume parts and a preform is required. This ability to fabricate highly functional parts, higher deposition rates, and an understandable ease for final part optimization are other advantages of this technique [2, 3].

Y. Farhaoui et al. (Eds.): ICAISE 2023, LNNS 837, pp. 17–25, 2024.
https://doi.org/10.1007/978-3-031-48465-0_3

Artificial intelligence (AI) is playing a critical role in Industry 4.0 as a driver of advanced automation, intelligent decision-making, and efficiency in industrial processes. AI offers promising solutions to address the challenges of additive manufacturing. For example, machine learning algorithms can analyze large datasets to identify patterns and automatically optimize part designs. AI can also improve the quality and reliability of manufactured products by predicting and correcting potential printing defects [4].The combination of AI and AM is opening up new possibilities for the design, process optimization, and quality of manufactured products. Interest in applying AI to AM to improve production efficiency and accuracy has grown in recent years [5]. The focus of this work is to compare the prediction performance of various algorithms for the tribological variables of the coefficient of friction and wear rate, since the characterization of these properties requires destructive studies to ensure the quality of the coatings.

2 State of the Art

Recently, the scientific community has been particularly interested in the monitoring and control of AM processes in order to overcome the defects such as dimensional variation, porosity, cracking, warping, and residual stress. A literature review on AM related to quality control methods used in metal AM is presented in [6]. The authors discuss key challenges associated with these techniques in metal AM, such as the lack of standardized quality protocols and the need to develop new materials and processes.

The melt pool is a term used to describe an area of molten material near the laser-material interface. Typically, critical information for the AM process is provided by the images obtained from this melt pool [7]. For example, monitoring the melt pool provides the understanding of microstructural changes by considering the thermal history [8]. In addition, the geometry of the coatings is strongly related to the changes observed in the melt pool, since the process parameters define the heat flux at which the coating will coalesce [9]. The methods studied for monitoring include segmentation algorithms such as *gravitational superpixels* [10] and the calibration of monitoring equipment [11, 12].

Some related studies [13, 14] have made predictions of mechanical properties based on monitoring variables, where the authors carry out a comparison between Random Forest (RF), artificial neural networks (ANN), support vectors for regression (SVR) and Gradient Enhancement (GBM), in which RF gets the best scores. A review of the application of intelligent algorithms in additive manufacturing processes is presented in [15]. It also provides an overview of supervised, unsupervised, and semi-supervised algorithms. More extensive work on applying AI to tribological data predictions is presented in [16, 17].

3 Materials and Methods

An ABB 6-axis robotic arm (model IRB 6620) coupled with a TRUMPF BEO D70 laser head was used to deposit Stellite-6 onto AISI 304 substrates by LMD. The deposition process was monitored with a scheme (Fig. 1) of two pyrometers, an Optris CT 2MH1 statically focused (fixed) to measure heat dissipation on the substrate and a dual-band pyrometer Metis H3M22 mounted (attached) on the laser head, also an IR camera FLIR

A6751SC. Four optimal configurations were tested, Table 1 shows the process parameter configured for printing Stellite 6 coatings (laser power, scanning speed, feeding rate, and laser spot size). The Stellite 6 is a Cobalt-based alloy, with a hardness of 36–45 HRC, 8.44 g/cm^3 of density, and a melting point between 1285 and 1410 °C.

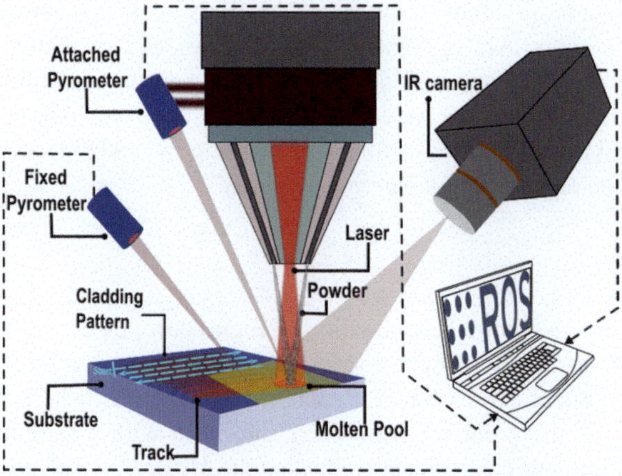

Fig. 1. Schematic setup of the instrumented-LMD cell

Table 1. LMD process parameters for print Stellite 6 coatings. Laser power (LP), scanning speed (SS), feeding rate (FR), and laser spot size.

	LP [W]	SS [mm/s]	FR [g/min]	SPOT [mm]
A	700	4	8	2.86
B	820	5	10	2.86
C	1000	5	9	2.86
D	830	6	11	2.86

3.1 Intelligent Algorithms

RF are an ensemble learning technique for regression, classification, and other applications that works by training a large number of decision trees. RF is a supervised learning method that builds decision trees on samples of data, makes and stores predictions on each one, and then votes on the best one. The RF hyperparameters configured for model optimization included the number of trees, maximum depth per branch, minimum number of samples to split an internal node, the minimum number of samples to be a leaf node, and the maximum number of features considered.

SVR use of the identical precept as Support Vector Machines (SVMs) but for regression problems. The difficulty with regression is finding a function that approaches the mapping from the input domain to the actual number in the training sample idea. The hyperparameters related to the performance of the SVR models are mainly focused on the use of different types of kernels (Linear, Polynomial, RBF), along with adjustments made to the regularization parameter C and varying the epsilon of the model.

The ANN model involves computations and mathematics that simulate the processes of the human brain. Many of the recent advances made in the research of artificial intelligence are based on ANNs. Neurons are interconnected by weighted connections. Several models were generated varying the number of neurons and hidden layers. Additionally, the activation functions were varied between "logistic" (sigmoid), "tanh", and "ReLu" (Rectified Linear Unit), as well as the learning rate, which can be "constant" or "adaptive", and the regularization parameter "alpha".

3.2 Metrics

The most common metrics in the evaluation of regression models are the *coefficient of determination* R^2, *mean absolute error* (MAE), *and mean squared error* (MSE). Table 2 shows the mathematical expressions for each of these metrics, where y_i represents the real value and $\widehat{y_i}$ as the predictor from each algorithm with n samples. The features importance in RF models can be measured using the *Mean Decrease in Impurity* (MDI) and *Mean Decrease in Accuracy* (MDA) metrics, even for prediction models, since scikitlern's methods are based on prediction variance reduction.

The MDI value represents the sum of the impurity reductions of all the split nodes. The average importance X_j is given by (1) over all trees φ_m (for $m = 1,...,M$), $p(t)$ is the portion $\frac{N_t}{N}$ of samples reaching t, j_t denotes the identifier of the variable used for splitting node t and $\Delta i(s_t, t)$ as a weighted impurity split.

Table 2. Metrics to evaluate the performance of prediction models.

Metrics	Expression
R^2	$1 - \frac{\sum_{i=1}^{n}\left(y_i-\widehat{y_i}\right)^2}{\sum_{i=1}^{n}(y_i-\overline{y_i})^2}$
MAE	$\frac{1}{n}\sum_{i=1}^{n}\left\|y_i - \widehat{y_i}\right\|^2$
MSE	$\frac{1}{n}\sum_{i=1}^{n}\left\|y_i - \widehat{y_i}\right\|^2$

$$MDI\left(X_j\right) = \frac{1}{M}\sum_{m=1}^{M}\sum_{t\in\varphi_m} 1(j_t = j)\left[p(t)\Delta i(s_t, t)\right] \quad (1)$$

MDA compares the predictions of each tree before and after eliminating a feature and considers importance as the contribution of each variable to the error reduction of the model. The importance per feature permutation is defined by (2), where L represents the feature permutation (with or without), and $m_{kl}, \ldots, k_{M^{-i}}$, denote the indices of the trees that have been built from bootstrap replicates with k random features.

$$MDA(X_j) = -\frac{1}{N} \sum_{(x_i, y_i) \in L} L\left(\frac{1}{M^{-i}} \sum_{l=1}^{M^{-i}} \varphi L^m kl(x_{i,}), y\right) \quad (2)$$

4 Results and Discussion

Once the deposition by LMD was completed, the coatings were polished and studied on a tribometer using Yttria-stabilized Zirconia as the counterface at an angle of 80° with reciprocating motions. The load was kept constant at 10N, and scratch tests were conducted at 5000, 12500, and 20000 cycles. Finally, wear rates and the coefficient of friction were calculated for all samples. Figure 2 shows how each coating was polished and worn for all 3 cycles.

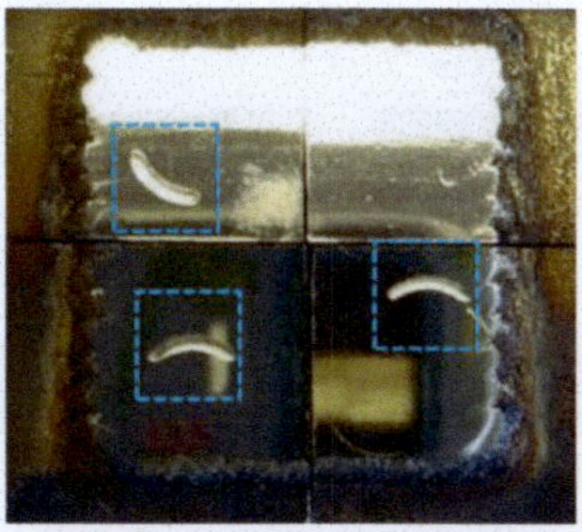

Fig. 2. Wear tracks with different cycles of the tribology study for a sample coating by LMD

All the features obtained and calculated from the data of the monitoring process are shown in Table 3. The information obtained from the master pyrometer (attached to the LMD head) was used to calibrate the other sensors and process the information. This resulted in a total of 4816 data points. This data set was divided into 482 (10%) samples for testing and 4334 (90%) as the training set. All features were measured from the melt pool. Orientation refers to the angle formed between the vertical and major axis of the melt pool, while eccentricity refers to its circularity, defined as $e = \frac{\sqrt{major^2 - minor^2}}{major}$. All of this information, along with the tribological test results, was integrated into a data set. The data were then processed using RF, SVR, and ANN algorithms. Model definition and hyper-parameter optimization were performed using Scikit-Learn tools and GridSearchCV.

4.1 Feature Importance

A comparison of the MDI metric for predicting friction coefficient and wear rate is shown in Fig. 3. It can be observed that geometric properties significantly affect the model

Table 3. Features computed from monitoring process. The minimum and maximum values for each feature are shown within the brackets.

Process parameter	Thermal	Geometric
Laser power (LP)	Profile temperature *[1100, 1827] [°C]*	Width *[1.15, 26.62] [px]*
Scan speed (SS)	Heating rate *[921,1465] [°C/s]*	Length *[1.15, 28.17] [px]*
Feeder rate (FR)	Cooling rate *[2218, 2805] [°C/s]*	Area *[1, 749.88] [px^2]*
–	–	Orientation *[-90, 90]*
–	–	Eccentricity *[0, 1]*

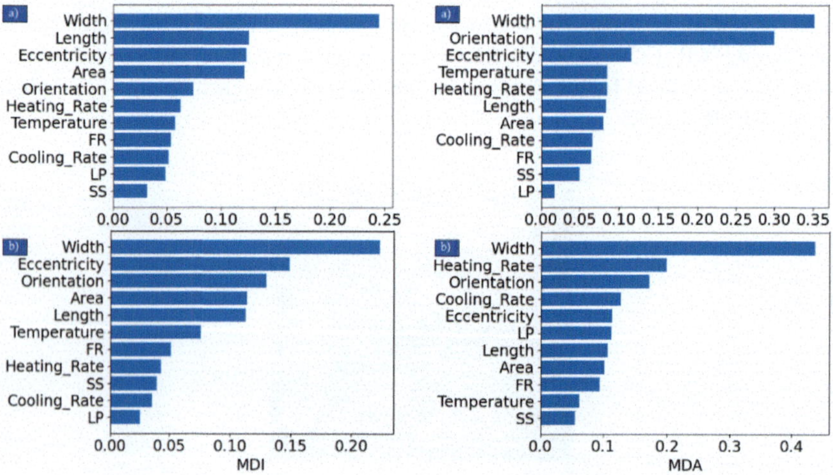

(a) Wear (b) Friction Coefficient.

Fig. 3. MDI and MDA metrics from RF. a) Wear. b) Friction Coefficient.

performance, with melt pool size(width/length),shape(area/eccentricity) and orientation being the main contributors in both cases. It can also be concluded that the inclusion of manufacturing parameters directly in the models does not have a significant impact on their performance. This is likely because including these variables affects the input energy flow to the substrate, modifying the temperatures and shape of the molten pool. As a result, the monitoring process information already includes this contribution. Also Fig. 3 shows the analysis from MDA,in both cases, the most important variable is again the width of the melt pool. However, in order to reduce the model error, the thermal heating and cooling rates of the melt pool become more important. Again, compared to other features, the manufacturing parameters make a smaller contribution.

4.2 Training Results

Figure 4.*a* shows a comparison of training runtime for all trained intelligent models, with RF having the lowest average time (*0.83s*), followed by SVR (*2.51s*), and finally ANN

(*6.32s*). The RF time results are in a range between [*0.02s, 3.42s*], SVR [*0.16s,16.08s*] and ANN [*3.31s,11.19s*]. The training runtime for SVR is strongly influenced by the parameter *C*, as it determines how the model fits to the training data. Higher values of *C* result in better fitting models, but take longer to train. On the other hand, the ANN models with two hidden layers were the ones that had a significant increase in the mean of this group.

As shown in Fig. 4.b RF obtained the best performance (*-0.16*) for all the trained models in the range [*– 0.33, – 0.11*]. ANN scored close (*– 0.25*) to RF range, between [*– 0.33, – 0.15*]. Within this group, the best predictions were obtained by the ReLu kernel, closely followed by the tanh kernel. The scores obtained by the logistic kernel were about 2 times lower than the others. SVR showed a lower overall performance (*–0.46*, from [*– 0.67, – 0.25*]). The best-performing kernel for these models was RBF, followed by polynomials. The linear kernel showed poor predictive performance due to the nature of the procedure.

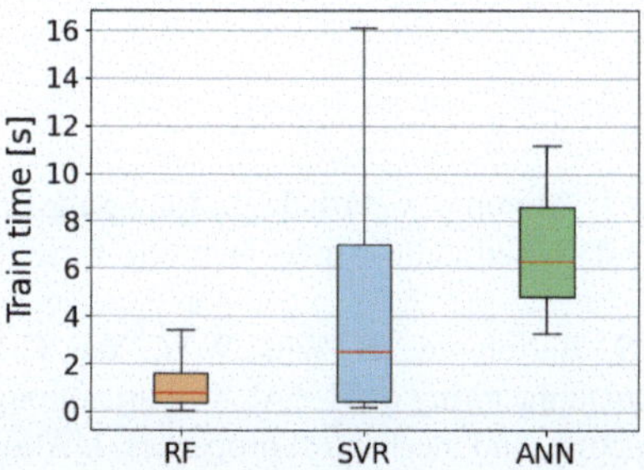

a)

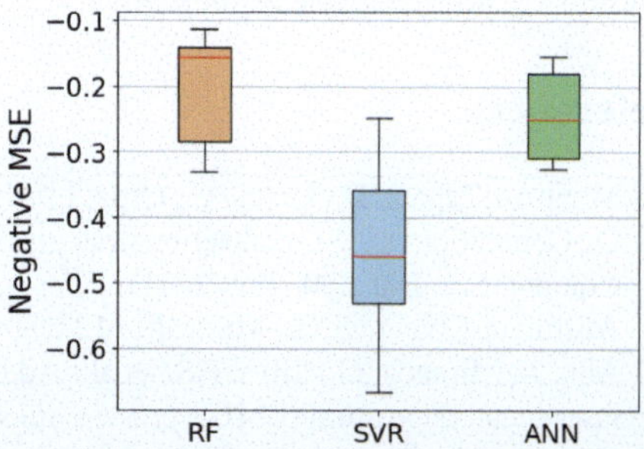

b)

Fig. 4. Comparative for all three evaluated algorithms. a) Training runtime. b) Negative MSE.

Finally, the models with the best adjustment of each algorithm were selected and their scores are presented according to the different metrics in Table 4. Random Forest with setting (max features: 5, min samples leaf: 1, min samples split: 2, n estimators: 190) got the best prediction for wear, while other RF architecture (max features: 5, min samples leaf: 1, min samples split: 2, n estimators: 130) was the best for friction.

Table 4. Accuracy on test data using the best models.

ML	WEAR			COF		
	$\mathbf{R^2}$	MAE	MSE	$\mathbf{R^2}$	MAE	MSE
RF	***0.93***	***0.1065***	***0.0689***	***0.82***	***0.2855***	***0.1761***
SVR	*0.82*	*0.2372*	*0.1780*	*0.69*	*0.4712*	*0.3032*
ANN	*0.89*	*0.1580*	*0.1135*	*0.77*	*0.3628*	*0.2291*

5 Conclusions

The paper presents the results of three algorithms (RF, SVR and ANN) for predicting the wear rate and friction coefficient of LMD's Stellite 6 coating. RF training has been used to identify the importance of the features studied. It was found that geometrical features such as melt pool size (width/length) and shape (area/eccentricity), together with its orientation and thermal heating and cooling rates, were the most important process variables. Furthermore, it can be concluded that including manufacturing parameters in training these intelligent models does not significantly alter their accuracy.

The RF algorithm showed the best fit, obtaining a R^2 of 0.93 for the wear rate and 0.82 for the friction coefficient, followed by the ANN models and finally by the SVR. The RF group also showed the shortest fitting runtime, followed by SVR and ANN. Overall, the intelligent algorithms proved to be viable for the prediction of complex material states, such as tribological variables, from experimental data in the deposition processes of additive manufacturing, which is very useful for the characterization of coatings by non-destructive methods.

References

1. Montoya-Zapata, D., Creus, C., Ortiz, I., Alvarez, P., Moreno, A., Posada, J., Ruiz-Salguero, O.: Generation of 2.5 D deposition strategies for LMD-based additive manufacturing. Procedia Comput. Sci. **180**, 280–289 (2021)
2. Azarniya, Colera, X.G., Mirzaali, M.J., Sovizi, S., Bartolomeu, F., Wits, W.W., Yap, C.Y., Ahn, J., Miranda, G., Silva, F.S., et al.: Additive manufacturing of Ti–6Al–4V parts through laser metal deposition (LMD): process, microstructure, and mechanical properties. J. Alloys Compd **804**, 163–191 (2019)
3. Moradi, M., Ashoori, A., Hasani, A.: Additive manufacturing of Stellite 6 superalloy by direct laser metal deposition–Part 1: effects of laser power and focal plane position. Opt. Laser Technol. **131**, 106328 (2020)
4. Li, X., Jia, X., Yang, Q., Lee, J.: Quality analysis in metal additive manufacturing with deep learning. J. Intell. Manuf. **31**, 2003–2017 (2020)
5. Gao, W., Zhang, Y., Ramanujan, D., Ramani, K., Chen, Y, Williams, C.B., Wang, C.C.L., Shin, Y.C., Zhang, S., Zavattieri, P.D.: The status, challenges, and future of additive manufacturing in engineering. Comput. Aided Des. **69**, 65–89 (2015)
6. Lee, J., Park, H.J., Chai, S., Kim, G.R., Yong, H., Bae, S.J., Kwon, D.: Review on quality control methods in metal additive manufacturing. Appl. Sci. **11**, 1966 (2021)
7. Seifi S.H., Yadollahi, A., Tian, W., Doude, H., Hammond, V.H., Bian, L.: In situ nondestructive fatigue-life prediction of additive manufactured parts by establishing a process–defect–property relationship. Adv. Intell. Syst. **3**, 2000268 (2021)
8. García-Moreno, A.-I.: A fast method for monitoring molten pool in infrared image streams using gravitational superpixels. J. Intell. Manuf. **33**, 1779–1794 (2022)
9. Nair, A., Khan, A.: Studies on effect of laser processed satellite 6 material and its electrochemical behavior. Optik **220**, 165221 (2020)
10. García-Moreno, A.-I., Alvarado-Orozco, J.-M., Ibarra-Medina, J., Martínez-Franco, E.: In-process monitoring of the melt-pool motion during continuous-wave laser metal deposition. J. Manuf. Process. **65**, 42–50 (2021)
11. Santhanakrishnan, S., Kovacevic, R.: Hardness prediction in multi-pass direct diode laser heat treatment by on-line surface temperature monitoring. J. Mater. Process. Technol. **212**, 2261–2271 (2012)

12. Feng, W., Mao, Z., Yang, Y., Ma, H., Zhao, K., Qi, C., Hao, C., Liu, C., Xie, H., Liu, S.: Online defect detection method and system based on similarity of the temperature field in the melt pool. Additive Manuf. **54**, 102760 (2022)
13. Wu, D., Jennings, C., Terpenny, J., Gao, R.X., Kumara, S.: A comparative study on machine learning algorithms for smart manufacturing: tool wear prediction using random forests. J. Manuf. Sci. Eng. **139** (2017)
14. Hasan, M.S., Kordijazi, A., Rohatgi, P.K., Nosonovsky, M.: Triboinformatics approach for friction and wear prediction of Al-graphite composites using machine learning methods. J. Tribol. **144** (2022)
15. Wang, Tan, X.P., Tor, S.B., Lim, C.S.: Machine learning in additive manufacturing: state-of-the-art and perspectives. Additive Manuf. **36**, 101538 (2020)
16. Yin, N., Xing, Z., He, K., Zhang, Z.: Tribo-informatics approaches in tribology research: a review. Friction **11**, 1–22 (2023)
17. Valizadeh, M., Wolff, S.J.: Convolutional Neural Network applications in additive manufacturing: a review. In: Advances in Industrial and Manufacturing Engineering, pp. 100072 (2022)

Predicting Future Sales: A Machine Learning Algorithm Showdown

Manal Loukili[1(✉)], Fayçal Messaoudi[1], Mohammed El Ghazi[1], and Hanane Azirar[2]

[1] Sidi Mohamed Ben Abdellah University, National School of Applied Sciences, Fes, Morocco
manal.loukili@usmba.ac.ma

[2] Faculty of Juridical, Economic and Social Sciences, Sidi Mohamed Ben Abdellah University, Fes, Morocco

Abstract. In the internet era, handling vast data volumes manually is impractical, while accurate sales prediction remains crucial for organizations. Machine learning techniques offer powerful tools to extract hidden patterns from extensive datasets, enhancing prediction accuracy. This paper uses machine learning models to forecast future sales based on historical data from the "Store Item Demand Forecasting" dataset, comprising five years of sales data for 50 items across ten stores. Regression techniques, including linear regression, Random Forest regressor, and XGBoost, were employed, along with the LSTM algorithm. Results evaluated using MAE, RMSE, and R-squared, indicate that the XGBoost model outperformed other models in predicting sales with higher accuracy, closely followed by linear regression.

Keywords: Supervised Learning · Sales Forecasting · Regression · Random Forest · Linear Regression · LSTM · XGBoost

1 Introduction

Customer satisfaction is vital in the business world, and organizations constantly seek to meet demands and increase profits through smart investments [1]. Predicting future sales plays a crucial role in making informed decisions to boost profits [2]. Sales forecasting, based on historical data, is essential for companies venturing into new markets, launching new products, or expanding significantly. Efficient and accurate sales forecasting enables resource optimization, cost reduction, and enhanced customer satisfaction. Machine learning has proven to be highly effective in several tasks in the world of e-business including predicting churn [3], sentiment analysis [4], recommendations [5], and sales prediction. With the help of machine learning algorithms, companies can analyze vast amounts of customer data to identify patterns and trends that can be used to predict customer behavior. This information can be used to develop strategies for customer retention and churn reduction, as well as for making personalized product recommendations. Machine learning can also help companies to predict sales volumes accurately, enabling them to optimize resource utilization and inventory control [6]. This

Y. Farhaoui et al. (Eds.): ICAISE 2023, LNNS 837, pp. 26–31, 2024.
https://doi.org/10.1007/978-3-031-48465-0_4

paper explores the use of machine learning algorithms for sales forecasting, emphasizing the importance of this process for businesses. It proceeds with methodology description, model comparison, and conclusion in subsequent sections [7, 8].

2 Methodology

The focus of this paper is to compare four machine learning algorithms for sales prediction namely: linear regression, Random Forest Regressor, XGBoost, and LSTM. The reasons for choosing these algorithms are as follows:

Linear Regression is simple to understand and explain and can be regularized to avoid overfitting. Random Forest provides strong and easily understandable predictions, efficiently processes large data sets, and achieves a higher level of accuracy than the decision tree algorithm for predicting outcomes. XGBoost uses the power of parallel processing, is highly flexible, and is able to handle missing data thanks to its built-in functions, as well as being faster than Gradient Boosting. Finally, LSTM cells are used in recurrent neural networks that learn to predict the future from sequences of varying lengths. Recurrent neural networks work with any type of sequential data and, unlike ARIMA and Prophet, are not limited to time series.

With the aim of developing and evaluating the set of prediction models used, we adopted the methodology presented in Fig. 1. In addition, we developed a range of functions listed in Table 1.

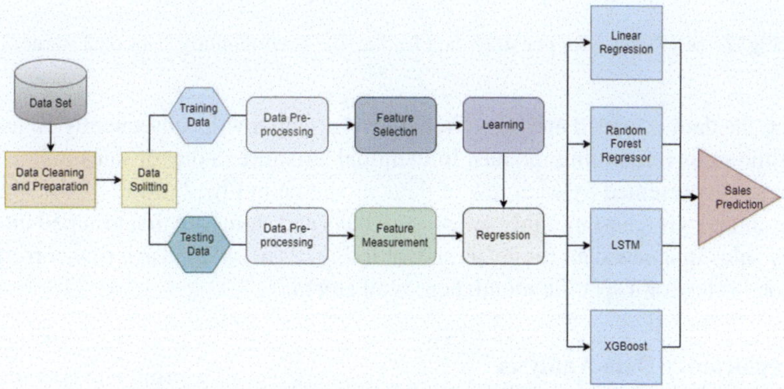

Fig. 1. The architecture of the adopted methodology.

2.1 Data Description and Visualization

The dataset used is "Store Item Demand Forecasting Challenge" which contains the sales data of a chain of stores. The dataset consists of four features, namely "date", "store", "item" and "sales". It contains a total of 913,000 rows, with each row representing sales data for a specific date, store and item.

Table 1. The functions used.

Function	Description
Train test split	The data is split in such a way that the last 12 months are part of the test set, and the rest of the data serves to form our model
Scaling the data	The data is scaled so that all variables are between -1 and 1 via a min-max scaler
Inverting the scale	The objective of this function is to invert the scaling of the previous function
Create prediction data frame	The data frame comprises the actual sales collected in our test set as well as the results predicted by our model to measure the performance of each model
Score the models	Evaluate the models based on the selected metrics

	date	store	item	sales
0	2013-01-01	1	1	13
1	2013-01-02	1	1	11
2	2013-01-03	1	1	14
3	2013-01-04	1	1	13
4	2013-01-05	1	1	10

(a)

	date	sales
0	2013-01-01	454904
1	2013-02-01	459417
2	2013-03-01	617382
3	2013-04-01	682274
4	2013-05-01	763242

(b)

Fig. 2. (a): Daily sales per store data frame. (b): Total monthly sales data frame.

First, the data is loaded and converted into a form that will subsequently be used for each of the sales forecasting models. In its initial structure, a day of sales in one of the ten stores is represented by each row of data, as shown in Fig. 2 (a).

The aim is to forecast monthly sales, so all days and stores are first grouped into total monthly sales in a new data frame, as shown in Fig. 2 (b), where each row corresponds to the total sales for a specific month across all stores.

2.2 Exploratory Data Analysis

The figure below (Fig. 3) represents graphically the distribution of sales per day and the distribution of total sales per store.

2.3 Determination of Stationarity

The plot of total monthly sales against time in Fig. 4 shows that mean monthly sales are rising with time, indicating that the data is not stationary. To make the data stationary, the difference between the sales of each month was computed and added to the data frame as a new feature (sales_diff), as shown in Fig. 6.

The following figure is a graphical representation of the appearance of the data before and after the differencing transformation (Fig. 4).

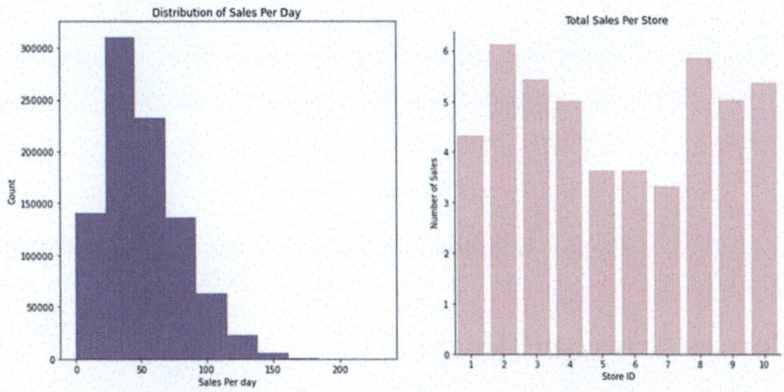

Fig. 3. Distribution of sales per day and total sales per store.

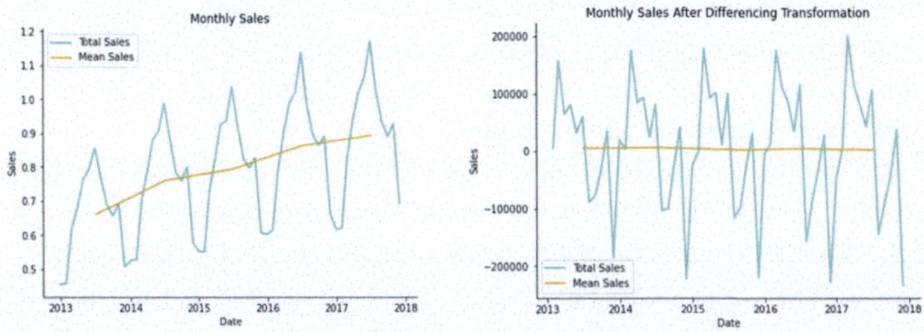

Fig. 4. Monthly sales plot before and after differencing transformation.

2.4 Lags Observing

A coherent look-back period for the regression models is provided by the autocorrelation and partial autocorrelation plots below (Fig. 5).

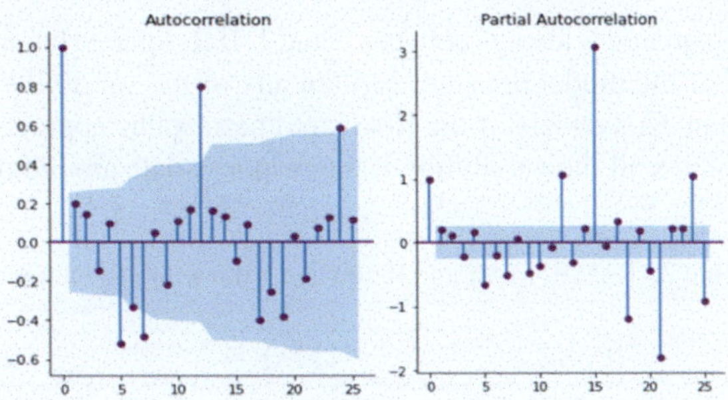

Fig. 5. Autocorrelation and partial autocorrelation graphs.

2.5 Data Set Regressive Modeling

Based on the above graphs, a 12-month look-back period was chosen for inclusion in our feature set. Next, a data frame was generated consisting of 13 columns, with 12 columns representing the months and one column for the dependent variable "sales_diff" (Fig. 6).

	date	sales	sales_diff	lag_1	lag_2	lag_3	lag_4	lag_5	lag_6	lag_7	lag_8	lag_9	lag_10	lag_11	lag_12
0	2014-02-01	529117	3130.0	19380.0	-186036.0	36056.0	-33320.0	-76854.0	-89161.0	60325.0	32355.0	80968.0	64892.0	157965.0	4513.0
1	2014-03-01	704301	175184.0	3130.0	19380.0	-186036.0	36056.0	-33320.0	-76854.0	-89161.0	60325.0	32355.0	80968.0	64892.0	157965.0
2	2014-04-01	788914	84613.0	175184.0	3130.0	19380.0	-186036.0	36056.0	-33320.0	-76854.0	-89161.0	60325.0	32355.0	80968.0	64892.0
3	2014-05-01	882877	93963.0	84613.0	175184.0	3130.0	19380.0	-186036.0	36056.0	-33320.0	-76854.0	-89161.0	60325.0	32355.0	80968.0

Fig. 6. Total monthly sales data frame.

3 Model Comparison and Results

The indicators MAE, RMSE, and R-squared were used to compare sales forecasting models, as shown in Table 2.

MAE or Mean Absolute Error provides a measure of the average magnitude of errors in a range of predictions, regardless of their direction. MAE is the average over the tested sample of the absolute differences of the prediction and the real observation where all individual differences have the same weight.

RMSE or Root Mean Square Error is a quadratic scoring method that also provides a measure of the average magnitude of the error. It corresponds to the square root of the average of the square differences of the prediction and the real observation.

MAE and RMSE values can be varied from 0 to ∞, and are negatively oriented scores, implying that the lower values are preferable.

R-squared or R2, is the coefficient of determination that serves to evaluate the quality of a linear regression. It corresponds mathematically to the proportion of the variance of a dependent variable that is explained by one or more independent variables in the regression model. It is the percentage of variation in the response variable that is explained by a linear model. Where:

$$R - squared = Explained\ variance / Total\ variance \tag{1}$$

The value of R-squared is always between 0 and 1. If R-squared has a value approaching 0, it implies that the model does not explain any of the variability of the response data around its mean. Conversely, if the R-squared has a value approaching 1, it means that the model explains all the variability of the response data around its mean.

Table 2. RMSE, MAE and R-squared values for each model.

Model	RMSE	MAE	R-Squared
LSTM	19281.054287	16007.250000	0.976882
Random forest	19227.637970	16531.583333	0.976955
Linear regression	16221.040791	15433.000000	0.980716
XGBoost	13701.003360	12342.666667	0.991301

From the comparison of the performance measures in Table 2, it can be seen that the XGBoost model is the best performing one, with the highest determination coefficient of 99.13% and relatively low errors. The Linear Regression algorithm is in second place with a determination coefficient of 98.07%. The Random Forest and LSTM algorithms are in third place with higher errors and relatively low coefficients of determination.

4 Conclusion and Outlook

In this paper, the following supervised machine learning algorithms were implemented: Linear Regression, Random Forest, XGBoost, and LSTM, to solve a regression problem that is the prediction of future sales. The XGBoost model was found to perform better in our sales prediction case, followed closely by the linear regression model. In this study, the simplified form of each model was used to illustrate how these models can be used for sales prediction. In order to minimize complexity, the models were only partially adjusted. In contrast, the LSTM model, in particular, can be enhanced by adding many additional nodes and layers. The current study can be improved by incorporating Big Data analytics, which is increasingly essential in today's business world for accurate sales forecasting. Further research could explore more complex model configurations and investigate additional features that could impact sales prediction.

References

1. Mediavilla, M.A., Dietrich, F., Palm, D.: Review and analysis of artificial intelligence methods for demand forecasting in supply chain management. In: CIRP Conference on Manufacturing Systems, pp. 1126–1131 (2022)
2. Zhang, C., Tian, Y.X., Fan, Z.P., Liu, Y., Fan, L.W.: Product sales forecasting using macroeconomic indicators and online reviews: a method combining prospect theory and sentiment analysis. Soft. Comput. (2019). https://doi.org/10.1007/s00500-018-03742-110.1007/s0
3. Loukili, M., Messaoudi, F., El Ghazi, M.: Supervised learning algorithms for predicting customer churn with hyperparameter optimization. Int. J. Adv. Soft Comput. Appl. **14**(3), 49–63 (2022). https://doi.org/10.15849/IJASCA.221128.04
4. Loukili, M., Messaoudi, F., El Ghazi, M.: Sentiment analysis of product reviews for e-commerce recommendation based on machine learning. Int. J. Adv. Soft Comput. Appl. **15**(1), 1–13 (2023). https://doi.org/10.15849/IJASCA.230320.01
5. Loukili, M., Messaoudi, F.: Machine learning, deep neural network and natural language processing based recommendation system. In: Kacprzyk, J., Ezziyyani, M., Balas, V.E. (eds.) International Conference on Advanced Intelligent Systems for Sustainable Development. AI2SD 2022. Lecture Notes in Networks and Systems, vol. 637. Springer, Cham (2023). https://doi.org/10.1007/978-3-031-26384-2_7
6. Kohli, S., Godwin, G.T., Urolagin, S: Sales prediction using linear and KNN regression. In: Advances in Machine Learning and Computational Intelligence: Proceedings of ICMLCI 2019, pp. 321–329. Springer, Singapore (2020)
7. Farhaoui, Y., et al.: Big data mining and analytics **5**(4), I–II (2022). https://doi.org/10.26599/BDMA.2022.9020004
8. Farhaoui, Y., et al.: Big data mining and analytics **6**(3), I–II (2023). https://doi.org/10.26599/BDMA.2022.9020045

A Detailed Study on the Game of Life

Serafeim A. Triantafyllou(✉)

Greek Ministry of Education and Religious Affairs, Athens, Greece
ser.triant@gmail.com

Abstract. In 1952, Alan Turing who is considered as a father of Computer Science, based on his previous scientific research on the theory of computation, he emphasized how important is the analysis of pattern formation in nature and developed a theory. In his theory, he described specific patterns in nature that could be formed from basic chemical systems. Turing in his previous studies in the theory of computation, he had constantly worked on symmetrical patterns that could be formed simultaneously and realized the necessity for further analysis of pattern formation in biological problems. However, it was until the late 1960s, when John Conway was the first to introduce the "Game of Life", an innovative mathematical game based on cellular automata, having a purpose to utilize the fundamental entities, called as cells, in two possible states described as "dead" or "alive". This paper tries to contribute to a better understanding of the "Game of Life" by implementing algorithmic approaches of this problem in PASCAL programming language.

Keywords: Game of Life · Cellular automata · Algorithms

1 Introduction

The "Game of Life" was first proposed by John Conway, and it is a cellular automaton which rules were firstly introduced by him in the late 1960s [1–3, 10]. The detailed description of Conway's work by Martin Gardner in the March 1970 Mathematical Games column in Scientific American, made the "Game of Life" popular to the public [7, 11]. Apparently, the "living" is considered as an example of metaphor in the "Game of Life" and this mathematical game has the potential of a universal Turing machine, which means that it is constructed in a way to be computed algorithmically [4, 8, 9]. This Turing machine is constructed from patterns in the Conway's "Game of Life" cellular automaton [5, 6]. Conway described a method of developing a register machine which can simulate a Turing machine [5, 9]. The Game of Life, invented by him is essentially a cellular automaton, where a selection of entities called as cells are arranged in a grid of an explicitly predetermined shape. Each cell gets a state from a binary set and updates its state according to a set of strict rules. Every cell transforms its state as a function of time and for Conway's "Game of Life", the cells include two states, named as "live" and "dead", and the necessary rules are driven by the states of the 8 neighboring cells, that are in the "alive" state [1, 4, 7]. The cells adjacent to another cell form its neighborhood. Cells in a neighborhood interact, and every cell on the cellular automaton grid has a neighborhood.

Y. Farhaoui et al. (Eds.): ICAISE 2023, LNNS 837, pp. 32–38, 2024.
https://doi.org/10.1007/978-3-031-48465-0_5

2 Background

The first cellular automaton which was developed by John von Neumann in 1966 was complex with its 29 states [13, 14], but there are much simpler ones which have only two states. John von Neumann's cellular automaton was a complex automaton with 29 different states based on a Universal Turing Machine that can recreate itself. Later, Christopher G. Langton in 1984, made the things simpler when he created a cellular automaton with only eight states, which gives up the universality, but still can execute a program (Langton's Loops) that can recreate itself [15]. A cellular automaton is a dynamic system arranged in a specified grid of cells each of which has a specific number of connected discrete states n, and in this paper, we consider n = 2 (for the states set as "live" and "dead") [12, 17–26]. Each state is updated in discrete time steps with a time function according to specific state rules [12, 14]. "Game of Life" is a game played on an infinite grid, where at any moment some cells are dead and some are live. Every cell is adjacent to 8 neighbors, creating the Moore neighborhood [14].

Melanie Mitchell-professor of complexity in the Santa Fe Institute mentions that [27]:

> Given that Conway's proof that the Game of Life can be made to simulate a Universal Computer—that is, it could be "programmed" to carry out any computation that a traditional computer can do—the extremely simple rules can give rise to the most complex and most unpredictable behavior possible.

Dr. Mitchell emphasizes that we should learn to live with uncertainty and inherent unpredictability.

The basic purpose of this study is to contribute to a better understanding of the "Game of Life" by implementing algorithmic approaches of this problem in PASCAL programming language.

3 Proposed Method

3.1 Detailed Algorithm Analysis

We consider a rectangular axis system with integer subdivisions and specifically we have chosen 9 x 9 subdivisions. Every position could be empty or possessed by a "living" organism with the symbol "*". Supposing that we are in a first state called as (FIRST GENERATION), we can move to the next state by following the basic rules:

1. Every position and for example the x position has 8 neighbors, the (1,2,3,4,5,6,7,8)
2. A "living" organism ("*") could be extinct (death state) and its position to be possessed by a blank symbol, if and only if, there are lesser than 2 or more than 3 "living" organisms as neighbors.
3. In an empty position an organism can be "born" (birth state), if and only if, in this position there are exactly 3 "living" organisms as neighbors.

3.2 Implementation of the Algorithmic Approaches in the PASCAL Programming Language

The decision to implement the algorithmic approaches in Pascal programming language was taken after careful consideration, because Pascal is a procedural programming language for good programming practices by using structured programming [16]. In this section we program in PASCAL programming language and the source code that is presented in detail, implements the following actions:

1. Create a rectangular axis system 9 × 9 (subroutine—procedure SCREEN).
2. Create a first state (subroutine—procedure FIRST GENERATION).
3. Create a next state starting from an existing state (subroutine—procedure NEXTGENERATION).
4. Finish the generation processes (subroutine—function FINISH).
5. With the help of the subroutines 1–4, a main program that starts from a first state moves to the creation of new generations, and terminates the overall process when needed, by using the function FINISH.

The source code in PASCAL programming language is the following (see the code in PASCAL programming language):

```
program PROBLEM(input,output);
uses Crt;
const size = 9;
type mchar = array[0..size+1, 0..size+1] of char;
var chessboard: mchar; generation: integer;
procedure SCREEN(var chessboard:mchar;generation:integer);
var i,j : 1..size;
BEGIN
    ClrScr;
    WRITELN;
    WRITELN( 'generation: ', generation);
    WRITELN;
    WRITE(' 1 2 3 4 5 6 7 8 9 ');
    WRITELN;
    FOR i : = 1 TO size DO
                WRITE (chessboard[i,j]:3);
            WRITELN
    END
END;
procedure FIRSTGENERATION (var chessboard: mchar; var generation:
integer);
                var i, j: integer; flag: boolean;
                BEGIN
                generation: = 1;
    FOR i: = 0 TO size + 1 DO
                FOR j: = 0 TO size + 1 DO
                         chessboard [i,j]: = ' ';
                flag: = FALSE;
                WRITE('to terminate the process give
```

```
:        ');
                 WRITELN (' (i < 1) OR (i >', size: 2, ' OR j < 1 OR
j > ', size: 2, ') ');
                WRITELN;
                REPEAT
                        WRITE ('give the coordinates i and j of the
organism: ');
                        READLN (i,j);
                          IF (i > 0) AND (i < = size) AND (j >0) AND
(j < = size)
                        THEN chessboard[i,j] := '*';
                        ELSE flag:= TRUE
                UNTIL flag
END;
procedure NEXTGENERATION(var chessboard:mchar);
var i,j: integer; helpboard : mchar;
i1, j1: 0..size+1;
neighbors: integer;
BEGIN
   neighbors : = 0;
   FOR i1:= i - 1 TO i + 1 DO
                FOR j1:= j - 1 TO j + 1 DO
                        IF chessboard[i1,j1] = '*' THEN
                              neighbors: = neighbors + 1;
   IF chessboard[i,j] = '*'THEN
   neighbors:= neighbors - 1;
helpboard:= chessboard;
FOR i:=1 TO size DO
   FOR j:=1 TO size DO
   BEGIN
       neighbors:= 0;
       FOR i1:= i - 1 TO i + 1 DO
       FOR j1:= j - 1 TO j + 1 DO
          IF chessboard[i1,j1]='*' THEN
               neighbors:= neighbors + 1;
   IF chessboard[i,j] = '*' THEN
      neighbors := neighbors - 1;
   IF helpboard[i,j] = ' ' THEN BEGIN
               IF neighbors = 3THEN
                  helpboard[i,j]:= '*';
               END
   ELSE IF (neighbors < 2) OR(neighbors > 3)
   THEN helpboard[i,j]:= ' ';
   chessboard := helpboard;
   END
END;
function FINISH: boolean;
var a: char;
BEGIN
WRITE ('next generation?(for yes press: y):');
```

```
READLN(a);
FINISH:= (a='y')
END;
BEGIN
ClrScr;
FIRSTGENERATION(chessboard, generation);
SCREEN(chessboard, generation);
WHILE FINISH DO
                BEGIN
                         WRITELN;
                         NEXTGENERATION(chessboard);
                         generation:= generation + 1;
                         SCREEN(chessboard, generation)
                END
END.
```

4 Presentation of Data and Outcomes

In this section, we present the data and final outcomes after running all the subroutines and the main program in PASCAL programming language (see the following code and Fig. 1).

to terminate the process give: (i < 1) OR (i > 9 OR j < 1 OR j > 9)
give the coordinates i and j of the organism: 4 5
give the coordinates i and j of the organism: 5 6
give the coordinates i and j of the organism: 6 6
give the coordinates i and j of the organism: 4 6
give the coordinates i and j of the organism: 0 0

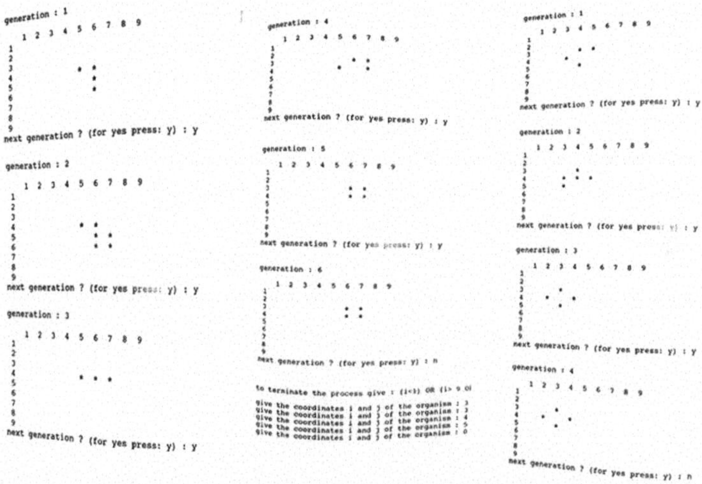

Fig. 1. Data and outcomes

5 Conclusions and Future Work

The "Game of Life" cellular automaton is a characteristic paradigm of a parallel collision-based computing machine, and its rules were invented by John Conway in the late 1960s. Nowadays, we are still getting inspiration from Conway's discovery. Dr. Conway was among one of the pioneers in the field of cellular automata and by introducing the "Game of Life", he introduced new scientific achievements in the domain of complexity science, with simulations that can be used to identify patterns and model problems like ants to traffic and many others such as clouds to galaxies. This paper tried to contribute to a better understanding of the "Game of Life" by implementing algorithmic approaches of this problem in PASCAL programming language. Future studies aim to study further the "Game of Life" and its dynamics to identify patterns and model many important problems.

References

1. Bays, C.: Introduction to cellular automata and Conway's game of life. In: Adamatzky, A. (eds.) Game of Life Cellular Automata. Springer, London (2010). https://doi.org/10.1007/978-1-84996-217-9_1
2. Caballero, L., Hodge, B., Hernandez, S.: Conway's 'Game of life' and the epigenetic principle. Front. Cell. Infect. Microbiol. **6** (2016). Available at: https://doi.org/10.3389/fcimb.2016.00057
3. Wainwright, R.: Conway's game of life: early personal recollections. In: Adamatzky, A. (eds.) Game of Life Cellular Automata. Springer, London (2010). https://doi.org/10.1007/978-1-84996-217-9_2
4. Rendell, P.: Turing universality of the game of life. In: Adamatzky, A. (eds.) Collision-Based Computing. Springer, London (2002). https://doi.org/10.1007/978-1-4471-0129-1_18
5. Rendell, P.: Conway's game life turing machine. http://www.rendell.uk.co/gol
6. Turing, A.M.: On computable numbers, with applications to the entscheidungsproblem. Proc. London Math. Soc. **42**, 230–265 (1937)
7. Gardner, M.: Mathematical games: the fantastic combinations of John Conway's new solitaire Game 'Life.' Sci. Am. **5**, 11 (1970)
8. Rogozhin, Y.: Small universal turing machines. Theor. Comput. Sci. **168**(2), 215–240 (1996). ISSN 0304-3975. 2, 13, 46
9. Smith, A.: Universality of Wolfram's 2, 3 Turing Machine, 2007. the Wolfram 2,3 Turing Machine Research Prize. 55, 185
10. Wolfram, S.: Universality and complexity in cellular automata. Physica **10D**, 1–35(1984). 4, 6, 13, 55, 185, 200
11. Gardner, M.: Mathematical games: the fantastic combinations of john conway's new solitaire game "life." Sci. Am. **223**, 120–123 (1970)
12. Kazakov, D., Sweet, M.: Evolving the game of life. In: Lecture Notes in Computer Science, vol. 3394 (2004). https://doi.org/10.1007/978-3-540-32274-0_9
13. Hirte, R.: John Horton Conway's game of life. An overview and examples (2022)
14. von Neumann, J.: Theory of self-reproducing automata. In: Burks, A.W. (1966)
15. Langton, C.G.: Self-reproduction in cellular automata. Physica D: Nonlinear Phenomena **10**(1–2), 135–144 (1984)
16. Zaks, R.: Introduction to Pascal (including UCSD Pascal). SYBEX Inc. (1981)

17. Triantafyllou, S.A.: A quantitative research about MOOCs and EdTech Tools for distance learning. In: Auer, M.E., El-Seoud, S.A., Karam, O.H. (eds.) Artificial Intelligence and Online Engineering. REV 2022. Lecture Notes in Networks and Systems, vol. 524. Springer, Cham (2023). https://doi.org/10.1007/978-3-031-17091-1_52
18. Triantafyllou, S.A.: TPACK and toondoo digital storytelling tool transform teaching and learning. In: Florez, H., Gomez, H. (eds.) Applied Informatics. ICAI 2022. Communications in Computer and Information Science, vol. 1643. Springer, Cham (2022). https://doi.org/10.1007/978-3-031-19647-8_24
19. Triantafyllou, S.A.: Use of Business Information Systems to Achieve Competitiveness. In: 2022 13th National Conference with International Participation (ELECTRONICA), pp. 1–4 (2022). https://doi.org/10.1109/ELECTRONICA55578.2022.9874433
20. Triantafyllou, S.A.: Work in progress: Educational Technology and Knowledge Tracing Models. In: 2022 IEEE World Engineering Education Conference (EDUNINE), pp. 1–4 (2022). https://doi.org/10.1109/EDUNINE53672.2022.9782335
21. Triantafyllou, S.A.: Magic squares in order 4K+2. In: 2022 30th National Conference with International Participation (TELECOM) (2022). https://doi.org/10.1109/TELECOM56127.2022.10017312
22. Triantafyllou, S.A.: Constructivist learning environments. In: Proceedings of the 5th International Conference on Advanced Research in Teaching and Education (2022). https://doi.org/10.33422/5th.icate.2022.04.10
23. Triantafyllou, S.A.: A Detailed study on the 8 queens problem based on algorithmic approaches implemented in PASCAL programming language. In: Silhavy, R., Silhavy, P. (eds.) Software Engineering Research in System Science. CSOC 2023. Lecture Notes in Networks and Systems, vol. 722. Springer, Cham (2023). https://doi.org/10.1007/978-3-031-35311-6_18
24. Triantafyllou, S.A., Sapounidis, T.: Game-based Learning approach and Serious Games to learn while you play. In: 2023 IEEE World Engineering Education Conference (EDUNINE) (2023). https://doi.org/10.1109/EDUNINE57531.2023.10102872
25. Triantafyllou, S.A.: What philosophy can teach us about games? In: Proceedings of the 7th International Conference on Social Sciences, Humanities and Education (2022). https://doi.org/10.33422/7th.icshe.2022.12.20
26. Triantafyllou, S.A.: Game-Based Learning and interactive educational games for learners—an educational paradigm from Greece. In: Proceedings of The 6th International Conference on Modern Research in Social Sciences (2022). https://doi.org/10.33422/6th.icmrss.2022.10.20
27. The lasting lessons of john Conway's game of life. The New York Times (2020)

A Decision-Making Model for Selecting Solutions and Improvement Actions Using Fuzzy Logic

Anass Mortada[1](✉) and Aziz Soulhi[2](✉)

[1] University Mohammed V-Agdal, Mohammedia School of Engineers, Rabat, Morocco
anass.mortada@gmail.com
[2] Higher National School of Mines, Rabat, Morocco
soulhi@enim.ac.ma

Abstract. The ability to solve problems and implement effective and relevant improvement actions is one of the most important factors in a company's success. Companies often find it difficult to determine which actions are worth implementing and prioritizing, and whether they are worth the effort and cost.

The objective of this article is to develop a fuzzy logic model that facilitates the decision and selection of relevant actions and solutions by estimating the value of action relevance based on four input indicators: "Efficiency", "Feasibility", "Cost" and "Implementation time".

This model enables companies and managers to follow up the solutions proposed during brainstorming sessions and analyze each one with precision, thanks to the detailed values of the input parameters that enable decisions to be made on the relevance of the actions and the extent to which they need to be implemented in order to be able to solve the problems with maximum efficiency and with the least resources.

Keywords: Decision Making · Fuzzy Logic · Artificial intelligence · Brainstorming · Action relevance · Solution selection

1 Introduction

Companies are striving to maintain their economic advantages as the industrial sector becomes more competitive by finding solutions to their challenges, improving their indicators and implementing the best ideas and activities [1], to stand out from the competition and offer high-quality goods and services at competitive costs and on time [2], this can ensure that the company remains competitive from the costumer's point of view and prevent any complaints [3].

Brainstorming involves bringing a group of people together to find a solution or produce a list of ideas that could solve a problem [4]. It is a commonly utilized technique for idea generating in a range of organizational environments [5].

However, due to the interaction of various aspects likely to affect the effectiveness of the solution and limit its relevance, companies often find it difficult to choose the best ideas and validate the most relevant activities.

Y. Farhaoui et al. (Eds.): ICAISE 2023, LNNS 837, pp. 39–45, 2024.
https://doi.org/10.1007/978-3-031-48465-0_6

Since artificial intelligence is mainly focused on creating models with intelligent behavior, Professor Lotfi Zadeh has developed fuzzy logic as an artificial intelligence logic system designed to codify natural human reasoning. Fuzzy logic is a very powerful tool for streamlining management and decision-making [6].

The boundaries of fuzzy sets are ambiguous, so it is possible for elements to belong partially to the same set. Unlike the conventional notion of binary sets, fuzzy logic modelling is based on the theory of fuzzy sets [7] (Fig. 1).

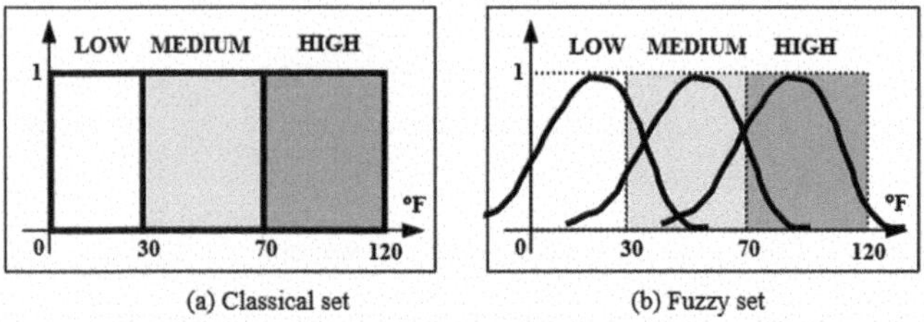

Fig. 1. Examples from the classical and fuzzy sets [8]

Fuzzy logic allows linguistic expressions to be converted into mathematical formulas, making it possible to move from a qualitative description based on an expert in the field to a quantitative description via the mathematical model. The two main elements of fuzzy logic are membership functions and fuzzy rules [6].

When modelling a process using fuzzy logic, the variables in the model must belong to fuzzy classes and be managed by IF…THEN rules so that a conclusion can be drawn for any combination of fuzzy classes of variables [7].

The process of "fuzzification" converts traditional, crisp data into fuzzy data [8], by specifying the membership functions for the input and output variables. By establishing the form of the membership functions and the degree of membership of each of the states to be defined, the numerical data is converted into linguistic variables [9]. Domain experts must define the membership functions, and the model then produces the output variable using the center of gravity method [9].

The control rules and membership functions already defined are combined in the fuzzy inference engine stage to produce the fuzzy output data [8].

Defuzzification is used to identify the set of fuzzy outputs once inference is complete. Accurate application of the model results requires a transfer from the "fuzzy world" to the "real world" [9].

A summary of fuzzy logic modeling is provided in (Fig. 2).

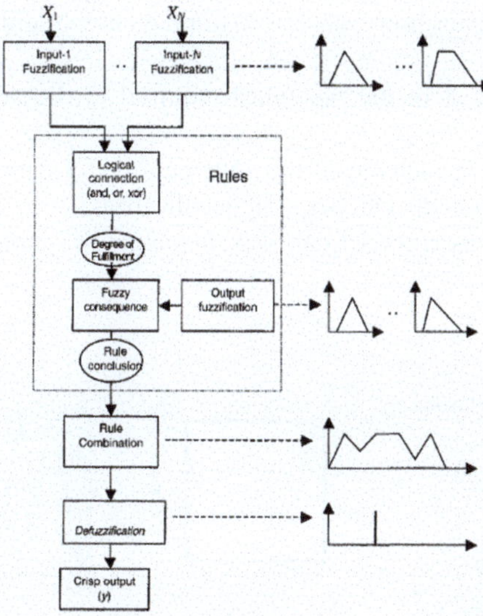

Fig. 2. A fuzzy logic-based model's schematic [7]

We propose a method based on a mathematical fuzzy logic model that calculates and evaluates the relevance of actions according to the values of four parameters: "Efficiency", "Feasibility", "Cost" and "Implementation time" in order to solve the problem of choosing appropriate activities.

2 Proposed Model

2.1 Action Relevance Estimation

The most important thing about an action is its ability to solve the problem or improve performance, which is what we call the action's efficiency, however, other parameters are also very important, such as feasibility, which indicates the level of constraints that can disrupt the implementation of the solution, the cost and also the time required to implement the action.

The interaction between all these parameters somewhat complicates the decision on action relevance, hence the importance of fuzzy logic.

In this article, we present a new method based on a fuzzy logic model for calculating "action relevance", based on the estimation of the input variables "Efficiency", "Feasibility", "Cost" and "Implementation time".

2.2 Indicators Definition

The following four indicators will be used to evaluate the action relevance as an output indicator:

Efficiency: i.e. the extent to which the action is capable of solving the problem or improving performance.

Feasibility: which means how easy it is to implement the solution and whether there are any constraints.

Cost: includes the cost of the resources required to develop and implement the solution.

Implementation time: which means the time required for the action to be definitively completed so that the company can begin to benefit from it.

As a result, the proposed model can be schematized as indicated in the following (Fig. 3).

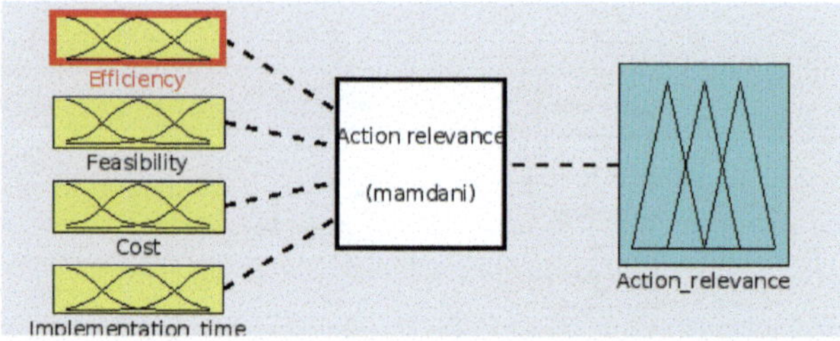

Fig. 3. Proposed fuzzy model

2.3 Modeling of Indicators

It is now time to model the proposed approach and the input and output indicators by determining the membership functions for each variable, as illustrated in (Figs. 4, 5, 6, 7 and 8).

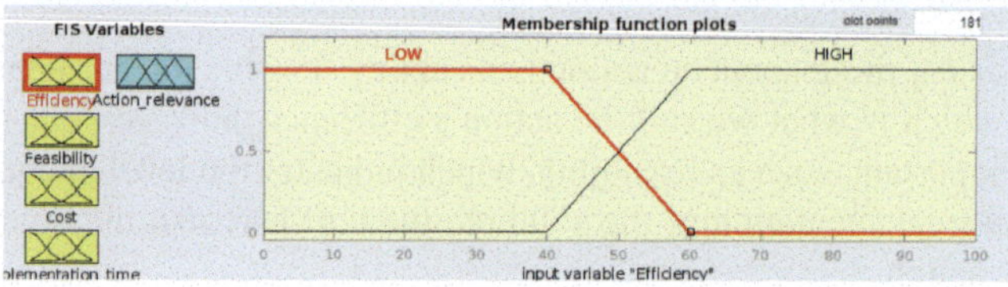

Fig. 4. Membership function for "Efficiency"

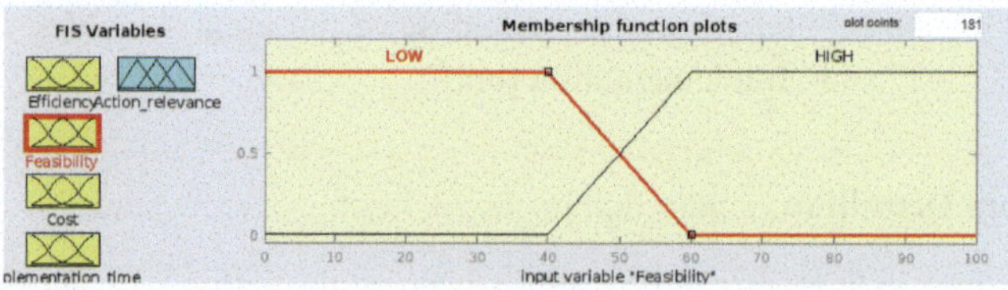

Fig. 5. Membership function for "Feasibility"

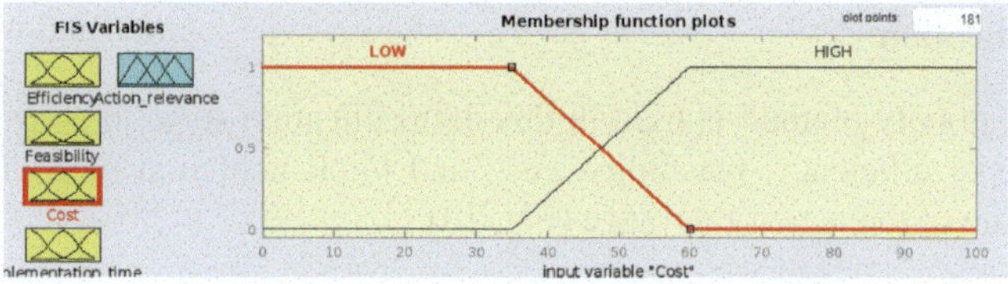

Fig. 6. Membership function for "Cost"

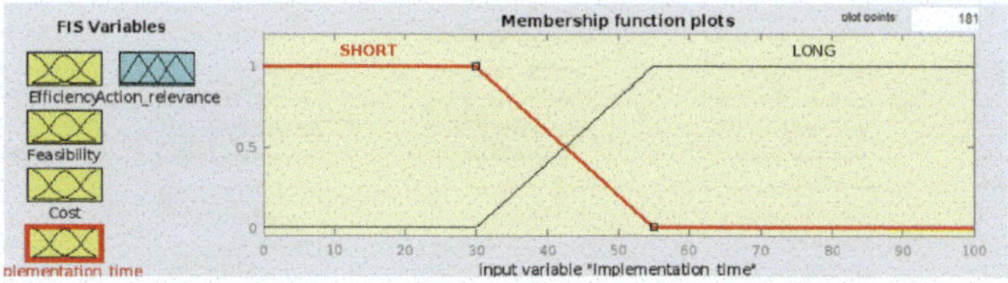

Fig. 7. Membership function for "Implementation time"

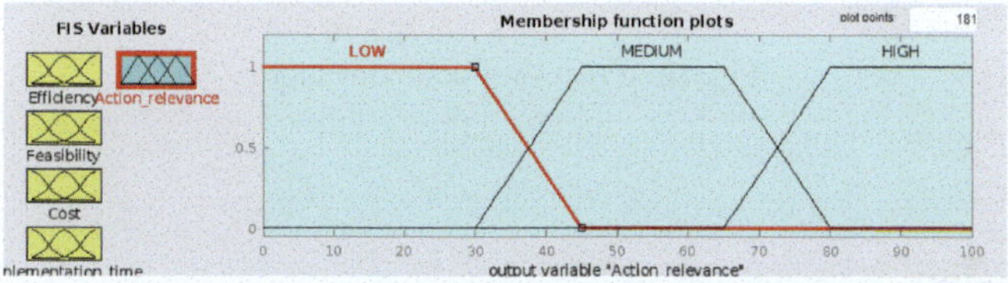

Fig. 8. Membership function for "Action relevance"

2.4 Fuzzy Inference

To manage the interaction between the many input variables, we will define fuzzy rules based on the expertise acquired in the field at this stage.

Using the <<AND>> operator, the following 16 fuzzy rules (2 * 2 * 2 * 2) are created (Fig. 9).

1. If (Efficiency is LOW) and (Feasibility is LOW) and (Cost is LOW) and (Implementation_time is SHORT) then (Action_relevance is LOW) (1)
2. If (Efficiency is LOW) and (Feasibility is LOW) and (Cost is LOW) and (Implementation_time is LONG) then (Action_relevance is LOW) (1)
3. If (Efficiency is LOW) and (Feasibility is LOW) and (Cost is HIGH) and (Implementation_time is SHORT) then (Action_relevance is LOW) (1)
4. If (Efficiency is LOW) and (Feasibility is LOW) and (Cost is HIGH) and (Implementation_time is LONG) then (Action_relevance is LOW) (1)
5. If (Efficiency is LOW) and (Feasibility is HIGH) and (Cost is LOW) and (Implementation_time is SHORT) then (Action_relevance is MEDIUM) (1)
6. If (Efficiency is LOW) and (Feasibility is HIGH) and (Cost is LOW) and (Implementation_time is LONG) then (Action_relevance is LOW) (1)
7. If (Efficiency is LOW) and (Feasibility is HIGH) and (Cost is HIGH) and (Implementation_time is SHORT) then (Action_relevance is LOW) (1)
8. If (Efficiency is LOW) and (Feasibility is HIGH) and (Cost is HIGH) and (Implementation_time is LONG) then (Action_relevance is LOW) (1)
9. If (Efficiency is HIGH) and (Feasibility is LOW) and (Cost is LOW) and (Implementation_time is SHORT) then (Action_relevance is MEDIUM) (1)
10. If (Efficiency is HIGH) and (Feasibility is LOW) and (Cost is LOW) and (Implementation_time is LONG) then (Action_relevance is MEDIUM) (1)
11. If (Efficiency is HIGH) and (Feasibility is LOW) and (Cost is HIGH) and (Implementation_time is SHORT) then (Action_relevance is LOW) (1)
12. If (Efficiency is HIGH) and (Feasibility is LOW) and (Cost is HIGH) and (Implementation_time is LONG) then (Action_relevance is LOW) (1)
13. If (Efficiency is HIGH) and (Feasibility is HIGH) and (Cost is LOW) and (Implementation_time is SHORT) then (Action_relevance is HIGH) (1)
14. If (Efficiency is HIGH) and (Feasibility is HIGH) and (Cost is LOW) and (Implementation_time is LONG) then (Action_relevance is HIGH) (1)
15. If (Efficiency is HIGH) and (Feasibility is HIGH) and (Cost is HIGH) and (Implementation_time is SHORT) then (Action_relevance is MEDIUM) (1)
16. If (Efficiency is HIGH) and (Feasibility is HIGH) and (Cost is HIGH) and (Implementation_time is LONG) then (Action_relevance is MEDIUM) (1)

Fig. 9. Presentation of fuzzy rules

2.5 Defuzzification

The center of gravity method is used in this defuzzification stage to convert the fuzzy set consisting of efficiency, feasibility, cost, and implementation time into a precise numerical value of the action relevance (Fig. 10).

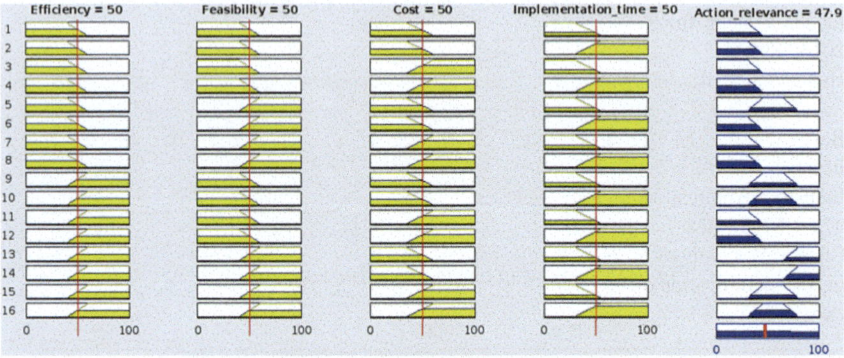

Fig. 10. Process of deffuzification

3 Conclusion

Improving performance and solving problems are major concerns for companies. They seek to implement solutions and actions capable of correcting the various anomalies. However, the many actions often proposed may not be adequate or relevant enough, due to lack of effectiveness or difficulty in implementation or even the high cost of deployment, which makes it difficult for managers to select the right solutions.

In this paper, we have developed a fuzzy logic model that calculates the output indicator: action relevance, based on four input indicators: ‘Efficiency’, ‘Feasibility’, ‘Cost’ and ‘Implementation time’.

This mathematical model makes it possible to indicate the degree of relevance of each action or solution raised during the brainstorming, in order to implement only those that can ensure maximum efficiency at minimum cost and with minimum constraints and implementation time.

References

1. Mortada, A., Soulhi, A.: A decision-making model based on fuzzy logic to support maintenance strategies and improve production lines productivity and availability (2023)
2. Mortada, A., Soulhi, A.: A decision-making model for quality improvement using fuzzy logic (2023)
3. Mortada, A., Soulhi, A.: A fuzzy logic model for ensuring customer satisfaction and preventing complaints about quality defects. (2023)

4. Mohd-Nassir, M.-D., Mohd-Sanusi, Z., Ghani, E.K.: Effect of brainstorming and expertise on fraud risk assessment **6** (2016)
5. Rowatt, W.C., Nesselroade, K.P., Beggan, J.K., Allison, S.T.: Perceptions of brainstorming in groups: the quality over quantity hypothesis. J. Creat. Behav. **31**, 131–150 (1997). https://doi.org/10.1002/j.2162-6057.1997.tb00786.x
6. Aguilar Lasserre, A.A., Lafarja Solabac, M.V., Hernandez-Torres, R., Posada-Gomez, R., Juárez-Martínez, U., Fernández Lambert, G.: Expert system for competences evaluation 360° feedback using fuzzy logic. Math. Probl. Eng. **2014**, 1–18 (2014). https://doi.org/10.1155/2014/789234
7. Hundecha, Y., Bardossy, A., Werner, H.-W.: Development of a fuzzy logic-based rainfall-runoff model. Hydrol. Sci. J. **46**, 363–376 (2001). https://doi.org/10.1080/02626660109492832
8. Bai, Y., Wang, D.: Fundamentals of fuzzy logic control—fuzzy sets, fuzzy rules and defuzzifications. In: Bai, Y., Zhuang, H., Wang, D. (eds.) Advanced Fuzzy Logic Technologies in Industrial Applications; Advances in Industrial Control, pp. 17–36. Springer, London, ISBN 978-1-84628-468-7
9. Chaabi, Y., Lekdioui, K., Boumediane, M.: Semantic analysis of conversations and fuzzy logic for the identification of behavioral profiles on facebook social network. Int. J. Emerg. Technol. Learn. IJET **14**, 144 (2019). https://doi.org/10.3991/ijet.v14i07.8832

Sensored Brushless DC Motor Control Based on an Artificial Neural Network Controller

Meriem Megrini(✉), Ahmed Gaga, and Youness Mehdaoui

Research Laboratory of Physics and Engineers Sciences (LRPSI), Research Team in Embedded Systems, Engineering, Automation, Signal, Telecommunications and Intelligent Materials (ISASTM), Polydisciplinary Faculty (FPBM), Sultan Moulay Slimane University (USMS), Beni Mellal, Morocco

megrinimeriem1@gmail.com

Abstract. Because of its high speed, low maintenance, and great torque capability, the BLDC motor is finding increasing uses. This motor is preferred over other motors due to its superior performance and ease of speed control using Power Converters. And enhanced artificial intelligence-based controllers. The purpose of this research is to use an Artificial Neural Network controller (ANNC) and a PID controller to manage the speed of a brushless DC motor. A detailed analysis is carried out based on the simulation results of both methods in the MATLAB/SIMILUNK environment. According to the results of the comparative investigation, the ANNC-based speed control method removes overshoot and peak time while also reducing the settling time of the system response. The ANNC-based simulation results are shown to be closer to the ideal reference control model.

Keywords: BLDC · ANNC · PID

1 Introduction

Brushless direct current (BLDC) motors have rapidly gained popularity in recent years due to their excellent characteristics and structure, such as high torque, compactness, and high efficiency, which make them suitable for a wide range of low to high speed applications, such as automotive, computer, safety critical sys- tems for the aerospace industry, and military [1–5]. Permanent magnet synchronous motors are categorised according to the wave shape of their induce emf, which might be sinusoidal or trapezoidal [1]. The sinusoidal form is referred to as a permanent magnet synchronous motor, while the trapezoidal type is re- ferred to as a PM Brushless dc (BLDC) machine [6]. Brushes are not used, as the name implies, therefore commutation is done electronically with a drive amplifier that uses semiconductor switches to adjust current in the windings depending on rotor position feedback [7]. Detected by incorporating external or internal position sensors [8]. In comparison to the classic permanent magnet DC motor, the brushless DC motor not only lacks mechanical brushes, but also swaps the stator and rotor structures. Its structure is diametrically opposed to that of a permanent magnet DC motor, in that the rotor of the brushless DC motor is made of permanent magnets, while

Y. Farhaoui et al. (Eds.): ICAISE 2023, LNNS 837, pp. 46–51, 2024.
https://doi.org/10.1007/978-3-031-48465-0_7

the stator is made of an armature wind- ing [9]. Many industrial applications utilizing BLDC motors necessitate precise speed and position data for rotor phase commutation. Traditionally, this action is carried out with sensors installed inside the motor, as hall-effect elements, or externally attached to the shaft, like encoders [10]. And employ artificial intel- ligence approaches for control [11]. Artificial Neural Networks (ANN), Genetic Algorithms, and Fuzzy Logic (FLC) are examples of branches in this category [12]. Because of its capacity to pick up complicated nonlinear processes with precision, reliability, and adaptability, ANNs appear to have the most meaning- ful impact in the field of motor drives in general and bldc motor in particular [13]. ANNs represent a viable study area for creating new position and speed estimation algorithms for BLDC motor drives [14]. And the most extensively used and praised control technique. Because of their high learning attribute and nonlinear mapping of multiple inputs and related outputs [15, 16]. This paper describes how to regulate the speed of a BLDC motor using a PID controller and an ANN controller. Hand tuning was employed in the PID controller. Following that, ANN was utilized to analyze the reaction of the motor drive system and compare the two controllers. The simulation of PID and ANN findings show that the ANN controller outperforms the PID controller in terms of rise time, settling time, and steady state error. The proposed work is structured as follows. Section II describes the materials and methods utilized, including the building of the BLDC motor, a description of the pid, and an introduction to ANN. Section III presents the results of the simulation in the Matlab/Simulink environment, as well as a discussion. Section IV summarizes the work and highlights the benefits of each controller.

2 Materials and Methods

2.1 BLDC Motor System

In comparison to a traditional brushed motor, the BLDC motor has a number of exceptional benefits, including improved performance, better static and dynamic characteristics, silent operation, long durability, and improved control ability [17]. The core components of a BLDC motor are the stator, rotor, and hall sensors. The stator, which is immobile, has slots along its inner perimeter where stator winding is located. The north and south poles of the rotor are permanent magnets that alternate. The magnetic field produced by the stator and rotor revolves at the same frequency in BLDC motors, which eliminates the slip that is typically observed in induction machines. Controlling the stator windings' energizing order regulates the rotor speed. The rotor position is sensed using three hall sensors placed into the stator. A brushless dc motor often needs an inverter and a position sensor to carry out "commutation" and transform electrical energy into mechanical energy.

2.2 Overview and Use of Controllers

PID controller PID control respectively stands for proportional, integral and derivative control, and is the most commonly used control technique in industry. The proportional control action is proportionate to the present control error. The integral term takes into account the error's history, or how long and how far the measured process variable has deviated from the fixed point over time. The derivative action is based on the expected future values of the control error.

Artificial neural network controller The ANN system is similar to the human organic nervous system. It consists of an input layer, an output layer, and a collection of hidden layers [16, 18]. A neural network is typically trained from a given example by determining the distinction between the network's processed output (often a prediction) and a target output. This distinction is the error. The network then uses this error value and a learning strategy to update its weighted associations. With each change, the neural network will create an increasingly comparable output that is increasingly comparable to the goal output.in brief, It learns (or is trained) by analyzing samples with known "input" and "result," establishing probability-weighted connections between the two, which are stored within the network's data structure.

3 Results and Discussions

The speed and electromagnetic torque of the proposed inverter driven BLDC motor system operating in a closed loop system with PID controller are shown in Fig. 1 as well as a reference speed of 68 rad/s.

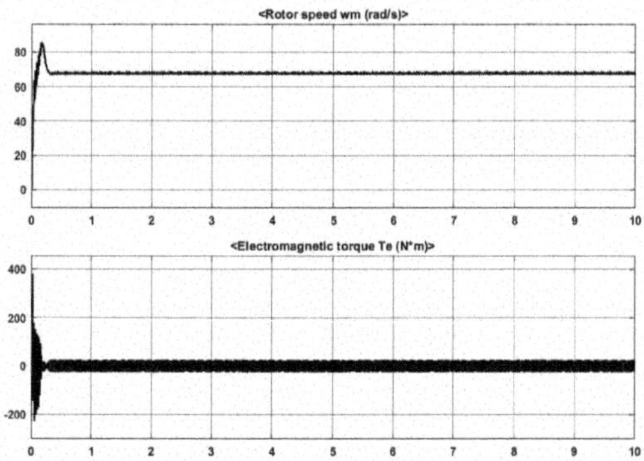

Fig. 1. Speed and torque response using PID with 68 rad/s speed condition.

The speed and electromagnetic torque of the proposed inverter driven BLDC motor system operating in a closed loop system with ANN controller are shown in Fig. 2 as well as a reference speed of 68 rad/s.

The speed and electromagnetic torque of the proposed inverter driven BLDC motor system operating in a closed loop system with PID controller are shown in Fig. 3 as well as a reference speed of 68 rad/s in the first 5 s, and 100rad/s in the last second.

The speed and electromagnetic torque of the proposed inverter driven BLDC motor system operating in a closed loop system with ANN controller are shown in Fig. 4 as well as a reference speed of 68 rad/s in the first 5 s, and 100rad/s in the last second.

According to the data obtained, it is apparent that the answers obtained by ANN controller are more valued than the PID because it eliminates overshoot and renders the

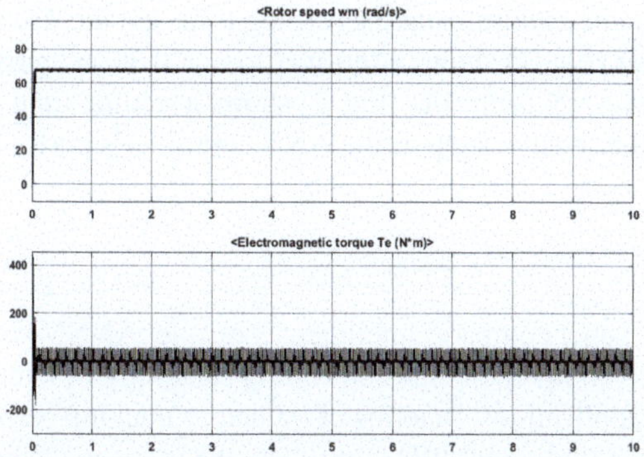

Fig. 2. Speed and torque response using ANNC with 68 rad/s speed condition.

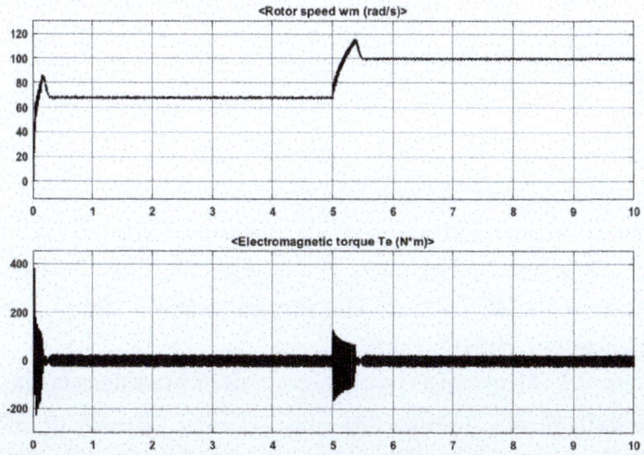

Fig. 3. Speed and torque response using PID with 68 rad/s in 5s and 100 rad/s after.

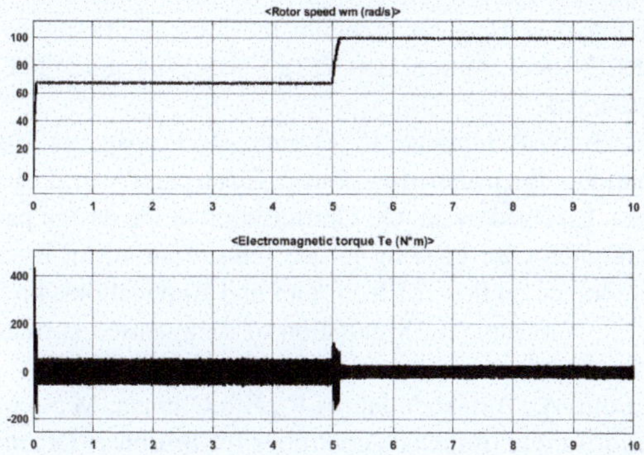

Fig. 4. Speed and torque response using ANNC with 68 rad/s in 5 s and 100 rad/s after.

system more quick, hence minimizing the settling time and the rise time. The response using a PID controller (Fig. 1) shows a settling time of 0.3s and a rise time of 0.1s, but the response using an ANN controller (Fig. 2) shows a settling time of 0.1s equal to the rise time. And the other results support the ANN controller's performance.

4 Conclusions

In the meantime, this paper compares intelligent controller ANNM with tradi- tional PID controllers in order to comprehend the controlling capability of intelli- gent controllers in industrial control applications. In terms of trajectory tracking performance, the ANN controller outperforms traditional PID controllers when given the reference speed trajectory. ANN control technique also produced quick reaction with minimal settling time, less overshoot, and zero steady-state error using a step reference input signal. The great performance of the sensored drive technique using the ANN speed controller function has been effectively validated. As a result, it is advised that it be used in a variety of industrial applications.

References

1. Gamazo-Real, J.-C., Mart´ınez-Mart´ınez, V., Gomez-Gil, J.: ANN-based position and speed sensorless estimation for BLDC motors. Measurement **188**, 110602, janv. (2022). https://doi.org/10.1016/j.measurement.2021.110602
2. Becerra, R.C., Ehsani, M.: High-speed torque control of brushless permanent magnet motors. IEEE Trans. Ind. Electron. **35**(3), 402–406, aouˆt (1988). https://doi.org/10.1109/41.3113
3. Kim, T.-Y., Lee, B.-K., Ehsani, M.: Sensorless control of the BLDC motors from near zero to high speed. In: Eighteenth Annual IEEE Applied Power Electronics Conference and Exposition, 2003. APEC '03, pp. 306–312. IEEE, Miami Beach, FL, USA (2003). https://doi.org/10.1109/APEC.2003.1179231
4. Bernth, J.E., Arezzo, A., Liu, H.: A novel robotic meshworm with segment-bending anchoring for colonoscopy. IEEE Robot. Autom. Lett. **2**(3), 1718–1724, juill. (2017). https://doi.org/10.1109/LRA.2017.2678540
5. Gamazo-Real, J.-C., Blas, J., Lorenzo, R., G´omez, J.: Propagation study of GSMpower in two dimensions in indoor environments—Part 2. Electronics World **116**, 30–34, mai (2010)
6. Lefley, P., Petkovska, L., Cvetkovski, G.: Optimisation of the design parameters of an asymmetric brushless DC motor for cogging torque minimization. In: Proceedings of the 2011 14th European Conference on Power Electronics and Applications, pp. 1–8 aouˆt (2011)
7. Vadla, V., Suresh, C., Naragani, R.: Simulation of fuzzy based current control strategy for BLDC motor drive
8. Bahari, N.B., Bin Jidin, A., Bin Abdullah, A.R., Bin Othman, M.N., Manap, M.B.: Modeling and simulation of torque hysteresis controller for brushless DC motor drives. In: 2012 IEEE Symposium on Industrial Electronics and Applications, pp. 152–155. IEEE, Bandung, Indonesia (2012). https://doi.org/10.1109/ISIEA.2012.6496618
9. Zhang, R., Gao, L.: The brushless DC motor control system Based on neural network fuzzy PID control of power electronics technology. Optik **271**, 169879, d´ec. (2022). https://doi.org/10.1016/j.ijleo.2022.169879
10. Celikel, R.: ANN based angle tracking technique for shaft resolver. Measurement **148**, 106910, d´ec. (2019). https://doi.org/10.1016/j.measurement.2019.106910

11. Vas, P.: Artificial-intelligence-based electrical machines and drives: application of fuzzy, neural, fuzzy-neural, and genetic-algorithm-based techniques, avr. (1999). Disponible sur: https://www.semanticscholar.org/paper/Artificial-Intelligence-Based-Electrical-Machines-Vas/4033915e28e737ed3b27f69fa1378c5a851fea40
12. Bose, B.K.: Neural network applications in power electronics and motor drives—an introduction and perspective. IEEE Trans. Ind. Electron. **54**(1), 14–33, f´evr. (2007). https://doi.org/10.1109/TIE.2006.888683
13. El-Sharkawi, M.A.: Neural network application to high performance electric drives systems. In: Proceedings of IECON '95—21st Annual Conference on IEEE Industrial Electronics, pp. 44–49. IEEE, Orlando, FL, USA (1995). https://doi.org/10.1109/IECON.1995.483331
14. Huang, F., Tien, D.: A neural network approach to position sensorless control of brushless DC motors. In: Proceedings of the 1996 IEEE IECON. 22nd International Conference on Industrial Electronics, Control, and Instrumentation, pp. 1167–1170. IEEE, Taipei, Taiwan (1996). https://doi.org/10.1109/IECON.1996.566044
15. Leena, N., Shanmugasundaram, R.: Artificial neural network controller for improved performance of brushless DC motor. In: 2014 International Conference on Power Signals Control and Computations (EPSCICON), pp. 1–6. IEEE, Thrissur, India, janv. (2014). https://doi.org/10.1109/EPSCICON.2014.6887513
16. Thayumanavan, P., Cs, S.G., B, A.,Sr, Y.: Artificial neural networks based analysis of BLDC motor speed control (2021)
17. Zhang, Q., Cheng, S., Wang, D., Jia, Z.: Multi-objective design optimization of high-power circular winding brushless DC motor. IEEE Trans. Ind. Electron. 1 (2017). https://doi.org/10.1109/TIE.2017.2745456
18. Farhaoui, Y., et al.: Big data mining and analytics **6**(3), pp. I–II 2023. https://doi.org/10.26599/BDMA.2022.9020045

A Machine Learning Based Approach to Analyze the Relationship Between Process Variables and Process Alarms

Sarafudheen M. Tharayil[1](✉), Rodrigues Paul[2], Ayman Qahmash[2], Sajeer Karattil[3], and M. A. Krishnapriya[4]

[1] Enterprise Analytics Group, CAD, Saudi Aramco, Saudi Aramco, Dhahran, Saudi Arabia
sharafudheendr@gmail.com

[2] Computer Science Department, King Khalid University, Abha, Saudi Arabia
drpaulprof@gmail.com, a.qahmash@kku.edu.sa

[3] Computer Science Department, MES College of Engineering, Kuttipuram, India
sajeerkarattil@gmail.com

[4] Computer Science Department, Hindustan Institute of Technology and Science, Kancheepuram, Chennai, India
priya_krishi@yahoo.co.in

Abstract. With the advancement of technology, industrial plants are adopting advanced process control systems and sensors to optimize operations. Alarm systems are utilized to manage the process and ensure safe and reliable operations. Although alarm systems can sometimes produce overhead to the operators if not managed properly, alarm systems can contribute to economic, environmental, and safety added value in industrial sectors. The relationship between alarm systems and process sensors may not be clearly known in complex plants. In this paper we propose an approach to iteratively improve the plant process management by quantifying the influence of process variables on the triggered alarms using machine learning and AI tools. This approach thus contributes to root cause analysis, continuous improvements and effective process controls. The proposed approach groups related alarms by analyzing times series alarm data and process data. In addition, the approach quantifies the influence between the process variables and related alarms triggered.

Keywords: AI · Machine Learning · Process Alarm · Control Systems · Plant Alarms · Classification · Data Science

1 Introduction

Plant alarms in industrial process plants consist of sophisticated sets of hardware and software systems connected in a layered structure. The functionality of the alarm management system is to alert process operators about any deviations from defined normal operations and to record any unwanted changes [1–4]. The indications provided by alarm systems should not be misleading, overloading, or disturbing to the plant operator. Traditional process control systems have improved over time and the sophistication in alarm

Y. Farhaoui et al. (Eds.): ICAISE 2023, LNNS 837, pp. 52–67, 2024.
https://doi.org/10.1007/978-3-031-48465-0_8

systems implementations brings the challenge of managing an increased number of triggered alarms. Most industrial plants have complex connections between the sensors and alarm systems which cause propagation of alarms. A flood of alarms can be overhead for operators, where the high number is more than what the operator can effectively manage and, therefore, can be a major reason for accidents. Organizations usually rely on operator or engineering subject matter experts for alarm resolution, but the human error and operator lag might impact the efficiency of the system. In contrast, manually checking each alarm for the related alarms can have significant operation downtime. Sophisticated tools are required to empower the human to predict and/or react to abnormalities in the plant alarms [5, 6].

Plant alarms and its supporting systems usually consist of Distributed Control Systems (DCS), Supervisory Control and Data Acquisition (SCADA), generated signals and Human-Machine Interface (HMI). SCADA helps the operators with user interfaces for managing devices present on the field by sending predefined commands for the industrial devices such as sensors and alarms. Signals are generated from sensors and associated controls such as Basic Process Control System (BPCS) which generate signals for control systems and Safety Instrumented System (SIS) that generate alarm signals. Human-Machine Interface (HMI) makes an interface and represents generated signals to the operators. Operators also use HMI to respond to alarm as well. Studies show poor performance of alarm systems in practice in many industries including Oil and Gas. So, more proactive solutions and intelligent mechanisms still need to be evolved. For example, Table 1 shows the summary of the survey regarding the performance measurements of alarm systems in the Oil and Gas sector showing the need for improvement in managing alarm systems [2, 7, 8].

Table 1. The performance measurements of alarm systems

Performance measurement	Oil and gas industry
Average alarm per hour	36
Average standing alarms	50
Peak alarm per hour	1300
Priority distribution (Low/Med/High)	25/40/35

The huge amount of data captured from sensors and alarm logs provides an opportunity for mining the data (DM) and apply machine learning (ML) algorithms, to better understand the relationship between the entities and thus lead to better decision making [9]. This article proposes an intelligent data driven machine learning technique based on historical plant and alarm data for improving the understanding of the relationship between triggered alarms and process control variables. The fundamental motivation is that some alarms may be triggered due to some abnormality caused by changes in processes, and if effectively traced, can lead to better process control, minimizing future triggered alarms and operating within safe thresholds. The proposed method attempts to identify related alarms that are likely to be triggered together and influential process control variables that are likely to contribute to triggering those alarms. This can help the operation team and plant analysts to take precautionary measures and to predict alarm behavior when specific process variables change or pinpoint most likely responsible process variables when alarms are triggered to reduce situations resulting in alarm floods.

2 Background

The alarm system works with layers of protection that guards from abnormal events such a catastrophe failure. The oil and gas industry is subject to risks due to the hazardous nature of hydrocarbon and process materials. Basic Process Control System (BPCS) is an independent layer which controls processes and monitors the facility. The sensor data feeds into process instruments and provide results based on the control functions suitable for approved design rule and set points [10].

The alarm management standards such as ISA 18.2 provides an outline for the successful design, implementation, operation and management of alarm systems in a process plant. The starting points of the alarm lifecycle are philosophy, monitoring and assessment, and audit.

Philosophy is conceptualized during initial setup of the facility, monitoring and assessment or audit could be applicable for an existing system. Such standards provide guidance to resolve and avoid the most common process control setbacks and bring a stable performance for the facility. The safety instrumented systems standard ISA-84 is a performance-based standard that specifies the activities and requirements to ensure the Safety Instrument Function (SIF) which provides the desirable consistency of guard from hazards. It also addresses the concept of Safety Integrity Level (SIL) quantifying a level of risk reduction. In spite of all the standards and protection layers, intelligent data driven tools are needed to improve decision making and operations [4] (Fig. 1).

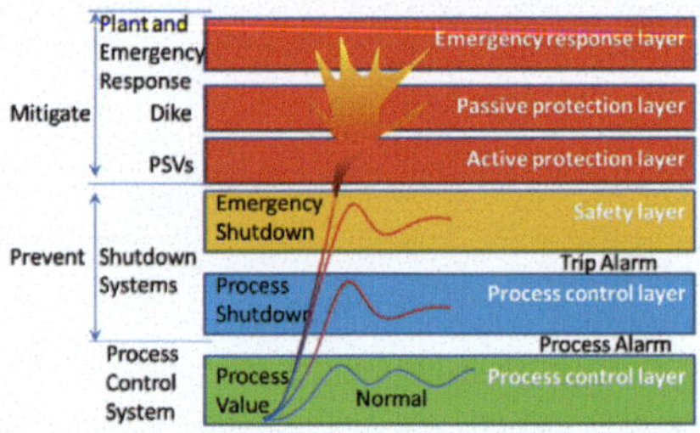

Fig. 1. Layers of protection in alarm management systems [11]

3 Related Work

Recent studies on alarm management systems shows the need of a Centralized Alarm Management System (CAMS) and its specifications [2]. The forecast of upcoming alarm events has attracted attention in both academia and engineering disciplines [12]. A data-driven approach is proposed by Haniyeh, et al., to address the early classification problem with unlabeled historical data using a semi-supervised approach based on vector representation and Gaussian mixture model [13]. A data driven alarm management system for preventing alarm flooding is discussed where a machine learning technique is used to gather the most important data associated with risk [7]. The historical alarm flood situations are analyzed using alarm subsequences and different studies use clustering approaches for a causal analysis to detect root cause [14, 15]. The frequent pattern mining used to find the most significant groupings of alarms in historical alarm data. But these methods have the issue of minimum support having the limitation of absolute or relative frequency with limited prediction in complex scenarios [16, 17].

Alarm analytics based on similarity measures such as Jaccard distance [1], term frequency-inverse document frequency (TF-IDF) representation and Levenshtein distance [12, 18], Pearson's correlation coefficient [15] are conducted in different researches. A recent efficient method could be the pairwise similarity measurement by Cheng et al. [19] which gives similarities of alarms based on two time-stamped alarm sequence patterns. Apart from these online mechanisms, different online pattern matching mechanisms are also discussed [4, 20, 21] using tools such as highly computationally modified proactive Smith-Waterman (MSW). The TF-IDF based unlabeled historical alarm flood data is unique in online alarm flood classification and provides support for online operators [22]. The implementation of the threshold alarm method combined with multiple classifiers in the decision set effectively improves alarm performance in regard to both accuracy and efficiency [23]. Related applications ranging from military surveillance [24] to health care [25] have been widely studied in the field of anomaly detection in alarm management systems. Some of the challenging factors that need to be considered when performing abnormalities of alarm for time-series data are, changing the pattern of the data and the temporal dependencies on the data the influence of noise [24].

4 Problem Statement

The industrial plants collect the information from sensors and make it available as process variable data. The alarm systems collect information related to alarm data and stores in different alarm management systems. The plant engineers and operators derive the relationship between the process variables and process alarms based on their industrial experience and the resolution of alarms are executed based on the semantics already defined. Human errors in deriving the meaning between process and alarms, and operator lags could impact the efficiency of the system and sometimes even catastrophic failures. With more numbers of alarms and process variables, alarm resolution becomes quite time consuming and each alarm process needs to be looked into individually. Manual checking each alarm code for the related alarm can have significant operation downtime and could lead to alarm floods. Motivated from this background, the problem here is to identify the alarms that are correlated in multiple alarm scenarios and derive possible hidden relationships between alarms and process tags and prioritize the alarms based on the importance of its relevance with process variables. These new insights will be then used by the plant operators and engineers to re-define the new priorities and semantics with in the process variables and process alarms.

5 Preliminaries

5.1 Process Variables and Alarm Tags

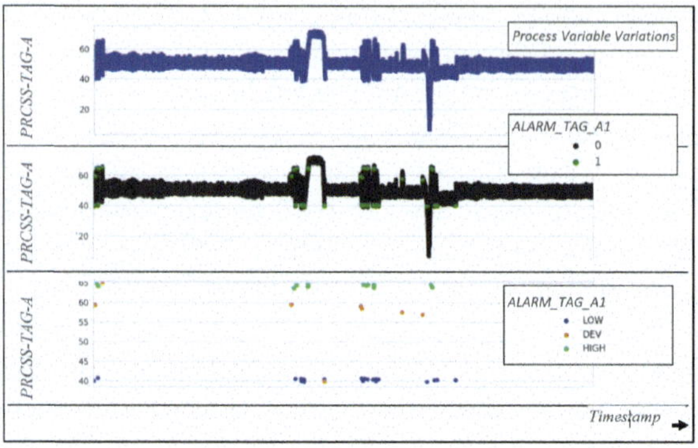

Fig. 2. Analysis on selected process tag and alarm

There is a clear relationship to the linked process variables. For example, Fig. 2 shows a thirty-day view of a specific process variable and its related alarm variable, with a clear indication of state changes such as LOW, deviation (DEV) and HIGH.

In a complex alarm system with multiple dependencies, multiple alarms can make alarm handling time consuming. The historical data of alarms and processes are utilized to determine relationships between process variables and alarm events which could be used to prioritize which of the alarms to resolve first and understand the alarm correlations for better alarm targeting. This way, the root cause of the issue could be resolved by looking into influential process variables, which helps the operator manage the process more effectively and prevent unwanted events.

5.2 Notions of Alarm and Process Variables

An alarm is triggered if there is a variation in sensor or process variable from a threshold set. When an alarm is triggered, it can be said that it is in an alarm state, or ALM. When an alarm is acknowledged by the operator, it can be said that it is in an acknowledgment state, or ACL. Finally, when an alarm stopped by the operator or related sensor values returned to normal, it can be said that it is in a returned state or an RTN. We define a binary value for an alarm a at time t as follows:

$$a(t) = \begin{cases} 0, & if\ state(a, t) = RTN \\ 1, & otherwise \end{cases}$$

where, state is the state of the alarm for alarm a at time t

An alarm does not necessarily have an associated process variable, but rather can be triggered according to its sensor readings. The frequency of recording data in the alarm system and process variables may differ. One of the challenges is to combine the two data from the two systems into one dataset for machine learning development. For example, process variables are sampled at one-minute intervals whereas alarm state change is provided with timestamp at one-second intervals. An alarm can be triggered at any time irrespective of process variable timestamp. Also, the alarm state is sorted by tags and state changes in ascending order and observe how the state changes are taking place deriving a state-change synthetic variable (Table 4). The actual count of alarms needs to be filtered based on the state changes. In this study, the focus is to find significant alarms and process variables that likely influence the triggered alarm. Therefore, we consider alarms having NA or more instances in the system, and disregard other entries such as noise and non-state changing alarms such as alarms standing for seven days or more (Table 2).

Table 2. State change derived from alarm variable

State	State	Change flag
ALM	ALM	1
ALM	ACK	0
ALM	RTN	0
ACK	ALM	1
ACK	ACK	1
ACK	RTN	0
RTN	ALM	1
RTN	ACK	1
RTN	RTN	1

5.3 Experimental Environment and Dataset

The study is conducted on one Gas Oil Separation Plant (GOSP) with a one-year time frame. Two datasets are provided, where, firstly, the process tags are captured for every time interval and the process variables are given with the value readings in each column in the dataset. A sample process variable dataset is shown in Table 2, where both data and process tag names are anonymized for security reasons (Table 3).

Secondly the alarm dataset is captured based on the timestamp of the event that happened, along with the important details such as event category, priority and alarm type. The experiments are conducted on the 300 process tags and its 100 related alarms while the tag names are anonymized (Table 4).

During the data pre-processing stage, the alarm data is carefully blended into the process variable dataset and treated as a single dataset as explained in the methodology section in this document.

Table 3. Sample process variable dataset

Timestamp	PNT1:FIC-4444	PNT1:FI-3333	PNT1:TI-5555	–	PNT1:FIC-8888
4/1/2020 16:40	220.20918	14.3111	18.11	–	5.4335
4/1/2020 16:41	220.21332	14.7432	18.10	–	5.4307
4/1/2020 16:42	220.21746	15.4962	18.09	–	5.4278
4/1/2020 16:43	220.22160	15.8937	18.08	–	5.4250
4/1/2020 16:44	220.22572	15.8588	18.07	–	5.4221
4/1/2020 16:45	220.22986	15.8104	18.06	–	5.4193
4/1/2020 16:46	220.23400	15.7839	18.05	–	5.4164
4/1/2020 16:47	220.23814	15.7574	18.07	–	5.4220
–	–	–	–	–	–

Table 4. Sample alarm dataset

Timestamp	ALARM	Priority	Tag	Event category
4/1/2020 1.55 PM	DISC_ALARM	Warning	FA-9999	PROCESS
6/1/2020 1.55 PM	TRIP	Critical	XL-2222-1	PROCESS
6/1/2020 1.59 PM	TRIP	Critical	XL-2233-C	PROCESS
6/1/2020 2.15 PM	TRIP	Critical	XL-4444-AA	PROCESS
7/1/2020 1.30 PM	TRIP	Critical	XL-4444-AA	PROCESS
–	–	–	–	–
–	–	–	–	–
7/1/2022 1.15 PM	DIAGNOSTIC	Sys Alarm	ESD-DIXXX	SYSTEM
7/1/2022 1.30 PM	LO_ALM	Warning	LIC-2222	PROCESS

A Summary of the data used is shown in Table 3. As we are dealing with a huge volume of the data, a random sample of a smaller dataset is used for development, which helps to scale-up the solution to the full dataset (Table 5).

Table 5. Summary of the dataset

Dataset name	#records	#process tags	# alarms	#days records	Periodicity of process	Periodicity of alarm
PL1_2020_SMALL	50,000	310	90	30	1 min	5 min
PL1_2020_FULL	861,000	310	90	640	1 min	5 min

We apply a temporal split on the data and use, train and test datasets on a 80:20 ratio and use cross validation on the training set. The plant down time data, and invalid process variables readings are removed based on subject matter knowledge. The data is anonymized and in order to get a feel of the data here is the statistics on the few of the process tags, after anonymizing the process tags and its values, which shows a faulty process variable which needs to be removed (Table 6).

Table 6. Statistics of the dataset

Sl	Tag	Min	Max	Avg	Std	Distribution
1	PROCESS_TAG_1	– 474.00	22879.40	18478.13	8297.12	
2	PROCESS_TAG_2	– 70.79	22837.12	18539.99	8162.15	
3	PROCESS_TAG_3	– 201.48	2280.77	18536.42	8240.00	
4	PROCESS_TAG_4	– 171.60	22726.48	14585.23	6507.78	
5	PROCESS_TAG_5	– 235.67	15607.20	12626.32	5092.15	
6	PROCESS_TAG_6	– 466.81	22578.88	14275.66	6636.25	
7	PROCESS_TAG_7	– 11.11	3130.94	1695.14	763.79	
8	PROCESS_TAG_8	0.74	235.45	127.84	64.36	
9	PROCESS_TAG_9	0.05	100.00	32.65	29.70	
10	PROCESS_TAG_10	– 0.59	21.18	2.61	1.46	
11	PROCESS_TAG_11					
12	PROCESS_TAG_12	77.16	159.25	131.01	16.92	

5.4 Machine Learning and Algorithms Used

Random forest classifier, gradient boost classifier and CatBoost classifier are used in the experiments. After different experiments, it is found that the CatBoost classifier gave robust results on different combinations of datasets. There are two runs executed for the classifier. Initially, a binary classifier is used to check whether there is an alarm present for 300 + process variables, as an elimination phase. From this run, the feature importance reveals the most influential process variables. For the experimental setup, a set of process variables varying from 25 to 50 are used for different experiments. The selected process variables are then identified as the independent variables for the second model.

The second classifier, execute the classifier for each alarm provided the most relevant alarms are already identified. Now, a CatBoost model is created for each alarm which reveals the most influencing process variable for the selected CatBoost model for specific alarm.

6 Methodology

6.1 Overall Learning Process

Two datasets are considered in this study, the process variables data and the alarm management system data. Process variables normally contain information from different parts of the plant such as temperature, voltage, pressure, flow and so on which influence the downstream alarms. Therefore, it is valuable to study the influence of those upstream process variables on the downstream alarms and quantify the relationship between them. Using a supervised machine learning approach, model feature importance is used to quantify the influence of process variables on accurately predicting alarms triggered, is an indicator that these variables are responsible for triggering the alarm (Fig. 3).

Given the number of process variable present in the study *(P1, P2, …, PN)*, there are many related process variables, a machine learning based classifier first learns the relation among various process loops. Now, after removing the redundant PI tag data, another machine learning classifier algorithm is applied to predict the alarm based on the most relevant PI tag. The overall process is shown in Fig. 2.

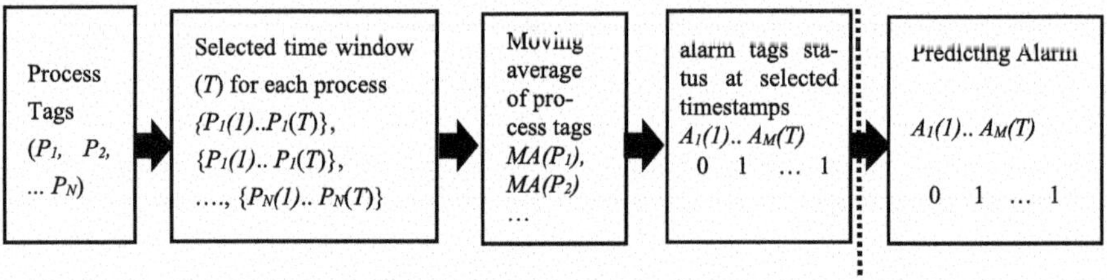

Fig. 3. The approach at high level.

6.2 Pre-processing

Preparing the data for machine learning usually requires a pre-processing step which involves activities such as excluding invalid data, handling missing values and outliers, and handling features multi-collinearity, which eliminates highly correlated features from being included in the training process. The algorithm used in this step can be summarised as following:

Algorithm (1): Pre-processing

Input: Values from process variables *(P1, P2,... P_N)*, alarm tags (*A1.. A_M*), plant downtime $\{T_{d1}, T_{d2} \dots T_{dN}\}$
Output: cleaner datasets

Steps:

1. Drop plant downtime from *(P1, P2,... P_N) and* tags (*A1.. A_M*) considering list $\{T_{d1}, \dots T_{dN}\}$
2. Drop missing values for plant variables and alarm variables
3. Calculate correlation coefficient between process variables (P_i, P_j) and remove highly correlated variables

$$r_P = \frac{\sum(P_i - \overline{P}_i)(P_j - \overline{P}_j)}{\sqrt{\sum(P_i - \overline{P}_i)^2 (P_j - \overline{P}_j)^2}}$$

4. Drop the highly correlated process variables if $r_P > r_{THRESHOLD}$

6.3 Merging Process Tag Data and Alarm Data

As alarm and process data in our study comes from independent systems, the data need to be merged in order to be used in a supervised machine learning approach. The merging process is explained in Algorithm 2.

Algorithm (2): Merging alarm and process tags

Input: preprocessed variables *(P1, P2,... P_N)* and alarm tags (*A1.. A_M*),, merge interval i_{MERGE}
Output: merged dataset $(\{p1_1, p2_1 \dots pN_1; a1_1 \dots aM_1\} \dots \{p1_x, p2_x \dots pN_x; a1_x \dots aM_x\}$ where x is the number of records in the new dataset

Steps:

1. For each process tag P_2 from $(P_1, P_2, \dots P_N)$; repeat

1.1 create an alarm subset A_{Pi} focusing on P_i
1.2 if alarms occurred during the time-period
1.3 Mark alarm as "1" or "0" based on state change logic as per in Table 2
1.4 Merge the process tags *(P1, P2,... P_N)* and alarm tags (*A1.. A_M*), and produce $MergedData = (\{p1_1, p2_1 \dots pN_1; a1_1 \dots aM_1\} \dots \{p1_x, p2_x \dots pN_x; a1_x \dots aM_x\}$

6.4 Train Machine Learning Model

After the data is prepared for machine learning, we select models which can generate feature importance such as random forests and boosted trees. Hyperparameter tuning and cross validation is applied on all models. The models are then trained for each alarm tag and their performance is compared using common classification metrics such as precision, recall and F1 score. Then, the best performing model is selected to generate feature importance.

Algorithm (3): Algorithm for process tag selection

Input: *MergedData, ({p1₁, p2₁... pN₁;a1₁... aM₁}...{p1ₓ, p2ₓ... pNₓ;a1ₓ... aMₓ}*

Output: *({p1₁, p2₁... p N̂ ₁;a1₁... aM₁}...{p1ₓ, p2ₓ... pNₓ;a1ₓ... aMₓ})*, where $\hat{N}$ is the new count of process variables

Steps:

1. For each alarm Tags, on Historical dataset, for any given time point (*T0*) we will look for next *L* minutes of lead time if an alarm is raised or not
2. From time point (*T0*) we will look back for *H* minutes of history of process tag values
3. Get moving average for each process tag over last *H* minutes of history as input and associate them with Alarm flag ("0" or "1") as output on *L* minutes lead time from *T0*
4. Train simple classifier using random sample from dataset (input & output)
5. Identify and $\hat{N}$ ranked process tag based on variable importance from the trained classifier and choose the Top $\hat{N}$ process tags for each alarm tags, giving output as

({p1₁, p2₁... p N̂ ₁;a1₁... aM₁}...{p1ₓ, p2ₓ... pNₓ;a1ₓ... aMₓ})

6.5 Generate Feature Importance

As discussed in Sect. 5.4, the trained machine learning model is used to calculate feature importance which measures the impact a process variable has in predicting an alarm accurately.

Algorithm (3): Finding linkage between alarm and process tags

Input: the merged & selective process tags *({p1₁, p2₁... pÑ₁;a1₁... aM₁}...{p1ₓ, p2ₓ... pNₓ;a1ₓ... aMₓ})*

Output: a) weight matrix *Wij*, b) evaluation matrix for machine learning algorithm

Steps:

1. Combine all the alarm data as a single output variable
   - 1.1 Y_{train} = binary class flag ("0" or "1") and Y_{train} is set to "1" if any of the relevant alarm is active. Otherwise "0"
2. Check the span of alarm to avoid chatter, nuisance, standing alarms
   - 2.1 Remove the alarm if the alarm is identified as chatter, nuisance or standing alarms
3. Taking one alarm find its most influencing process variable
   - 3.1 First find the most influencing process tags for this Alarm
   - 3.2 For a given time window, find the influence of the selected alarm on the most prominent process variables for the alarm
   - 3.3 Hyperparameter tuning on the model to improve the model accuracy
4. Considering one alarm at a time
   - 4.1 Consolidate the feature importance and find the most influencing process variables
   - 4.2 Construct the weight matrix *Wij* where *i = process variable* and *j = alarm*
   - 4.3 Use *Wij* to identify the influencing process variables for a specific alarm

7 Experiments and Results

7.1 Analysis of the Process Data and Alarm Data

The similarity analysis on the process data revealed the need for removal of 26 process tags that have high covariance (Fig. 4). The exploratory data analysis showed the need for removal of 74 tags which are not having quality data. For the alarm data, 55 alarms are identified as relevant alarms after removing chatter, standing and nuisance alarms. The plant data and the alarm data are then merged together to produce a single dataset with the internal time of 5 min. As per the logic explained in the previous section, for each alarm, a sliding window of 15 intervals is created based on the most influencing process variables based on the feature importance of the cat-boost results. This sliding window is then used for finding the most influencing process tag for the selected alarm.

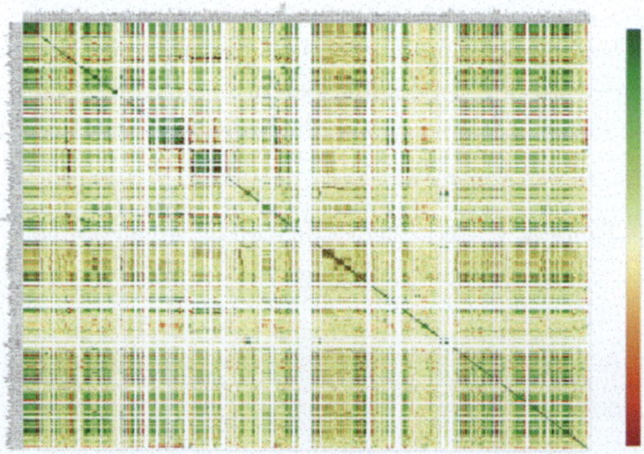

Fig. 4. Covariance of process tags

7.2 Deriving the Relationship Between Alarm and Process Tags

As discussed before, if we already know the relation between the process tag and alarm tag, it is easy to derive the one-to-one relation between the alarm tag and process tags as shown in Fig. 5. The mechanism derived here brings more insights into the alarm and process system deriving multiple process variables which could be contributing to a specific alarm event.

7.3 Performance Evaluation of the Machine Learning Algorithms

In our study a cat boost classifier is used with the log loss as the loss function. The table below shows average performance metrics.

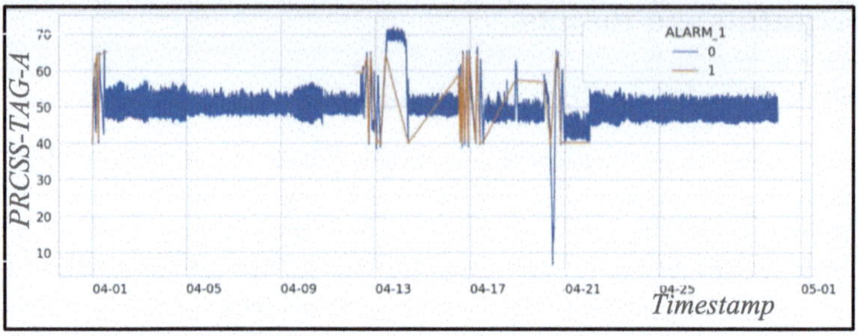

Fig. 5. One to one relation between the alarm tag and process tags

A flavor of performance evaluation of the machine learning algorithm for identifying the linkage between process variables shown in Table 5. The actual names of the alarm and process variables are anonymized for privacy reasons as per NDA. Among the 55 alarms considered, 7 of them provide excellent performance (F1 Score > 0.8) while another set of 9 alarms provide acceptable performance (F1 Score > 0.5). 16 of these alarms provide some meaningful insights to the process to alarm relationships (Table 7).

Table 7. Performance evaluation of ML algorithm

	Performance measures			
Alarm_Tag	Accuracy (%)	Precision (%)	Recall (%)	F1
ALARM_100	99.992	95.679	86.592	0.909
ALARM_55	99.986	85.714	95.094	0.902
ALARM_93	99.993	89.041	90.909	0.900
ALARM_96	99.849	84.950	92.788	0.887
ALARM_20	99.981	90.476	84.300	0.873
ALARM_58	99.811	77.442	90.362	0.834
–	–	–	–	–

	ALARM_42	ALARM_44	ALARM_29	ALARM_41	ALARM_16	ALARM_54	ALARM_45	ALARM_34	ALARM_8	ALARM_43	ALARM_43	ALARM_51	ALARM_17	ALARM_17	ALARM_19	ALARM_46	ALARM_43	ALARM_50	ALARM_19	ALARM_33	ALARM_12	ALARM_2	ALARM_45	ALARM_31	ALARM_53	ALARM_45
PLANT_TAG_345	3.6	11.9	0.2	0.9	3.3	0.8	7.9	2.4	4.5	4.5	5.5	1.2	6.3	7.8	0.6	2.5	0.2	3.2	0.0	0.2	1.0	8.5	6.6	2.6	0.1	0.2
PLANT_TAG_251	4.3	1.4	4.0	3.9	0.8	0.4	0.8	3.9	0.9	1.2	1.0	1.2	1.5	1.0	4.1	5.6	4.4	12.3	1.0	5.7	5.9	1.1	9.8	2.5	1.5	1.0
PLANT_TAG_194	0.7	2.5	8.1	2.5	9.6	15.8	2.7	2.0	0.5	1.6	0.7	1.1	2.4	3.1	1.6	1.2	2.2	1.0	1.5	0.7	1.5	1.5	0.4	1.8	0.9	6.5
PLANT_TAG_34	0.0	5.3	0.4	0.0	0.4	0.3	4.8	3.1	2.0	3.3	1.7	3.1	9.9	2.0	0.5	0.6	0.2	0.1	4.5	0.8	2.1	1.5	0.0	5.5	4.8	2.8
PLANT_TAG_293	1.7	2.2	1.0	1.1	2.3	4.3	2.3	2.6	1.3	0.5	1.7	0.9	2.8	0.4	0.6	0.6	0.4	1.0	4.4	2.9	0.4	1.4	0.5	1.2	2.7	1.2
PLANT_TAG_72	0.9	1.4	0.5	1.5	0.4	2.0	2.6	0.3	2.8	1.4	2.2	1.1	2.0	1.7	1.2	0.7	1.5	1.8	2.0	0.1	0.6	0.5	0.4	2.5	2.0	0.8
PLANT_TAG_211	1.0	0.6	0.0	0.0	0.0	0.0	3.3	0.5	0.0	1.9	4.3	4.2	0.5	0.0	5.4	0.4	0.0	0.0	0.0	3.6	0.0	0.0	3.0	0.2	2.7	0.4
PLANT_TAG_267	0.3	0.9	0.6	2.9	1.5	4.7	2.3	0.8	0.5	1.8	0.3	1.2	0.7	0.6	2.9	0.5	1.6	0.2	0.0	0.2	0.5	1.4	0.5	0.3	1.0	2.5
PLANT_TAG_124	2.3	1.5	0.3	2.9	1.3	0.9	1.6	0.4	0.3	1.2	0.8	1.0	0.5	0.5	0.4	0.5	1.2	0.2	0.8	1.9	1.8	3.1	1.5	2.1	1.2	0.5
PLANT_TAG_103	0.3	0.4	2.7	4.1	0.1	0.9	0.6	2.1	2.0	0.3	0.1	1.6	0.4	1.3	0.7	0.0	5.2	0.9	0.0	0.2	0.3	0.5	0.0	2.3	0.2	1.7
PLANT_TAG_72	1.0	0.7	2.9	0.3	0.9	0.4	0.7	1.0	0.5	1.4	0.3	1.0	1.1	0.7	1.5	1.5	2.5	1.4	0.4	0.6	0.7	0.5	2.4	0.6	1.8	2.1
..	..	..	..	..	..	..	..	..	..	..	..	..	..	..	..	..	..	..	..	..	..	..	..	..	..	..
..	..	..	..	..	..	..	..	..	..	..	..	..	..	..	..	..	..	..	..	..	..	..	..	..	..	..

Fig. 6. Sample of importance of alarm tag over process tags

The weight matrix Wij matrix helps to find the importance of the variable for each alarm as shown in Fig. 6 and graphical display of selected alarm is given in Fig. 7. As shown in figure, there are a few process variables that have a direct linkage to the alarm while the other process variables are less relevant. As the values in each column shows the importance of process variables in percentage each column will sum to 100.

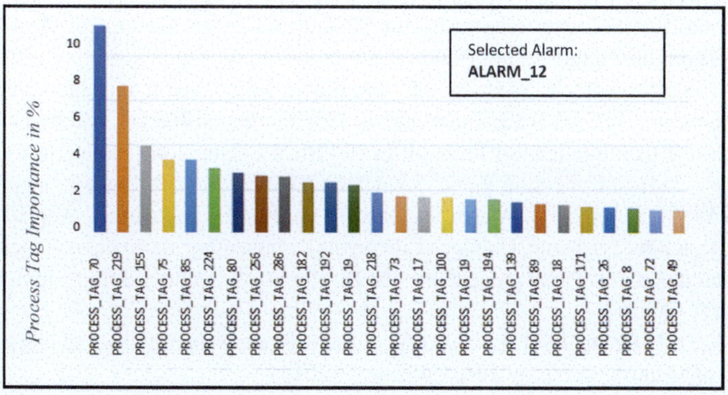

Fig. 7. Graphical display of importance of process tags for selected alarm

8 Summary and Future Work

The experiments show that, it is not possible to derive the relationship between alarms and process tags for all the alarms in consideration due to the limitations of classical machine learning algorithms and its tuneable hyperparameters. As the next step of the research, more sophisticated tools such as deep learning could be adopted to learn based on the data which was unable to get satisfactory results in this experiment. LSTM could be one such option for further research. Moreover, more data mining tools could be developed such as a graph-based sequencing which can show the linkage between alarms and processes which could show the relationship between the processes and alarms and this might help the operator enabling quick decision making.

Further, the current research is limited to a single plant data for a specific time period. The research can be expanded to different types of plant alarms in order to capture the characteristics of different functional areas.

References

1. Jiandong, W., Fan, Y., Tongwen, C., Sirish, S.L.: An overview of industrial alarm systems: main causes for alarm overloading research status and open problems. IEEE Trans. Autom. Sci. Eng. **13**(2), 1045 (2016)
2. Mahmoud, A.H., Joseph, S.P.D., Askar, J.: Achieving operational efficiencies from a centralized alarm management system. In: Abu Dhabi International Petroleum Exhibition & Conference, Abu Dhabi, UAE (2021)
3. ISA, ANSI/ISA-18.2, Management of Alarm Systems for the Process Industries. International Society of Automation. ISA, Durham, USA (2009)
4. M, L., C, M., Grimholt, C., Hollender, M.: Advances in alarm data analysis with a practical application to online alarm flood classification. J. Process Control **79**(Jul), 56–71 (2019)
5. Rubinstein, E., Mason, J.F.: An analysis of Three Mile Island: the accident that shouldn't have happened. IEEE Spectrum (1979)
6. Executive, H.A.S.: The explosion and fires at the Texaco Refinery, Milford Haven, 24 July 1994: a report of the investigation by the Health and Safety Executive, Texaco Refinery: Milford Haven on 24 July 1994 (1997)
7. Ahillya, B.S., Deepak, K., Ashish, M.: Intelligent based alarm management system for plant automation. In: 3rd IEEE International Conference on Recent Trends in Electronics, Information & Communication Technology (RTEICT), Bangalore, India (2020)
8. Rothenberg, D.H.: Momentum Press, NY, USA (2009)
9. ANSI/ISA, "Management of Alarm Systems for the Process Industries," ANSI/ISA, 2016
10. Stauffer, T., Sands, N., Dunn, D.: Get a Life (cycle)! connecting alarm management and safety instrumented systems. Paper presented at the ISA Safety & Security Symposium, Sellersville. EXIDA, PA (2010)
11. Ahmed, K., I, I., Chen, T., D, J., Burto, T.: Similarity analysis of industrial alarm flood data. IEEE Trans. Autom. Sci. Eng. **10**(2), 452–457 (2013)
12. Haniyeh, A.S., Jun, S., Tongwen, C.: Early classification of industrial alarm floods based on semisupervised learning. IEEE Trans. Industr. Inf. **18**(3), 1845–1853 (2021)
13. Fahimipirehgalin, M., Weiss, I., Vogel-Heuser, B.: The historical alarm flood situations are analyzed using alarm subsequences and different studies uses clustering approaches for a causal analysis to detect root cause. Proc. Eur. Control Conf. (ECC) 2056–2061 (2020)
14. Rodrigo, V., Chiou, M., Hagglund, T., Hollende, M.: Causal analysis for alarm flood reduction. IFAC **49**(7), 723–728 (2016)
15. Folmer, J., Vogel-Heuser, B.: Computing dependent industrial alarms for alarm flood reduction. Proc. Int. Multi-Conf. Syst. Sygnals Devices (Mar. 2012), 2012 (2012)
16. Vogel-Heuser, B., Schütz, D., Folmer, J.: Criteria-based alarm flood pattern recognition using historical data from automated production systems (aPS). Mechatronics **31**, 89–100 (2015)
17. Fullen, M., Schüller, P., Niggemann, O.: Validation of similarity measures for industrial alarm flood analysis, pp. 93–109. Springer, Berlin, Germany (2018)
18. Cheng, Y., Izadi, I., Chen, T.: Pattern matching of alarm flood sequences by a modified Smith-Waterman algorithm. Chem. Eng. Res. Des. **91**(6), 1085–1094 (2013)
19. Lai, S., Yang, F., Chen, T.: Online pattern matching and prediction of incoming alarm floods. J. Process. Control. **56**, 69–78 (2017)

20. Shang, J., Chen, T.: Early classification of alarm floods via exponentially attenuated component analysis. IEEE Trans. Ind. Electron. **67**(10), 8702–8712 (2020)
21. Fullen, M., Schüller, P., Niggemann, O.: Semi-supervised case-based reasoning approach to alarm flood analysis. Proc. Mach. Learn. Cyber-Phys. Syst. 53–61 (2020)
22. Haque, S., Aziz, S.: False alarm detection in cyber-physical systems for healthcare applications. AASRI Procedia **5**(AASRI ProcediaAASRI Procedia), 54–61 (2013)
23. Patcha, A., Park, J.: An overview of anomaly detection techniques: existing solutions and latest technological trends. Comput. Netw. **51**(12), 3448–3470 (2007)
24. Lin, J., Keogh, E., Fu, A., Herle, H.: Approximations to magic: finding unusual medical time series. In: IEEE, CBMS'05, 18th IEEE Symposium on Computer-Based Medical Systems
25. Rajavellu, S.: Layers of protection, zoomdots (2017)
26. Sarafudheen, Tharayil, M., Rodrigues, P.: Utility based anonymization for big data systems using apache spark. J. Adv. Res. Dynam. Control Syst. 12(Special Issue) (2017). Elsevier/Scopus

Improving the Resolution of Images Using Super-Resolution Generative Adversarial Networks

Maryam J. Manaa[1](✉), Ayad R. Abbas[1], and Wasim A. Shakur[2]

[1] Computer Science Department, University of Technology, Baghdad, Iraq
cs.20.28@grad.uotechnology.edu.iq,
ayad.r.abbas@uotechnology.edu.iq

[2] Computer Science Department, College of Education for Pure Sciences/Ibn Al-Haitham, University of Baghdad, Baghdad, Iraq
wisam.a.s@ihcoedu.uobaghdad.edu.iq

Abstract. This article introduces an innovative strategy for improving image super-resolution through the utilization of Super-Resolution Generative Adversarial Networks (SRGANs). By harmoniously incorporating perceptual loss functions and refining the model's structure, the proposed technique strives to strike an equilibrium between quantitative measurements and perceptual authenticity. Empirical assessments conducted on benchmark datasets demonstrate that the resulting high-resolution images, generated through this method, showcase exceptional quality and perceptual fidelity in comparison to conventional methods. The integration of SRGANs represents a noteworthy leap in the domain of image resolution enhancement, holding the potential to deliver visually captivating and perceptually plausible high-resolution images across a wide spectrum of applications.

Keywords: Deep Learning · Generative Adversarial Network (GAN) · Image Super-Resolution · Image Quality · Super-Resolution GAN

1 Introduction

The pursuit of enhancing image resolution has been a longstanding challenge in the field of computer vision and image processing. The demand for high-quality images spans various applications, including medical imaging, surveillance, entertainment, and remote sensing. As technology continues to evolve, the need to convert low-resolution images into high-resolution counterparts becomes increasingly essential [1]. These algorithms work to transform existing low-resolution images into high-resolution counterparts, aiming to recover the latent true information. The term "image super-resolution technology" denotes the application of technical approaches to generate high-quality images from one or more low-resolution sources [2]. The advancement of deep learning techniques, particularly Convolutional Neural Networks (CNNs) and Generative Adversarial Networks (GANs), has opened new avenues for addressing the challenge of image super-resolution. CNNs have significantly improved image resolution by directly mapping low-resolution

Y. Farhaoui et al. (Eds.): ICAISE 2023, LNNS 837, pp. 68–77, 2024.
https://doi.org/10.1007/978-3-031-48465-0_9

images to higher-resolution versions. However, conventional CNN-based methods often struggle to capture fine details and intricate textures, limiting their ability to produce perceptually realistic results [3, 4]. In recent years, the emergence of Generative Adversarial Networks (GANs) has marked a paradigm shift in image synthesis tasks, including image super-resolution. GANs introduce a novel adversarial framework where a generator network generates images and a discriminator network evaluates their authenticity. The interplay between these networks leads to the generation of increasingly realistic images. This adversarial approach has shown remarkable success in enhancing the perceptual quality of super-resolved images [6]. The paper introduces a novel approach to improving image resolution through the utilization of Super-Resolution Generative Adversarial Networks (SRGANs) in Fig. 1. SRGANs harness the power of GANs and combine them with perceptual loss functions, emphasizing the perceptual quality of generated images over traditional pixel-wise reconstruction. By incorporating perceptual loss, SRGANs aim to produce images that not only exhibit higher quantitative fidelity but also align closely with human perceptual judgments [5]. Within the framework of SRGANs, this study delves into the architecture, training strategies, and loss functions that contribute to the superior performance of the proposed approach. This paper investigates modifications to the SRGAN architecture, including the removal of batch normalization layers from residual blocks, utilization of a deeper network with 16 residual blocks, and optimization of the loss function to achieve more realistic and visually appealing results. To validate the effectiveness of the proposed SRGAN approach, comprehensive experiments are conducted on benchmark datasets commonly used in image super-resolution research. The experimental results demonstrate the capability of SRGANs to restore intricate textures and generate high-resolution images that exhibit enhanced perceptual realism. The subsequent sections of this paper are organized as follows: Sect. 2 offers Related Works of SRGAN, deep learning, and super-resolution image images. Section 3 develops the framework of SRGAN based on this. The experimental findings are presented in Sect. 4, and a discussion of the methodology and potential future research areas is included in Sect. 5. Section 6 presents the conclusion.

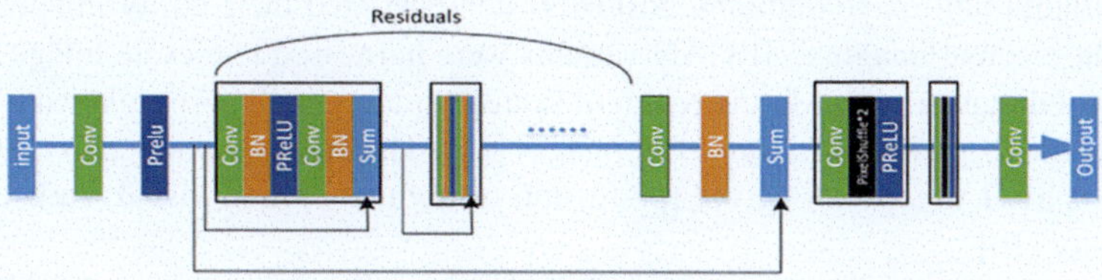

Fig. 1. General structure of generator network in SRGAN

2 Related Work

Over the last decade, the field of image super-resolution has garnered significant attention and has found numerous practical applications, spanning from enhancing the readability of license plates and signs in low-resolution CCTV footage to identifying faces in surveillance videos, processing low-resolution medical images, improving satellite and aerial imagery, and analyzing text within images. Moreover, its potential extends to augmenting the resolution of films, both old and new video games, and astronomical photographs. The pursuit of refining image resolution has evolved from classical methods to the emergence of innovative deep-learning techniques. Traditional approaches, such as bicubic interpolation and Lanczos resampling, aimed to amplify image dimensions but frequently led to the degradation of intricate details and authentic textures. These methods struggled to restore finer and more intricate textural characteristics, often resulting in a synthetic rather than genuine appearance [7, 8]. The rise of deep learning introduced Convolutional Neural Networks (CNNs) as a breakthrough. The SRCNN (Super-Resolution Convolutional Neural Network) brought a revolutionary change to image processing by allowing a direct translation from low-resolution images to high-resolution images using convolutional layers [3]. However, these initial CNN-based methods encountered challenges in capturing nuanced textures and details, which limited their capacity to generate perceptually realistic outputs. A pivotal transformation occurred with the introduction of Generative Adversarial Networks (GANs) in the context of super-resolution. Ledig et al. introduced the Super-Resolution GAN (SRGAN) in 2017, fusing adversarial networks with perceptual loss functions [5]. This integration bridged the gap between quantitative metrics and perceptual quality, resulting in super-resolved images that surpassed the capabilities of conventional CNN-based techniques. Subsequent research endeavored to refine the architecture and training strategies of SRGANs. Variations in generator depth, configurations of residual blocks, and incorporation of skip connections were explored to optimize performance [9]. Attention mechanisms also emerged as a significant advancement. Techniques such as self-attention and spatial attention were introduced to capture long-range dependencies and enhance the extraction of image features [4, 10]. Additional methods, including Wavelet Transform and Laplacian Pyramid Fusion, were employed to adeptly manage multi-scale information [11]. In response to the computational challenges of training GANs, techniques like progressive growing and network pruning were proposed. Progressive growing introduced a stepwise approach to expanding the generator and discriminator layers, facilitating stable training [12]. Network pruning aimed to alleviate computational demands while preserving performance [13]. Beyond the confines of traditional image domains, SRGANs demonstrated their utility in specialized sectors. In medical imaging, SRGAN variants were harnessed to elevate image quality for medical diagnoses [14]. In the realm of satellite imagery, SRGANs played a pivotal role in enhancing low-resolution aerial images [15]. The creative world also reaped the benefits of SRGANs, enabling the production of high-resolution visuals imbued with authenticity [16].

In summation, the development of SRGANs, merging the prowess of GANs with perceptual loss functions, has ushered in a new epoch of super-resolution techniques. This amalgamation bridges the gap between quantitative metrics and perceptual quality, redefining the landscape of generating visually captivating high-resolution images across a wide spectrum of applications [22–24].

3 Proposed Method

The major objective of the SRGAN is to generate a new version with super-resolution (SR) from the input image of low resolution (LR). The main contribution of each approach focused on this field of art is how to modify the standard SRGAN in a specific manner that leads to generating the image with more enhancements, rather than keeping this image with the same clear viewing when enlarge at a high resolution. In this section, the architecture of the proposed network is described in more detail in addition to the evaluation metrics that are considered to evaluate the robustness of the proposed approach.

3.1 Improved Model Structure

In this research, the general structure of SRGAN was used with some modifications to the general structure The generator network is designed to up-sample the low-resolution input images to higher resolution while preserving fine details and textures. To achieve this, the generator employs a series of residual blocks. The elimination of batch normalization (BN) within the residual blocks of the Super-Resolution Generative Adversarial Network (SRGAN) involves the normalization of activations. In tasks involving super-resolution, the primary objective is the accurate restoration of intricate details and high-frequency information within images. The inclusion of batch normalization within the residual blocks can unintentionally diminish the impact of high-frequency components, resulting in the forfeiture of nuanced textures and realistic fine points. This effect could lead to the creation of images characterized by an excessively smooth and artificial appearance, lacking the subtleties that contribute to visual authenticity. The decision to exclude batch normalization from the residual blocks within the SRGAN architecture is geared towards the preservation of the intricate textures and fine particulars that hold the utmost significance for perceptual authenticity. This strategic choice is aligned with the overarching objective of generating high-resolution images that fulfill quantitative standards while concurrently exhibiting perceptual realism. The absence of batch normalization empowers the model to concentrate on capturing the distinct attributes of the input image, thereby generating high-resolution outcomes that retain their visual integrity Shown in Fig. 2.

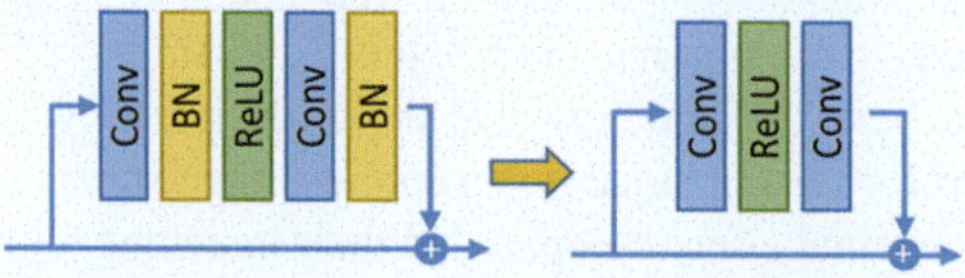

Fig. 2. The effect of removing the BN Layer from the standard structure of RB in SRGAN.

And the use of 16 residual blocks in SRGAN is a design choice made to enhance the depth and expressive power of the generator network. The inclusion of 16 residual blocks within the architecture of the Super-Resolution Generative Adversarial Network

(SRGAN) is a deliberate strategy aimed at substantially enhancing the quality and genuineness of the resulting high-resolution images. This strategic decision is rooted in a comprehensive understanding of the intricate nature of super-resolution tasks and the critical role that network depth plays in accurately capturing the intricate nuances present in images. Super-resolution involves the recovery of fine textures and high-frequency details that tend to be lost during the process of downscaling. Each individual residual block contributes by introducing a residual mapping that aids in the restoration of these essential elements. The collective impact of integrating 16 such blocks into the SRGAN framework enables an iterative mechanism of refining and improving the representation of these vital high-frequency features.

Moreover, the intentional selection of employing 16 residual blocks empowers the network to proficiently capture features that span varying scales. The initial blocks are dedicated to capturing fundamental attributes, while subsequent blocks progressively fine-tune and amalgamate these attributes into a coherent and intricate representation. This hierarchical methodology of feature extraction equips the SRGAN to accurately recreate intricate textures, sharp edges, and intricate patterns, culminating in the generation of high-resolution images that exude heightened perceptual authenticity shown in Fig. 3.

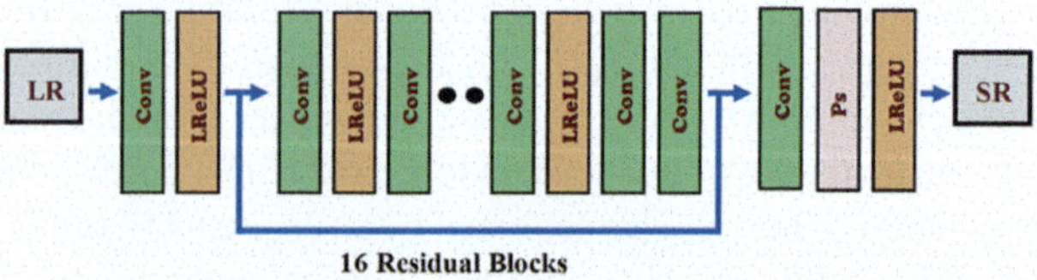

Fig. 3. A network model of the proposed SRGAN generator.

4 Experimental

To evaluate the effectiveness of our proposed approach, we conducted extensive experiments on benchmark datasets commonly used for image super-resolution tasks. We compared our method with traditional CNN-based approaches and the original SRGAN architecture. The experimental results demonstrate that our proposed method yields images with enhanced visual quality, finer details, and improved perceptual realism. The used two datasets, Set5 [18] and Set14 [2] are used for testing. Only one image from each dataset was used for testing since it is a single-image super-resolution model. All dataset's super-resolved images are compared to their matching high-resolution images.

Based on the provided assessment metrics and the identical testing datasets, the proposed model is then contrasted with the prior SR models.

4.1 Evaluation Metrics

In this research, two evaluation metrics are considered to evaluate the performance of the proposed approach to generate the image with super-resolution. The first metric is the peak signal-to-noise ratio (PSNR) The perceptual quality of the image often declines when the image becomes extremely smooth. Peak Signal Noise Ratio (PSNR), a widely used metric to assess picture quality, was later discovered to be of little benefit because it only considers MSE rather than the real perceptual quality of an image.

$$\boldsymbol{PSNR} = \mathbf{10log10}\boldsymbol{MAXI2MSE} \tag{1}$$

where

MAX_I is the maximum possible pixel value of the image.
MSE is the mean squared error between corresponding.

which is determined using definition (1) and the obtained value of this metric can describe the quality of the resulting image according to the average summation of differences between the corresponding pixels in the LR and SR images. The second metric is structural similarity (SSIM) which is determined using definition (2) and evaluates the similarity of the LR and SR images based on brightness, contrast, and structure properties.

$$\boldsymbol{SSIM} = \mathbf{2}\boldsymbol{\mu X \mu Y} + \boldsymbol{C}\mathbf{1}\boldsymbol{\sigma XY} + \boldsymbol{C}\mathbf{2}\boldsymbol{\mu X}\mathbf{2} + \boldsymbol{\mu Y}\mathbf{2} + \boldsymbol{C}\mathbf{1}\boldsymbol{\sigma X}\mathbf{2} + \boldsymbol{\sigma Y}\mathbf{2} + \boldsymbol{C}\mathbf{2}' \tag{2}$$

were

$\boldsymbol{\sigma}_{\boldsymbol{X}}$ is the average value of X, and $\boldsymbol{\sigma}_{\boldsymbol{y}}$ represents the average value of Y; $\boldsymbol{\sigma}_{\boldsymbol{X}}^2$ represents the variance of X,

$\boldsymbol{\sigma}_{\boldsymbol{Y}}^2$ represents the variance of Y, and σ_{XY} represents the covariance between X and Y; $\boldsymbol{C}_1 = (0.001 * \mathrm{L})^2$, $\boldsymbol{C}_2 = (0.003 * \mathrm{L})^2$ are the two variables used to maintain stability, and L represents the dynamic range of image pixels.

Finally, the large value of PSNR and the closer value to 1 of SSIM reflect the significant percentage of clearness for the resulting image.

5 Result

In this research, the performance of the proposed approach is evaluated by a set of images in Set5 and Set14 datasets which are used mainly by similar works thus all the outcomes can be validated easily with the corresponding results from these studies.

5.1 Qualitative Evaluation

In this section, the proposed approach is evaluated according to the visual view of the reconstructed image segment which is generated from applying the proposed improving method to a specific area in the LR image. The qualitative results of four images from Set5 and Set14 Datasets are mentioned in Fig. 4.

Fig. 4. The qualitative results of enlarge segments from applying the proposed approach to four images with different resolutions.

5.2 Quantitative Evaluation

In this section, another evaluation scenario is applied to validate the performance of the proposed approach to generate super-resolution images using two standard evaluation metrics (PSNR and SSIM). The images in the Set5 and Set14 Datasets are considered in this evaluation. The determined values of these evaluation metrics using the proposed and some well-known existing works: DVDR-SRGAN [19], SRGAN [5], ESRGAN [11], Beby-GAN [20], and SPSR [21] are illustrated in Table 1.

Table 1. Validation results of PSNR and SSIM obtained by the proposed improving approach and other five existing works.

#	Method description	Set5		Set14	
		PSNR	SSIM	PSNR	SSIM
1	DVDR-SRGAN	28.52	0.825	24.81	0.702
2	SRGAN	26.91	0.804	23.87	0.677
3	ESRGAN	27.35	0.806	23.61	0.650
4	Beby-GAN	27.82	0.801	24.69	0.701
5	SPSR	28.44	0.824	24.75	0.696
6	Proposed model	28.64	0.892	24.69	0.698

The validation results in Table 1 view that the DVDR-SRGAN technique scores the highest PSNR values among the first five techniques for both Set5 and Set14 Datasets. At the same time, the average PSNR value which is obtained by the proposed approach exceeds the corresponding value which is obtained by the DVDR-SRGAN technique using Set5 Dataset and also stands nearest the average PSNR value using Set14 Dataset. The same comparison result is repeated for SSIM except that the recorded average value of SSIM using Set5 is more powerful (nearest to 1) than the same SSIM value in the DVDR-SRGAN technique. Thus, the significant validation results of PSNR and SSIM in the proposed approach prove its ability to resolve the trouble of the general lack of authenticity in adversarial perception techniques and save the visual quality of the resulting image in a case of enlarged details.

6 Conclusion and Future Work

In summary, this study has introduced a revolutionary approach to enhancing image super-resolution using Super-Resolution Generative Adversarial Networks (SRGANs). Through the seamless integration of perceptual loss functions and optimization of the model architecture, our proposed methodology attains a harmonious equilibrium between quantitative metrics and perceptual fidelity. Empirical findings from benchmark datasets underscore the superior quality and perceptual authenticity of the generated high-resolution images, surpassing conventional techniques. The integration of

SRGANs represents a substantial leap forward, holding the promise of delivering visually captivating and perceptually authentic high-resolution images across a diverse array of applications.

Looking ahead, this study offers several promising directions for future exploration. Firstly, there is room for further architectural refinement to push the boundaries of image super-resolution. Exploring diverse configurations of residual blocks, skip connections, and network depths could yield even more remarkable outcomes. Moreover, delving into alternative loss functions, such as variations of adversarial loss or modifications to perceptual loss, may contribute to advancing the perceptual realism of the generated images.

References

1. Nasrollahi, K., Moeslund, T.B.: Super-resolution: a comprehensive survey **25**(6) (2014)
2. Bevilacqua, M., Roumy, A., Guillemot, C., Morel, M.L.A.: Low-complexity single-image super-resolution based on nonnegative neighbor embedding. BMVC 2012—Electron. Proc. Br. Mach. Vis. Conf. (Ml), 1–10 (2012). https://doi.org/10.5244/C.26.135
3. Dong, C., Loy, C.C., He, K., Tang, X.: Image super-resolution using deep convolutional networks. IEEE Trans. Pattern Anal. Mach. Intell. **38**(2), 295–307 (2016). https://doi.org/10.1109/TPAMI.2015.2439281
4. Chen, R., Qu, Y., Li, C., Zeng, K., Xie, Y., Li, C.: Single-image super-resolution via joint statistic models-guided deep auto-encoder network. Neural Comput. Appl. **32**(9), 4885–4896 (2020). https://doi.org/10.1007/s00521-018-3886-2
5. Ledig, C., et al.: Photo-realistic single image super-resolution using a generative adversarial network. In: Proceedings of the IEEE Conference on Computer Vision and Pattern Recognition, CVPR, vol. 2017, pp. 105–114 (2017). https://doi.org/10.1109/CVPR.2017.19
6. Goodfellow, I.J., et al.: Generative adversarial nets. Adv. Neural. Inf. Process. Syst. **3**(January), 2672–2680 (2014). https://doi.org/10.3156/jsoft.29.5_177_2
7. October, M.: A Wavelet Tour of Signal Processing (2008)
8. "Ten Lectures of Wavelets"
9. Tai, Y., Yang, J., Liu, X., Xu, C.: MemNet: a persistent memory network for image restoration
10. Woo, S., Park, J., Lee, J., Kweon, S.: CBAM: convolutional block attention module
11. Wang, X., Yu, K., Wu, S., Gu, J., Liu, Y.: ESRGAN : enhanced super-resolution generative adversarial networks, pp. 1–16
12. Karras, T., Aila, T., Laine, S., Lehtinen, J.: Progressive growing of GANs for improved quality, stability, and variation (2017), [Online]. Available: http://arxiv.org/abs/1710.10196
13. Durdanovic, I., Graf, H.P.: P f e c n (2016), 1–13 (2017)
14. Bai, L., et al.: Multispectral U-Net : a semantic segmentation model using multispectral bands fusion mechanism for landslide detection, pp. 73–76 (2022)
15. Tao, Y.: Super-resolution restoration of spaceborne ultra-high-resolution images using the UCL OpTiGAN system (2021)
16. Agustsson, E., Timofte, R.: NTIRE 2017 challenge on single image super-resolution: dataset and study. IEEE Comput. Soc. Conf. Comput. Vis. Pattern Recognit. Work. **2017-July**, 1122–1131 (2017). https://doi.org/10.1109/CVPRW.2017.150
17. Martin, D., Fowlkes, C., Tal, D., Malik, J.: A database of human segmented natural images and its application to evaluating segmentation algorithms and measuring ecological statistics. Proc. IEEE Int. Conf. Comput. Vis. **2**, 416–423 (2001). https://doi.org/10.1109/ICCV.2001.937655

18. Zeyde, R., Elad, M., Protter, M.: On single image scale-up using sparse-representations. Lect. Notes Comput. Sci. (including Subser. Lect. Notes Artif. Intell. Lect. Notes Bioinformatics) **6920** LNCS(1), 711–730, 2012. https://doi.org/10.1007/978-3-642-27413-8_47
19. Qu, H., Yi, H., Shi, Y., Lan, J.: DVDR-SRGAN : differential value dense residual (2023)
20. Li, W., Zhou, K., Qi, L., Lu, L., Lu, J.: Best-buddy GANs for Highly detailed image super-resolution (2019)
21. Ma, C., Rao, Y., Cheng, Y., Chen, C., Lu, J., Zhou, J.: Structure-preserving super resolution with gradient guidance, pp. 7769–7778
22. Farhaoui, Y.: All, big data mining and analytics **6**(3), I–II (2023). https://doi.org/10.26599/BDMA.2022.9020045
23. Farhaoui, Y.: Big data analytics applied for control systems. In: Lecture Notes in Networks and Systems, vol. 25, pp. 408–415 (2018). https://doi.org/10.1007/978-3-319-69137-4_36
24. Farhaoui, Y., et al.: Big data mining and analytics **5**(4), I–II (2022). https://doi.org/10.26599/BDMA.2022.9020004

Three Levels of Security Including Scrambling, Encryption and Embedding Data as Row in Cover Image with DNA Reference Sequence

Asraa Abdullah Hussein(✉), Rafeef M. Al Baity, and Sahar Adill Al-Bawee

College of Science for Women, University of Babylon, Hillah, Iraq
wsci.israa.abdullah@uobabylon.edu.iq

Abstract. Data security has become a more important and pressing issue as a result of the growing use of computers in everyday life and the unchecked growth of the internet. The unauthorized access to data is prevented by the use of new secure communication methods including cryptography and steganography. The proposed system consists of three levels to hide data inside colored images. The first level is to scatter the data through the use of the seed. The second level is to encrypt the scattered data into deoxyribonucleic acid (DNA) and match each two segments of the string with DNA reference sequence to produce set of index. The last level is to embedding the index set as a row inside the cover image. The results of the experiments proved the efficiency of the proposed system where the PSNR values ranged between (69.5616–82.9210).

Keywords: DNA · Cryptography · Steganography · RGB · Security · Image

1 Introduction

With the rapid advancement of communication technology and the internet, sharing of information over the internet and mobile networks has become a common form of communication [1]. It is very essential that the communication is made in an extremely secure manner with the primary concern being how to transmit the information securely and prevent the data from hacking, unauthorized access, or modification; as a result, information security became crucial to facilitating the confidential exchange of information between any sender and receiver [2, 3]. Preserve this information a security techniques invented such as data hiding and cryptography that used the most frequently in the sectors of communication and computer security [4].

Cryptography is the science of using mathematics to encrypt and decrypt data to keep messages secured by transforming intelligible data form into an unintelligible form [5]. Steganography used to disguise data so thoroughly that others who aren't intended recipients don't recognize the stego-graphic medium that carrying secret information. The process of protecting private and confidential information without arousing the attention or suspicion of others in a digital media (images, audio, video, and text) is called Steganography [6]. In contrast to cryptography which focuses on hiding information

Y. Farhaoui et al. (Eds.): ICAISE 2023, LNNS 837, pp. 78–83, 2024.
https://doi.org/10.1007/978-3-031-48465-0_10

within messages, steganography is more concerned with concealing information itself. However, steganography focuses on making the message go undetected while both methods are useful in protecting sensitive information from prying eyes, they also have flaws that can be exploited. Integrating steganography and cryptography in a single system improves security [7 8]. The DNA reference sequence is a fundamental resource used in genetics and genomics research presents in European Bioinformatics Institute (which is known as EBI Database) [9]. It serves as a standardized representation of the human genome or the genome of any other organism.

2 Related Work

In 2019 [10] this study was divided into two parts: The first part included encoding the text inside an image where the encoding was performed after performing an XOR operation between the text and the shared password between the sender and receiver. The second part involves encoding a random text inside the image. The results showed that the proposed method is characterized by difficulty in detecting the text.

In 2021 [11] a 3D chaotic cat map was proposed based on the output of the irregular 3D cat map, the pixel coordinates and color component were determined to include the secret message randomly in the less significant bit (LSB) value of the cover image. Therefore, the proposed method provides a high level of security by eliminating various attacks. In 2020 [12] presented a hiding method consisting of two levels of confidentiality where the first level includes hiding the data inside an image based on a random key and the second level includes encrypting the stego image with the encryption key before sending it to the recipient in order to increase confidentiality.

In 2022 [13–15] hyper chaotic map with modified LSB are combined and used in conjunction with cryptography and steganography to enhance security. Performing message encryption by convert data into binary representation to achieve different string lengths before hiding. In the steganography a random key stream is generated using a combination of two low-complexity chaotic maps: tent map and Ikeda map. Finally using LSB color image masking a low-complexity XOR operation is applied to the most significant bits in the 24-bit color-envelope images. In 2020 [16] a two-stage technique suggested a secret text message in a cover image. The two phases included a chaotic map encryption technique to encrypt the text and an image-masking technique using LSB to hide the cipher text in an image. The results showed that depending on the chaotic map and the process of hiding the text in the image using LAB, with the help of XOR can provide more security and complexity.

In 2020 [17] new technology was used to mask the text in the file speech without any visible fuss. The idea of the system consists of a scrambled data using a chaos map and then using Zaslavsky map to encoded text file. Hiding encoded data in the speech cover based on using K-means indexing. The results showed that the system is a good method because a data file cannot be detected or retrieved. In 2022 [18] The LSB algorithm was used to hide confidential information within the cover photo. The algorithm was applied to insert a binary code of information into the LSB for each color component in the image where the LSB algorithm is embedded in the color image. This method is effective for hiding the data.

3 Proposed System

This Section reviewing details of the steps taken by both the sender and the recipient. The proposed system as shown in the following Fig. 1.

Part one:" Sender side details"

Step 1: Read the text and convert it into ASCII and then Binary form.
Step 2: Scatter the binary form by using seed with one of the Shuffle functions.
Step 3: The stage of encryption the sender and receiver have the same DNA reference sequence so perform the encryption process by convert scrambling binary form into DNA and then generate index by matching it with DNA reference sequence.
Step 4: Hide these indexes inside the color image where the proposed method adopts an unconventional embedding idea instead of converting the cipher text into binary and including it inside the cover image the index series (representing the cipher text) is represent as a row for the hiding sites in the cover image and the column is selected in sequence from 1 to the NO. Of columns as needed. The recipient is facing the problem of which sites were chosen by the sender when embedding; to facilitate this process the following is followed:

A. Change the 8th bit of the value for the locations that precede the locations (represents the hidden information) into (0).
B. Change the 8th bit of the value for the locations that represents the hidden information into (1).

The following example shows the stages of the process of hiding text at the sender: the plain text: 'hi' , Original binary text: 0110100001101001.

Key scrambling = 7 16 2 14 11 6 9 13 8 4 1 12 15 10 5 3.

Scrambling binary text = 0110100100000111. DNA form = CGGCAACT.

Encryption form (matching with DNA reference sequence) = 12 19 8 27.

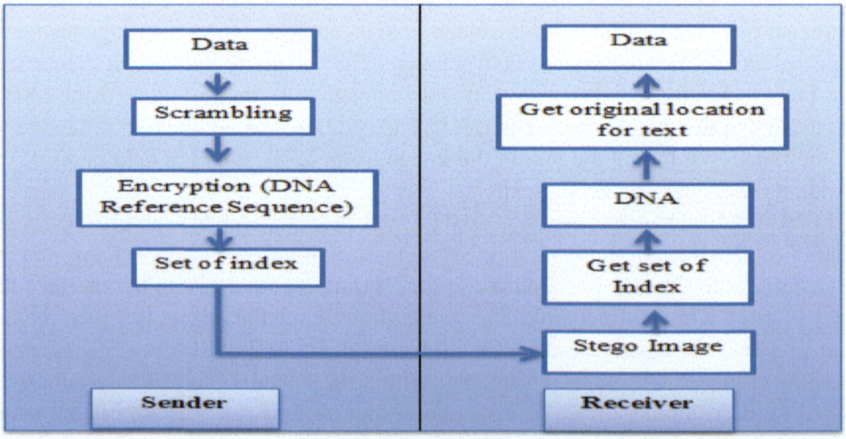

Fig. 1. Proposed system.

Part two:" Recipient side details"

The hidden image is received by the recipient and begins to extract the hidden information by following the steps:

A. Check column by column and takes the index and puts it in a vector for the first pixel which the value of the 8th bit is equal to (1) and so on for the whole image so that the outputs of this step are a set of indexes.
B. Find matches index in DNA reference sequence and the process continues until all the sequences are matched to obtain DNA form and convert into binary.
C. Return the binary form to its original positions by using the seed.
D. Convert binary into ASCII and then character to obtain original text.

4 Discussion and Results

Testing the proposed system on a set of texts of different sizes and RGB standard cover images (Baboon, Lena and Peppers) of dimensions 512 × 512. Measuring the performance of the proposed system by applying peak signal-to-noise ratio (PSNR) as shown in the (Tables 1 and 2). The Fig. 2 shows the cover image Lena after embedding the data in three different experiments with the PSNR of each experiment.

Table 1. Set of experiments(1) for proposed system

Cover image	Length of text	Bits to be hidden	PSNR
Peppers	25	200	82.9210
	77	616	77.8914
	451	3608	70.1784
	529	4232	69.5616

Bits to be hidden=200 PSNR= 82.4039

Bits to be hidden= 616 PSNR= 77.8137

Bits to be hidden=4232 PSNR= 69.5516

Fig. 2. Stego Lena after hiding data.

The strength of the proposed system comes from placing three obstacles in front of hackers and people who are not authorized to prevent them from accessing confidential data. The first challenge facing hackers and attackers is scatter the binary sites, the second challenge is to convert it into DNA and take every two segments and match them with data that is chosen from among of thousand bases located on the EBI site to be transformed into the index set. The third and final challenge is dealing with the index as a row to include it inside cover image without the need to convert it to binary.

Table 2. Set of experiments(2) for proposed system

Cover image	Length of text	Bits to be hidden	PSNR
Baboon	25	200	82.9210
	77	616	78.0511
	451	3608	70.1607
	529	4232	69.6017

5 Conclusion

It is not possible to overlook the danger associated with the technological development which has become dominant in all aspects of life therefore researchers are in a continuous effort to enrich the field of information security and protection with a lot of research and studies. The proposed system offers three levels of confidentiality represented first by scattering the original text through the use of the scrambling; the second step includes encrypting the scattered text by converting it into DNA and then searching in DNA reference sequence for a match to convert it into set of index and finally hiding theses index as row in cover color image. It is worth noting that the nature of the hiding in the proposed system is considered random and this step makes it difficult for the hacker to know the concealment sites that represent confidential information. The efficiency of the proposed system was measured using PSNR and good ratios were obtained.

References

1. Mohammed, A.N., Saif, M.A., Ahmed, M., Majid, J.: Steganography and cryptography techniques based secure data transferring through public network channel. Baghdad. Sci. J. **19**(6), 1362–1368 (2022). https://doi.org/10.21123/bsj.2022.6142
2. Rusul, M.N., Jinan Ali, A., Elaf Ali, A.: Hide text depending on the three channels of pixels in color images using the modified LSB algorithm. Int. J. Electric. Comput. Eng. **10**(1), 809–815 (2020). https://doi.org/10.11591/ijece.v10i1.pp809-815
3. Ahmed, T., Thahab, A.: Secure image steganography based on burrows wheeler transform and dynamic bit embedding. Int. J. Electric. Comput. Eng. (IJECE) **9**(1), 460–467 (2019).https://doi.org/10.11591/ijece.v9i1.pp460-467
4. Nabi, S.H., Sarosh, P., Parah, S., Mohiuddin Bhat, G.: Information embedding using DNA sequences for covert communication, in multimedia security, pp. 111–129. Springer, Singapore (2021). https://doi.org/10.1007/978-981-15-8711-5_6

5. Zou, C., Wang, X., Zhou, C., Xu, S., Huang, C.: A novel image encryption algorithm based on DNA strand exchange and diffusion. Appl. Math. Comput. **430**, 127291 (2022). https://doi.org/10.1016/j.amc.2022.127291
6. Yasir Ahmed, H., Nada Elya, T., Mohammed Qasim, A.: An enhanced approach of image steganographic using discrete shearlet transform and secret sharing. Baghdad. Sci. J. **19**(1), 197–207 (2022). https://doi.org/10.21123/bsj.2022.19.1.0197
7. Zhiguo, Q., Zhenwen, C., Xiaojun, W.: Matrix coding-based quantum image steganography algorithm. IEEE. **7,** 35684–35698 (2019). https://doi.org/10.1109/ACCESS.2019.2894295
8. Taha, M.S., Mohd Rahim, M.S., Lafta, S.A., Hashim, M.M., Alzuabidi, H.M.: Combination of steganography and cryptography: a short survey, 2nd international conference on sustainable engineering techniques (ICSET 2019). Baghdad. Iraq. IOP Conf. Ser. Mater. Sci. Eng. **518**, 1–14 (2019). https://doi.org/10.1088/1757-899X/518/5/052003
9. European Bioinformatics Institute. http://www.ebi.ac.uk/
10. Azal, H.: A new method for hiding text in a digital image. J. Southwest Jiaotong. Univ. **55**(2), 1–6 (2020). https://doi.org/10.35741/issn.0258-2724.55.2.4
11. Sarab, M.H., Zuhair Hussein, A., Ghadah, K.A., Safa, A.: Chaos-based color image steganography method using 3 d cat map. Iraqi. J. Sci. **62**(9), 3220–3227 (2021). https://doi.org/10.24996/ijs.2021.62.9.34
12. Hassanain, R.K., Hadi Hussein, M., Keyan Abdul-Aziz, M.: Hiding encrypted text in image steganography. Periodic. Eng. Nat. Sci. **8**(2), 703–707 (2020). https://doi.org/10.21533/pen.v8i2.1302
13. Iman, Q., Zaid Ameen, A., Mustafa, A., Mudhafar, J., Junchao, M., Vincent, O.: A lightweight hybrid scheme for hiding text messages in colour images using LSB, LAH transform and chaotic techniques. J. Actut. Netw. Sens. **11**, 66 (2022). https://doi.org/10.3390/jsan11040066
14. Farhaoui, Y.: All big data mining and analytics. **6**(3), I–II (2023). https://doi.org/10.26599/BDMA.2022.9020045
15. Farhaoui, Y.: All big data mining and analytics. **5**(4), III (2022). https://doi.org/10.26599/BDMA.2022.9020004
16. Zahraa, S., Raniah, A., Amal, A.: Hiding encrypted text in image using least significant bit image steganography technique. Int. J. Eng. Res. Adv. Technol. (IJERAT) **6**, 8 (2020). https://doi.org/10.31695/IJERAT.2020.3642
17. Iman, Q.A., Amal, H.K.: Hiding text in speech signal using k-means, LSB techniques and chaotic maps. Int. J. Elect. Comput. Eng. (IJECE) **10**(6), 5726–5735 (2020). https://doi.org/10.11591/ijece.v10i6.pp5726-5735
18. Taleb, A.S.: Obaid embedding secret data in color image using LSB. J. Educ. Pure Sci. Univ. Thi-Qar. **12**(2) (2022)

Acceptance and Barriers of ICT Integration in Language Learning: In the Context of Teacher Aspirants from a Third World Country

Kristine May C. Marasigan[1](✉), Bernadeth Abequibel[1], Gadzfar Haradji Dammang[2], John Ryan Cepeda[3], Izar U. Laput[1], Marisol Tubo[1], and Jovannie Sarona[1]

[1] Western Mindanao State University, 7000 Zamboanga City, Philippines
xt202001476@wmsu.edu.ph

[2] Mindanao State University-Sulu, Capitol Site, 7400 Jolo, Sulu, Philippines

[3] Ateneo de Zamboanga University, 7000 Zamboanga City, Philippines

Abstract. Technology plays an important role in 21st century education. In equal importance, its role is also striking in the field of English language learning. However, to ensure the effectiveness and efficiency of its integration, major stakeholders such as students who are directly influenced by technology incorporation in their learning process must foster favorable attitudes toward it. Hence, it is important to examine learners' level of technology acceptance, barriers that hinder ICT incorporation, and variables that could influence it. Nevertheless, there is a scarcity of literature that delves into the case of language major prospective teachers from the Philippines, a linguistically heterogeneous third world country. Therefore, the present study, by adhering to descriptive and quantitative approaches, surveyed a total of 92 language major aspirants from all years, whose ages ranged from 19 to 23. One questionnaire consisting of two dimensions (acceptance and barriers of ICT integration) with a reliability score of 0.927 was adopted from the study of Hashemi et al. (JAMA 9:1–20, 2022). Hence, this study reveals that aspirants who are language majors have a *'positive attitude'* toward ICT integration in language learning. Moreover, the study also discloses that internet-related issues prevail as the primary barrier that could hamper their intention to utilize ICT tools. Furthermore, in terms of the association between their acceptance and gender, the study reveals that the latter does not influence the former. The analyses of the results are further deliberated in the study.

Keyword: English learning · ICT integration · Third world country · Gender

1 Introduction

As technology continuously advances, its integration in different fields has become more prevalent. This progress is more specifically evident in the domain of education, as there seems to be concerted efforts to incorporate ICT in the teaching-learning process [1, 2]. Such efforts include investing in technological tools given that they can support education, which is an idea anchored by a long-term transformative vision of what education

Y. Farhaoui et al. (Eds.): ICAISE 2023, LNNS 837, pp. 84–92, 2024.
https://doi.org/10.1007/978-3-031-48465-0_11

should be [3, 4]. Undoubtedly, the integration of ICT has become an integral component of education [5]; hence, it comes as no surprise that it eventually alters how education is delivered, which is evident from the manner of how teachers teach, and learners learn [6].

Moreover, according to Sabti and Chaichan [7], the advent of technological development has altered not only education but also language learning. Supportive of this is the claim of Rahmati et al. [8] that the application and utilization of technological tools are also striking in the field of English language learning, which has been extensively empowered by the prominence of ICT integration [9]. Several studies with varying contexts and constructs have similarly reported that English language learning has been substantially enhanced and optimized by means of incorporating ICT-related tools [10–16]. However, the success of its incorporation still depends on many factors. As anchored to the Technological Acceptance Model, the acceptability of technology influences one's attitude. According to de la Rama et al. [17] and Cabangcala et al. [18], attitude acts as a determinant of the success or failure of something, as it influences behavior and actions. In this sense, learners' intention to integrate ICT is directly influenced by their technology acceptance, which impacts their attitude. Having stated this, it is worth noting that the present study adopted the *'perceived usefulness'* and *'perceived ease of use'* as the two determinants of identifying one's acceptance toward the incorporation of ICT.

Additionally, provided that digital technology has developed into an integral factor in language learning [9], it becomes implicit for aspirants of language major teachers to develop a favorable attitude toward its integration. However, although several studies have already explored the case of learners [10–12] and teachers [13–16], it should be noted that most of these studies are from foreign countries, and limited investigations were situated in the Philippines. Therefore, this warrants the necessity to delve into the case of Filipino teacher aspirants in terms of their acceptability level toward ICT integration. Furthermore, taking into consideration that the integration of ICT could be influenced by other factors apart from attitude, this investigation also aims to determine which barrier impedes its integration in language learning. Additionally, gender as a variable will also be investigated given that there are inconclusive findings in the literature, which revealed that while some studies disclosed that gender is not a factor [10, 13, 15, 19, 20], there are also other studies that declared otherwise [16]. Hence, the scope of this present study could provide further understanding regarding the status of ICT integration in language learning from a third world context.

Research question:

This study explores the acceptance of and barriers to ICT integration among English language major teacher aspirants. The following questions guide the present investigation:

1. What is the perception of English language major aspirants regarding ICT usefulness and ease of use in English language learning?
2. What common barriers do the respondents experience that hinder them from utilizing ICT in learning English?
3. Is gender a contributing factor that influences their acceptance of ICT integration?

2 Methodology

2.1 Research Design

This research study utilized a quantitative-descriptive design to determine the perceptions of language major aspirants regarding their acceptance of ICT and barriers that could influence their intention to employ it in English language learning. According to Cresswell [21], quantitative research design involves the process of gathering quantifiable information that can be subjected to statistical treatment. Moreover, provided that this study also intended to describe the respondents' perceptions through the gathered information, it also made use of a descriptive research design [22]. In addition, the responses were gathered in one shot through a survey questionnaire to ensure an efficient way of collecting information [23], hence implying that this study is also cross-sectional.

2.2 Respondents

The selection for the respondents of this study was determined using the following specifications: (1) respondents must be language majors who are enrolled in the College of Teacher Education and (2) they must have taken at least one pedagogical technology-related subject, in this case, the Technology for Teaching and Learning (TTL). With the specifications provided, a total of 92 English language prospective teachers from all year levels, whose age range was from 18 to 22 with a mean age of 20.04, consented to respond in this study. It is worth noting that more than half of the samples (81.1%) were female.

2.3 Instrument

To quantify the perceptions of teacher aspirants regarding the acceptance of ICT, one questionnaire with two dimensions with an original Cronbach's alpha score of 0.813 was adopted from the research instrument of Hashemi et al. [10]. The questionnaire is composed of 20 closed-ended items that aim to collect the respondents' perceptions regarding ICT usefulness and ease of use in English language learning. Moreover, in terms of the respondents' perception of the barriers to ICT integration, they were asked to choose from a given list of barriers that they thought hinder them from using ICT. The declared overall reliability score of the instrument is 0.927, which is noticeably higher than the adopted instrument's original reliability score. According to the general rule of thumb, a very good level of internal consistency is $\alpha = 0.8$ or greater [24], hence implying that the instrument used in the study has excellent reliability.

2.4 Data Gathering Procedure

The digitation of the adopted instrument was carried out after the validation process. The link to the digitized instrument was subsequently disseminated among the target respondents using a Messenger account dedicated to data gathering. Moreover, it is worth noting that the first section of the survey link consists of a consent form to guarantee that the respondents voluntarily participated in the study; hence, 68.65% of the target population agreed with the said consent. In this section, confidentiality and anonymity were guaranteed by assuring the respondents that no identifying information would be collected and that the gathered data would not be disclosed to third parties; hence, the retrieved file was encrypted to keep the information secured.

2.5 Data Analysis Technique

The collected data were subjected to statistical treatment using IBM SPSS version 25. According to Ozgur et al. [25], because this application could conveniently handle statistical treatment even for studies with enormous data, it has been the top platform used in social sciences research. To determine the respondents' perceptions, mean scores were provided with corresponding interpretations based on computed equal intervals. The provided corresponding interpretations are as follows: 1.00–1.79 for very negative attitude (VNA); 1.80–2.59 for negative attitude (NA); 2.60–3.39 for neutral attitude (neA); 3.40–4.19 for positive attitude (PA); and 4.20–5.00 for very positive attitude (VPA). Moreover, to identify which barriers hinder the respondents from using ICT, the percentage and frequency of responses were used. On the other hand, to determine whether there is a gender difference in their perception, the responses underwent statistical treatment using one-way ANOVA.

3 Results and Discussion

Teacher Aspirants' Acceptance of ICT in English Language Learning.

The responses were retrieved from Google Forms by converting the gathered data into an Excel spreadsheet. The collected information was then subjected to descriptive statistical treatment using SPPS, which entailed the utilization of both the mean and standard deviation. Moreover, equal intervals were also computed to serve as the basis for the interpretation of the mean scores.

Table 1. Teacher aspirants' acceptance of ICT

Dimensions	M	SD	Interp
Perceived usefulness	4.08	0.52	PA
Perceived ease of use	3.96	0.50	PA
Overall acceptance	4.02	0.49	PA

Scale 1.00–1.79, *Very Negative Attitude (VNA)* 1.80 to 2.59, *Negative Attitude (NP)* 2.60 to 3.39, *Neutral Attitude (neA)* 3.40 to 4.19, *Positive Attitude (PA)* and 4.20 to 5.0- *Very Positive Attitude (VPA)* N = 92

Table 1 provides the results of language major teacher aspirants' level of ICT acceptance in their process of English language learning. As such, this study reveals the respondents' overall '*positive attitude*' in terms of utilizing ICT (M = 4.0272, SD = 0.49600). It specifically entails that they view the two determinants of technology acceptance favorably and positively, as they likewise have positive attitudes toward its perceived usefulness (M = 4.0848, SD = 0.52955) and perceived ease of use (M = 3.9696, SD = 0.50877). Their positive attitude toward the two determinants equates to their acceptance of ICT, which is a finding that resonates with some studies in the literature, such as the investigations of Hashemi et al. [10], Ketmuni [26], Sulistilyo et al. [12], and Sabti and Chaichan [7].

In closer scrutiny, the respondents of this study believe that the incorporation of ICT is beneficial and effective for their English language learning. Moreover, the data also reveal that aspirants with language majors do not have difficulty in terms of utilizing ICT-related tools, as they could maneuver them with ease, which is a quality expected from 21st century prospective teachers.

Barriers to ICT Integration.

To determine which barrier commonly hinders the language teacher aspirants from integrating ICT in their language learning process, they were asked to select one barrier that they experience from the list provided. To decipher the respondents' perceptions, the frequency and percentage of responses were used. The analysis of the data is provided below.

Table 2. Frequency report of barriers to ICT Integration in English Language Learning

Ranking	Barrier	Response rate	
		Frequency	Percentage (%)
1	Lack of internet	48	52.2
2	Lack of institutional support	10	10.9
3	Lack of pedagogical training	10	10.9
4	Lack of time	10	10.9
5	Lack of technological devices	8	8.7
6	Lack of ICT infrastructure	4	4.3
7	Lack of electricity	2	2.2
8	Lack of confidence in using ICT	0	0
Total		92	100

Table 2 presents a list of barriers that could possibly hamper ICT integration, in which the descriptive result divulges that as per the perceptions of language major teacher aspirants, lack of internet acts as the main barrier, noting that 52.2% of the sample disclosed that it is the primary hindrance that encumbers them from incorporating ICT in their language learning. This denotes that their intention to exploit ICT-related tools is dependent on their internet connectivity. This finding corroborates the studies of Asnawi et al. [27], Hashemi et al. [10], and Ketmuni [26], in which they likewise reported that the incorporation of ICT in learning is impeded by internet-related issues. However, this result is explicable by the fact that problems with internet connection continuously prevail in the country.

Interestingly, none of the respondents believed that *'lack of confidence in ICT'* hinders them from incorporating it in their learning. This therefore contradicts the studies of Goktas et al. [28] and Sabti and Chaichan [7], which revealed skill-related issues as a common hindrance that hampers ICT integration. Such a finding could allude to the idea that as 21st century learners, it is an implicit expectation for them to carry the competence of being digital natives who are adept, proficient, and skilled with technology. Hence, along with this is the expectation of their confidence in their skills in terms of utilizing ICT-related tools.

Differences in the Acceptance of ICT in English Language Learning across Genders.

The data were statistically analyzed to determine whether the respondents' attitudes would significantly differ when gender was considered as a variable. Specifically, their responses were subjected to statistical treatment by employing one-way ANOVA. The results are reported below.

Table 3. Language major teacher aspirants' acceptance of ICT across genders

Dimensions	Male		Female		Sig. (2 tailed)	Interp
	M	SD	M	SD		
Perceived usefulness	3.99	0.71	4.10	0.48	0.437	Not significant
Perceived ease of ICT use	3.79	0.58	4.00	0.48	0.116	Not significant
Overall ICT acceptance	3.89	0.63	4.05	0.46	0.222	Not significant

**Difference is significant at level 0.05

The results presented in Table 3 disclose that gender does not influence the language major teacher aspirants' ICT acceptance in English language learning, as the difference between the male respondents and their female counterparts is not significant ($p = 0.222$). This finding implies that the respondents accept ICT integration and foster a positive attitude toward it irrespective of their gender, which therefore deflates a certain contention that technology is a male-dominated area. The results of the present study are in parallel with the findings of Bećirović et al. [9], Hashemi et al. [10], Ketmuni [26], and Teo et al. [29] and in contradiction with the study of Mahdi and Al-Dera [13]. Moreover, it is worth noting that the results of this study could be explained by several reasons. First, it is important to acknowledge that both male and female respondents of this study, who are language major teacher aspirants, are classified as digital natives who are likewise well versed with technology. Second, both male and female respondents are required to take technology-related pedagogy classes as prescribed in their curriculum. Hence, these reasons suggest that they were exposed to the same opportunity to utilize and explore technology in English language learning irrespective of whether they were male or female.

4 Conclusion

The success or failure of ICT incorporation into language learning is determined by various factors, including one's acceptance of it, which in turn could determine one's attitude. Having cited that most of the studies were conducted from contexts in which the English language has a status of being a foreign language, it is necessary to conduct the present investigation in a context of a third-world country that characterized English as not only one of its official languages but also a second language. Moreover, given that learners are one of the major stakeholders who are directly affected by the incorporation of ICT-related tools in their English language learning process, it becomes imperative to delve into their level of acceptance of technology, barriers that hinder ICT incorporation, and variables that could influence it. Nevertheless, there exists a scarcity of studies that examine the cases of prospective teachers from the Philippines. Therefore, to provide insights from learners who are situated in a third world context, the present study explored the attitude of language major teacher aspirants regarding their acceptance of ICT. This study discloses that the respondents have an overall favorable attitude toward ICT incorporation, as they garnered a reasonable level of acceptance toward the two identified determinants of technology acceptance (perceived usefulness and perceived ease of use). Moreover, this paper also reveals that in terms of barriers that could hinder them from incorporating ICT, Internet-related issues persist as the primary hindrance, which deeply reflects the status of Internet connectivity in the country. Furthermore, when gender is considered as a variable, this study discloses that it does not influence the language major teacher aspirants' ICT acceptance, as the difference between the male respondents and their female counterparts is not significant.

In reference to the enumerated results, this paper implies that while language major aspirants have a favorable view of ICT incorporation, they nonetheless encounter some obstacles. With this, this paper also suggests that to further understand this construct, future investigations should cover a wider scope. It is also interesting to find the attitude of other prospective teachers who are majoring in various disciplines but are nonetheless expected to both utilize ICT and carry out the competence of the language of instruction employed. Such future studies could aid stakeholders in education to make the process of incorporating technology in language learning more beneficial.

References

1. Kumar, S., Daniel, B.: Integration of learning technologies into teaching within Fijian polytechnic institutions. Int. J. Educ. Technol. **13**, 1–17 (2016). https://doi.org/10.1186/s41239-016-0036-8
2. Yang, S., Kwok, D.A.: Study of students' attitude towards using ICT in a social contructivist environment. Aust. J. Educ. Technol. **33**, 50–62 (2017). https://doi.org/10.14742/ajet.2890
3. Ramdhony, D., Mooneeapan, O., Dooshila, M., Kokil, K.: A study of university students' attitude towards integration of information technology in higher education in Mauritius. High. Educ. Q. **75**, 1–16 (2020). https://doi.org/10.1111/hequ.12288
4. Williamson, B., Komljenovic, J.: Investing in imagined digital futures: the techno-financial 'futuring' of edtech investors in higher education. Critic. Stud. Educ. 1–16 (2022). https://doi.org/10.1080/17508487.2022.2081587

5. Goriss-Hunter, A., Sellings, P., Echter, A.: Information communication technology in schools: students exercise "digital agency" to engage with learning. Technol. Knowl. Learn. 1–16 (2022).https://doi.org/10.1007/s10758-021-09509-2
6. Al-Adwan, A., Li, A., Al-Adwan, A., Abbasi, G., Albelbis, N., Habibi, A.: Extending the Technology Acceptance Model (TAM) to predict university students' intentions to use metaverse-based learning platforms. Educ. Inform. Technol. 1–33 (2023).https://doi.org/10.1007/s10639-023-11816-3
7. Sabti, A., Chaichan, R.S.: Saudi high school students; attitudes and barriers toward the use of computer technologies in learning English. SpingerPlus 1–8 (2014). https://doi.org/10.1186/2193-1801-3-460
8. Rahmati, J., Izadpanah, S., Shahnavaz, A.: A meta-analysis on educational technology in English language teaching. Lang. Test. Asia **11**, 1–20 (2021). https://doi.org/10.1186/s40468-021-00121-w
9. Bećirović, S., Brdarević-Čeljo, A., Delić, H.: The use of digital technology in foreign language learning. SN Soc. Sci. **1**, 1–21 (2021). https://doi.org/10.1007/s43545-021-00254-y
10. Hashemi, A., Na, S., Noori, A., Orfan, S.: Gender difference on the acceptance and barriers of ICT use in English language learning: students' perspectives. Cogent Human. **9**, 1–20 (2022). https://doi.org/10.1080/23311983.2022.2085381
11. Sarimanah, E., Soeharto, S., Dewi, F., Efendi, R.: Investigating the relationship between students' reading performance, attitudes toward ICT, and economic ability. Heliyon **8**, 1 (2022). https://doi.org/10.1016/j.heliyon.2022.e09794
12. Sulistilyo, U., Al Ariz, T., Handayani, R., Ubaidillah, M., Wiryotinoyo, M.: Determinants of technology acceptance model (TAM) towards ICT use for English language learning. J. Lang. Educ. **8**, 18–31 (2022). https://doi.org/10.17323/jle.2022.12467
13. Lee, C., Yeung, A.S., Ip, T.: Use of computer technology for English language learning: do learning styles, gender, and age matter? Comput. Assist. Lang. Learn. **29**, 1035–1051 (2016). https://doi.org/10.1080/09588221.2016.1140655
14. Li, B.: Ready for online? Exploring EFL teachers' ICT Acceptance and ICT literacy during COVID-19 in Mainland China. J. Educ. Comput. Res. **60**, 196–219 (2021). https://doi.org/10.1177/07356331211028934
15. Qaddumi, H., Smith, M., Masd, K., Bakeer, A., Abu-ulbeh, W.: Investigating Palestininan in-service teachers' beliefs about the integration of information and communication technology (ICT) into teaching English. Educ. Inform. Technol. 1–21 (2023). https://doi.org/10.1007/s10639-023-11689-6
16. Mahdi, H.S., Al-Dera, A.S.: The impact of teachers' age, gender, and experience on the use of information and communication technology in EFL teaching. Engl. Lang. Teach. **6**, 57–67 (2013). https://doi.org/10.5539/elt.v6n6p57
17. de la Rama, J.M., et al.: Virtual teaching as the 'new norm': analyzing science teachers' attitude toward online teaching, technological competence, and access. Int. J. Adv. Sci. Technol. **29**, 12705–12715 (2020). https://doi.org/10.2139/ssrn.3654236
18. Cabangacala, R., Alieto, E., Estigoy, E., Delos Santos, M., Torres, J.: When language learning suddenly becomes online: analyzing English as second language learners' (ELLs) attitude and technological competence. TESOL Int. J. **16**, 115–131 (2021)
19. Islahi, F., Nasrin,: Exploring teacer attitude toward information technology with a Gender Prospective. Contemp. Educ. Technol. **10**, 37–54 (2019). https://doi.org/10.30935/cet.512527
20. Farhaoui, Y.: Teaching computer sciences in Morocco: an overview. IT Profess. **19**(4), 12–15, 8012307 (2017). https://doi.org/10.1109/MITP.2017.3051325
21. Creswell, J.: Research design: qualitative, quantitative, and mixed methods, 2nd edn. SAGE Publications, Thousand Oaks, CA (2003)
22. Somblingo, R., Alieto, E.: English language attitude among Filipino prospective language teachers: an analysis through the mentalist theoretical lens. Asian ESP J. 23–41 (2020)

23. Dillman, D., Smith, J., Christian, C.: Internet, mail and mixed-modesurveys: the tailored design. John Wiley and Son, Hoboken, NJ (2009)
24. Ursachi, G., Horodnic, I.A., Zait, A.: How reliable are measurement scales? external factors with indirect influence on reliability estimators. Procedia Econ. Finan. **20**, 679–686 (2015). https://doi.org/10.1016/S2212-5671(15)00123-9
25. Ozgur, C., Kleckner, M., Li, Y.: Selection of statistical software for solving big data problem: a guide for business, students, and universities. SAGE Open 1–12 (2015).https://doi.org/10.1177/2158244015584379
26. Ketmuni, M.: The acceptance of online English language learning of undergraduate students at Rajamangala University of Technology Thanyaburi. Psychol. Educ. **58**, 1464–1470 (2021). https://doi.org/10.17762/pae.v58i1.930
27. Asnawi, M., Quismullah, Y., Rena, J.: Perceptions and barriers to ICT use among English teachers in Indonesian. Teach. Engl. Technol. **18**, 2–23 (2018)
28. Goktas, Y., Yildirim, Z., Yildirim, S.: Main barriers and possible enablers of ICT integration into pre-service teacher education programs. Educ. Technol. Soc. **12**, 193–204 (2009)
29. Teo, T., Fan, X., Du, J.: Technology acceptance among pre-service teachers: does gender matter? Aust. J. Educ. Technol. **21**, 235–251 (2015). https://doi.org/10.14742/ajet.1672

Detection of Pesticide Responsible of Intoxication: An Artificial Intelligence Based Method

Rajae Ghanimi(✉), Fadoua Ghanimi(✉), Ilyas Ghanimi, and Abdelmajid Soulaymani

Ibn Tofail University, Av. de L'Université, Kénitra, Morocco
ghanimi.rajae@gmail.com, {fadoua.ghanimi, Abdelmajid.Soulaymani}@uit.ac.ma, ghanimiilyas@gmail.com

Abstract. According to WHO surveys, African countries import less than 10% of the pesticides used in the world, but they account for half of accidental poisonings and more than 75% of fatal cases. Faced with this alarming figure, and in order to prevent death and improve emergency treatment of cases of pesticide poisoning, it becomes important to use the potential of artificial intelligence in the diagnosis, prognosis and processing of these admissions. Our approach is essentially based on algorithms, in particular decision support software capable of predicting, based on major clinical signs, the most probable pesticide in the triage room. Before moving on to the confirmation stage based on biological and toxicological investigations, which are often costly and time-consuming.

Keywords: Pesticide · Machine learning · Artificial intelligence · Diagnostics · Emergencies · Triage room

1 Introduction

Poisoning is a public health problem worldwide and is one of the most common reasons for frequenting hospital emergency departments [1]. Although the incidence of poisonings is difficult to estimate accurately, the wide availability and accessibility of chemicals and their widespread use in various applications including medicine, agriculture and industry have increased the risk of poisoning. The acute toxicity of pesticides results from misuse, accidental use of pesticides (domestic accidents) or intentional poisoning, often extremely serious. Organophosphate pesticides and carbamates cause the most frequent cases of pesticide poisoning [2]. Exposure is mainly via the mucocutaneous and respiratory route (inhalation), the oral route of exposure would concern the general population more by accidental or intentional ingestion of pesticides. According to the World Health Organization (WHO), there are one million serious pesticide poisonings worldwide each year, causing approximately 220,000 deaths per year. The early diagnosis of these cases of poisoning is decisive on the prognostic level [3].

Y. Farhaoui et al. (Eds.): ICAISE 2023, LNNS 837, pp. 93–98, 2024.
https://doi.org/10.1007/978-3-031-48465-0_12

Our work consists of a Machine Learning (ML) algorithm capable of predicting the type of pesticide responsible for the poisoning based on the patient's clinical signs in emergency room.

To develop our algorithm. We worked on the database of the national poison control center of MOROCCO, including 8107 cases of poisoning by pesticides from March 1980 to December 2014, in order to extract the data necessary for training an algorithm.Our patients come from both urban and rural areas our data is composed of different ages: 1 newborn (maternal-fetal intoxication);12 baby walkers, 417 children, 2473 teenagers;5180 adults and 23 elderly people.The distribution by sex showed a female predominance, 5150 women (i.e. 64%) against 2957 men (i.e. 36%).

And the predominance of the suicidal cause (98%), the other causes (drug addiction, professional, criminal and for abortion) represent less than 2%.these etiologies were detected through history taking and family interview, clinical symptoms, and laboratory examination of gastric lavage products.

2 Sate of Art

Chavan et al. [4] introduced a classification model using the k-nearest neighbor (k-NN) algorithm. Their model focused on 118 chemicals from the NEDO (New Energy and Industrial Technology Development Organization) RTD database, now known as the Hazard Evaluation Support System (HESS). They employed both acute toxicity classes (LD50) as responses and eight PaDEL-derived fingerprints as predictors. The developed model effectively predicted LD50 classes for 70 out of 94 chemicals in the training set and 19 out of 24 chemicals in the test set [4]. Additional studies, notably by JC Carvaillo, R Barouki, X Coumoul, and K Audouze [5], introduced the AOP-helpFinder program, designed to predict product toxicity by analyzing extensive data from scientific literature. The program relied on two methods: the first involved text analysis, targeting relevant terms including chemical names (e.g., bisphenol S, pesticides) and descriptions of pathological biological processes. The second method focused on identifying critical keywords. Researchers compiled dictionaries containing known substance names (e.g., bisphenol S from PubChem) and numerous adverse effects pathways (Adverse Outcome Pathways, AOPs).In the context of paraquat poisoning, Huiling Chen et al. [6] explored the utilization of common blood indices for diagnosing PQ toxicity and assessing its severity using a machine learning approach. They employed a support vector machine-based method along with feature selection to predict PQ poisoning risk. The approach was validated on 79 individuals, distinguishing between living and deceased patients, and was rigorously evaluated for accuracy, sensitivity, and specificity. Despite numerous studies on technology and artificial intelligence in healthcare, diagnosis, and pharmacology, the researchers noted a lack of publications discussing the use of artificial intelligence specifically for diagnosing poisoning cases. This identified gap serves as the motivation for their upcoming research in this innovative field.

3 The Proposed Method

The approach is essentially based on Machine Learning (ML) algorithms, capable of predicting, based on major clinical signs, the most probable pesticide in the triage room. Hopefully, this prediction by the software has to occur before moving on to the confirmation stage based on biological and toxicological investigations, which are often costly and time-consuming. Based on the patient's clinical signs reported by the doctor in charge in emergency room, the work consisted of predicting the type of pesticide responsible for the poisoning.

3.1 Database Description

The approach is essentially based on Machine Learning (ML) algorithms, capable of predicting, based on major clinical signs, the most probable pesticide in the triage room. Hopefully, this prediction by the software has to occur before moving on to the confirmation stage based on biological and toxicological investigations, which are often costly and time-consuming. Based on the patient's clinical signs reported by the doctor in charge in emergency room, the work consisted of predicting the type of pesticide responsible for the poisoning.

The data used is collected from the national poison control center of Morocco. The Database included 8107 cases of pesticide poisoning from March 1980 to December 2014, in order to extract the data necessary for training an algorithm. Pesticides represent the second cause of poisoning in our database after drugs (Figs. 1, 2).

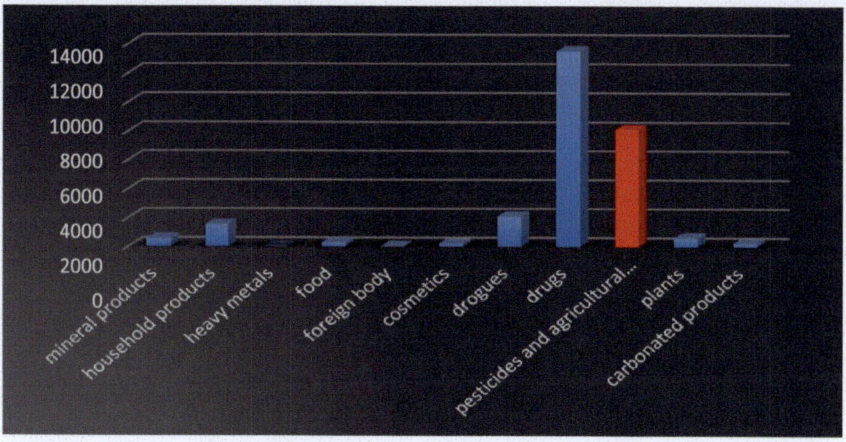

Fig. 1. Database by category of toxic substance

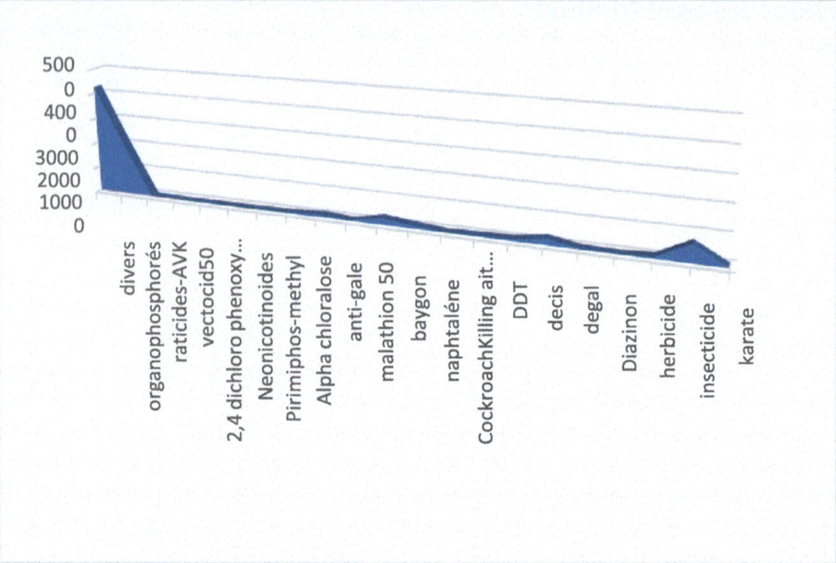

Fig. 2. Graphical representation of pesticide poisoning cases

3.2 Machine Learning Algorithms

Three different ML models were tested in this study for prediction: SVM, Random Forest and XGBoost. GridSearchCV function from scikit-learn library was used to perform the search for model's parameters. The evaluation metrics used to evaluate the models were precision, recall and accuracy.

***Support vector machine* SVM.**

Support Vector Machine (known as SVM) is a supervised learning method designed for binary classification "either positive or negative class". It is applied for the purpose of finding patterns from the collection of data [7]. Generally, the pattern classification applies an activity to be involved in two main steps; the first is mapping the input to higher dimension feature space, this is done due to the fact of the SVM that usually depends on geometrics characteristics of inputted data, and, the second is finding the most suite hyper plane that classifies the mapped features in the higher dimensional space. It's beneficial for classification and regression approaches. The margin is the space between the hyperplane and the nearby data point. The core target is discovering the hyperplane with the greatest database split; into two classes for gaining novel vectors with proper classification.

Random forest.

Random Forest are supervised ensemble learning models used for classification and regression. The Random Forest consists of several decisional trees to classify a new object from an input vector while the random forest's input vector is the input of every tree [8]. The scientist proves that this method achieves accurate and stable prediction with high performance.

XGBoost.

XGBoost [9] is a decision tree ensemble based on gradient boosting designed to be highly scalable. XGboost uses randomization techniques to reduce overfitting and to increment training speed. In addition, XGBoost implements several methods to increment the training speed of decision trees not directly related to ensemble accuracy. Specifically, XGBoost focuses on reducing the computational complexity for finding the best split, which is the most time-consuming part of decision tree construction algorithms.

4 Result and Discussion

To analysis of the result in this experiment, we considered different analyses to examine the three machine learning models for the prediction of the nature of poisoning pesticides applied to the dataset. In terms of accuracy, XGBoost achieved the highest accuracy of 75% and SVM achieved the worst performance 62%. With respect to F1-score, XGBoost achieved the highest score 84%, Random Forest achieved 83% and SVM performs 74%. (see Table 1).

According to these results we conclude that XGBoost Algorithm is more effective than the other models for predicting the nature of the pesticide responsible of poisoning case. Especially since XGBoost has demonstrated in several studies its performance in the field of medical diagnostics [10–12].

Table 1. Results for the three machine learning algorithms with the test dataset

	Accuracy score	Precision score	Recall score	F1-score
XGBoost	0.75	0.91	0.78	0.84
SVM	0.62	0.72	0.76	0.74
Random forest	0.74	0.85	0.81	0.83

5 Conclusion

It is recognized that it is difficult to predict the nature of the pesticide responsible of poisoning in time due to several risk factors.

Due to these factors, scientists have turned to modern approaches such machine learning to predict disease.

Three machine learning models, namely respectively:XGBoost, Random forest and SVM were developed in this work. XGBoost predicted the best with a performance of 84% in the F1-score used in this study. As a result, the objectives set at the introduction were achieved and this research responded to the expectations of emergency clinicians.

References

1. Mégarbane, B.: Présentation clinique des principales, intoxications et approche par les toxi dromes. Réanimation **21**, S482–S493 (2012)
2. El-Sarnagawy, G.N., Abdelnoor, A.A., Abuelfadl, A.A. et al.: Comparison between variousscoring systems in predicting the need for intensive care unit admission of acute pesticide-poisoned patients. Environ. Sci. Pollut. Res. **29**, 3399934009 (2022)
3. Thabet, H., Brahmi, N., Elghord, H., Kouraichi, N., Amamou, M.: Intoxications par les insecticides organophosphorés et carbamates. In: Intoxications aiguës. Références en ré animation. Collection de la SRLF. Springer, Paris (2013)
4. Chavan, S., Friedman, R., Nicholls, I.A.: Acute toxicity-supported chronic toxicity prediction: a k-nearest neighbor coupled read-across strategy. Int. J. Mol. Sci. **16**, 11659–11677 (2015)
5. Carvaillo, J.C., Barouki, R., Coumoul, X., Audouze, K.: Linking Bisphenol S to adverse outcome pathways using a combined text mining and systems biology approach. Environ. Health Perspect. **127**, 4 CID: 047005 (2019)
6. Chen, H., Hu, L., Li, H., Hong, G., Zhang, T., Ma, J., Lu, Z.: An effective machine learning approach for prognosis of paraquat poisoning patients using blood routine indexes. Basic Clinic. Pharmacol. Toxicol. **120**(1), 86–96 (2017)
7. Vapnik, V.N.: The nature of statistical learning theory. Springer-Verlag, New York (1995)
8. Breiman, L.: Mach. Learn. **45**(1), 5–32 (2001)
9. Chen, T., Guestrin, C.: Xgboost: A scalable tree boosting system. In: Proceedings of the 22Nd ACM SIGKDD International Conference on Knowledge Discovery and Data Mining, KDD '16, pp. 785–794, New York, NY, USA (2016)
10. Sami, S.M., Bhuiyan, M.I.H.: Power transformer fault diagnosis with intrinsic time-scale decomposition and XGBoost classifier. In: Arefin, M.S., Kaiser, M.S., Bandyopadhyay, A., Ahad, M.A.R., Ray, K. (eds) Proceedings of the International Conference on Big Data, IoT, and Machine Learning. Lecture Notes on Data Engineering and Communications Technologies, vol. 95. Springer, Singapore (2022)
11. Jiang, Y.Q., Cao, S.E., Cao, S., et al.: Preoperative identification of microvascular invasion in hepatocellular carcinoma by XGBoost and deep learning. J. Cancer Res. Clin. Oncol. **147**, 821–833 (2021)
12. Li, Q., Yang, H., Wang, P., et al.: XGBoost-based and tumor-immune characterized gene signature for the prediction of metastatic status in breast cancer. J. Transl. Med. **20**, 177 (2022). https://doi.org/10.1186/s12967-022-03369-9

A Literature Review of Tutoring Systems: Pedagogical Approach and Tools

Fatima-Zohra Hibbi(✉)

SMARTiLab, EMSI Rabat, Rabat, Morocco
f.hibbi@emsi.ma

Abstract. The use of intelligent tutoring systems or assistants (ITS) as support tools in the context of programming learning has many advantages since they are oriented towards the development of more effective personalised teaching and learning processes. From the point of view of the role of teachers, the development of this type of device offers them a great opportunity, among which is the reduction of time spent designing content and promoting appropriate knowledge. On the other hand, from the student's point of view, learning takes place autonomously, as these systems are designed to complement, not replace, the role of the teacher. In short, ITS is distinguished from other education systems by their ability to monitor the cognitive states of individual learners and respond appropriately. This system has received a great deal of attention from disciplines such as cognitive science, education and computer science. The ultimate goal is to emulate expert human tutors in the way they teach and interact with learners. In what follows, the authors present a state of the art of intelligent tutoring system.

Keywords: Intelligent tutoring system · Programmed learning · Computer assisted instruction

1 Introduction and Background

Programming raises many problems for students who are just starting to learn it [1]. Specifically, much of this content involves a degree of abstraction in students that they do not yet have, because they do not understand the real effect that modifying the source code of a program could have on its execution or because they do not provide appropriate solutions to problems using a programming language. In 1984, «Bloom» showed from a study that learners who were supervised and followed by tutors during the learning process through a combination of traditional assessment and correction instructions performed better than those who received a traditional teaching group [2]. When students have difficulty understanding concepts or exercises, the most effective option is to use a one-to-one tutor [3, 4].

Human tutors can offer students a whole range of services. A good human tutor allows students to do as much work as possible while guiding them to stay on track [4]. Of course, students who learn on their own can also increase their knowledge and

Y. Farhaoui et al. (Eds.): ICAISE 2023, LNNS 837, pp. 99–104, 2024.
https://doi.org/10.1007/978-3-031-48465-0_13

reasoning skills. They give students advice and suggestions rather than explicit solutions [5]. This motivates students to overcome difficulties. What's more, human tutors are very interactive in this sense. What they do is give constant feedback to the students while they are solving the problems. The assistance of a tutor enables a certain type of practice-led learning, in which learners reap the rewards of active problem-solving while tutors minimize the risks [5].

In order to enable a tutoring system to give feedback similar to that of a human tutor, researchers will need to ensure that it interacts with students in the same way as human tutors. This leads to the question of how can a System deal with students as effectively as human tutors? To answer this question we will look at the emergence of programmed learning. The second section will focus on the intervention of artificial intelligence in order to develop an intelligent tutoring system (ITS).

2 Programmed Learning

2.1 Definition of Programmed Learning

Programmed learning is a learning methodology or technique first proposed by B. At the end of the 60s and throughout the 70s, programmed learning was popular on the community. Nevertheless, its pedagogical value was lost in the early 80s through the difficulty to implement it and its limitations were not understood by specialists. Programmed learning remains popular in self-study manuals [1].

2.2 Principles of Programmed Learning

The presentation of information, the learner's question and response; this technique is based on the principle that the learner learns immediately whether he made a mistake or not and progresses by following the program. The different steps are organized in logical sequences [1]. Skinner's linear program is divided into small sequences, which each learner can complete without difficulty. According to Skinner, 95% of correct answers mean a successful program.

The linear programming style was developed by B. Skinner's. Linear program is divided into small sequences, which each learner is able to complete without difficulty. First, the training unit is divided into small instructional sequences, and then the information is presented, the learner responds, the correct answer is proposed, and the process is repeated. Early systems were linear presentations of information and guided exercises, with the lesson's progress monitored by a control system. In modern systems, particularly with visualisation systems and simulated environments, control often rests with the student or tutor CAI. The Fig. 1 presents the principle of linear programming [1].

2.3 Branch Programming by Crowder

The branched programming method was initiated by N. The learner progresses through the main pathway from number one to number two, then tree and so on. The figure below shows an example of programmed teaching using N. The Fig. 2 presents the branch programming by Crowder [2].

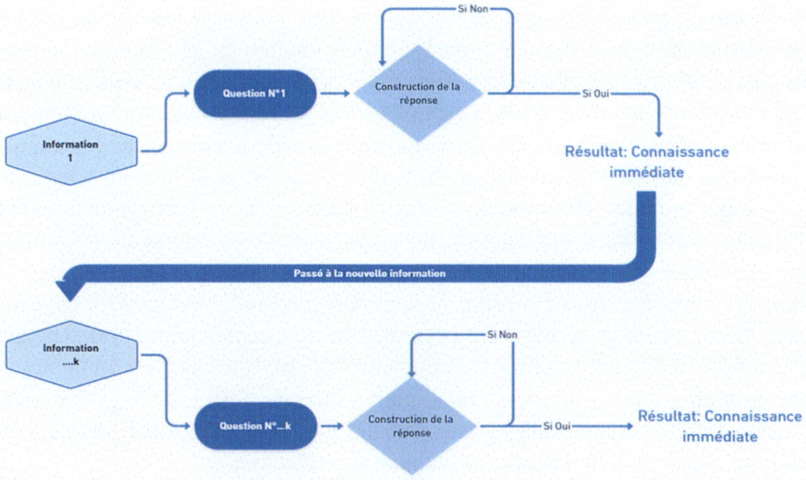

Fig. 1. Linear programming (skinner) [1]

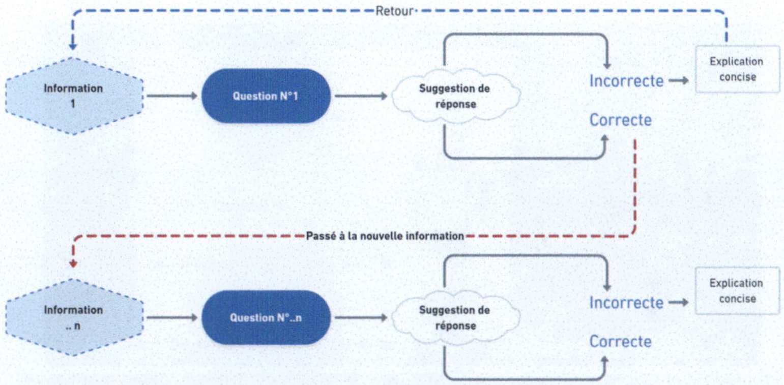

Fig. 2. Branch programming (crowder) [2]

3 Intelligent Tutors

3.1 Intelligent Tutoring System: Definition and Architecture

ITS are computer programs that use artificial intelligence techniques to provide intelligent tutors who can know what they are teaching, who they are teaching and how they are teaching [6]. AI is used to simulate human tutors to produce intelligent tutors. Most of the research in educational software involving AI has been conducted under the name of intelligent computer-assisted instruction. Although Intelligent Tutoring systems have not yet reached the same level of effectiveness as expert human tutors, numerous studies show the impact of these systems on significantly increasing learning.

The architecture of intelligent tutoring systems is very diverse [7]. It is very rare to find two systems based on the same architecture. In addition, an ITS must know about communication to present the desired information to the students. On the other hand, even though tutoring systems differ enormously in their internal structures and components and display a wide variety of characteristics, their behavior is similar in certain respects [8]. The outer loop mainly decides which of several tasks the students should perform next. The decision is made based on the student's knowledge history and background. The knowledge module is also used to assess the overall progress of the student against the knowledge module. The module evaluates each learner's performance based on their behavior when interacting with the tutoring system to determine their knowledge, perceptual abilities and reasoning skills [9].

There are many intelligent tutoring systems designed and developed for educational purposes. Some of these systems are presented in the second subsection. The author suggests a new intelligent tutoring system named Smart Tutoring System dedicated to learning programming using meta-collaborative learning [10]. The Fig. 3 presents a simulation of the smart tutoring system and the Fig. 4 illustrates the strength of the metacognitive approach in one area of collaborative discussion.

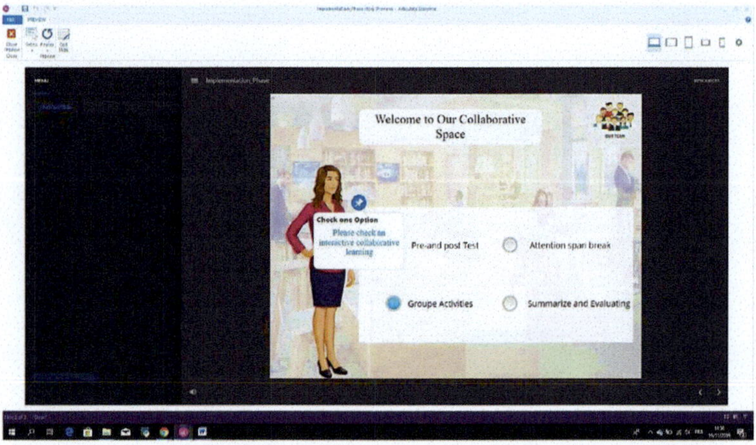

Fig. 3. Smart tutoring system [10]

3.2 Example of intelligent Tutoring Systems

- Computer science domain: such as teaching the object programming language Java [10, 11, 13] ITS to help computer science students learn debugging skills. ITS to help computer science students learn computer programming [13], an agent-based ITS for parameter passing in Java programming, Evaluation of Java Expressions [14], Linear Programming [15], Learner Performance Prediction using NT and ITS [11], and Intelligent Tutoring System for teaching advanced topics in Information Security [13] development and evaluation of Oracle Intelligent Tutoring System (OITS) [13]etc.

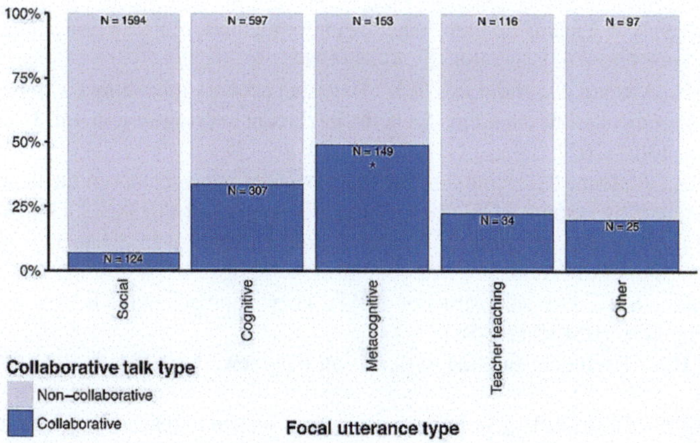

Fig. 4. The strength of the metacognitive approach in one area of collaborative discussion [10]

- Mathematics: ITS to help students aged 8–12 learn primary mathematics [12]. Researchers in the field have carried out a comparative study between animated intelligent tutoring systems (AITS) and video-based intelligent tutoring systems (VITS) [13], etc.
- Languages: there are English language tutoring systems, such as [14] and an intelligent tutoring system for teaching English grammar tenses [14], etc.

4 Conclusion

This chapter has presented a general overview of this thesis. It presents a review of the literature on the concepts of tutor and tutoring. In the first link, we presented the role of a human tutor during the learning process and its main missions. We then defined programmed learning and set out the principles of programmed learning. Secondly, we described the emergence of the new concept of Computer-Assisted Instruction (CAI), which are an open door to trigger intelligent tutoring systems using artificial intelligence techniques.

The gap between human tutors and ITS tutoring systems is narrowing, but not even remotely closed. There are many models of knowledge, teaching styles and learner knowledge. Each model has its advantages and disadvantages. Hybrid models have also been created to enhance and reinforce traditional models. The close partnership between ITS, AI and psychology holds a great promise for the advancement of ITS.

References

1. Figueiredo, J., García-Peñalvo, F.J.: Building skills in introductory programming. In: Proceedings of the Sixth International Conference on Technological Ecosystems for Enhancing Multiculturality, Salamanca, Spain, 24–26 October 2018; ACM, New York USA, pp. 46–50 (2018)

2. Bloom, B.S.: The 2 sigma problem: The search for methods of group instruction as effective as one-to-one tutoring. Educ. Res. **13**, 4–16 (1984)
3. Hibbi, F.Z., Abdoun, O., Haimoudi, E.K.: Bayesian network modelling for improved knowledge management of the expert model in the intelligent tutoring system. Int. J. Adv. Comput. Sci. Appl. **13**(6) (2022)
4. Hibbi, F.Z., Abdoun, O., Haimoudi, E.K.: Knowledge management in the expert model of the smart tutoring system. ACM (2020). https://doi.org/10.1145/3386723.3387895
5. Koti, M.S., Kumta, S.D.: Role of intelligent tutoring system in enhancing the quality of education. Int. J. Adv. Stud. Sci. Rese. **3**, 330–334 (2018)
6. Nazimuddin, S.K.: Assisted instruction (CAI): a new approach in the field of education. Int. J. Sci. Eng. Res. (IJSER) (2015)
7. Nwana, H.S.: Intelligent tutoring systems: an overview. Artif. Intell. Rev. **4**(4), 251–277 (1990)
8. Elizabeth, C., Glenn, B.D.: An intelligent tutoring system to teach debugging. In: Artificial intelligence in education: 16th international conference, AIED 2013, Memphis, TN, USA, July 9–13 (2013)
9. Naser, S.A., Ahmed, A., Al-Masri, N., Abu Sultan, Y.: Human computer interaction design of the LP-ITS: linear programming intelligent tutoring systems. Int. J. Artific. Intell. Appl. (IJAIA) **2**(3), 60–70 (2011)
10. Fatima-Zohra, H., Otman, A., Elkhatir, H.: Collaborative metacognitive technique in smart tutoring system. In: Motahhir, S., Bossoufi, B. (eds) Digital technologies and applications. ICDTA 2021. lecture notes in networks and systems, vol. 211. Springer, Cham (2021). https://doi.org/10.1007/978-3-030-73882-2_58
11. Naser, S.S.A.: Intelligent tutoring system for teaching database to sophomore students in Gaza and its effect on their performance. Inform. Technol. J. Scialert **5**(5), 916–922 (2006)
12. Hennessy, S., O'Shea, T., Evertsz, R., Floyd, A.: An intelligent tutoring system approach to teaching primary mathematics. Educ. Stud. Math. **20**(3), 273–292 (1989)
13. Mahmoud, M.H., Abo El-Hamayed, S.H.: An intelligent tutoring system for teaching the grammar of the Arabic language. J. Electric. Syst. Inform. Technol. (2016)
14. Alhabbash, M.I., Mahdi, A.O., Abu Naser, S.S.: An intelligent tutoring system for teaching grammar English tenses. Euro. Acad. Res. **4** (2016)

IoT in Agriculture: Security Challenges and Solutions

Khaoula Taji(✉), Ilyas Ghanimi, and Fadoua Ghanimi

Electronic Systems, Information Processing, Mechanics and Energy Laboratory, IbnTofail University, Kenitra, Morocco
taji.khaoula95@gmail.com, Fadoua.Ghanimi@uit.ac.ma

Abstract. The concept of the Internet of Things (IoT) pertains to employing conventional Internet protocols to facilitate interactions between humans and objects, as well as between objects themselves. Functioning with machines and entities, IoT enables computation, communication, and collaboration among interconnected elements. This integration has extended to various facets of agriculture such as cultivation, surveillance, processing, marketing, and consumer engagement. While recognized security needs for IoT systems have prompted standardization efforts and proposed security mechanisms, persisting challenges and gaps remain due to both unresolved use cases and inadequate security provisions in many deployed IoT devices and systems. This study offers a cybersecurity-oriented overview of IoT-based agriculture, exploring potential applications of IoT devices, classifying agriculture's IoT architecture into four layers, scrutinizing security threats, and proposing effective countermeasures based on IoT architectural layers. An innovative IoT-driven smart agriculture model, fortified by security measures, is also proposed to mitigate cyber threats and enhance agricultural productivity.

Keywords: Internet of Thing (IoT) · IoT-based agriculture · Secure (IoT) devices in farming · Meta-Model · Platform-specific model (PSM) · Asset · Physical object · Denial-of-service (DOS)

1 Introduction

The agricultural industry is receiving increased attention in many developing countries as sophisticated computers and the internet of Things are gradually integrated into farming operations such land cultivation, farm tracking, water management, soil pH positioning, method of procedure, and food marketing. This is an effort to address issues associated with food security throughout the world, agricultural export requirements, economic diversification, and the rise of the digital economy [1]. Security concerns have grown more relevant as agricultural systems in developing nations adopt more sophisticated IoT for applications including water and soil quality monitoring, smart vegetable gardens, the milk and egg business, and scientific monitoring of diseases and pests. IoT connects computers, smart devices, structural machinery, artifacts, and people through the internet and mobile applications. In an IoT network, sensors and actuators interact to perceive, intelligently place, monitor, and track [2]. This vivid remote control on an IoT

Y. Farhaoui et al. (Eds.): ICAISE 2023, LNNS 837, pp. 105–111, 2024.
https://doi.org/10.1007/978-3-031-48465-0_14

network transforms traditional control into intelligent control, improving efficiency and productivity. IoT ecosystems need all devices to interact, making device, network, and data security the biggest concern. Internet of things agriculture boosts food production and quality through sensors, wireless networking, cloud computing, machine learning, big data, and IoT [1]. Based on iot technologies and remedies, underground smart sensors for blueberry irrigation have reduced water waste by 70% and information models to estimate and prevent crop diseases are being integrated into farmland processes to reduce waste, maximize yields, and improve operational efficiency [3]. IoT-based farming might help solve food shortages, butit may raise cyber-security concerns.

This study focuses on the issue of data security in the context of Internet of Things-based farming. With the objective of increasing agricultural productivity by addressing cybersecurity challenges, it explores the possible applications of IoT devices in agriculture, analyses security threats, and offers counter measures. Recent investigations on IoT in farming have used a variety of methods. Some have focused on using blockchain technology to enhance data integrity and traceability. Machine learning methods such as anomaly detection have been used to the study of irregularities in agricultural systems. Satellite imagery and remote sensing technologies have also made real-time surveillance and early threat identification feasible. Collaborative intrusion detection systems, which pool the data from several sensors, have also shown to be effective. The variety of solutions to security concerns in IoT-based agriculture is reflected in the combination of various methods.In the next parts of this study, we'll look at: 2. a complete review of the literature on IoT-based agriculture; 3. The possible security risks and challenges this field faces; and 4. a suggested mixed metamodel as a way to deal with these problems. Through this study, we hope to help people understand and solve security problems related to IoT-driven agriculture practices. This will help make the farming sector's technology more safe and long-lasting.

2 Literature Review and Discussion

Many research have examined IoT-agriculture security issues. To protect IoT-enabled agricultural systems, [4] studied encryption, authentication, and intrusion detection. Also rising is blockchain technology for data integrity and transparency. Researchers recommend anomaly detection for hacking and other undesirable activities. Agricultural IoT predictive security analytics using machine learning and AI are popular [5]. These studies illustrate the need to address security challenges in IoT-based agriculture and stimulate innovative ideas to safely and sustainably integrate technology into agricultural operations [6]. In [7], researchers undertook a thorough examination of IoT-based intelligent agricultural systems, exploring their potential integration into agriculture to enhance productivity and efficiency. They evaluated diverse IoT technologies and devices employed for computing, transmission, and storage. The potential of IoT for precision farming is illuminated in [8], encompassing technologies like smart devices and intelligent UAVs, resulting in a cyber-physical system susceptible to security vulnerabilities due to its diverse devices and technologies. The authors scrutinized security concerns and proposed mitigation techniques. Sinha et al. [9] explored IoT advancements and security challenges in smart agriculture, delivering a comprehensive overview. In [4], a five-layered edge-based design, integrating IoT and LoRA for smart farming, aimed at energy efficiency, data quality, and system safety. The role of blockchain in smart farming was investigated in [10], suggesting an IoT and blockchain-based product tracking system. Awan et al. employed smart contracts to enhance trust and evaluated their architecture's performance. Hydroponic farming monitoring was realized in [11], unveiling a fully automated web-based system while overlooking security measures, affecting system reliability. In [9], an IoT-based crop and irrigation tracking system was devised, utilizing detectors for various measurements, albeit omitting security considerations. The complete literature assessment of the aforementioned research shows that security concerns permeate IoT-based smart agricultural systems at every architectural tier. Some threats have been reported, while others have not. Application layer issues including access control, data leaks, and eavesdropping are addressed [12]. However, denial of service and phishing threats remain on this tier. Forgery, man-in-the-middle attacks, and unauthorized access remain edge layer security vulnerabilities without remedies [13]. Routing attacks, DDOS, signal manipulation, interruption, and DOS assaults are network layer vulnerabilities that consumers have yet to experience. The perception layer needs solutions for fraudulent nodes, vulnerable sensors, and node capture.

3 IoT-Based Agriculture Threats

The Internet of Things–based agricultural risk research found many limitations. Cross-layer security procedures and solutions were developed to combat these attacks [14]. Key-based authentication, secure wireless connections, encryption, and segregated application containers reduced risks and protected agricultural systems.

Smart farming is susceptible to cyberattacks due of heterogeneous IoT devices generating dynamic and geographical data. Regularly monitoring agricultural activities with data. Accessing this data without authorization may harm agricultural [15]. Attackers may employ third-party agronomy analytics to break security, gain competitive advantages, and lose money. Physical damage, theft, tampering, and animal interference may destroy IoT devices. Some devices are vulnerable to DoS attacks due to low memory and power usage. Precision agricultural IoT drones may be captured, revealing vital data. DoS, wireless channel jamming, and man-in-the-middle attacks may affect agriculture networks. Interconnected IoT farms distribute malware faster. Present malware detection approaches may not function in IoT agricultural architecture [16]. AI-enhanced malware detection is suggested, but not IoT-based agricultural malware. Man-in-the-middle exploits, DoS, DDoS, and cloud/edge layer threats such illegal services, data manipulation, and DoS attacks may hinder IoT-based farming [15, 17]. Fake IoT device IDs undermine automated activities in these attacks. Hearing IoT communications harms privacy. DoS and DDoS attacks [10] harm IoT agricultural systems. Agriculture pH may be altered by unauthorized entry. Trojan horses and MITM compromise biometric data. Finally, IoT-based farming faces cyber threats such data tampering, unauthorised access, and process interruptions [15–17] (Fig. 1).

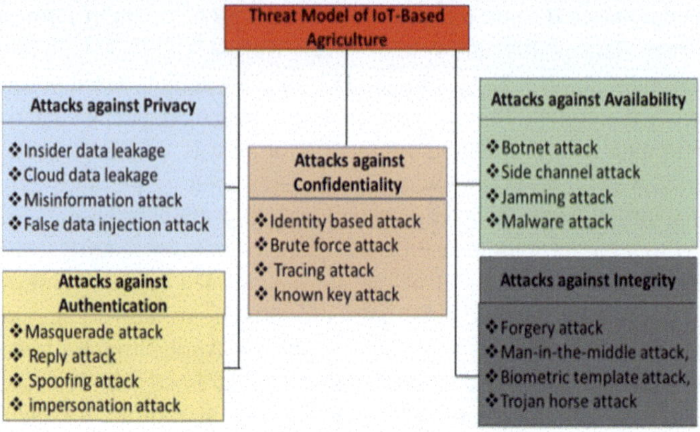

Fig. 1. Smart farming cyber attacks

Aspects of IoT-based agriculture under attack

This research categorizes assaults into five (5) distinct types, including privacy, authentication, confidentiality, availability, and integrity attack, taking into account various literatures.

Intrusions Into Privacy: An attacker used the position and authentication of smart devices in agriculture to get illegal access to the data and violate device privacy. Accessing agricultural data generated at the IoT sensing layer may expose a farmer's everyday actions and provide a competition an unfair advantage [6]. Illegal farm data access can lead to agricultural espionage. Secrecy, anonymity, and autonomy define farm privacy.

Attempts to Break Authentication: Identity forgery is used in this attack to get access to systems and nodes. There at sensor, network, fog, or cloud layer, the attacker assumes the identity of a genuine device user to obtain access to agricultural devices. Masked, retaliatory, spoofing, and impersonation attacks are all examples of identity-based attacks. **Assaults on Confidentiality:** As a sort of hostile eavesdropping, the attacker uses an IoT-based farming gadget and net-work with an entry point at the perception layer to listen in on conversations and take actions that aren't intended by the intended recipients. An identity-based, brute force, tracing, or known key assault can all be used in this approach [18].**Threats Affecting Availability:** This attack's purpose is to disrupt the service of IoT devices in agriculture. This type of Denial of Service attack occurs when an attacker floods a server with a significant volume of undesired data, updates IoT software remotely by injecting fake data, or launches an attack on the accurate localization of UAVs with a rogue 5G station. Botnet attacks are another type of availability attack. **Anti-Integrity Attacks:** Misinformation, trojan horses, and biometric attacks are all possible methods of assault. The attacker's goal is to obtain network access devices and alter their settings and data without the owner's permission. As a result, data generation may be compromised, possibly leading to errors [7].

4 The Proposed Hybrid Metamodel for Smart Agriculture

In our pursuit of advancing smart agriculture, we present a novel hybrid meta-model. This combination of Internet of Things (IoT) [9] and security [10] frameworks is carefully intended to improve IoT solutions in agriculture while protecting agricultural data [19]. Smart agriculture relies on the IoT meta-model [11, 20]. A "Physical Object" class represents virtual entities with unique IDs, services, and physical features. Under the "Domain" meta-class, these "Physical Objects" are essential to the IoT Ecosystem. This ecosystem captures agricultural system dynamics using IoT nodes with microcontrollers, microprocessors, sensors, and actuators, layers, levels, and protocols.In smart agriculture, our hybrid concept is essential for security. It connects the 'Actor,' 'Attacker,' 'Asset,' and 'Role.' The 'Asset' class is directly linked with the 'Physical Object' class, underscoring its significance in safeguarding agricultural ecosystem components and data streams. This connection underscores the importance of protecting agricultural assets and data. The security meta-models used in [11, 21] can secure critical agricultural components. This linkage safeguards the 'Physical Objects or IoT Nodes' and their data against unauthorized access (Fig. 2).

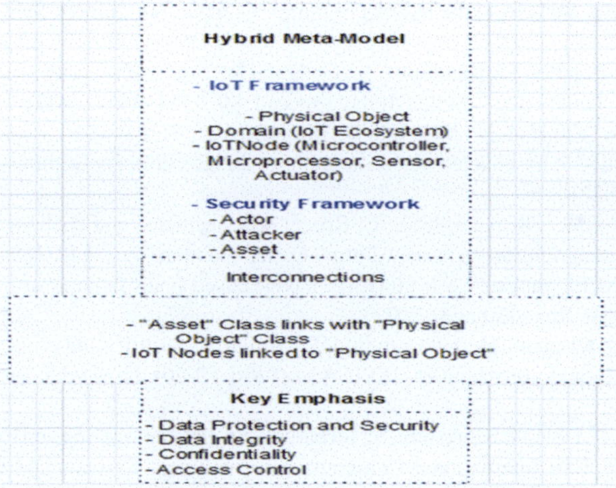

Fig. 2. Proposed hybrid meta-model for smart agriculture security

5 Conclusion and Future Works

Establishing and managing a security mechanism is essential in IoT-based designs so that such technologies may be extensively used without concern for cyber security incidents. In almost every area, researchers have primarily focused on developing IoT-based systems, although many have also highlighted the need of security in those systems. Throughout this research, we focused on the IoT-based intelligent agricultural domain and reviewed several publications on security domain. We also showed off a cutting-edge

IoT-based smart agriculture system with security features on each IoT layer. Increased productivity will surely emerge from this helping to narrow the security weakness in the IoT-based intelligent agricultural sector. In future works, we will also validate our secure model for smart agriculture by implementing it with specific algorithms, strategies and security techniques to ensure the highest level of security for the system. We will also look at security vulnerabilities in other IoT domains and try to solve them by providing a tighter security framework in those areas.

References

1. Mingjun, W., Zhen, Y., Wei, Z., Xi-shang, D., Xiaofei, Y., Chenggang, S., Xuhong, L., Fang, W., Jinghai, H.: A research on experimental system for internet of things major and Design and Manufacturing Informatization, 1:261–263, 2012. application project. 2012 3rd International Conference on System Science, Engineering
2. Rose, K., Eldridge, S.D., Chapin, L.: The internet of things: an overview understanding the issues and challenges of a more connected world (2015)
3. Lee, A., Wang, X., Nguyen, H., Ra, I.: A hybrid software defined networking architecture for next-generation iots. KSII Trans. Internet Inf. Syst. **12**, 932–945 (2018)
4. Orestis, M., Mouratidis, H., Fish, A., Panaousis, E., Kalloniatis, C.: A conceptual model to support security analysis in the internet of things. Comput. Sci. Inform. Syst. **14**(2), 557–578 (2017)
5. Rekha, S., Thirupathi, L., Renikunta, S., Gangula, R.: Study of security issues and solutions in Internet of Things (IoT). Mater. Today Proceed. **80**, 3554–3559 (2023)
6. KhanhQuy, V., Van Hau, N., Van Anh, D., Minh Quy, N., Tien Ban, N., Lanza, S., Randazzo, G., Muzirafuti, A.: Iot-enabled smart agriculture: architecture, applications, and challenges. Appl. Sci. (2022)
7. Bahadur Sinha, B., Dhanalakshmi, R.: Recent advancements and challenges of internet of things in smart agriculture: a survey. Future Gener. Comput. Syst. **126**, 169–184
8. Raja Gopal, S., Prabhakar, V.S.V.: Intelligent edge based smart farming with lora and iot. Int. J. Syst. Assuran. Eng. Manage. (2022)
9. Gupta, M., Abdelsalam, M., Khorsandroo, S., Mittal, S.: Security and privacy in smart farming: challenges and opportunities. IEEE Access **8**, 1 (2020). https://doi.org/10.1109/access.2020.2975142
10. Rajalakshmi, P., Devi Mahalakshmi, S.: Iot based crop-field monitoring and irrigation automation. In: 2016 10th International Conference on Intelligent Systems and Control (ISCO), pp. 1–6 (2016)
11. Allae, E., Belangour, A.: A big data security layer meta-model proposition. Adv. Sci. Technol. Eng. Syst. **4**(5), 409–418 (2019)
12. Khader, R., Eleyan, D.: Survey of dos/ddos attacks in IOT. Sustain. Eng. Innov. **3**(1), 23–28 (2021)
13. Su, J., He, S., Wu, Y.: Features selection and prediction for IoT attacks. High-Confid. Comput. **2**(2), 100047 (2022)
14. Anthi, E., Williams, L., Javed, A., Burnap, P.: Hardening machine learning denial of service (DoS) defences against adversarial attacks in IoT smart home networks. Comput. Secur. **108**, 102352 (2021)
15. Olivier, F., Carlos, G., Florent, N.: New security architecture for Iot network. Procedia Comput. Sci. **52**, 1028–1033 (2015)
16. Raj, H., Kumar, M., Kumar, P., Singh, A., Verma, O.P.: Issues and challenges related to privacy and security in healthcare using iot, fog, and cloud computing. Adv. Healthcare Syst. (2022)

17. Bikash, K.P., Saugat, B., Pal, K.: Iot-based applications in healthcare devices. J. Healthcare Eng. (2021)
18. Kariri, E.: Iot powered agricultural cyber-physical system: security issue assessment. IETE J. Res. (2022)
19. Yaacoub, J.-P.A., Noura, H.N., Salman, O., Chehab, A.: Ethical hacking for IoT: Security issues, challenges, solutions and recommendations. Int. Things Cyber-Physic. Syst. **3**, 280–308 (2023)
20. Sushanth, G., Sujatha, S.: IOT based smart agriculture system. In: 2018 international conference on wireless communications, signal processing and networking (WiSPNET), pp. 1–4. IEEE (2018)
21. Ruengittinun, S., Phongsamsuan, S., Sureeratanakorn, P.: Applied internet of thing for smart hydroponic farming ecosystem (hfe). In: 2017 10th International Conference on Ubi-media Computing and Workshops (Ubi-Media), pp. 1–4 (2017)
22. Sfar, A.S., Chtourou, Z., Challal, Y. A systemic and cognitive vision for Iot security: a case study of military live simulation and security challenges. In: 2017 International Conference on Smart, Monitored and Controlled Cities (SM2C), pp. 101–105 (2017)
23. Singh, D., Kumar Mishra, M., Lamba, A.K.: Security issues in different layers of iot and their possible mitigation (2020)
24. Khan, K., Khan, S.U., Zaheer, R., Khan, S.: Future internet: The internet of things architecture, possible applications and key challenges. In: 2012 10th International Conference on Frontiers of Information Technology, pp. 257–260 (2012)
25. Gondchawar, N., Kawitkar, R.S.: IoT based smart agriculture. Int. J. Adv. Res. Comput. Commun. Eng. **5**(6), 838–842 (2016)
26. Zanella, A., Bui, N., Paolo Castellani, P.: Lorenzo Vangelista, and Michele Zorzi. Internet of things for smart cities. IEEE Int. Things J. **1**, 22–32 (2014)
27. HosseinAlavi, A., Jiao, P., Buttlar, W.G., Lajnef, N.: Internet of things-enabled smart cities: state-of-the-art and future trends. Measurement (2018)
28. Khajenasiri, I., Estebsari, A., Verhelst, A., Gielen, G.E.E.: A review on internet of things solutions for intelligent energy control in buildings for smart city applications. Energy Procedia **111**, 770–779 (2017)
29. Li, Y., Alqahtani, A., Solaiman, E., Perera, C., Jayaraman, P.P., Buyya, R., Morgan, G., Ranjan, R.: Iot-cane: a unified knowledge management system for data-centric internet of things application systems. J. Parallel Distribut. Comput. **131**, 161–172 (2019)
30. Rettore de AraujoZanella, A., Germano da Silva, E., Carlos Pessoa Albini, L.: Security challenges to smart agriculture: Current state, key issues, and future directions. Array **8**, 100048 (2020)
31. Vangala, A., Das, A.K., Chamola, V., Korotaev, V., Rodrigues, J.J.: Security in IoT-enabled smart agriculture: architecture, security solutions and challenges. Clust. Comput. **26**(2), 879–902 (2023)
32. Kamalov, F., Pourghebleh, B., Gheisari, M., Liu, Y., and Moussa, S.: Internet of medical things privacy and security: Challenges, solutions, and future trends from a new perspective. Sustainability **15**(4), 3317 (2023)
33. Karankar, N., Seth, A.: A comprehensive survey on internet of things security: challenges and solutions. Mobile Comput. Sustain. Inform. Proceed. ICMCSI **2023**, 711–728 (2023)
34. Jazzar, M., Hamad, M.: An analysis study of IoT and DoS attack perspective. In: Proceedings of International Conference on Intelligent Cyber-Physical Systems: ICPS 2021 (2022)
35. Iván, A., Garcés, K., Castro, H., Cabot, J.: Modeling self-adaptative IoT architectures. In: 2021 ACM/IEEE International Conference on Model Driven Engineering Languages and Systems Companion (MODELS-C), pp. 761–766. IEEE (2021)

Big Data Analytics in Business Process: Insights and Implications

Swati Sharma(✉)

Jindal Global Business School, O. P. Jindal Global University, Sonipat, India
swati@jgu.edu.in

Abstract. The study explores the role of big data analytics (BDA) in business process by presenting a comprehensive review of listed literature on Scopus database. 1122 studies are analyzed keyword-wise to identify the trend and future scope of BDA in business process. The study suggests that none of business process aspects are untouched by big data analytics. From planning to execution, manufacturing to customer satisfaction, costing to performance management and corporate governance to public relation, all aspect of business process has employed big data analytics tools in one way or other. BDA has also admitted its presence in different business domain like supply chain, service-industry, industry 4.0 and sustainable business practices. Internet of Things (IoT), Artificial Intelligence (AI), Deep learning, Decision support system, Neural network, Predictive analysis, Cloud computing and Machine learning are mostly used big data analytics tools. Keyword analysis also provide insights of currently researched topics and future scope for under researched topics.

Keywords: BDA · Big data analytics · Business process · Network visualization · IoT · AI

1 Introduction

The idea of integrating big data engineering with business process has been visited many times by scholars as such integration brings efficiency and agility in business process [1–4]. Several studies have also reviewed the literature on topic by employing different literature review methods including systematic review and bibliometric analysis [5–8]. These studies have focused on different area of business process such as supply chain, e-commerce, banking, real-estate, smart governance etc. [9–12]. The present study contributes to the existing literature by analyzing keywords indexed in underlying literature and presenting comprehensive insights and implications for researchers, businesspeople, governing agencies and other stakeholder.

Y. Farhaoui et al. (Eds.): ICAISE 2023, LNNS 837, pp. 112–118, 2024.
https://doi.org/10.1007/978-3-031-48465-0_15

2 Research Methodology

We use Scopus database to extract all studies on Big Data Analytics (BDA) in Business Process. We employ different keywords i.e., business process, business innovation, business process engineering, business process re-engineering, big data, big data technology, big data analytics to broaden the search horizon. These seven keywords have generated 5224 articles. Further we limit the article by putting subject area filter i.e., Business, management and Accounting. This filter has limited the number of studies to 1136. Further we have excluded non-English language articles i.e., 4 German, 3 Spanish, 3 Portuguese, 3 Chinese, 2 Lithuanian, 1 Serbian, 1 Russian and 1 Italian. This extraction has left us with 1122 articles. Table 1 present the descriptive statistics of these 1,122 articles. The underlying studies ranges from 1983 to July 2023. Year 1985 to 1991, 1994 and 95, and 1998 and 99 has 0 published articles. Year 2019 has highest number of published articles i.e., 158. On average 96 articles are published excluding the 0 publication years. Range of publication is also very high as till 2016, only 237 articles are published and remaining 885 articles are published 2017–2023. Last 8 years have been accounted for more than 75% of total publications.

Table 1. Descriptive Statistics of published articles on BDA in business process

Statistics	Value	Statistics	Value
Mean	96.09	Skewness	−0.29
Standard error	13.90	Range	135
Median	106	Minimum	23
Mode	#N/A	Maximum	158
Standard deviation	46.11	Sum	1057
Kurtosis	−1.42	Count	11

This study investigates the keywords used in literature year-wise and frequency wise to draw inference about current trends and future scope from extant literature. The study includes both author keywords as well as indexed keywords for analysis. Keywords are analysed to find under-researched and promising topics, and VOSViewer software is used to graphically represent the keyword analysis. Table 2 represent keywords and its co-occurrence in literature.

3 Results and Discussion

Out of 5216 keywords indexed in 1122 articles, this study categories the keywords into three classifications i.e., business process, big data analytics tools and others. Table 3 lists all business process related keywords as per their frequency (F) and total link strength (TLS). As per VOSviewer manual, each keyword has link strength, represented by positive numerical value. Total strength link represents number of documents which

Table 2. Keyword frequency and co-occurrence

Frequency of keywords	Co-occurrence	Frequency of keywords	Co-occurrence
1	5216	11	91
2	1074	12	83
3	580	13	75
4	378	14	67
5	277	15	62
6	216	16	59
7	166	17	54
8	133	18	56
9	119	19	46
10	101	20	41

have co-occurrence of two keywords [13]. Total 40 such business process keywords have been identified. Big data analytics have been used for business operations from decision making to ERP, from marketing to human resource management, from cost management to performance management. None of business process mechanism is untouched by big data analytic.

Table 4 shows which big data analytics tools are employed in business process. IoT has highest total strength link with the most frequently used keyword followed by AI.

Figure 1 shows the network visualization of the keywords which have been used minimum 5 times. Such 277 keywords are presented in Fig. 1 with their link to other keyword. The bigger the circle, the higher the frequency of co-occurrence. The line represents link of keyword co-occurrence e.g., how many time keyword Internet of Things and Big Data appear together is represented by the line connecting these two keywords.

Figure 2 on BDA tool overlay visualization of keywords used between 2018 and 2020 shows how over the years frequency of keywords used have been changed. Keywords like data mining, process control, decision support systems have frequented the literature in 2017–18 whereas keywords like AI, Machine learning, deep learning, blockchain and IoT has been indexed in literature heavily from 2018 onwards.

Table 3. Business Process related keyword analysis

Keywords	F	TLS	Keywords	F	TLS
Decision making	99	697	Social networking	16	115
Data analytics	103	675	E-commerce	19	109
Information management	77	508	Forecasting	13	82
Data mining	65	413	Costs	11	81
Competition	47	390	Survey	13	81
Advance analytics	45	388	Strategic planning	10	79
Information system	48	330	Risk management	14	78
Industry 4.0	55	289	Customer satisfaction	7	73
Enterprise resource management	25	193	ERP	9	73
Knowledge management	33	191	Performance management	12	72
Sales	24	180	Risk assessment	11	61
Sustainable development	25	171	Manufacturing	6	54
Commerce	25	170	Public relation	5	54
Industrial research	19	156	Smart city	6	53
Competitive intelligence	18	152	International trade	7	50
Innovation	36	149	Automotive industry	5	40
Supply Chain management	29	146	Budget control	6	40
Human Resource management	31	139	Governance	12	35
Administrative data processing	16	122	Banking	7	34
Marketing	23	117	Finance	8	27

Table 4. BDA keyword frequency and Total link strength

Keywords	F	TLS	Keywords	F	TLS
IoT	97	623	Network architecture	8	63
AI	82	404	Cyber physical system	6	60
Meta data	40	323	Block chain	19	55
Decision support system	33	256	Classification of information	7	51
Machine learning	46	205	Sentiment analysis	11	46
Cloud computation	28	156	System engineering	6	46
Learning system	14	112	Structural equation modelling	5	45
Embedded system	12	111	Six Sigma	6	40
Predictive analysis	17	110	Decision tree	6	26
Process mining	15	92	Requirement engineering	5	25
Process control	9	80	Text mining	7	25
Deep learning	15	75	Cluster analysis	5	20
Data visualization	7	66	Neural network	5	19

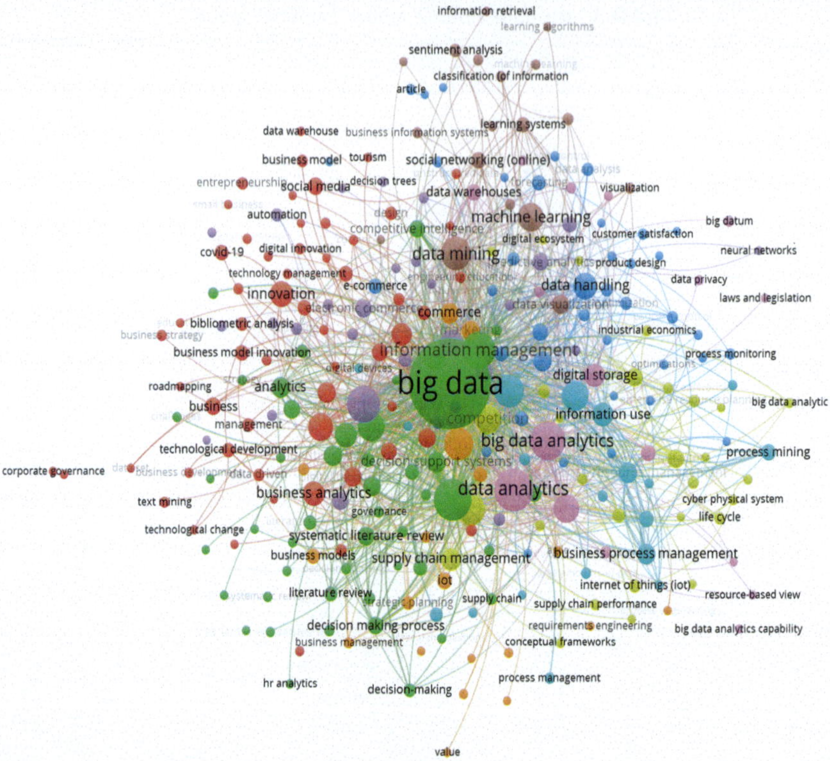

Fig. 1. Network visualization of keyword co-occurrences

4 Scope and Limitation of Study

Though present study does keyword analysis of 1122 articles for presenting a comprehensive view on big data in business process, there are few limitations that exists with the study. We search only one research engine for finding literature i.e., Scopus, that may limit the view of the present paper. We also put the subject filter whereas other subject may have studies based on big data analytics and business process. These limitations can be overcome by eliminating subject filter and exploring other research databases for finding the articles on the topic.

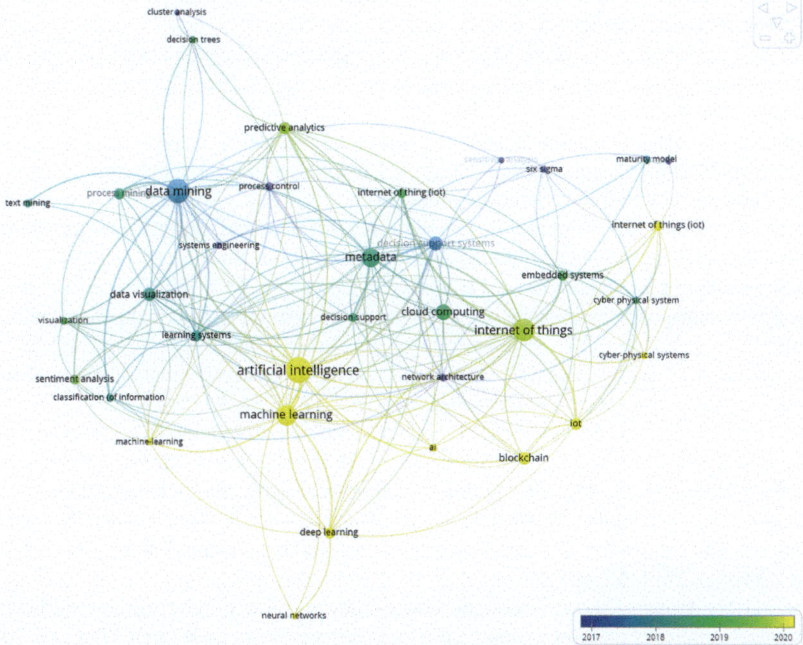

Fig. 2. Overlay visualization of big data analytics keyword

5 Conclusion

Big data analytics has received substantial attention from scholars due to its important role in promoting businesses with respect to smart technologies and as a competitive advantage in digital economy [11, 14–16]. However, the integration of big data analytics to the business process still remains a challenge [15]. Hence, for smooth transition, the researchers will continue to explore the underlying topic as the present study suggest that all aspect of business process has employed big data analytics tools in one way or other. IoT, AI, Deep learning, Decision support system, Neural network, Predictive analysis, Cloud computing and Machine learning are mostly used big data analytics tools. The surge in published article in last ten years also establishes importance of BDA in business process.

References

1. Choi, T.M., Chan, H.K., Yue, X.: Recent development in big data analytics for business operations and risk management. IEEE Transact. Cybernet. **47**(1), 81–92 (2016)
2. Rialti, R., Marzi, G., Silic, M., Ciappei, C.: Ambidextrous organization and agility in big data era: the role of business process management systems. Bus. Process Manage. J. **24**(5), 1091–1109 (2018)
3. Vera-Baquero, A., Colomo Palacios, R., Stantchev, V., Molloy, O.: Leveraging big-data for business process analytics. Learn. Organ. **22**(4), 215–228 (2015)

4. Fosso Wamba, P.S.: Big data analytics and business process innovation. Bus. Process Manage. J. **23**(3), 470–476 (2017)
5. Ancillai, C., Sabatini, A., Gatti, M., Perna, A.: Digital technology and business model innovation: a systematic literature review and future research agenda. Technol. Forecast. Soc. Change. **188**, 122307 (2023)
6. Yordanova, Z.: Big data in the innovation process–a bibliometric analysis and discussion. In: European, Mediterranean, and Middle Eastern Conference on Information Systems, pp. 122–133. Springer Nature Switzerland, Cham (2022)
7. Sardi, A., Sorano, E., Cantino, V., Garengo, P.: Big data and performance measurement research: trends, evolution and future opportunities. Measur. Bus. Excell. (2020)
8. López-Robles, J.R., Otegi-Olaso, J.R., Gómez, I.P., Cobo, M.J.: 30 years of intelligence models in management and business: a bibliometric review. Int. J. Inform. Manage. **48**, 22–38 (2019)
9. Fengchen, W.: The present and future of the digital transformation of real estate: a systematic review of smart real estate. Бизнес-информатика. **17**(2), 85–97 (2023)
10. Durão, M., Veríssimo, M., Moraes, M.: Social Media research in the Hotel Industry: a bibliometric analysis. In: Digital Transformation of the Hotel Industry: Theories, Practices, and Global Challenges, pp. 153–171. Springer International Publishing, Cham (2023)
11. Alsmadi, A.A., Shuhaiber, A., Al-Okaily, M., Al-Gasaymeh, A., Alrawashdeh, N.: Big data analytics and innovation in e-commerce: current insights and future directions. J. Financ. Serv. Market. **26**, 1–8 (2023)
12. Purba, F.N., Arman, A.A.: A systematic literature review of smart governance. In: 2022 International Conference on Information Technology Systems and Innovation (ICITSI), pp. 70–75. IEEE (2022)
13. Guo, Y.M., Huang, Z.L., Guo, J., Li, H., Guo, X.R., Nkeli, M.J.: Bibliometric analysis on smart cities research. Sustainability. **11**(13), 3606 (2019)
14. Sultana, S., Akter, S., Kyriazis, E., Wamba, S.F.: Architecting and developing big data-driven innovation (DDI) in the digital economy. J. Global Inform. Manage. (JGIM). **29**(3), 165–187 (2021)
15. Akter, S., Bandara, R., Hani, U., Wamba, S.F., Foropon, C., Papadopoulos, T.: Analytics-based decision-making for service systems: a qualitative study and agenda for future research. Int. J. Inform. Manag. **48**, 85–95 (2019)
16. Farhaoui, Y.: All big data mining and analytics **6**(3), I–II (2023).https://doi.org/10.26599/BDMA.2022.9020045

Chatbots for Medical Students Exploring Medical Students' Attitudes and Concerns Towards Artificial Intelligence and Medical Chatbots

Berrami Hind[1(✉)], Zineb Serhier[1,2], Manar Jallal[1], and Mohammed Bennani Othmani[1,2]

[1] Medical Informatic Department, Hospital August 20, Casablanca, Morocco
hindberrami.lechgar@gmail.com

[2] Neuroscience and Mental Health Laboratory, Hassan II University, Casablanca, Morocco

Abstract. Introduction: Artificial intelligence (AI) encompasses the concept of automated machines that can perform tasks typically carried out by humans, doctor-patient communication will increasingly rely on the integration of artificial intelligence (AI) in healthcare, especially in medicine and digital assistant systems like chatbots. The objective of this study is to explore the understanding, utilization, and apprehensions of future doctors at the Faculty of Medicine in Casablanca regarding the adoption of artificial intelligence, particularly intelligent chatbots. Methods: A cross-sectional study was conducted among students from the 1st to 5th year at the Faculty of Medicine and Pharmacy in Casablanca. Probability sampling was implemented using a clustered and stratified approach based on the year of study. Electronic forms were distributed to randomly selected groups of students. Results: Among the participants, 52% of students fully agreed to utilize chatbots capable of answering health-related queries, while 39% partially agreed to use chatbots for providing diagnoses regarding health conditions. About concerns, 77% of the respondents expressed fear regarding reduced transparency regarding the utilization of personal data, and 66% expressed concerns about diminished professional autonomy. Conclusion: Moroccan Medical students are open to embracing AI in the field of medicine. The study highlights their ability to grasp the fundamental aspects of how AI and chatbots will impact their daily work, while the overall attitude towards the use of clinical AI was positive, participants also expressed certain concerns.

Keywords: Medical students · Chatbots · Artificial intelligence · Concerns

1 Introduction

The healthcare system is currently undergoing a digital revolution, and artificial intelligence (AI) will play a crucial role in shaping the everyday practices of medical professionals [1]. The advent of digital applications, which can be accessed irrespective of

Y. Farhaoui et al. (Eds.): ICAISE 2023, LNNS 837, pp. 119–124, 2024.
https://doi.org/10.1007/978-3-031-48465-0_16

location and time, has opened up new possibilities in medicine and health communication, consequently altering the doctor-patient dynamic [2]. The increasing significance of e-health applications, wearable devices, and AI tools like chatbots empowers patients to gather their own health data [3, 4]. Additionally, the digital integration of patients, hospitals, physicians, and other healthcare services is driving a shift from a doctor-centered approach to a more patient-centered model of treatment [5]. In order to harness the potential of these technological advancements and ensure optimal care for patients, future medical practitioners must possess the appropriate skills [6].

Machine learning (ML) is a branch of artificial intelligence (AI) that enhances its capabilities by learning from data and experiences, rather than relying on predefined rules as in traditional approaches [7]. Progress in ML has yielded numerous advantages in terms of accuracy, decision-making, fast processing, cost-effectiveness, and the ability to handle intricate data [8]. Chatbots is an example of AI systems that have evolved through the application of ML techniques.

The objective of this study is to explore the understanding, utilization, and apprehensions of future doctors at the Faculty of Medicine in Casablanca regarding the adoption of artificial intelligence, particularly intelligent chatbots.

2 Methods

In April 2023, a cross-sectional study was conducted among students from the 1st to 5th year at the Faculty of Medicine and Pharmacy in Casablanca. The sample size was determined using Epi Info software, taking into account a percentage of 52% of medical students taken from a European article who had knowledge in terms of AI in radiology, with a precision of 5%. The calculated sample size was 335 students.

Probability sampling was implemented using a clustered and stratified approach based on the year of study. Electronic forms were distributed to randomly selected groups of students.

The data analysis was performed using R software, and the association was assessed using the chi-square test with a significance level of 5%.

3 Results

3.1 Socio-demographic Characteristics

A total of 393 responses were received from participating students, with a mean age of 20.6 ± 1.84 years. Among the respondents, women comprised 67% of the sample, indicating a significant female predominance.

We note a predominance of 5th year students 22.4% followed by 4th year medical students 20.1%.

3.2 Attitudes of Medical Student Toward Intelligent Chatbots in Medicine

Among the participants, 52% of students fully agreed to utilize chatbots capable of answering health-related queries, while 39% partially agreed to use chatbots for providing diagnoses regarding health conditions. Most respondents, accounting for 67%, expressed agreement with the use of chatbots for scheduling appointments for patients.

Additionally, 52% strongly supported the utilization of chatbots to assist them in addressing health concerns and to provide supplementary information about their well-being (Table 1).

Table 1. Attitudes toward intelligent chatbots in medicine

	Completely ready N (%)	Partially ready N (%)	Not quite ready N (°%)	Not at all ready N (%)	I don't know N (%)
To respond to your health-related questions	204 (52.0)	140 (35.6)	24 (6.0)	17 (4.4)	5 (5.0)
To give diagnoses on your health condition	81 (20.6)	152 (38.7)	75 (19.1)	70 (17.8)	15 (3.8)
To you propose care	86 (21.9)	155 (39.4)	67 (17.0)	71 (18.1)	14 (3.6)
To conduct certain tests or examinations	126 (32.1)	156 (39.7)	53 (13.5)	35 (8.9)	23 (5.8)
Planning your appointments with patients	263 (66.9)	99 (25.2)	9 (2.3)	12 (3.1)	10 (2.5)
To Support you in matters of health and provide you with additional information	205 (52.2)	148 (37.7)	20 (5.1)	10 (2.5)	10 (2.5)

3.3 Potential Applications of AI in Medicine Perceived by Medical Student

Among students, 68% noted the importance of AI in automating diagnostics, followed by the optimization of continuing education for 17% of respondents (Fig. 1).

3.4 Medical Students Concerned About the Use of AI in Clinical Practice

When it comes to apprehensions surrounding the integration of artificial intelligence into their clinical practice, 77% of the respondents expressed fear regarding reduced transparency regarding the utilization of personal data.

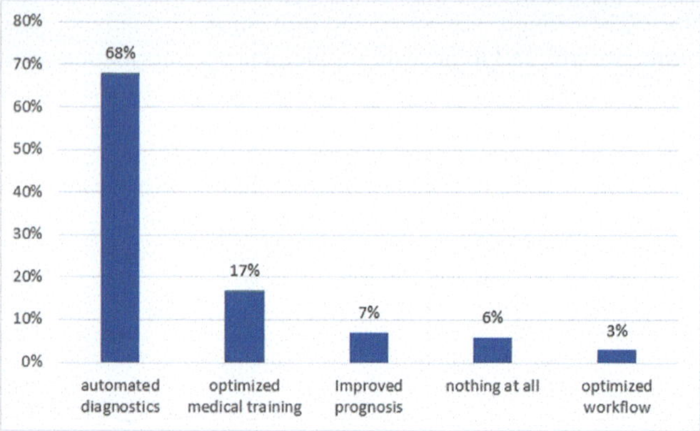

Fig. 1. Perceived potential applications of AI in medicine

Additionally, 66% expressed concerns about diminished professional autonomy, while an equal percentage of 62% were worried about being overwhelmed by the increased reliance on artificial intelligence in their daily work. In terms of job security, 56% of the participants agreed with the notion that they would face negative consequences as a result of the widespread adoption of AI (Table 2).

Table 2. Medical students concerned about the use of AI in medicine.

Questions	Agree	Neutral	Disagree
I am concerned that there is less transparency about how personal data is used	298 (77.0)	51 (13.2)	38 (9.8)
I'm afraid of having less autonomy at work	257 (66.6)	71 (18.4)	58 (15.1)
I fear that my qualifications are no longer sufficient for the requirements of my field of work	184 (61.5)	80 (20.7)	69 (17.8)
I am afraid of being overloaded by the use of artificial intelligence in my daily work	241 (62.5)	77 (19.9)	68 (17.6)
I am afraid to lose my job after the widespread use of AI	216 (56.0)	65 (16.8)	105 (27.2)

4 Discussion

Artificial intelligence (AI) encompasses the concept of automated machines that can perform tasks typically carried out by humans [9]. The field of AI is rapidly evolving, and numerous applications have already become part of our daily lives, such as speech and text recognition and email spam filters [10]. However, the potential of AI to revolutionize global health in low- and middle-income countries is also being explored due to its ability to apply advanced analytical methods to large datasets involved in complex diagnostic tasks [11].

In our study, we found that 52% of medical student completely agreed to utilize chatbots that can answer health-related questions, while 39% partially agreed to use chatbots for diagnosing health conditions. In a separate study of medical students, 50% showed partial willingness to use chatbots for answering health-related questions, while 59% were not willing to use them for health diagnoses [12].

Furthermore, the majority of respondents (67%) agreed with using chatbots to schedule appointments for patients, and 52% strongly supported their use in providing support for health issues and delivering additional health information. Concerning Potential applications of AI in medicine perceived by medical student, 68% noted the importance of AI in automating diagnostics, followed by the optimization of continuing education for 17% of respondents.

Regarding concerns about AI use in clinical practice, 56% expressed fear of job loss due to its generalization, while 77% were concerned about the lack of transparency regarding the use of personal data. Additionally, 67% expressed fears of reduced autonomy and being overloaded by the use of AI in their daily work. These findings align with an article that highlighted concerns about computerized chatbots, including decreased human presence, which can lead to increased distrust in healthcare services. Healthcare professionals and patients often lack confidence in chatbot capabilities, which can raise concerns about clinical care risks, liability, and increased workload rather than reduced workload [13].

5 Conclusion

According to this research, it is evident that medical students are open to embracing AI in the field of medicine. The study highlights their ability to grasp the fundamental aspects of how AI and chatbots will impact their daily work. While the overall attitude towards the use of clinical AI was positive, participants also expressed certain concerns. These concerns mainly revolved around the lack of transparency regarding the utilization of personal data and the potential loss of autonomy when working with AI in medical practice.

Considering the future advancements in the healthcare workplace, it becomes crucial to underscore the importance of incorporating these new core competencies into medical curricula. This will enable physicians to actively participate in shaping the technological trajectory of patient treatment, equipped with comprehensive knowledge and confidence in utilizing AI tools.

References

1. Digitale Medienprodukte in der Arzt-Patienten-Kommunikation. springerprofessional.de. https://www.springerprofessional.de/digitale-medienprodukte-in-der-arzt-patienten-kommunikation/12057004 (n.d). Accessed 12 June 2023
2. Kundu, S.: How will artificial intelligence change medical training? Commun. Med. **1**, 8 (2021). https://doi.org/10.1038/s43856-021-00003-5
3. Lin, B., Wu, S.: Digital transformation in personalized medicine with artificial intelligence and the internet of medical things. Omics J. Integr. Biol. **26**, 77–81 (2022). https://doi.org/10.1089/omi.2021.0037

4. Bates, M.: Health care Chatbots are here to help. IEEE Pulse **10**, 12–14 (2019). https://doi.org/10.1109/MPULS.2019.2911816
5. Rajpurkar, P., Chen, E., Banerjee, O., Topol, E.J.: AI in health and medicine. Nat. Med. **28**, 31–38 (2022). https://doi.org/10.1038/s41591-021-01614-0
6. Ng, K.H., Wong, J.H.D.: A clarion call to introduce artificial intelligence (AI) in postgraduate medical physics curriculum. Phys. Eng. Sci. Med. **45**, 1–2 (2022). https://doi.org/10.1007/s13246-022-01099-2
7. Kersting, K.: Machine learning and artificial intelligence: two fellow travelers on the quest for intelligent behavior in machines. Front. Big Data **1**, 6 (2018). https://doi.org/10.3389/fdata.2018.00006
8. Velayutham, S.: Handbook of Research on Applications and Implementations of Machine Learning Techniques. IGI Global (2019)
9. Amisha, N., Malik, P., Pathania, M., Rathaur, V.K.: Overview of artificial intelligence in medicine. J. Fam. Med. Prim. Care. **8**, 2328–31 (2019). https://doi.org/10.4103/jfmpc.jfmpc_440_19
10. Taulli, T.: Artificial intelligence basics: a non-technical introduction. (2019). https://doi.org/10.1007/978-1-4842-5028-0
11. Schwalbe, N., Wahl, B.: Artificial intelligence and the future of global health. Lancet Lond. Engl. **395**, 1579–1586 (2020). https://doi.org/10.1016/S0140-6736(20)30226-9
12. Moldt, J-A., Festl-Wietek, T., Madany Mamlouk, A., Nieselt, K., Fuhl, W., Herrmann-Werner, A.: Chatbots for future docs: exploring medical students' attitudes and knowledge towards artificial intelligence and medical chatbots. Med. Educ. Online **28**, 2182659 (n.d.). https://doi.org/10.1080/10872981.2023.2182659
13. Parviainen, J., Rantala, J.: Chatbot breakthrough in the 2020s? an ethical reflection on the trend of automated consultations in health care. Med. Health Care Philos. **25**, 61–71 (2022). https://doi.org/10.1007/s11019-021-10049-w

Design and Analysis of the Rectangular Microstrip Patch for 5G Application

Karima Benkhadda(✉), Fatehi A. L. Talqi, Samia Zarrik, Zahra Sahel, Sanae Habibi, Abdelhak Bendali, Mohamed Habibi, and Abdelkader Hadjoudja

Laboratory of Electronics Treatment Information, Mechanic and Energetic, Department of Physics, Faculty of Science, Ibn Tofail University, Kenitra, Morocco
Karima.benkhadda@uit.ac.ma

Abstract. This paper is introducing a design for a patch antenna for 5G applications. The prototype suggested of antenna is design to operate at a frequency of 24 GHz, utilizing Rogers RT5880 with a permittivity equal to 2.2 and a loss tangent of 0.0009. The simulation of the suggested patch is conducted using the CST Studio Suite software. The primary objectives of this research are to attain return loss, elevated gain, reduced VSWR, improved directivity, and overall enhanced efficiency in operation. The simulation results reveal promising performance metrics: a return loss of −31.54 dB, a gain of 8.093 dB, a VSWR of 1.05, a directivity of 8.127 dBi, and a bandwidth featuring a −10 dB reflection coefficient, encompassing a span of 2.8046 GHz (from 22.705 to 25.51 GHz). Additionally, the antenna system demonstrates an impressive efficiency of 99.58%. Overall, the presented antenna exhibits good gain characteristics across the frequency band.

Keywords: Antenna · 5G · Gain

1 Introduction

The realm of microwave and radio frequency communications is experiencing a growing demand for compact transceiver applications [1]. Microstrip antennas play a pivotal role as critical components in these systems [2, 3]. Among them, microstrip patch antennas (MPAs) be garnered significant attention in recent years due to their low profile, lightweight, and cost-effective manufacturing advantages [4]. The significance of microstrip patch antennas has been magnified in the advancement of modern wireless communication systems, driven by the increasing interest in diverse wireless applications [5]. This field has attracted active participation from researchers and scientists due to the manifold possibilities that wireless technology offers [6].The microstrip antenna design comprises three main elements: a patch, a dielectric substrate, and a ground plane [7]. Microstrip patch antennas are available in various shapes, including square, circular, rectangular, triangular, elliptical, and dipole [8]. Among these shapes, circular and rectangular antennas stand out as the most widely used and preferred options in demanding wireless applications due to their versatility and excellent performance capabilities [9]. Throughout the design process, special emphasis is placed on the electrical

Y. Farhaoui et al. (Eds.): ICAISE 2023, LNNS 837, pp. 125–133, 2024.
https://doi.org/10.1007/978-3-031-48465-0_17

characteristics of these antennas, encompassing center frequency (f_0), voltage standing wave ratio (VSWR), return loss, gain, and directivity [10]. These considerations are crucial for achieving high-speed data transfer capabilities. Patch antennas can be undergo modification to match the desired resonant frequency, and radiation pattern [11].

The main focus of this study is on the design and performance evaluation of a rectangular patch antenna operating at 24 GHz, custom-made for 5G applications [12]. The key goals of this antenna design are to enhance return loss and achieve a standard Voltage Standing Wave Ratio (VSWR) and attain a suitable bandwidth to ensure its effective usage in 5G wireless systems [13]. To address the inherent narrow bandwidth of microstrip antennas, the researchers adopted adjustments such as proximity-coupled feeding [14]. The design process involves initial dimension calculations, followed by simulation and analysis. Initially, the antenna is replicated with its starting dimensions, and then proximity-coupled feeding is introduced to enhance its performance. At times, we add slots [15] or a quarter-wave [16] structure to elevate diverse attributes of an antenna. These enhancements encompass factors such as gain, radiation, along with the widening of bandwidth and the refinement of directivity. However, our proposed microstrip antenna yields satisfactory outcomes. This research article presents a comprehensive overview of the entire process involved in designing and analyzing the rectangular Microstrip patch antenna, which is intended for various wireless and broadband communication applications [17]. The simulations carried out in this study utilized computer simulation technology (CST) Microwave Studio, allowing for a precise evaluation of the antenna's design and performance. The study sheds light on the potential application of the proposed antenna in 5G wireless communication scenarios, offering promising results for future 5G deployments.

2 Design Parameters for Rectangular Patch Antenna

The construction of a Microstrip patch antenna involves three critical parameters, as detailed below [18]:

- Frequency of operation (f_0): The antenna is designed to operate with a resonant frequency of 24 GHz for this research setup.
- Dielectric constant of the substrate ($\boldsymbol{\varepsilon_r}$): The dielectric constant of substrate plays a significant role in influencing the Microstrip antenna's performance.
- Height of the dielectric substrate ($\boldsymbol{h_s}$): Determining the antenna dimensions involves using specific equations.

For this study, we have opted for a rectangular patch shape due to its ease of design and analysis, as well as its broader bandwidth compared to other shapes (Fig. 1).

The choice substrate material is Rogers RT5880, possessing a copper metal thickness of 1.575 mm. The Table 1 showcases different substrate thicknesses available in the market.

The antenna's performance is evaluated through various metrics, including bandwidth, directivity, return loss, gain, and radiation efficiency. To achieve our objectives, we have employed techniques such as optimization of antenna dimensions.

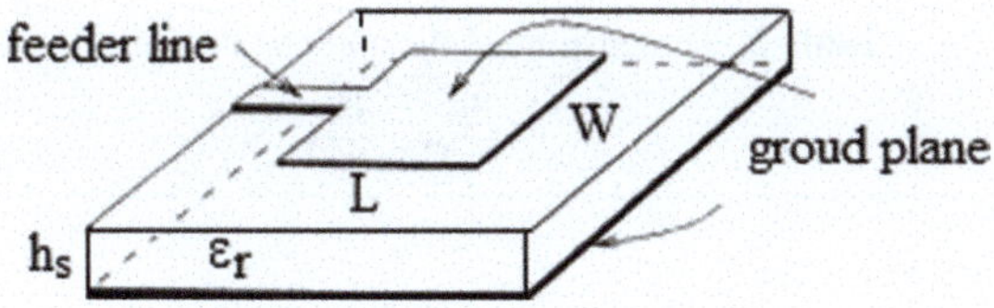

Fig. 1. Microstrip patch antenna

Table 1. Dimensions of the antenna design.

Material	Substrate thickness (mm)
ROGER RT 5880	0.127
	0.252
	0.508
	0.787
	1.575

The formulas used to size the Microstrip antenna according to the literature are [19–21]:

$$W = \frac{c}{2f_r\sqrt{\frac{(\varepsilon_r+1)}{2}}} \tag{1}$$

$$L = \frac{c}{2f\sqrt{\varepsilon_{reff}}} - 2\Delta l \tag{2}$$

$$\varepsilon_{eff} = \frac{\varepsilon_r+1}{2} + \frac{\varepsilon_r-1}{2}\left[1+12\frac{h}{w}\right]^{-\frac{1}{2}} \tag{3}$$

$$\frac{\Delta l}{h} = 0.412\frac{(\varepsilon_{reff}+0.3)\left(\frac{W}{h}+0.264\right)}{(\varepsilon_{reff}+0.258)\left(\frac{W}{h}+0.8\right)} \tag{4}$$

The W represents the width, L is the actual length, L_{eff} is the effective length, ϵ_{reff} is the effective dielectric constant and $\boldsymbol{\Delta L}$ is Fringe length.

L = the actual length.

Initially, we employ the EM Talk website to compute the dimensions of the patch and microstrip line for the microstrip antenna, through numerous iterations, the dimensions of both the patch and the line underwent alterations. The dimensions of the antenna design used in this work are shown in (Table 2, Fig. 2).

Table 2. Dimensions of the antenna design

Parameters	Values
Wg	10.5
Ws	10.5
Lg	12
Ls	12
Wp	4.3
Lp	2.3
Wf	1.5
Lf	5
Hs	1.575
t	0.035

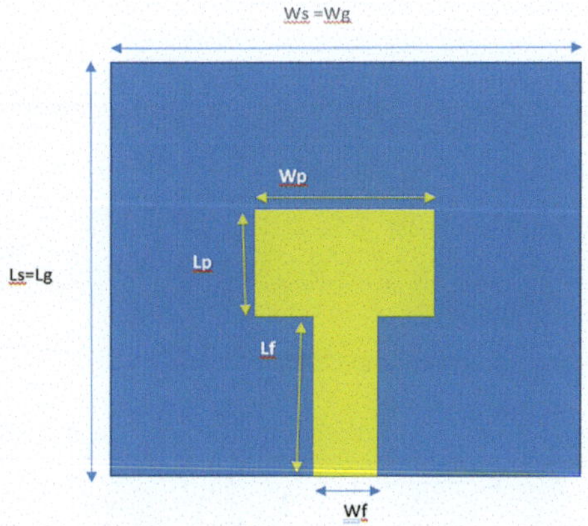

Fig. 2. Geometry of microstrip patch antenna in CST

3 Results and Discussions

The Microstrip patch antenna is meticulously designed and simulated through CST, a powerful software adept at analyzing intricate 3D and multilayer configurations. This versatile program is widely utilized in design of different antenna depending on her application and the resonance frequency, enabling precise calculations and graphical representations of essential parameters. Among the critical metrics it can derive and visualize are S-Parameters, VSWR, gain, and directivity.

3.1 Return Loss

For the proposed antenna, the parameter S_{11} is computed, and the corresponding simulated return loss is graphically represented in Fig. 3, exhibiting an outstanding value of

−31.54 dB, which is excellent. It is evident that the antenna's resonance occurs precisely at 24.02 GHz.

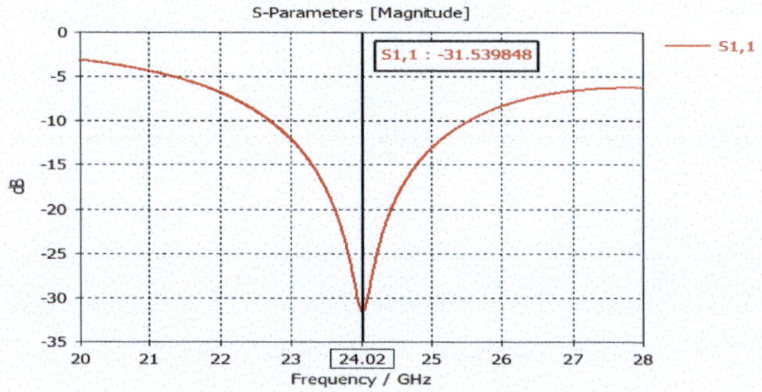

Fig. 3. S-parameters of antenna patch in CST

3.2 Voltage Standing Wave Ratio (VSWR)

The VSWR must ideally be between 1 and 2. In Fig. 4, the VSWR (Voltage Standing Wave Ratio) value of 1.05 at a frequency 24.02 GHz. This value is small, so he has a good efficacient for coverage.

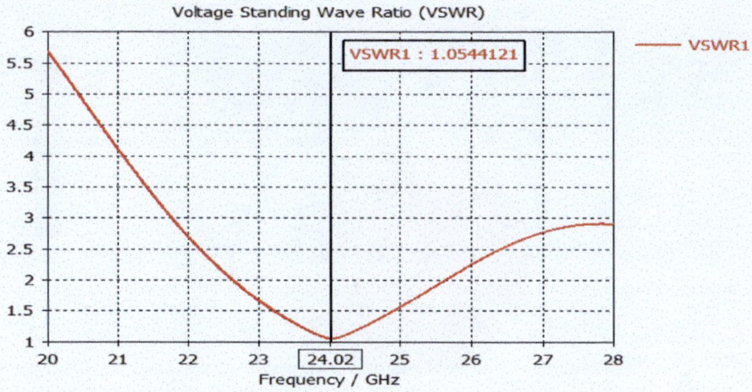

Fig. 4. VSWR of antenna patch in CST

3.3 Bandwidth

The antenna's bandwidth was determined from the Fig. 5 taking the values of f_1 and f_2 at −10 dB, the obtained bandwidth for the designed antenna is 2.8046 GHz.

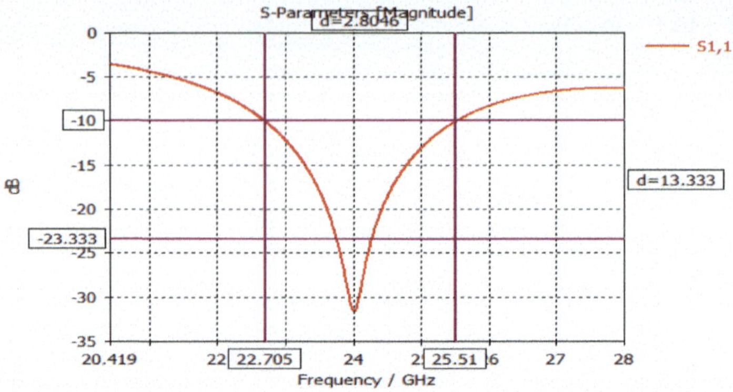

Fig. 5. Bandwidth of antenna patch in CST

3.4 3D Radiation Pattern for Gain and Directivity

3.4.1 Gain

Gain holds immense significance in assessing an antenna's performance, as it represents the ratio of the radiated field intensity of the specific antenna to that of a reference antenna. Usually measured in decibels (dB), antenna gain also provides insights into the direction of maximum radiation. Within the context of this research, the proposed antenna's gain is precisely measured art a frequency of 24 GHz, resulting in a value of 8.093 dB, as shown in Fig. 6.

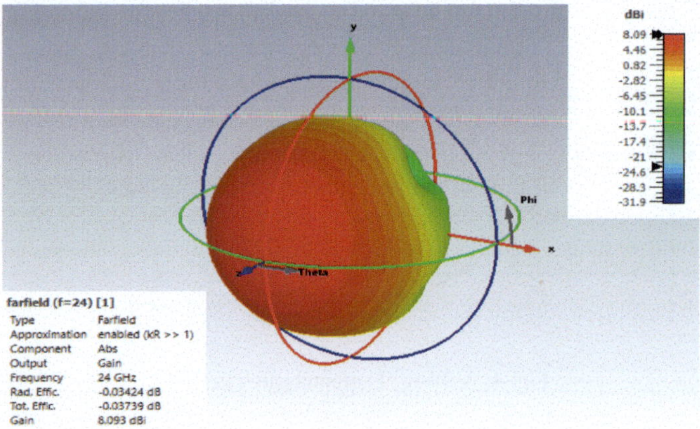

Fig. 6. 3-D radiation pattern gain of antenna patch in CST (gain)

3.4.2 Directivity

The designed antenna exhibits a remarkable directivity of 8.127 dBi at a frequency of 24 GHz, clearly depicted in Fig. 7. The primary goal is to optimize the antenna's radiation pattern, directing its response towards a specific direction to enhance power transmission or reception. The directivity of the antenna is directly influenced by the shape of its radiation pattern.

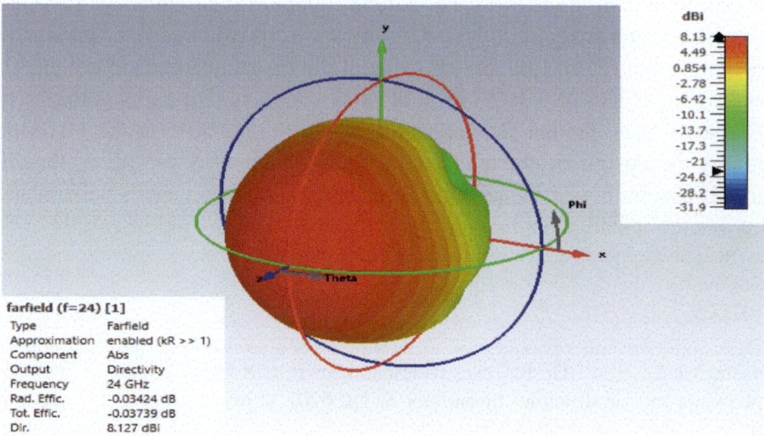

Fig. 7. 3-D radiation pattern directivity of antenna patch in CST

Table 3 presents a comparison of the parameter values for the suggested microstrip antenna and various designs found in the existing literature. Clearly, our antenna, as well as antennas [22–24], employ the Rogers RT5880 substrate material. Notably, antenna [25] stands apart by using FR-4 material. Every adjustment to a microstrip antenna parameter brings about a shift in antenna performance. Ultimately, the convergence of the gain and directivity values is important for enhancing antenna efficiency.

Table 3. Comparison the proposed antenna with various other antennas

Parmeter references	Size	Resonnance frequency (GHZ)	Substrate material	Dielectric constant	S-parameters (dB)	VSWR	Gain (dB)	Directivity	Efficiency %
[22]	7.14 × 8.52	27.988	Taconic TLY-5	2.2	−36.18	1.03	6.72	–	–
[25]	7.9 × 14.7	28	FR-4	4.4	−14.15	1.48	6.06	6.74	–
[23]	6 × 15	25.75	Rogers RT5880	–	−47.65	1.008	5.58	–	–
[24]	6.285 × 7.235	27.954	Rogers RT5880	2.2	−13.48	1.54	6.63	8.37	70.18
Proposed	10.5 × 12	24.02	Rogers RT5880	2.2	−31.54	1.05	8.093	8.127	99.58

4 Conclusion

In this paper, we showcase the design and simulation of a basic rectangular microstrip patch antenna tailored for 5G applications, including weather radar, surface ship radar, and select communication satellites. The antenna operates at a frequency of 24 GHz. The study encompasses an analysis of the radiation pattern, as well as crucial parameters like gain, efficiency, and return loss. Specifically, at the resonant frequency of 24.02 GHz, the measured return loss of -31.54 dB indicating excellent mat-ching at that frequency point. The suggested antenna showcases a bandwidth of 2.8046 at the -10 dB level. Additionally, the designed antenna demonstrates a gain of 8.093 dB at the 24 GHz frequency. These findings provide valuable insights into the antenna's performance and suitability for the intended applications.

References

1. Al-yasir, Y.I.A., et al.: Design of a wide-band microstrip filtering antenna with modified shaped slots and sir structure. Inventions. **5**, 1 (2020). https://doi.org/10.3390/inventions5010011
2. Mohsen, M.K., Isa, M.S.M., Isa, A.A.M., Abdulhameed, M.K., Attiah, M.L., Dinar, A.M.: Enhancement of boresight radiation for leaky wave antenna array. Telkomnika (Telecommunication Comput. Electron. Control. **17**, 2179–2185 (2019). https://doi.org/10.12928/TELKOMNIKA.v17i5.12631
3. Colaco, J.: Antenna for 5G applications. 682–685 (2020)
4. Tegegn Gemeda, M., Fante, K.A., Goshu, H.L., Goshu, A.L.: Design and analysis of a 28 GHz microstrip patch antenna for 5G communication systems. Int. Res. J. Eng. Technol. 881–886 (2021)
5. Przesmycki, R., Bugaj, M., Nowosielski, L.: Broadband microstrip antenna for 5g wireless systems operating at 28 ghz. Electron. **10**, 1–19 (2021). https://doi.org/10.3390/electronics10010001
6. Rana, S., Rahman, M.: Study of microstrip patch antenna for wireless communication system. Int. Conf. Adv. Technol. ICONAT 1–5 (2022). https://doi.org/10.1109/ICONAT53423.2022.9726110
7. Darimireddy, N.K., Ramana Reddy, R., Mallikarjuna Prasad, A.: Design of triple-layer double U-slot patch antenna for wireless applications. J. Appl. Res. Technol. **13**, 526–534 (2015). https://doi.org/10.1016/j.jart.2015.10.006
8. B, A.D.S., Cahyono, F.B., Bagus, H.B.: Proceedings of the International Conference on Advance Transportation, Engineering, and Applied Science (ICATEAS 2022). Atlantis Press International BV (2023). https://doi.org/10.2991/978-94-6463-092-3
9. Rana, M.S., Islam, S.I., Al Mamun, S., Mondal, L.K., Ahmed, M.T., Rahman, M.M.: An S-Band microstrip patch antenna design and simulation for wireless communication systems. Indones. J. Electr. Eng. Informatics. **10**, 945–954 (2022). https://doi.org/10.52549/ijeei.v10i4.4141
10. Abdel-Jabbar, H., Kadhim, A.S., Saleh, A.L., Al-Yasir, Y.I.A., Parchin, N.O., Abd-Alhameed, R.A.: Design and optimization of microstrip filtering antenna with modified shaped slots and SIR filter to improve the impedance bandwidth. Telkomnika (Telecommunication Comput. Electron. Control. **18**, 545–551 (2020). https://doi.org/10.12928/TELKOMNIKA.v18i1.13532

11. Mushaib, M., Anil Kumar, D.: Designing of microstrip patch antenna using artificial neural network: a review. **11**, 193–199 (2020)
12. Kim, J., Oh, J.: Liquid-crystal-embedded aperture-coupled microstrip antenna for 5g applications. IEEE Antennas Wirel. Propag. Lett. **19**, 1958–1962 (2020). https://doi.org/10.1109/LAWP.2020.3014715
13. Colaco, J., Lohani, R.: Design and implementation of microstrip circular patch antenna for 5g applications. In: 2nd International Conference Electronics Communication Computer Engineering ICECCE (2020). https://doi.org/10.1109/ICECCE49384.2020.9179263
14. Chaitanya, G., Arora, A., Khemchandani, A., Rawat, Y., Singhai, S.: Comparative study of different feeding techniques for rectangular microstrip patch antenna. Int. J. Innov. Res. Electr. Electron. Instrum. Control Eng. **3**, 32–35 (2015). https://doi.org/10.17148/IJIREEICE.2015.3509
15. Bhargava, A., Sinha, P.: Multi rectangular slotted hexa band micro-strip patch antenna for multiple wireless applications. In: 2nd International Conference Electronics Communication Computer Engineering ICECCE 2018, pp. 902–904 (2018). https://doi.org/10.1109/CESYS.2018.8723965
16. Banuprakash, R., Hebbar, H.G., Janani, N., Neha, R., Raghav, K.K., Sudha, M.: Microstrip array antenna for 24GHz automotive RADAR. In: 2020 7th International Conference Smart Structure System ICSSS, pp. 5–10 (2020). https://doi.org/10.1109/ICSSS49621.2020.9202360
17. Rana, M.S., Smieee, M.M.R.: Design and analysis of microstrip patch antenna for 5G wireless communication systems. Bull. Electr. Eng. Informatics. **11**, 3329–3337 (2022). https://doi.org/10.11591/eei.v11i6.3955
18. Rahman, Z., Mynuddin, M.: Design and Simulation of High Performance Rectangular Microstrip Patch Antenna Using CST microwave studio design and construction of a magnetic levitation system using programmable logic controller view project design and simulation of high performance Re. **8**, 2225–2229 (2020)
19. Upender, P., Harsha Vardhini, P.A.: Design analysis of rectangular and circular microstrip patch antenna with coaxial feed at s-band for wireless applications. In: Proceeding 4th International Conference IoT Social Mobile, Analog Cloud, ISMAC 2020, pp. 274–279 (2020). https://doi.org/10.1109/I-SMAC49090.2020.9243445
20. Wang, Q., Mu, N., Wang, L., Safavi-Naeini, S., Liu, J.: 5G MIMO Conformal Microstrip Antenna Design. Wirel. Commun. Mob. Comput. 2017, (2017). https://doi.org/10.1155/2017/7616825
21. Sarade, S.S., Ruikar, S.D., Bhaldar, H.K.: Design of microstrip patch antenna for 5G application. Techno-Societal 2018 Proc. 2nd Int. Conf. Adv. Technol. Soc. Appl. **1**, 253–261 (2020). https://doi.org/10.1007/978-3-030-16848-3_24
22. Darsono, M., Wijaya, A.R.: Design and simulation of a rectangular patch microstrip antenna for the frequency of 28 GHz in 5G technology. J. Phys. Conf. Ser. 1469 (2020). https://doi.org/10.1088/1742-6596/1469/1/012107
23. El-Wazzan, M.M., Ghouz, H.H., El-Diasty, S.K., Aboul-Dahab, M.A.: Compact and integrated microstrip antenna modules for mm-wave and microwave bands applications. IEEE Access. **10**, 70724–70736 (2022). https://doi.org/10.1109/ACCESS.2022.3187035
24. Darboe, O., Manene, F., Konditi, D., Bernard, D., Konditi, O.: A 28 GHz rectangular microstrip patch antenna for 5g applications. Int. J. Eng. Res. Technol. **12**, 854–857 (2019)
25. Kavitha, M., Dinesh Kumar, T., Gayathri, A., Koushick, V.: 28 GHZ printed antenna for 5G communication with improved gain using array. Int. J. Sci. Technol. Res. **9**, 5127–5133 (2020)

Reading in the 21st Century: Digital Reading Habit of Prospective Elementary Language Teachers

Loise Izza Gonzales[1(✉)], Radam Jumadil Yusop[2], Manilyn Miñoza[1], Arvin Casimiro[1], Aprilette Devanadera[3], and Alexandhrea Hiedie Dumagay[1]

[1] Western Mindanao State University, Zamboanga City, Philippines
xt202000932@wmsu.edu.ph
[2] Mindanao State University-Sulu, 7400 Jolo, Philippines
[3] Southern Luzon State University, Lucban, Quezon, Philippines

Abstract. The contemporary world is dominated by technologies and can be seen in various fields, such as education or even for daily purposes. Furthermore, the existence of computer-generated tools gives rise to the demand for digital reading, which has solely become a competitor of traditional printed reading. Although the growth of digital breakthroughs is evident, the literacy rate of students tends to decline, especially in the case of the Philippines. Previous studies delve into reviewing the digital reading habits of students and are limited in terms of preservice teachers. Thus, this investigation aimed to distinguish the digital reading habits of preservice elementary language educators. This study accumulated a total of 95 respondents with an age range of 18–27 years old. The tool utilized in this study was the Digital Reading Habit Questionnaire (Cronbach's Alpha = 0.89). The participants were found to have *'satisfactory'* digital reading habits. Furthermore, there was no variation between the male and female prospective elementary language teachers.

Keywords: Digital reading · Reading habits · Technology · Gender

1 Introduction

The utilization of technology has been widespread and deemed perpetual in the contemporary era. The twenty-first century exposed people to digital breakthroughs that are applicable to their daily lives [1]. It was noted that modern civilization applies it as a means for communication or getting in touch with each other [2] or even a way to have a form of indulgence in a digital sense [3]. Moreover, it has been saliently available in education in which assimilation of information and communication technologies (ICT) were present in teaching strategies and activities in the classroom [4, 5]. It was even attested that education is dynamic due to its adaptability with the presence of technology and global proliferation during this computer-generated or virtually inclined era [6]. Furthermore, institutions tend to establish expenditures with ICT equipment for them

Y. Farhaoui et al. (Eds.): ICAISE 2023, LNNS 837, pp. 134–141, 2024.
https://doi.org/10.1007/978-3-031-48465-0_18

to anticipate the exemplary quality of education that produces internationally competitive minds [7]. Additionally, the outbreak of the COVID-19 pandemic has stimulated e-learning platforms since the modification of face-to-face learning to an online modality is a necessity to combat the odds that unceasingly make education possible [8]. Hence, it was noted that there is an intensification of ICT tools in the aspect of learning for students and teaching strategies for educators [9].

Reading as a macro skill is indispensable in academia and even in daily interactions, especially during this digital age. Through various advancements, digital media has become popular, replacing traditional books or hard-copy documents [10]. The exceptionally swift growth of technologies such as smartphones, laptops, computers, and tablets contributed to the paradigm shift in education [11]. At present, learners favor reading or searching for information through websites, e-books, portable document format (PDF) files, audiobooks, and even social media platforms [12, 13]. It was also asserted that the accessibility of these digital tools tends to be facile since they can be utilized through the Internet, which is embedded in a plethora of resources [14]. Through this, it is implied that students or adolescents in general have an abundance of online references that they can use for requirements in school. Additionally, reading digitally has *'significant versatility'* since it can be done anywhere or at any time [13]. This pertains to its capacity to store information online, save space, and even replace bulky books, which makes it economical [15, 16].

However, even though the progress of ICT tools seems to be efficient, problems in terms of reading have become rampant. It was even said that students tend to be distracted regarding reading online, which leads them to read less or more direct text than wordy text [17]. It was also claimed that some might lead to an extreme case in which younger ones tend to reduce their time for reading or avoid it at all due to the demand for other activities found on the Internet [18, 19]. On the other hand, the predicament about reading comprehension has become a major dilemma and often the topic of scholarly studies [20–23]. Developing countries such as the Philippines also experience debacles in terms of literacy rate. In fact, this country turned out to have the poorest rating in terms of performance for reading and writing [24]. Through this, it is essential to take note that even if the digitization of platforms in learning was present, there might be other learners who struggle in adapting to reading with digital text.

Given that electronic equipment and other computer-based tools for reading are in demand during these contemporary times, it is truly a necessity for teachers to be flexible regarding handling classes, especially with language courses. In the pool of studies, it was seen that there is a scarcity of content regarding discussing the digital reading habits of prospective teachers. Nonetheless, there are only two studies, Abequibel et al. [14] and Maden [16], that focus on high school language educators. Thus, this investigation centers on examining the digital reading habits of preservice elementary language educators at a nonmetropolitan university. The findings of this study will provide essential aspects in terms of acquiring adequate knowledge in assessing the reading habits of aspiring reading mentors. It was also noted that the participants in this study were chosen since they can evaluate the digital reading habits of their future learners as well. Gender has been widely studied in language studies [25, 26]. Thus, this study also endeavors to examine digital reading habits about gender to additionally contribute to or contextualize it in the vast field of literature.

Research question:

The preceding questions will aid in examining preservice elementary language teachers' digital reading habits:

(1) What is the prospective elementary language teacher's digital reading habit?
(2) Is there a significant difference in digital reading habits between male and female participants?

2 Methodology

2.1 Research Design

This investigation employed a descriptive-quantitative design. The study aimed to quantify digital reading habits, which showcased that it possesses the qualities of quantitative research [27]. Moreover, this study is a descriptive type of research because it analyzes both mean and standard deviation [28]. Additionally, this research study is considered cross-sectional because it was conducted within a short period of time, which was attainable through the utilization of a survey tool [29]. Correspondingly, employing survey tools embodied efficiency, particularly in investigations with large numbers of participants [30]. Likewise, this study is nonexperimental since controlled setups were not employed [31].

2.2 Respondents

This study focused on aspiring elementary language educators from a nonmetropolitan university. The requirement of the following participants conformed with the inclusion criteria: (1) the prospective teacher must be assigned with language courses such as English or Filipino and (2) the prospective teacher must be under elementary and general education. This study accumulated a sample size of 95. The majority of the females (81.1%) responded to the survey. Notably, this confirms that females dominate the teaching profession [14, 32, 41]. Furthermore, the participants' ages ranged from 18 to 27, with an average age of 20.33 and a standard deviation of 1.55.

2.3 Research Instrument

This investigation employed a survey questionnaire. Furthermore, the tool entitled Digital Reading Habit Questionnaire was adapted from the perusal of Maden [16]. Furthermore, it consists of two dimensions: reading psychology and daily use. Additionally, the former has nine items, and the latter contains seventeen items in total. The overall questionnaire earned a Cronbach's alpha score of 0.89, which indicates that it has a reliability that can be interpreted as '*good*' [33]. On the other hand, it can also imply to be '*very reliable*' [34]. Notably, the items adhere to a four-point Likert scale: '*never*', '*seldom*', '*mostly*', and '*always*.'

2.4 Data Gathering Procedure

After being subjected to validity, the tool was finalized through a Google form. An online survey was conducted in hopes of reducing waste in terms of hard copies and saving time through its automaticity. The instrument was employed from June 24 until July 7, 2023. Moreover, the respondents were recognized prior to being communicated through online social media platforms, specifically Facebook and Messenger. A total of 280 students were contacted. However, only 95 responded just in time for the data analysis.

2.5 Data Analysis Technique

For the analysis of data to be possible, answers of the survey tool were coded in SPSS. For the demographics, the gender utilized its dichotomous aspect in which male is coded as 1 and female as 2. The positive statements were coded as 1 for '*never*', 2 for '*seldom*', 3 for '*mostly*', and 4 for '*always*.' Additionally, two items in the questionnaire were deemed negative. Hence, it was subjected to reverse coding. Moreover, the mean or average scores were elucidated through a certain scale: 1.0–1.74 '*poor*', 1.75–2.4 '*fair*', 2.5–3.24 '*satisfactory*', and 3.25–4.0 '*very satisfactory.*' To distinguish the prospective elementary language teachers' digital reading habits, descriptive statistics were utilized. Specifically, the mean and standard deviation. In addition, one-way analysis of variance (one-way ANOVA) was used to classify the difference in digital reading habits between the variable gender.

2.6 Ethical Consideration

A consent letter was distributed to the participants by employing a Google form link access together with the survey tool. Consequently, the letter explained that the investigation is deemed volitional, and it occurred during a short span of time that lasted approximately 15 min or less. Furthermore, it was noted that the questionnaire was subjective since there were no standard answers that were anticipated. Furthermore, it was highlighted that all the data, especially names and email addresses, were held confidential.

3 Results and Discussion

Responses from the Digital Reading Habit Questionnaire.

To disinter the respondents' digital reading habits, the questionnaire was administered, and the data were analyzed through the application of descriptive statistics such as the mean (M) and standard deviation (SD). It was deciphered through a scale: 1.0–1.74 '*poor*' (P), 1.75–2.4 '*fair*' (F), 2.5–3.24 '*satisfactory*' (S), and 3.25–4.0 '*very satisfactory*' (VS).

Table 1 divulges that the prospective elementary language teachers have a '*satisfactory*' overall digital reading habit. Moreover, the result was consistent with the case of Abequibel et al. [14], who also conducted a study with preservice teachers. Similarly, both the reading psychology dimension and the daily use dimension were discovered to have an interpretation of '*satisfactory*', which agrees with the investigation that was

Table 1. Prospective elementary language teachers' digital reading habits

Dimension	Mean	SD	Descriptor	Interpretation
Reading psychology	2.84	0.37	Mostly	S
Daily use	3.01	0.38	Mostly	S
General digital reading habit	2.96	0.34	Mostly	S

also conducted with preservice educators by Maden [16]. Thus, prospective language educators tend to read digital text often and adapt it in their daily routines.

Furthermore, it was evident that the daily use dimension garnered a higher score (M-3.01; SD-0.38) compared to reading psychology (M-2.84; SD-0.37). Indeed, digital reading habits flourished due to being exposed to electronic platforms daily. This claim is accepted since digital media has been part of the everyday routine of people, especially in keeping in touch with friends, family, acquaintances, and even in the workplace or school. Over the years, technology has developed, and individuals have tended to adapt it to make it as efficient and convenient as possible. Even in the educational field, instructors started to adapt more digital text as part of their lectures. It was noted that the experience of online learning heightened the demand for reading through digital screens. Hence, with the outcome of this study, it is evident that digital reading will establish great interest or importance, especially in academia.

Digital Reading Habit across gender.

To distinguish whether male and female respondents vary in terms of digital reading habits, the data gathered were analyzed. One-way analysis of variance (ANOVA) was employed to ascertain the difference in the variable gender.

Table 2. Digital reading habit and gender

Dependent	Independent	M	SD	p value	Interpretation
Digital reading habit	Male	2.99	0.34	0.626	Not significant
	Female	2.95	0.34		

*Significant at alpha = 0.05

Table 2 shows that the p value surpassed the alpha level of 0.05, which indicates that gender does not affect the digital reading habits of prospective elementary language educators. Furthermore, the findings in this empirical research indicate that the construct of gender does not manipulate or influence digital reading habits. Through this, it is asserted that the roles of gender were seen as neutral or impartial. Thus, males and females share the same abilities or routines regarding reading digital text.

The result of this study has another point of view compared to the studies of Abequibel et al. [14], Maden [16], and Horton-Ramos [35] since these investigations claimed that the construct of gender is truly an influential contributor to digital reading habits. Furthermore, it debunks the claim that females were more inclined to read digitally than males [36]. Likewise, it does not agree that males establish superiority when reading online text [10]. Nevertheless, the current study corroborates the outcome of Abidin et al. [37], which indicated that there was fairness in terms of gender regarding reading digitally.

The explanation behind this relates to the fact that teachers were exposed to digital media throughout their lives and even applied to their curriculum in school. It was even attested that neither males nor females can outstand one another regarding their outlook toward reading through screens [15]. Moreover, it was discovered that both males and females embodied expertise in manipulating electronic devices [38]. Additionally, it was even noted that in terms of digital competence, gender garnered neutral results [39, 40]. Hence, these findings in various studies support the outcome of the study, which indicates that both male and female prospective elementary language educators have the same level of digital reading habits.

4 Conclusion and Implications

This study focused on determining the digital reading habits of prospective elementary language teachers. It was discovered that the respondents have a *'satisfactory'* overall digital reading habit. Additionally, it was perceived that gender differences in this case are not present since the statistical findings were deemed *'neutral.'*

Even though it was asserted that the digital reading habits of the participants are acceptable, there is still room for improvement regarding this matter that can help to assuage and achieve a high level of digital reading habits. Through this, it can be implied that incorporating more activities done online or with the aid of technological tools can boost digital reading habits. Furthermore, it was noted that both genders have the capacity to compete in this technologically competitive world. The findings of this research regarding the variation of gender toward digital reading habits are essential since they address a diverse point of view that adds to the limited data in the literature. Through this, future research should delve more deeply into assessing gender differences about digital reading habits. As future language teachers at the primary level, assessing the respondents' digital reading habits can absolutely pave the way for combating literacy deficiency in the Philippines. The outcome of this study shall set standards or guidelines for the betterment of the educational system in the country.

References

1. Pond, W.K.: Twenty-first century education and training: implications for quality assurance. Int. High. Educ. **4**, 185–192 (2001). https://doi.org/10.1016/S1096-7516(01)00065-3
2. Chien, T.C., Chen, Z.H., Chan, T.W.: Exploring long-term behavior patterns in a book recommendation system for reading. Educ. Technol. Soc. **20**, 2027–2036 (2017)
3. Sun, B., Loh, C.E., Nie, Y.: The COVID-19 school closure effect on students' print and digital leisure reading. Comput. Educ. **2**, 100033 (2021). https://doi.org/10.1016/j.caeo.2021.10003
4. Aditya, B.R., Andrisyah, I.A.N., Atika, A.R., Permadi, A.: Digital disruption in early childhood education: a qualitative research from teachers' perspective. Procedia Comput. Sci. **197**, 521–528 (2022)

5. Coman, C., Țîru, L.G., Meseșan-Schmitz, L., Stanciu, C., Bularca, M.C.: Online teaching and learning in higher education during the coronavirus pandemic: students' perspective. Sustainability **12**, 10367 (2020). https://doi.org/10.3390/su122410367
6. Rillo, R., Alieto, E.: Indirectness markers in Korean and Persian English essays: implications for teaching writing to EFL learners. English Int. J. **13**, 165–184 (2018)
7. Wu, B.: Identifying the influential factors of knowledge sharing in e-learning 2.0 systems. Int. J. Enterp. Inf. Syst. **12**, 85–102 (2016)
8. Mumbing, A.L.L., Abequibel, B.T., Buslon, J.B., Alieto, E.O.: Digital education, the new frontier: determining attitude and technological competence of language teachers from a developing country. Asian ESP J. **17**, 300–328 (2021)
9. Ratten, V.: The post COVID -19 pandemic era: changes in teaching and learning methods for management educators. Int. J. Manage. Educ. **21**, 100777 (2023)
10. Karim, N.S.A., Hasan, A.: Reading habits and attitude in the digital age: analysis of gender and academic program differences in Malaysia. Electron. Libr. **25**, 285–298 (2007). https://doi.org/10.1108/02640470710754805
11. Wu, W.H., Wu, Y.C.J., Chen, C.Y., Kao, H.Y., Lin, C.H., Huang, S.H.: Review of trends from mobile learning studies: a meta-analysis. Comput. Educ. **59**, 817–827 (2012)
12. Hu, J., Yu, R.: The effects of ICT-based social media on adolescents' digital reading performance: a longitudinal study of PISA 2009, PISA 2012, PISA 2015 and PISA 2018. Comput. Educ. **175**, 104342 (2021). https://doi.org/10.1016/j.compedu.2021.104342
13. Spjeldnæs, K., Karlsen, F.: How digital devices transform literary reading: The impact of e-books, audiobooks and online life on reading habits. New Media Soc. **0**, 1–17 (2022).https://doi.org/10.1177/14614448221126168
14. Abequibel, B., Ricohermoso, C., Alieto, E., Barredo, C., Lucas, R.: Prospective reading teachers' digital reading habit: a cross-sectional design. TESOL Int. J. **16**, 246–260 (2021)
15. Eijansantos, A.M., Alieto, E.O., Morgia, J.D., Ricohermoso, C.D.: Print-based texts or digitized versions: an attitudinal investigation among senior high school students. Asian EFL J. **27**, 308–339 (2020)
16. Maden, S.: Digital reading habit of pre-service Turkish language teachers. S. Afr. J. Educ. **38**, 1–10 (2018). https://doi.org/10.15700/saje.v38ns2a1641
17. Mangen, A., Van der Weel, A.: The evolution of reading in the age of digitisation: an integrative framework for reading research. Literacy **50**, 116–124 (2016)
18. Chalari, M., Vryonides, M.: Adolescents' reading habits during COVID-19 protracted lockdown: to what extent do they still contribute to the perpetuation of cultural reproduction? Int. J. Educ. Res. **115**, 102012 (2022). https://doi.org/10.1016/j.ijer.2022.102012
19. Rutherford, L., Waller, L., Merga, M., McRae, M., Bullen, E., Johanson, K.: Contours of teenagers' reading in the digital era: scoping the research. New Rev. Child. Lit. Librarian. **23**, 27–46 (2017)
20. Cho, B.Y., Hwang, H., Jang, B.G.: Predicting fourth grade digital reading comprehension: a secondary data analysis of (e)PIRLS 2016. Int. J. Educ. Res. **105**, 101696 (2021). https://doi.org/10.1016/j.ijer.2020.101696
21. Huang, H.C., Chern, C.L., Lin, C.C.: EFL learners' use of online reading strategies and comprehension of texts: an exploratory study. Comput. Educ. **52**, 13–26 (2009). https://doi.org/10.1016/j.compedu.2008.06.003
22. Srinivasan, V., Murthy, H.: Improving reading and comprehension in K-12: evidence from a large-scale AI technology intervention in India. Comput. Educ. Artific. Intell. **2**, 100019 (2021). https://doi.org/10.1016/j.caeai.2021.100019
23. Türk, E., Erçetin, G.: Effects of interactive versus simultaneous display of multimedia glosses on L2 reading comprehension and incidental vocabulary learning. Comput. Assist. Lang. Learn. **27**, 1–25 (2014). https://doi.org/10.1080/09588221.2012.692384

24. Rosales, S.S.: Seeing the 'hidden' disability: a quantitative analysis of the reading comprehension in English of learners suspected with dyslexia. Asian EFL J. **27**, 1–31 (2020)
25. Buslon, J., Alieto, E., Pahulaya, V., Reyes, A.: Gender divide in attitude towards Chavacano and cognition towards mother tongue among prospective language teachers. Asian EFL **27**, 41–64 (2020)
26. dela Rama, J.M., Sabasales, M., Antonio, A., Ricohermoso, C., Torres, J., Devanadera, A., Tulio, C., Alieto, E.: Virtual teaching as the 'new norm': analyzing science teachers' attitude toward online teaching, technological competence and access. Int. J. Adv. Sci. Technol. **29**, 12705–12715
27. Creswell, J.: Educational research: planning, conducting, and evaluating quantitative and qualitative research, 4th edn. Pearson Education, Boston, MA (2012)
28. Carretero, A.L., de la Rosa, J., Sanchez-Rodas, D.: Applying statistical tools systematically to determine industrial emission levels associated with the best available techniques. J. Clean. Prod. **112**, 4226–4236 (2016). https://doi.org/10.1016/j.jclepro.2015.06.037
29. Setia, M.: Methodology series module 3: Cross-sectional studies. Indian J. Dermatol. **61**, 261–264
30. Singh, Y.: Fundamental research methodology and statistics. New Age International, New Delhi (2006)
31. Torres, J.M.: Positioning Philippine English grammar and lexicon in four discourse quadrants. Asian EFL J. **22**, 253–276 (2019)
32. Alieto, E.: Cognition as predictor of willingness to teach in the Mother Tongue and the Mother Tongue as a subject among prospective language teachers. Sci. Int. (Lahore) **31**, 135–139 (2019)
33. George, D., Maller, P.: SPSS Windows step by step: a simple guide and reference, 11.0 (2nd Edition), Allyn & Bacon, Boston (2003)
34. Hair, J.F., Black, W.C., Babin, B.J., Anderson, R.E.: Multivariate data analysis seventh Edition, Pearson Education Upper Saddle River, New Jersey (2010)
35. Horton-Ramos, M.: Reading in the digitized era: analyzing ESL graduate students' e- reading habit. Asian EFL **27**, 67–85 (2020)
36. McKenna, M.C., Conradi, K., Lawrence, C., Jang, B.G., Meyer, J.P.: Reading attitudes of middle students: results of a US survey. Read. Res. Q. **47**, 283–306 (2012)
37. Abidin, M.J.B.Z., Pourmohammadi, M., Varasingam, N., Lean, O.C.: The online reading habits of Malaysian students. Read. Matrix Int. Online J. **14**, 164–172 (2014)
38. Fernández-Márquez, E., Vázquez-Cano, E., López Meneses, E.: Los mapas conceptuales multimedia en la educación universitaria: recursos para el aprendizaje significativo. Campus Virtuales **5**, 10–18 (2016)
39. Jacinto, M.J., Alieto, E.: Virtual teaching attitude and technological competence among English as second language (ESL) teachers: implications for the management of learning. Asian EFL J. **27**, 403–433 (2020)
40. Javier, C.: The shift towards new teaching modality: examining the attitude and technological competence among language teachers teaching Filipino. Asian ESP J. **16**, 210–244 (2020)
41. Farhaoui, Y.: Teaching computer sciences in Morocco: an overview. IT Profess. **19**(4), 12–15, 8012307 (2017). https://doi.org/10.1109/MITP.2017.3051325

Understanding and Designing Turing Machines with Applications to Computing

Serafeim A. Triantafyllou(✉)

Greek Ministry of Education and Religious Affairs, Athens, Greece
ser.triant@gmail.com

Abstract. Alan Turing was a pioneer in the computability theory. This theory has its origins from the fields of mathematical logic, theory of computation and computer science. In 1936, Turing introduced the concept of an automatic machine called since then as "Turing Machine". This concept set strong foundations by helping to define computable functions and algorithms. A Turing machine is a mathematical model of computation. This model describes an abstract machine that handles symbols on a computer tape according to a set of rules. With the help of Turing machines, it is achievable by researchers to design and implement any computer algorithm. This paper tries to contribute to a better understanding of Turing machines and their utilization in Algorithms Theory.

Keywords: Turing Machines · Algorithms · Mathematical models of computation

1 Introduction

Nowadays, the computing machines seem to have a complicated structure, but they remain basic machines that rely on the basic principles of the "*Turing Machines*". According to these principles, it is considered that every algorithm can be implemented with a proper type of a Turing machine [1–13]. Turing machine is a machine (automaton) capable of enumerating some arbitrary subset of valid strings of an alphabet. These strings are part of a recursively enumerable set. A Turing machine has a tape of infinite length on which it can perform read and write operations [1, 2, 14–17]. Supposing that we have a computer (human or machine), the following problem is set: With the help of a detailed process or program to solve a function f for a given value x of its variables. This problem is shown in Fig. 1 (where P is the so-called process or program). In addition, we consider that the computer will write its basic calculations to a storage medium (for example, a piece of paper that is divided into squares). In each square, the computer can place precisely one element of the alphabet. Although for many algorithms the two-dimensional storage medium of paper is used, it is more efficient to use a computer tape that is a one-dimensional storage medium [18–25]. The computer tape is divided into squares, and it is considered that the tape is unlimited and oriented from left to right for the two ends (see Fig. 2).

No academic titles or descriptions of academic positions should be included in the addresses.

Y. Farhaoui et al. (Eds.): ICAISE 2023, LNNS 837, pp. 142–148, 2024.
https://doi.org/10.1007/978-3-031-48465-0_19

Fig. 1. P process or program

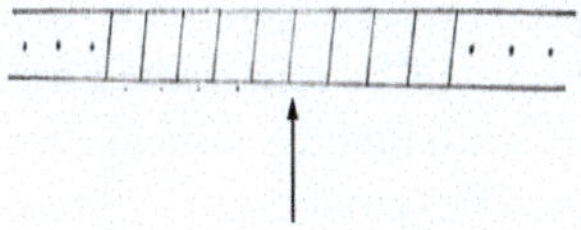

Fig. 2. The computer tape

We hypothesize the alphabet *S = {a1, a2, ..., an}* and the null sign *a0* to be represented by *. Approximately all the fields (except for a finite set of values) of the computer tape, contain the * symbol (that means they are empty-null). One field of the tape in every time moment is *active* field. It is accepted that the active field is found on a (*reading and writing*) *tape head*, where each moment we can write or read from the active field one symbol of the alphabet S U {*} (see in Fig. 1 the field with the arrow). In every time moment, the storage content of the computer tape is so called as *writing process* of the tape [1, 2, 14–18].

2 Computational Steps

There are four basic computational steps that we can follow (see Table 1): a_k: *write process* in the *active* field of the symbol a_k (the pre-existing symbol is deleted), *r*: transition of the active field to one position to the right, ℓ: transition of the *active* field to one position to the left, *s*: Stop. The machine stops (Stop) when data is written upon the computer tape and that means that the calculation is over. After every computational step the computer tape gets into a new state. Therefore, the initial state of the computer tape is called as *initial state*, and the final state as *final state*. Every state for practical reasons is divided into three basic components [19–25]: (i) The program, (ii) The content of the computer tape, (iii) The active field in every time moment. The above-mentioned steps will be called in this study as *basic computational steps* and will be symbolized as $U \in \{\alpha_\kappa, \ell, r, s\}$. To better understand the concept of program we divide the initial program to *basic subprograms*. A basic subprogram is every ordered triple in the form of α_i, U_i, k_i where: $\alpha_i \in \{\alpha_0, \alpha_1, \ldots, \alpha_\nu\}$, $U_i \in \{\alpha_\kappa, \ell, r, s\}$, k_i = the number of each subprogram. A typical subprogram executes in a simple way: if the active field contains the α_i execute the step U_i and go to the subprogram k_i. A matrix in the following form is called as *subprogram*.

$$\begin{bmatrix} \kappa & \alpha 0 & U0 & \kappa 0 \\ \kappa & \alpha 1 & U1 & \kappa 1 \\ \ldots & \ldots & \ldots & \ldots \\ \kappa & \alpha \nu & U\nu & \kappa \nu \end{bmatrix}$$

A successive number of m of the above-mentioned programs is called as *program* or *Turing matrix* or *machine matrix*.

3 Turing Machines-Background

3.1 Conceptualizing Turing Machines

Given the alphabet $A = \{\alpha_0, \alpha_1, \ldots, \alpha_n\}$ where $n \geq 1$, and α_0 as the null sign, we consider the symbols r, ℓ, s $\notin$ A. Let $k_i \in N$ where k_i and i = 1(1) m are natural numbers. The sets {r, ℓ, s} and $\{k_1, k_2, \ldots, k_m\}$ will be represented as: V = { r, ℓ, s} and $K = \{ k_1, k_2, \ldots, k_m\}$. By numbering next, with integer numbers $z \in Z$, the fields of the computer tape, where the z field is to the left next to the $z + 1$ field we can represent the computer tape with the following figure (see Fig. 3):

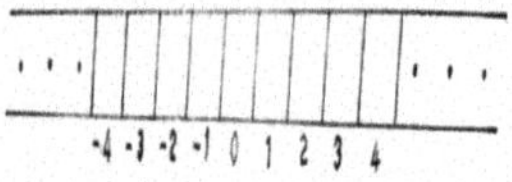

Fig. 3. The computer tape

Next, we present the following definitions:

Definition 1: Every function in the form: f: Z → A where f(z) ≠ α_0 for the finite set of z (for approximately every z ∈ Z, f(z) = α_0), is called as the writing function of the computer tape. The set {(z, f(z)), z ∈ Z} is called as the writing process of the computer tape and is represented by gr(f). The number z is the identifier number of the field and f(z) the storage content of the field z.

Definition 2: The functions: r: Z → Z where r(z) = z + 1, ℓ: Z → Z where l(z) = z – 1, s: Z → Z where s(z) = z will be called as shift operators.

Definition 3: Every ordered quartet b = (k, a, u, k') with k, k' ∈ K, a ∈ A and u ∈ A U V will be called as basic program or basic command. The k is called as a state and the k' is called as the next state.

Definition 4: The set b_i of the commands $b_{i,j}$, $b_i = \{ b_{i,j} = (k_i, a_j, u_{i,j}, k_{i,j}), j = 0(1)n\}$is called as subprogram or (full) command. A subprogram can be written as the following matrix:

$$\begin{bmatrix} ki & a0 & ui, 0 & ki, 0 \\ ki & a1 & ui, 1 & ki, 1 \\ ki & a2 & ui, 2 & ki, 2 \\ \ldots & \ldots & \ldots & \ldots \\ \ldots & \ldots & \ldots & \ldots \\ ki & an & ui, n & ki, n \end{bmatrix}$$

Definition 5: The b subprograms where b_i, i = 1(1)m is called as a Turing machine of the *alphabet A* and is represented as *M(A).*

$$M(A) = \begin{bmatrix} k1 & a0 & u1,0 & k1,0 \\ \ldots\ldots & \ldots & \ldots \\ k1 & an & u1,n & k1,n \\ k2 & a0 & u2,0 & k2,0 \\ k2 & an & u2,n & k2,n \\ \ldots\ldots & \ldots & \ldots \\ \ldots\ldots & \ldots & \ldots \\ km & a0 & um,0 & km,0 \\ \ldots\ldots & \ldots & \ldots \\ km & an & um,n & km,n \end{bmatrix}$$

Also, we have the *following symbolisms*: *A: external alphabet*, *K: set of states* A U V: *internal alphabet*. In the matrix M(A) for every element $(k_i, a_j) \in K \times A$ there is a precise command $(k_i, \alpha_j, u_{ij}, k_{ij})$ that starts with k_i a_j. In the Turing program k_1 is the first state that appears in the matrix M(A) and is called as *initial state* K_M of the M(A) matrix. If to a basic command $(k_i, \alpha_j, u_{ij}, k_{ij})$ let $u_{ij} = s$, then the k_i is called as the *final state.*

3.2 Result After Running a Command

For a basic command b = (k, α, u, k') we have the following result: If u = {s, ℓ, r} then the tape head transports to no position, or to one position to the left or to one position to the right. If u = {α_i} then the α can be replaced with α_i.

Definition 6: For each ordered triple (v, f, k) where v: the identifying number of a field of the computer tape, f: a writing function of the computer tape, k: a state, will be called as a cluster. A cluster *(v, f, k)* with $k = k_M$ will be called as the initial cluster. For every cluster *(v, f, k)* there is one precise command of M(A) that starts with k, f(v). This basic command is called as *basic cluster command.* A cluster *(v, f, k)* is called as *final cluster*, if the basic command starts with k, f(v), s. In every cluster *(v, f, k)* that is not final, we place a next cluster *(v',f', k')* that is represented as *F (v, f, k) = (v', f', k'). If (k, f (v), u, ậ)* the command of the *(v, f, k)* cluster, then *u' = { v, u ∈ A} or u' = {v + 1, u = r}*, or *u' = { v – 1, u = ℓ}* and *f'(x) = { f (x), x ≠ v or u ∈ V}* or *f'(x) = { u, x = v and u ∉ V}* and k = ậ. The transition from a cluster to its next will be called as a *computational step.* The initial cluster will be called as 0 (zero) step, its next as 1(first) step etc. The Turing machine will stop after the *n* step and the *n*- cluster will be the *final cluster*. Let an M(A) Turing machine and v_0 the number of the field and f_0 the writing function. We choose as the initial cluster the $(v_0, f_0, k_M = k_0)$. This option process will be called as '*process of Turing Machine M to the writing function* f_0 *in the field* v_0' and will be represented as $(v_0, f_0) \Rightarrow (v_0, f_0, k_M)$. Let (v_i, f_i, k_i), i = 1, 2, … are the next clusters of the (v_0, f_0, k_M). If *n* is the step and the (v_n, f_n, k_n) is the final cluster, then we state that the Turing Machine M stops after the *n* step in the position v_n that contains the string $f_n(v_n)$. If $f_n(v_n) \neq \alpha_0$ then there is a max word w (without spaces) upon the computer tape that contains the string f_n (v_n). In this case, the machine M stops to the word w.

4 Basic Concepts of the Theory of Algorithms with the Help of Turing Machines

At first, the Turing Machine stops exactly after the word w, if from the left of the tape head we find the word w and to the right there are only spaces, while the left next position to the tape head contains element $\alpha_i \neq \alpha_0$ (see Fig. 4).

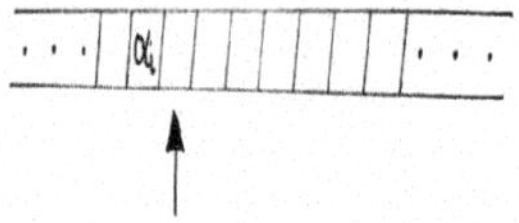

Fig. 4. Computer tape instance

Next, we consider the concept of algorithm to be equivalent to a Turing Machine and we go a step further to give precise definitions. Therefore, we consider the alphabet $A_0 = \{\alpha_1, \alpha_2, \ldots, \alpha_n\}$ and we choose only words $w \in A^*$ with $w \neq$ *empty word.*

Definition 7: Let the function $f: A0^* \to A0^*$. The function is computable by *Turing* $\Leftrightarrow \ni M(A)$ with $A_0 \subset A$. If we place to the empty computer tape one word $w \in A0^*$, and place next the machine to one field of the tape, then the machine M stops after a finite set of steps exactly after the word w' = f(w). The above-mentioned definition is the same for functions of k-positions: f: $(A_0{}^*)^k \to A_0{}^*$ where $w = w_1\alpha_0 w_2\alpha_0 \ldots \alpha_0 w_k$ with a_0 to be the null sign and $w_i \in A_0{}^*$, i = 1(1)k.

5 Examples of Turing Machines

Let the alphabet $A = \{a_0, a_1, \ldots, a_n\}$, and a_0 the null sign. The following paradigms of Turing programs are presented in detail to contribute to a better understanding of Turing Machines.

Paradigm 1: The right machine: (name) r: r: =

$$\begin{bmatrix} 0 & a0 & r & 1 \\ \cdots & \cdots & \cdots & \cdots \\ 0 & an & r & 1 \\ 1 & a0 & s & 1 \\ \cdots & \cdots & \cdots & \cdots \\ 1 & an & s & 1 \end{bmatrix}$$

If the right machine is set next to a random writing function and to a field *v*, it stops after a step to the next field *v + 1* without the initial write of the computer tape to be modified. Only the new active field of the computer tape is transposed to one position to the right.

Paradigm 2: The left machine: (name) ℓ:ℓ: $= \begin{bmatrix} 0 & a0 & \ell & 1 \\ \cdots & \cdots & \cdots & \cdots \\ 0 & an & \ell & 1 \\ 1 & a0 & s & 1 \\ \cdots & \cdots & \cdots & \cdots \\ 1 & an & s & 1 \end{bmatrix}$.

The functionality is the same with the r machine. The only difference is that the new active field of the computer tape is transposed to one position to the left.

6 Conclusions and Future Work

Turing machines in mathematics, are models that simulate a machine that mechanically operates on a tape. On this tape there are symbols, which the machine can read and write, one at a time, using a tape head. The mathematical description of a Turing machine set the foundations to prove properties of computation in general—and specifically, the uncomputability of the Entscheidungsproblem that is a decision problem. Decision problems are problems that can be expressed in yes/no form of the input values. The field of computability theory classifies the undecidable decision problems by Turing degree. Turing degree measures the noncomputability and unsolvability. In computer science and mathematical logic, it is a set of natural numbers that measures the level of algorithmic unsolvability of the set. Future studies aim to study in depth the Turing completeness, that means a computational system that can compute every Turing-computable function.

References

1. Eilenberg, S.: Automata, Languages and Machines. Academic Press. New York (1973)
2. Minsky, M: Computation: Finite and Infinite Machines. Prentice-Hall (1967)
3. Triantafyllou, S.A.: TPACK and Toondoo digital storytelling tool transform teaching and learning. In: Florez, H., Gomez, H. (eds.) Applied Informatics. ICAI 2022. Communications in Computer and Information Science, vol. 1643. Springer, Cham (2022). https://doi.org/10.1007/978-3-031-19647-8_24
4. Triantafyllou, S.A.: A quantitative research about MOOCs and EdTech tools for distance learning. In: Auer, M.E., El-Seoud, S.A., Karam, O.H. (eds.) Artificial Intelligence and Online Engineering. REV 2022. Lecture Notes in Networks and Systems, vol. 524. Springer, Cham (2023). https://doi.org/10.1007/978-3-031-17091-1_52
5. Triantafyllou, S.A.: Work in progress: educational technology and knowledge tracing models. In: 2022 IEEE World Engineering Education Conference (EDUNINE) (2022). https://doi.org/10.1109/edunine53672.2022.9782335
6. Triantafyllou, S.A.: Use of business information systems to achieve competitiveness. In: 2022 13th National Conference with International Participation (ELECTRONICA) (pp. 1–4). IEEE (2022). https://doi.org/10.1109/ELECTRONICA55578.2022.9874433
7. Triantafyllou, S.A., Sapounidis, T.: Game-based learning approach and serious games to learn while you play. 2023 IEEE World Engineering Education Conference (EDUNINE) (2023). https://doi.org/10.1109/EDUNINE57531.2023.10102872
8. Triantafyllou, S.A.: Magic squares in order 4K+2. In: 2022 30th National Conference with International Participation (TELECOM) (2022). https://doi.org/10.1109/TELECOM56127.2022.10017312

9. Triantafyllou, S.A., Georgiadis, C.K.: Gamification Design Patterns for user engagement. Inf. Educ., pp. 655–674 (2022). Available at: https://doi.org/10.15388/infedu.2022.27
10. Triantafyllou, S.A.: Game-based learning and interactive educational games for learners—an educational paradigm from Greece. In: Proceedings of The 6th International Conference on Modern Research in Social Sciences (2022). https://doi.org/10.33422/6th.icmrss.2022.10.20
11. Triantafyllou, S.A.: A detailed study on the 8 Queens problem based on algorithmic approaches implemented in PASCAL Programming Language. In: Silhavy, R., Silhavy, P. (eds.) Software Engineering Research in System Science. CSOC 2023. Lecture Notes in Networks and Systems, vol 722. Springer, Cham (2023). https://doi.org/10.1007/978-3-031-35311-6_18
12. Triantafyllou, S., Georgiadis, C.: Gamification of MOOCs and security awareness in corporate training. In: Proceedings of the 14th International Conference on Computer Supported Education (2022). Available at: https://doi.org/10.5220/0011103000003182
13. Barker-Plummer, D.: Turing Machines (2004)
14. Ackermann, W.: Zum Hilbertschen Aufbau der reellen Zahlen. Math. Annalen **99**, 118–133 (1928)
15. Knuth, E.D.: The Art of Computer Programming. Addison-Wesley. Publ. Co. (1969)
16. Korfhage, R.: Logic and Algorithms. John Wiley. New York (1966)
17. Maurer, H.: Theoretische Grundlagen der Programmiersprachen. Bibl. Institut. Manheim (1969)
18. Turing, A.M.: On computable numbers, with an application to the Entscheidungsproblem. Proc. Lond. Math. Soc., s2–42(1), 230–265 (1937). https://doi.org/10.1112/plms/s2-42.1.230
19. Whitehead, A.N., Russel, B.: Principia Mathematica. Cambridge University Press, London, vol. 1 (1910), vol. 2 (1912), vol. 3 (1913)
20. Hilbert, D.: Mathematical problems. Bull. Am. Math. Soc. **8**(437–445), 478–479 (1901)
21. Hilbert, D.: Uber das Unendliche. Math. Annalen **95**, 161–190 (1926)
22. Li, Y.: Some results of fuzzy Turing machines. In: 2006 6th World Congress on Intelligent Control and Automation (2006). https://doi.org/10.1109/wcica.2006.1713000
23. Vinayagam, G.S., Ezhilarasu, P., Prakash, J.: Applications of Turing machine as a string reverser for the three input characters—a review. In: 2016 10th International Conference on Intelligent Systems and Control (ISCO) (2016). https://doi.org/10.1109/isco.2016.7726890
24. Mallik, A., Khetarpal, A.: Turing machine based syllable splitter. In: 2021 Fourth International Conference on Computational Intelligence and Communication Technologies (CCICT) (2021). https://doi.org/10.1109/ccict53244.2021.00028
25. Gontumukkala, S.S., Godavarthi, Y.S., Gonugunta, B.R., M., S.: Implementation of tic tac toe game using multi-tape Turing machine. In: 2022 International Conference on Computational Intelligence and Sustainable Engineering Solutions (CISES) (2022). https://doi.org/10.1109/cises54857.2022.9844404

How Can Cloud BI Contribute to the Development of the Economy of SMEs? Morocco as Model

Najia Khouibiri and Yousef Farhaoui(✉)

STI Laboratory, T-IDMS, Department of Computer Science, Faculty of Sciences and Techniques Errachidia, Moulay Ismail University of Meknes, Meknes, Morocco
n.khouibiri@edu.umi.ac.ma, y.farhaoui@fste.umi.ac.ma

Abstract. Since business intelligence has played a crucial role in enhancing the competitiveness of organizations, providing insights that guide decision-making and drive business growth, and cloud technology has improved collaboration and data sharing, in this research paper we propose and emphasize the need for providing Cloud BI as a modern technology that can contribute to the growth of the economy of small and medium-sized enterprises in Morocco. We also discussed the extent to which the success of these enterprises is linked to the success of the Moroccan economy, and the importance of the relevant authorities paying special attention to these technologies, rather than relying solely on financial support, and that technological integration can help enhance the competitiveness of these companies. Finally, we concluded our research exploration by proposing a framework that adopts the migration of BI to the cloud as a special case of Cloud BI.

Keywords: Cloud Computing · Business Intelligence (BI) · Small and Medium-Sized Enterprises (SMEs) · Morocco · Migration · Homomorphic Encryption

1 Introduction

Over the last five years, the use of cloud Business Intelligence (BI) has been highly skilled at the level of economic growth of most organisations. In 2018, the level off adoption of cloud BI doubled to twice that of 2016, as the use of cloud BI is one of the key factors for success in business intelligence. (About 66% of leading enterprises in business intelligence are really cloud-based) [1], which was predicted by a global study years ago It was conducted by IBM in 2011, which stated that it is expected that more than 60% of companies will accelerate to embrace cloud computing in order to improve their competitive position in the job market. So If small and medium-sized enterprises form the backbone of any economy and the success of the latter is dependent on their success, then the success and prosperity of these companies will also only be achieved by adopting Cloud BI. However, most developing countries, such as Morocco, still lack the courage to fully finance cloud-based BI technology and focus on making it an integral and inseparable part of the information and communication technology sector,

Y. Farhaoui et al. (Eds.): ICAISE 2023, LNNS 837, pp. 149–159, 2024.
https://doi.org/10.1007/978-3-031-48465-0_20

and integrate it into all sectors (education, agriculture, tourism, economy, etc.). During 2020 the Royal Monarch of Morocco gave the launch of a program to support micro and medium enterprises "INTILAKA" this shows that Morocco recognizes the important role played by small enterprises in the country's economic growth. But this realization still requires technological integration based on modern technologies such as Cloud BI. Especially because this latter, is the combination of the term business intelligence that is very popular with organizations and refers to the use of a range of technologies to understand business data by officials of organizations with the aim of making smart business decisions and The term cloud computing as a platform for providing online computer services.so the cloud Business Intelligence offers a promising alternative by providing middle enterprises with access to powerful analytical tools and data storage capabilities.

In this paper, we take Morocco as a model for developing countries and propose the need to adopt Cloud BI for small and medium-sized enterprises with government support for their development, the latter of which seems to have been hit hard by the repercussions of the COVID crisis and the Ukrainian crisis [2, 3]. In Sect. 2, we discussed the concept of cloud computing and services, as well as the advantages provided by this technology. We also defined the concept of business intelligence and discussed its structure and steps. We concluded this section by presenting the concept of cloud BI and its benefits for small and medium-sized enterprises. In Sect. 3, we describe the relationship between cloud BI and Moroccan small and medium-sized enterprises, with a mention of these companies' concerns regarding the adoption of cloud BI. As for Sect. 4, it describes the proposed framework for transferring business information to the cloud with its main advantages taking into account the concerns mentioned earlier. Finally, we conclude in Sect. 5 with the expected conclusions and perspectives [4, 5].

2 State of the Art

2.1 Cloud Computing

Cloud computing is the provision and enablement of the computing power online It converts the software to a service in which clients do not pay for a license but for the amount they use; so that become computing power and storage space a merchandise, purchased as needed and expanded as required [6]. It is a cutting-edge technology that provides computing resources to organizations via the Internet [7].Or according to what Gartner that defined it "cloud computing" is a computing way that IT-related abilities are extensively evolved as a "service" utilizing Internet technologies to serve numerous external users [8].

2.1.1 Cloud Computing Services

Three types of services are offered by the cloud computing architecture [9], (Fig. 1):

- **Infrastructure as a service (IaaS):** In general, Iaas provides [9] everything related to infrastructure services (virtualisation, servers, storage and networking) while giving the customer the ability to change and manage his cloud infrastructure according to what he needs [9] among the providers of this service we find:Amazon EC2,Google Compute Engine GCE.

- **Platform as a service (PaaS):** In this service, the user it only manages the data and application resources, but the provider manages everything related to cloud infrastructure (like: virtualisation, servers, storage and networking) and (OS, middleware, runtime).As an example of some of the providers of this service we mention: Aws Elastic, Beans stalk, App Engine.
- **Software as a service (SaaS):** This service provides the customer with a ready-to-use application. The customer does not manage anything, only he has to pay the monthly or annual subscription fees that are determined according to the agreed contract. It is between the company or organization, the client and the service provider. Among the providers of this service, we mention:Office 365,Salesforce,Gmail,Google Drive, One Drive [10, 11].

2.1.2 Cloud Computing Features

The are several the features of cloud computing, we mention the following:

- **Resource pooling:** Cloud service providers can pool IT resources to serve multiple clients, with various physical and virtual resources dynamically assigned and re-allocated based on demand.
- **Quick Elasticity:** Cloud resources can be quickly increased or reduced to meet evolving demand, allowing users to pay only for the resources they need.
- **Wide network access:** Cloud services are accessible through the Internet from anywhere, using any device that has an Internet connection [12].
- **Deployment models**: There are four main deployment models in cloud: private cloud, public cloud, hybrid cloud, and multi-

In general, cloud computing offers organizations with a flexible, scalable and cost-effective means of accessing computing resources and services, without the need to invest in costly hardware and software.

2.2 Business Intelligence

Business intelligence (BI) is a combination of software tools and hardware solutions is a broad term that unites many technologies and operations It is both a method and a product that saves and analyses data to make decisions easier for customers (organisatons, academic associations…).The operation includes several ways that businesses employ to create their usable data, which may help them thrive and undoubtedly forecast the actions of their rivals [13, 14].

2.2.1 Business Intelligence Architecture

The three basic stratums (as the Fig. 3 shows) in the architecture of BI are [1, 6]:

1. **Data Layer:** This layer is in charge of gathering, managing, and storing raw data from various data sources (CRM,[1] ERP,[2] etc.). Typically, the data is kept in a data

[1] CRM: Customer Relationship Management.

[2] ERP: Enterprise Resource Planning.

warehouse or in its subset (data mart), that are often created (serve a specific department or business function) to give business users faster and more flexible access to data and also because it contains a subset of the data warehouse which is optimised for specific kinds of analysis [15, 16].
2. **Logic Layer:** This layer is in charge of turning unprocessed data into information that can be used for business purposes. To produce reports, dashboards and analytics. The data is cleaned, converted and aggregated.the data is accessed in this layer and analyzed by BI tools, which are programs offer a user-friendly interface so that business users may interact with the data and learn about the performance of the organisation.
3. **Presentation Layer:** layer is in charge of providing end users with business information. In order to assist users in making informed decisions, information is presented in reports, dashboards, charts and visualisations, which is displayed through the BI portal interface, this latter in turn offers a way for end-users to access and interact with the data.

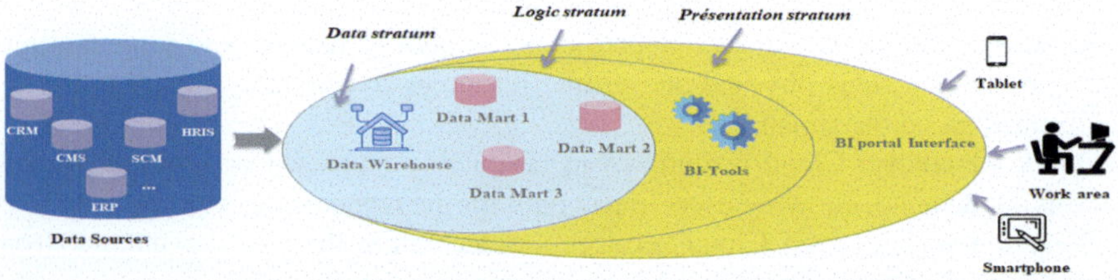

Fig. 1. Non-cloud-based BI system (the generalised system).

2.3 Cloud BI

2.3.1 Definition

Cloud BI is the current way to carry out BI. BI software is hosted in the Cloud rather than being installed and maintained as expensive and complex software on-premises. It may be accessed using any web browser. To utilize this service, there is no requirement to acquire hardware or install any software. Additionally, the system will automatically assign more resources as user computing needs grow. Customers just pay for what they use with Cloud BI, rather than needing to pay to continually prepare for high load utilization, thanks to this elasticity of scale. Since traditional BI is expensive and difficult to access, while BI on the Cloud changed the norm when the people became able to access easily on the BI and the latter become inexpensive [17].

2.3.2 Migration of BI to the Cloud

Migration of BI to the Cloud is the operation of transferring Business Intelligence (BI) applications and services from on-premises infrastructure to cloud-based infrastructure. This include Transferring data, applications, and other resources from local servers to cloud-based servers [18] which that can offers benefits like scalability, flexibility, and cost savings. Migrating from BI to the cloud may also involve adopting cloud-based BI tools and services, which can provide additional features and capabilities.

3 Morocco, SMEs and Cloud BI

3.1 Important Stats and Discussions

Small and medium-sized enterprises (SMEs) are the backbone of the country's economy and one of its main pillars. The health and economic well-being of citizens are closely linked to the performance of businesses. It seems that Morocco recognizes the important role played by SMEs, this prompted him to speed up the implementation of several programs, including the "Intilaka" program. This program aims to support these companies and provide them with loans so that they can overcome the difficulties caused by the COVID crisis [19, 20], which has literally destroyed the global economy and caused serious economic and social damage to most countries, including Morocco.

It should be noted that small and medium-sized enterprises (SMEs) are considered one of the main providers of employment, where these small projects account for 85.5%, employing nearly 73% of the workforce. To illustrate the role played by these institutions, we rely on statistics from 2018, where the study was conducted on a population of 208,919 institutions. It was found that 748,207 of them are small and medium-sized enterprises, representing 99.4% of the total population, with 85.5% being micro-enterprises and 8.1% being very small enterprises. Large companies represent only 0.6% of the total [21].

The figure below shows the distribution of institutions by business size (VSP,[3] SB,[4] MB,[5] LB,[6] Micro-business[7]) of the total number of ALE[8]:

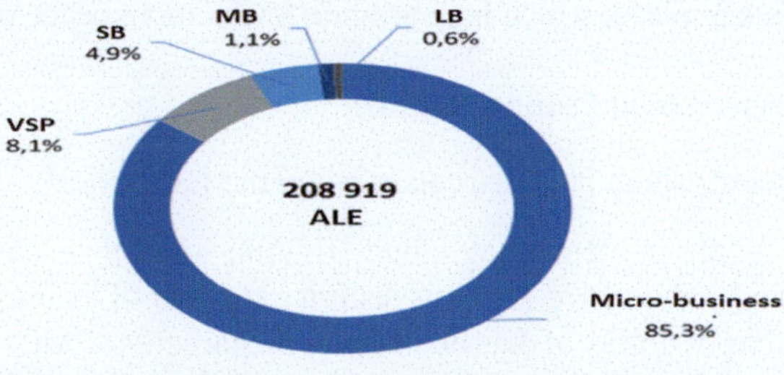

Fig. 2. The distribution of institutions by business size [21].

[3] VSP: Very Small Business.

[4] SB: Small Business.

[5] MB: Medium-sized Business.

[6] LB: Large Business.

[7] Micro-business: a very small company, usually with fewer than 10 employees and an annual turnover below a certain threshold.

[8] ALE: Active Legal Entities and it represents the total number of the total legal and active institutions.

This means that SMEs companies are the beating heart of the country's economy. And if we talk about finding effective solutions for the prosperity of these institutions and their economy, it will only be a technology solution, as it is one of the main factors that contribute to the development of the economies of countries. These small and medium enterprises are still able to compete, but they should adopt the latest technologiesto survive in time,otherwise they will be destined to failure [7], and of which Cloud BI is the most prominent. This latter which has helped many medium-sized companies [22] in other countries to develop their economy rapidly ((99% of all companies in North America and Europe) [6] because it relies on business intelligence as the main assistant and the basic technology in making critical decisions for the most prominent companies and cloud computing that helps to grant a set of privileges.Which should be with the support of the government and the concerned authorities, as well as with the help of regional telecom operators, as it is not enough just to make financing programs, but the auxiliary structures and technologies must be provided to implement this technology, which is Cloud BI [23, 24].

3.2 Benefits of Cloud BI in SMEs

BI cloud computing can offer a various of advantages for small and medium-sized enterprises (SMEs), we mention the following [25]:

1. **Cost savings:** By using cloud BI, SMEs are spared from purchasing pricey gear and software as well as the expenses related to maintaining and updating them. For SMEs, this can lead to considerable cost savings.
2. **Scalability:** Cloud BI enables SMEs to adjust the size of their BI infrastructure as needed without worrying about the constraints of on-premises infrastructure.
3. **Security:** To secure the data of their clients, cloud BI providers often have strong security measures in place, which may provide SMEs more peace of mind.

3.3 SMEs Concerns About Using Cloud BI

SMEs frequently worry about [13], we can mention the following:

1. **Data security:** while implementing cloud BI. And when their data is kept on the cloud, SMEs can be concerned about its security and fear of losing it.
2. **Data privacy:** the possibility of data breaches or illegal access.SMEs could also worry about how their information is utilized and whether it complies with applicable laws.
3. **Cost-effectiveness:** MEs must make sure that the advantages of implementing cloud BI exceed the expenses.

4 Proposed Framework

Our framework was inspired by a previous framework introduced in a study published in 2014 [26], which is a special case of cloud BI. This framework is based on the idea of migrating BI to the cloud. Our approach involves modifying the drawbacks of the previous framework and fully considering the concerns of the companies mentioned in this paper.

We include in our approach the factors that influence the decision to adopt Cloud BI in general and migrate BI to the cloud in particular. Our explanation depends on specific models and examples that show the possibility of the results of applying our proposed framework (Fig. 3):

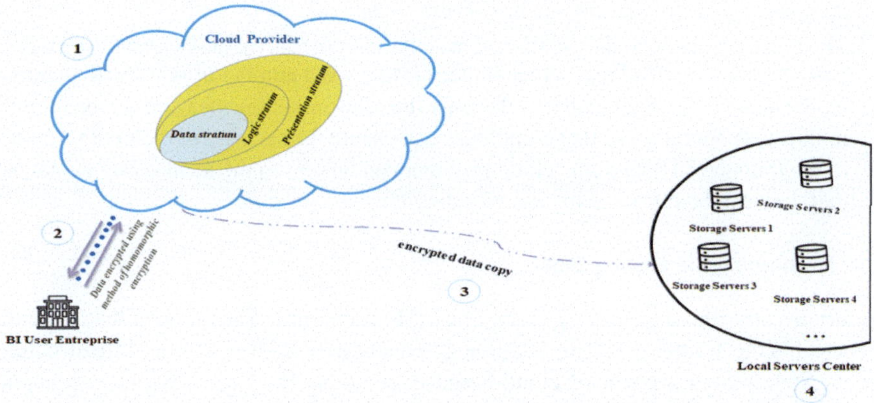

Fig. 3. Schematic architecture of proposed framework (cloud-based BI system).

After the BI user Enterprise transfers its data to the cloud, BI tools (logic stratum) work to operate the cloud environment and analyze the received and stored data at the data warehouse, so that it can be re-analyzed and easily accessed later through the presentation stratum (see Fig. 2). This presents the information in the form of reports and charts through a BI portal interface, which provides a means for end-users to access the data through various devices such as mobile phones, tablets, etc. Modern BI portal interfaces are often designed to be responsive and adaptable to different types of devices, providing users with a consistent experience across different devices.

(1) Our framework relies on a single provider instead of multiple providers because using multiple providers breaks several fundamental goals of integrating computing technology, which also raises concerns for organizations using these technologies, namely: - Cost: Using multiple providers means higher costs due to fees paid, depending on the type of service, so our approach respects this rule by relying on a single provider. - Data security and privacy: We do not deny that using multiple providers provides multiple backups of data and improves the ability to deal with failures and disasters, but this is considered a risk to privacy. Instead of data being exploitable by one provider, it will be exploitable by many, increasing the chances of errors and problems.

In a previous study from 2014, the issue of partial migration was raised based on migrating half of the data and retaining the other half, but this would also lead to increased costs and burden users with more than one infrastructure, especially if there are multiple providers. Instead of focusing on multiple infrastructures, the organization should only focus on a cloud infrastructure, which will force them to manage more than one infrastructure (local and other cloud infrastructures!!!).

(2) Our system supports secure data and cost reduction, which is a barrier for small and medium-sized companies, by using a single provider and fully migrating data, but with an emphasis on adopting the latest encryption technologies (Homomorphic Encryption), which we will discuss later, and the need to create a contract under the supervision of the relevant authorities between the provider and the beneficiary company to ensure that no one has access to the data.

(3) The proposed system also takes into account the possibility of data loss, so we propose the creation of a local server center **(4)** that has a number of servers (depending on the size of the organization's data) by the relevant authorities and regional communication operators so that companies can ensure the security of their data. Copies of the transferred data to the cloud provider can be sent to the local server center, so that they can be retrieved in case of a disaster.

4.1 Homomorphic Encryption

We briefly describe the security technology (Homomorphic Encryption) that is being proposed in our framework. This technology creates a high-level security infrastructure for the storage and transfer of cloud-based data bases. The Fig. 4 displays shematic architecture suitable with the suggested framework.

Two levels were used in the construction of the framework. The first is the level of base data service provider, which is located in a public, unreliable cloud. The second is the client level (Enterprise) that a proxy client deploys in the client environment. The client has a proxy that manages communication between the client apps and the cloud-based data base, which may be queried, when a client executes a request, the proxy converts it into a chiseled request that is executed directly in the cloud. When a query result is returned from the cloud, the proxy dechiffres it before sending it to the client. Le proxy is dependent on a metadata module that contains the underlying data models and the rounding and dechiffrement keys [2].

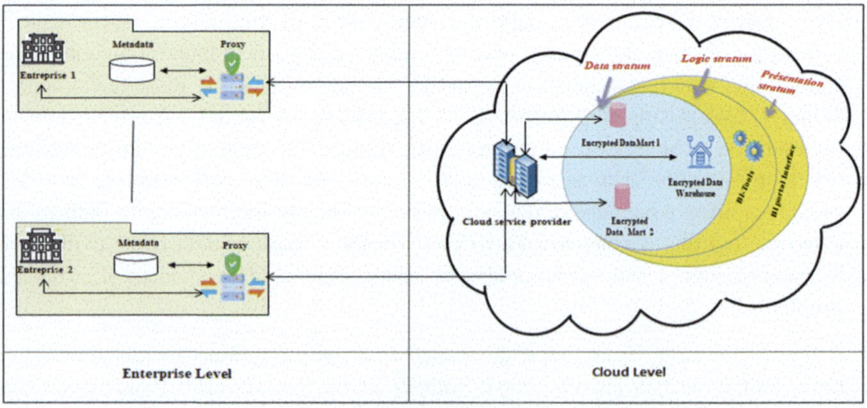

Fig. 4. Shematic architecture suitable with the proposed framework.

5 Conclusion and Perspectives

The paper aimed to provide cloud BI in general and to move BI to the cloud in particular. The study explored the extent of the relationship between SMEs and the national economy, and concluded that the success or failure of the Moroccan economy depends on the success or failure of small and medium-sized enterprises. The study proposed a planning framework for migrating BI to the cloud, in which we tried to take into account the concerns of SMEs. However, the proposed framework in this study cannot be guaranteed to actually overcome the concerns of companies, as concerns may increase or decrease depending on the real framework and the type of service that the company wants and that the provider offers. Therefore, it should be given special attention. It has become clear that financing alone is not sufficient to increase quality and improve the situation of these companies, it is necessary to help them integrate modern technologies such as Cloud BI, as it is one of the technologies that has helped many companies thrive.

Our study opens the door for contributions and other future prospects, such as making our framework more efficient and proposing it to several other developing countries, as well as generalizing the idea that these small companies are the center of economic success in other countries. Therefore, future research can be conducted on other countries and integrating them with other modern technological techniques beyond business intelligence and cloud computing. Regardless of the national character, the economic structure can be compared to that of other countries, and therefore the rule can be generalized.

References

1. Khan, S., Zhang, B., Khan, F., Chen, S.: Business intelligence in the cloud: a case of Pakistan. In: Dans 2011 IEEE International Conference on Cloud Computing and Intelligence Systems (pp. 536–540). IEEE, Beijing, China (2011). https://doi.org/10.1109/CCIS.2011.6045126
2. Farhaoui, Y.: Design and implementation of an intrusion prevention system. Int. J. Netw. Secur. **19**(5), 675–683 (2017). https://doi.org/10.6633/IJNS.201709.19(5).04
3. Farhaoui, Y., et al.: Big Data Min. Anal. **6**(3), I–II (2023). https://doi.org/10.26599/BDMA.2022.9020045
4. Farhaoui, Y.: Intrusion prevention system inspired immune systems. Indonesian J. Electr. Eng. Comput. Sci. **2**(1), 168–179 (2016)
5. Farhaoui, Y.: Big data analytics applied for control systems. Lect Notes Netw. Syst. **25**, 408–415 (2018). https://doi.org/10.1007/978-3-319-69137-4_36
6. Camargo-Perez, J.A., Puentes-Velasquez, A.M., et al.: Sanchez-Perilla, Integration of big data in small and medium organizations: Business intelligence and cloud computing. J. Phys.: Conf. Ser. **1388**(1), 012029 (2019). https://doi.org/10.1088/1742-6596/1388/1/012029
7. Kasem M., Hassanein, E.: Cloud business intelligence survey. Int. J. Comput. Appl. **90**(févr.) (2014). https://doi.org/10.5120/15540-4266
8. Kumar, V., Laghari, A.A., Karim, S., Shakir, M., Brohi, A.A.: Comparison of fog computing & cloud computing. Int. J. Math. Sci. Comput. **1**, 31–41 (2019)
9. Tole, A.A.: Cloud computing and business intelligence. Database Syst. J. 5(4) (2014)
10. Farhaoui, Y., et al.: Big Data Min. Anal. **5**(4), I II. https://doi.org/10.26599/BDMA.2022.9020004
11. Alaoui, S.S., Farhaoui, Y.: Hate speech detection using text mining and machine learning. Int. J. Decis. Support Syst. Technol. **14**(1), 80 (2022). https://doi.org/10.4018/IJDSST.286680

12. Dziembek, D. Ziora, L.: Cloud-based business intelligence solutions in the management of Polish Companies. In: Silaghi, G.C., Buchmann, R.A., Niculescu, V., Czibula, G., Barry, C., Lang, M., Linger, H., Schneider, C. (eds.) dans Advances in Information Systems Development, dans Lecture Notes in Information Systems and Organisation, vol. 63 (pp. 35–52). Springer International Publishing, Cham (2023). https://doi.org/10.1007/978-3-031-32418-5_3
13. El Ghalbzouri, H., El Bouhdidi, J.: Integrating business intelligence with cloud computing: state of the art and fundamental concepts. In: Ben Ahmed, M., Teodorescu, H.-N.L., Mazri, T., Subashini, P., Boudhir, A.A. (eds.) dans Networking, Intelligent Systems and Security, dans Smart Innovation, Systems and Technologies, (pp. 197–213). Springer, Singapore (2022). https://doi.org/10.1007/978-981-16-3637-0_14
14. Elmalah K., Nasr, M.: Cloud business intelligence. Int. J. Adv. Netw. Appl. **10**, 4120–4124 (2019). https://doi.org/10.35444/IJANA.2019.100612
15. Alaoui, S.S., Farhaoui, Y.: Data openness for efficient e-governance in the age of big data. Int. J. Cloud Comput. **10**(5–6), 522–532 (2021). https://doi.org/10.1504/IJCC.2021.120391
16. El Mouatasim, A., Farhaoui, Y.: Nesterov step reduced gradient algorithm for convex programming problems. Lect. Notes Netw. Syst. **81**, 140–148 (2020). https://doi.org/10.1007/978-3-030-23672-4_11
17. Elshibani, F.: Benefits of Using Cloud Business Intelligence to Improve Business Maturity - ProQuest
18. Kocaman, B., Gelper, S., Langerak, F.: Till the cloud do us part: technological disruption and brand retention in the enterprise software industry. Int. J. Res. Market. **40**(2), 316–341 (2023). https://doi.org/10.1016/j.ijresmar.2022.11.001
19. Tarik, A., Farhaoui, Y.: Recommender system for orientation student. Lecture Notes in Networks and Systems, **81**, 367–370 (2020). https://doi.org/10.1007/978-3-030-23672-4_27
20. Sossi Alaoui, S., Farhaoui, Y.: A comparative study of the four well-known classification algorithms in data mining. Lecture Notes Netw. Syst. **25**, 362–373 (2018). https://doi.org/10.1007/978-3-319-69137-4_32
21. Lemrajni, L.: Comportements des TPME marocaines, et mesures gouvernementales de soutien, face à la crise sanitaire. Int. J. Econ., Manag. Finance (IJEMF) 1(1), Art. no 1 (2022). https://doi.org/10.5281/zenodo.7536856
22. Owusu, A., Broni, F., Penu, O., Boateng, R.: Exploring the critical success factors for cloud BI adoption among Ghanaian SMEs (2020)
23. Farhaoui, Y.: Teaching computer sciences in Morocco: an overview. IT Prof. **19**(4), 12–15, 8012307 (2017). https://doi.org/10.1109/MITP.2017.3051325
24. Farhaoui, Y.: Securing a local area network by IDPS open source. Procedia Comput. Sci. **110**, 416–421 (2017). https://doi.org/10.1016/j.procs.2017.06.106
25. Mondal, S.: Cloud business intelligence as a solution for empowering SMEs. EPRA Int. J. Multi. Res. (IJMR) **8**(9), Art. no 9 (2022)
26. Yagoub, M.A., Laouid, A., Kazar, O., Bounceur, A., Euler, R., AlShaikh, M.: An adaptive and efficient fully homomorphic encryption technique. In: dans Proceedings of the 2nd International Conference on Future Networks and Distributed Systems, dans ICFNDS'18 (pp. 1–6). Association for Computing Machinery, New York, NY, USA (2018). https://doi.org/10.1145/3231053.3231088

27. Puthal, D., Sahoo, B.P.S., Mishra, S., Swain, S.: Cloud computing features, issues, and challenges: a Big picture. In: dans 2015 International Conference on Computational Intelligence and Networks (pp. 116–123). IEEE, Odisha, India (2015). https://doi.org/10.1109/CINE.2015.31
28. Juan-Verdejo, A., Surajbali, B., Baars, H., Kemper, H.-G.: Moving business intelligence to cloud environments. In: dans 2014 IEEE Conference on Computer Communications Workshops (INFOCOM WKSHPS), avr. 2014, p. 43–48. https://doi.org/10.1109/INFCOMW.2014.6849166

Deep Learning for Dynamic Content Adaptation: Enhancing User Engagement in E-commerce

Raouya El Youbi[1(✉)], Fayçal Messaoudi[2], and Manal Loukili[1]

[1] National School of Applied Sciences, Sidi Mohamed Ben Abdellah University, Fez, Morocco
{raouya.elyoubi,manal.loukili}@usmba.ac.ma

[2] National School of Business and Management, Sidi Mohamed Ben Abdellah University, Fez, Morocco
faycal.messaoudi@usmba.ac.ma

Abstract. In recent years, the landscape of online businesses has undergone a significant transformation. With the proliferation of e-commerce websites, providing a personalized user experience has become critical for success. This paper presents a study on deep learning for dynamic content adaptation in e-commerce, aimed at enhancing user engagement. The methodology involved collecting and preprocessing data from an e-commerce website, which included visitor behavior data. A recurrent neural network (RNN) with long short-term memory (LSTM) cells were chosen as the deep learning architecture. The model was trained and evaluated using various performance metrics, such as accuracy, precision, recall, F1-score, click-through rate (CTR), average session duration, and conversion rate. The results demonstrated that the deep learning model outperformed the baseline model in all evaluation metrics. The deep learning model achieved an accuracy of 92%, indicating its success in adapting content based on user behavior.

Keywords: Deep learning · Dynamic content adaptation · E-commerce · Neural networks

1 Introduction

In the realm of e-commerce, personalized user experiences have become pivotal for businesses seeking to engage customers and drive conversions [1]. With the vast amount of information available online, users expect tailored content that resonates with their preferences, needs, and behaviors [2]. Dynamic content adaptation, the process of dynamically customizing website content based on user profiles, has emerged as a powerful strategy to enhance user engagement in e-commerce [3]. By leveraging advanced technologies, such as deep learning, e-commerce websites can now provide highly personalized experiences that captivate users and foster long-term customer relationships [4].

Content adaptation in e-commerce refers to the process of customizing the presentation of products, recommendations, and user interfaces based on individual user characteristics and interactions [5].

Y. Farhaoui et al. (Eds.): ICAISE 2023, LNNS 837, pp. 160–165, 2024.
https://doi.org/10.1007/978-3-031-48465-0_21

Deep learning, a subfield of machine learning, has emerged as a powerful and promising approach for content adaptation in e-commerce [6]. It is a class of algorithms that leverage neural networks with multiple layers to learn hierarchical representations of data and capture intricate relationships [7]. Deep learning models, such as recurrent neural networks (RNNs) with long short-term memory (LSTM) cells, excel at processing sequential data and can effectively capture user behavior and preferences over time [8].

The primary motivation behind this study is to leverage the potential of deep learning algorithms, particularly recurrent neural networks (RNNs) with long short-term memory (LSTM) cells, to capture intricate user behavior patterns over time and adapt content dynamically and thus drive higher conversion rates and revenue generation.

The remainder of this paper is structured as follows: Sect. 2 provides the methodology, including data collection and preprocessing, the deep learning model architecture employed, outlining its components, and the model training and evaluation. Section 3 presents the experimental results and analysis. Section 4 concludes the paper. The references are provided at the end of the paper, listing the sources that have been cited throughout the paper.

2 Methodology

2.1 Data Collection and Preprocessing

The data used in this study was collected from an e-commerce website from June 2, 2015, to August 1, 2015. The dataset includes visitor behavior data, such as the items viewed, added to cart, and purchased. Initially, the data was preprocessed to remove any duplicate entries and handle missing values. The preprocessing steps also involved converting the timestamp data into a more readable format and creating new features to enrich the dataset. For example, additional features were created, such as the total number of items viewed and the total view count for each visitor. The final dataset comprised information on 1,407,580 unique visitors, out of which 11,719 visitors made a purchase during the study period.

The data preprocessing tasks were performed using Python and popular libraries such as pandas and NumPy. Duplicate entries were identified and removed, and missing values were handled appropriately. The timestamp data was transformed into a more human-readable format using the datetime module. To augment the da-taset, new features were engineered, such as aggregating the total number of items viewed and computing the total view count for each visitor.

2.2 Deep Learning Model for Dynamic Content Adaptation

For this study, a recurrent neural network (RNN) with long short-term memory (LSTM) cells were chosen as the deep learning architecture. The LSTM architecture was implemented using the Keras library in Python. The model consisted of an input layer, embedding, LSTM layers, a fully connected dense layer, and an output layer. The LSTM layers help capture the sequential nature of user behavior and adapt content accordingly (Fig. 1).

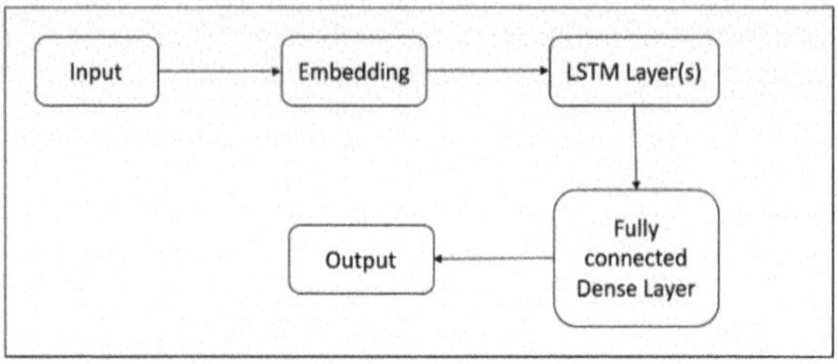

Fig. 1. The deep learning model architecture used for dynamic content adaptation.

- Input Layer:
  - Input shape: Sequential data representing user behavior.
  - No parameters to tune
- Embedding Layer:
  - Embedding size: 100
- LSTM Layer(s):
  - Number of layers: 2 LSTM layers
  - LSTM cell units: 128 units in each LSTM layer.
  - Activation function: Default activation (tanh) in LSTM cells.
  - Dropout: Applied between LSTM layers for regularization, with a dropout rate of 0.2.
- Fully-Connected Dense Layer:
  - No. of units: 64 units in the dense layer.
  - Activation function: ReLU (Rectified Linear Unit).
- Output Layer:
  - Output units: 1.
  - Activation function: Sigmoid.

2.3 Model Training

The preprocessed dataset was split into training, validation, and testing sets, with a split ratio of 70%, 15%, and 15% respectively. The deep learning model was initialized with the defined architecture, including the input layer, embedding layer, LSTM layer(s), fully connected dense layer, and output layer. The model was configured with a categorical cross-entropy loss function, the Adam optimizer, and accuracy as the evaluation metric. Subsequently, the model was compiled with the specified configuration.

To train the model, the fit() function was used on the training data. The training process involved specifying the batch size, the number of epochs, and the validation data. The model's training progress was monitored by observing the loss and accuracy metrics on both the training and validation sets.

2.4 Model Evaluation

After training, the model was evaluated using the testing set. The evaluate() function was employed to calculate several evaluation metrics to assess the performance of the deep learning model for dynamic content adaptation. The following metrics were used:

- Accuracy measures the proportion of correctly predicted instances out of the total instances in the testing set. It provides an overall assessment of the model's predictive performance.
- Precision calculates the proportion of true positive predictions out of all positive predictions made by the model. It indicates the model's ability to accurately identify relevant content and minimize false positives.
- Recall, also known as sensitivity or true positive rate, calculates the proportion of true positive predictions out of all actual positive instances in the dataset. It measures the model's ability to capture relevant content and minimize false negatives.
- F1-score is the harmonic mean of precision and recall. It provides a balanced measure of the model's performance, considering both precision and recall. The F1-score is particularly useful when there is an imbalance between positive and negative instances in the dataset.
- Click-through Rate (CTR) measures the proportion of users who clicked on the recommended items out of the total number of recommendations displayed. It reflects the effectiveness of the personalized content recommendations in capturing user interest and generating user engagement.
- Average Session Duration calculates the average length of time users spend on the e-commerce website per session. It is an indicator of user engagement and satisfaction with the recommended content.
- Conversion Rate measures the proportion of users who made a purchase out of the total number of users who interacted with the recommended content. It quantifies the impact of personalized content adaptation on driving actual sales and revenue generation.

3 Results and Discussion

See Table 1.

The deep learning model exhibited superior performance across all evaluation metrics compared to the baseline model. The accuracy of the deep learning model reached an impressive 92%, outperforming the baseline model's accuracy of 82%. This indicates that the deep learning model successfully adapted the content based on user behavior and preferences, resulting in more accurate predictions.

The precision and recall scores of the deep learning model were notably higher, at 91% and 93% respectively, compared to the baseline model's precision of 78% and

Table 1. Summarizes the performance metrics of the baseline model and the deep learning model.

Metric	Baseline model	Deep learning model
Accuracy	0.82	0.92
Precision	0.78	0.91
Recall	0.85	0.93
F1-score	0.81	0.92
Click-through Rate (CTR)	0.23	0.38
Average Session Duration	213 s	312 s
Conversion Rate	0.12	0.26

recall of 85%. This implies that the deep learning model effectively identified and recommended relevant content to users, resulting in a more personalized and engaging user experience. The higher precision indicates a reduced number of irrelevant recommendations, while the higher recall signifies a greater ability to capture relevant items.

Furthermore, the F1-score, which combines precision and recall, demonstrated a substantial improvement with the deep learning model achieving a score of 92% compared to the baseline model's score of 81%. This indicates a higher balance between precision and recall, highlighting the model's ability to effectively adapt content while minimizing false positives and false negatives.

The deep learning model also significantly increased the click-through rate by 15 percentage points, from 23% in the baseline model to 38%. This suggests that the personalized content recommendations generated by the deep learning model captured users' interests more accurately, resulting in a higher likelihood of user engagement and interactions. The higher CTR indicates that users were more inclined to click on recommended items, indicating increased interest and engagement.

Moreover, the average session duration increased from 213 s in the baseline model to 312 s in the deep learning model. This longer engagement duration implies that users found the personalized content more relevant and engaging, leading to increased exploration and interaction within the e-commerce website. The extended session duration indicates improved user satisfaction and a higher level of interest in the recommended content.

Additionally, the deep learning model achieved a conversion rate of 26%, a substantial improvement over the baseline model's conversion rate of 12%. This demonstrates that the personalized content adaptation facilitated by the deep learning model positively influenced user purchase decisions, ultimately driving higher conversion rates and revenue generation for the e-commerce website. The improved conversion rate indicates that the deep learning model effectively guided users towards making purchase decisions, resulting in increased revenue and business success.

4 Conclusion

In conclusion, this study examined the use of deep learning techniques for dynamic content adaptation in e-commerce websites with the goal of improving personalized user experiences. The deep learning model, trained on real-world e-commerce data, outperformed the baseline model across various metrics. It achieved a remarkable accuracy of 92%, indicating its precise adaptation to user behavior and preferences. The model demonstrated higher precision and recall scores, highlighting its effectiveness in identifying and recommending relevant content. The F1-score showed a substantial improvement, striking a better balance between precision and recall. Moreover, the model increased the click-through rate by 15 percentage points, leading to improved user engagement and satisfaction. It also achieved a conversion rate of 26%, positively influencing user purchase decisions and generating higher revenue for the e-commerce website.

This study has implications for online businesses seeking to enhance user experiences and drive conversions by leveraging deep learning techniques for content adaptation. Future research should explore advanced architectures, incorporate additional contextual information, and investigate the generalizability of the model across different e-commerce domains and datasets, providing valuable insights for broader applications.

References

1. Akter, S., Wamba, S.F.: Big data analytics in E-commerce: a systematic review and agenda for future research. Electron. Mark. **26**, 173–194 (2016)
2. Loukili, M., Messaoudi, F., El Ghazi, M.: Sentiment analysis of product reviews for E-commerce recommendation based on machine learning. Int. J. Adv. Soft Comput. Its Appl. **15**(1), 1–13 (2023)
3. Pereira, A.M., et al.: Customer models for artificial intelligence-based decision support in fashion online retail supply chains. Decis. Support. Syst. **158**, 113795 (2022)
4. Semerádová, T., Weinlich, P.: Website Quality and Shopping Behavior: Quantitative and Qualitative Evidence. Springer Nature (2020)
5. Loukili, M., Messaoudi, F.: Machine learning, deep neural network and natural language processing based recommendation system. In: Kacprzyk, J., Ezziyyani, M., Balas, V.E. (eds.) International Conference on Advanced Intelligent Systems for Sustainable Development. AI2SD 2022. Lecture Notes in Networks and Systems, vol. 637. Springer, Cham (2023). https://doi.org/10.1007/978-3-031-26384-2_7
6. Agarwal, B., Nayak, R., Mittal, N., Patnaik, S.: Deep Learning-Based Approaches for Sentiment Analysis (p. 4). Springer, Singapore (2020)
7. Li, B., Pi, D.: Learning deep neural networks for node classification. Expert Syst. Appl. **137**, 324–334 (2019)
8. Quadrana, M., Karatzoglou, A., Hidasi, B., Cremonesi, P.: Personalizing session-based recommendations with hierarchical recurrent neural networks. In: Proceedings of the Eleventh ACM Conference on Recommender Systems, pp. 130–137 (2017)

PID Versus Fuzzy Logic Controller Speed Control Comparison of DC Motor Using QUANSER QNET 2.0

Megrini Meriem(✉), Gaga Ahmed, and Mehdaoui Youness

Research Laboratory of Physics and Engineers Sciences (LRPSI), Research Team in Embedded Systems, Engineering, Automation, Signal, Telecommunications and Intelligent Materials (ISASTM), Polydisciplinary Faculty (FPBM), Sultan Moulay Slimane University (USMS), Beni Mellal, Morocco

megrinimeriem1@gmail.com

Abstract. This paper discusses the impact and significance of the PID controller and the fuzzy logic controller on the performance of a DC motor, particularly its speed regulation. On the one hand, the Proportional-Integral-Derivative (PID) controller and the fuzzy logic controller (FLC) are simulated in the MATLAB/SIMULINK environment. The simulation results, on the other hand, are validated in the experiment using LabVIEW software and a QUANSER QNET 2.0 DC motor. The LabVIEW software visualizes the system's response using virtual instruments and either stops or runs the process. A USB cable connects this software to the QUANSER QNET 2.0 DC motor. Although the PID controller is more widely used in industry, it still has some disadvantages, the most significant of which is that it cannot be more efficient with a non-linear and dynamic system. As a result, the fuzzy logic controller is presented in this work to be tested and compared to the PID controller. The fuzzy logic controller's inputs are error and change of error, and its output is armature voltage. A Mamdani engine system is used, with 7 membership functions for each input and output. The simulation and experiment results confirm that the fuzzy logic controller outperforms the PID controller in terms of stability and rapidity.

Keywords: Proportional-integral-derivative · Fuzzy logic · QNET 2.0 DC motor · LabVIEW software

1 Introduction

DC motor is a machine that converts electrical energy to mechanical energy [1]. It has a crucial role as an electromechanical energy converter in manufacturing operations, such as vehicles, material handling, and robotics [2]. To obtain its performance plenty of algorithms are developed the conventional one is the Proportional-Integral-Derivative (PID) controller, fuzzy logic, what's more, the controller uses artificial intelligence that made the system able to work automatically without human help as Artificial Neural Network (ANN) [3]. In this paper, both first controllers are used to control the speed of the QNET 2.0 DC motor.

Y. Farhaoui et al. (Eds.): ICAISE 2023, LNNS 837, pp. 166–171, 2024.
https://doi.org/10.1007/978-3-031-48465-0_22

QNET 2.0 DC motor is a board consisting of a direct drive motor, an encoder to measure angular position, an inertia disk, and a Pulse Width Modulation (PWM) power amplifier to drive the motor.

QNET 2.0 DC motor board necessitates the National Instrument Educational Laboratory Virtual Instrumentation Suite (NI ELVIS), as its name indicates is a virtual laboratory that has 12 common instruments used [4, 5].

LabVIEW is the software used to make the connection between the board NI ELVIS, QNET 2.0 DC motor and to do the control of this motor. Due to this software, the control is easier and more efficient where we can find the PID, and fuzzy logic blocks to do the test directly what's more, with this advanced software, the response will be shown in it more clearly [6]. The most important thing, this work focuses on some toolkits of QNET DC Motor and a driver named control and simulation where the blocks of control PID and Fuzzy logic could be used.

PID controller is the most popular controller in the domain of regulation but it has the problem of dynamic system's non-linearity [7] in contrast, Fuzzy logic avoids the non-linearity problem [8].

The fuzzy logic theory differs from the Boolean theory which is represented by 0 and 1. It can represent values from 0 to 1 that are fuzzy [9]. It uses the fuzzy membership functions to adjust the control signal. In this paper, a MISO system is treated; two inputs and one output are used. The fuzzy inputs have 7 memberships and the output also.

A fuzzy logic controller is a type of artificial intelligence [10] it has 3 parts, a fuzzifier as a digital-analog converter, which transforms the input signal to a fuzzy signal, a fuzzy inference engine, where the rules are mentioned to control the fuzzified signal, and the defuzzifier work as an analog-digital converter where the fuzzified signal transformed to a signal used to control the system, in this case, is the DC motor [11].

This paper presents the speed control of a DC motor using a Quanser QNET 2.0 DC motor. In the first, the closed loop was done by PID controller, it was simulated in MATLAB/SIMULINK environment and implemented using QNET 2.0 DC Motor hardware and the LabVIEW software to visualize the responses of the system and to control the motor. After that, the Fuzzy logic controller takes place of the PID controller in the closed loop of the system. This controller is either simulated in MATLAB/SIMULINK and, tested regarding the same software/ hardware. Finally, the simulation and experimental results are the same. Where, the fuzy logic controller gives performed results than the PID controller.

2 Materials and Methods

2.1 DC Motor Material

DC machine is an electric machine that has a bidirectional conversion of energy. When the conversion is from mechanical to electrical energy, the machine is a gener-ator. In contrast, when the conversion is from electrical to mechanical energy. The machine is a motor [1]. A DC motor will be presented in this work, where it is more used in industry, especially in robotics [12]. The QNET 2.0 DC motor can be approximated by a 1st order transfer function eliminating the insignificant pole. A bump test was done to identify the parameters; steady state gain K and time constant τ:

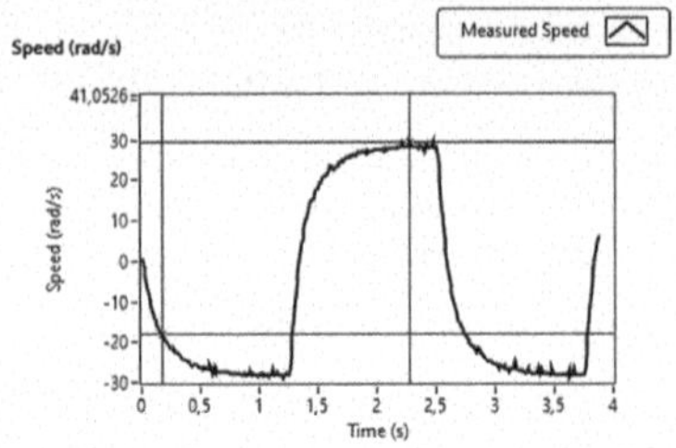

Fig. 1. Measurement graphs for identification of QNET DC motor model

From the figure (Fig. 1): K = 29 rad/v.s, and $\tau = 0.17$s.Where, the system model of this work is presented by this transfer function:

$$H(S) = \frac{29}{0.17S + 1} \tag{1}$$

2.2 PID Controller Method

PID controller is a robust and simple controller. It is mostly used in closed loops and it gives good responses by a simple method called the tunning method. As mentioned before the QNET 2.0 DC motor is set by the voltage thus the regulation of error between the desired speed and the measured acquired necessitates voltage that gives a good result.

2.3 Fuzzy Logic Controller Method

A fuzzy logic controller is inspired by a fuzzy set method presented by L. Zadeh in 1965 to present a flow values [13]. Membership functions presented by linguistic variables is the main idea of this method [13]. The fuzzy logic controller is widely used in the last center in industry for making processes more intelligent. It consists by 4 crucial blocks; fuzzification (convert crisp inputs to flow variables), defuzzification (convert flow variables to crisp output), between them there is inference engine that use Mamdani system, and the part of rule base that present the relationship between the inputs and the output. A MISO (Multiple Input Single Output) systems is treated in this work. Where; error and change of error are the inputs and the voltage applied to the DC motor is the output. Both inputs and output are presented by triangular membership functions based on Mamdani fuzzy inference system. The error, change error, and voltage has 7 membership functions. These membership functions generates 49 rules base that has the form:

IF(error (e) is membership function 1) and (change error (DE) is membership function 2) THEN (voltage (V) is membership function 3).

3 Simulation and Experimental Results

3.1 Simulation Results

The system's reaction to a speed of 100 rad/s using the proportional-integral-derivative controller is depicted in (Fig. 2), and fuzzy logic controller is dipected in (Fig. 3):

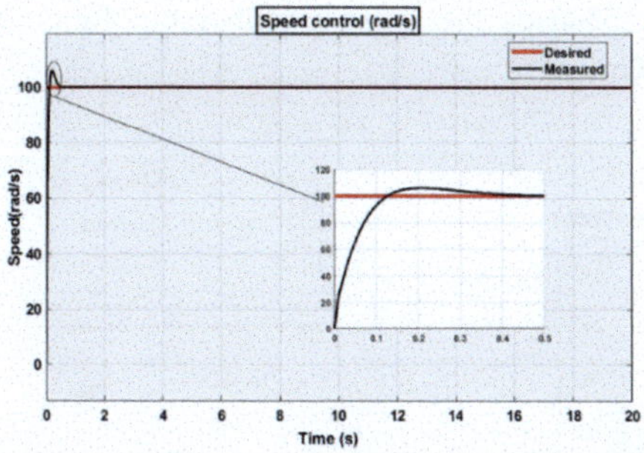

Fig. 2. SIMULINK response of speed control using the PID controller

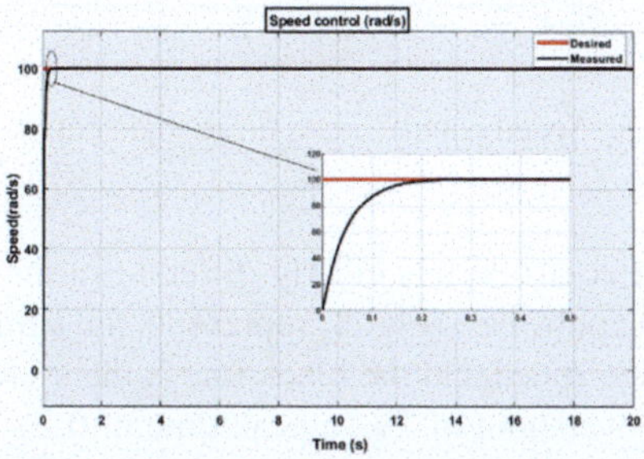

Fig. 3. SIMULINK response of speed control using fuzzy logic controller

3.2 Experimental Results

The result in (Fig. 4) uses the same values of Kp, Ki, Kd parameters founded by MATLAB/SIMULINK simulation.

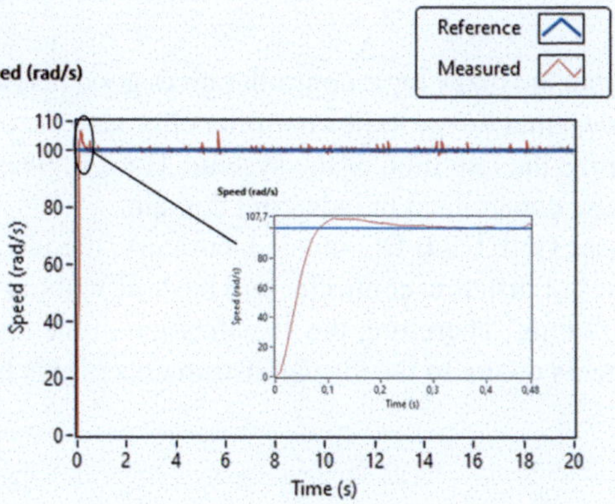

Fig. 4. Speed control using the PID controller.

The result in (Fig. 5) uses the same membership functions founded on MATLAB/SIMULINK simulation.

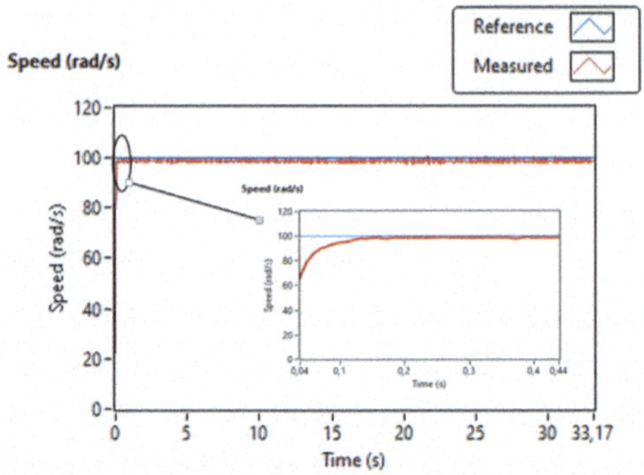

Fig. 5. Speed control using fuzzy logic controller

The response of both controllers is clearly different in the simulation and experimental tests. For the PID controller, (Fig. 2) represents the simulation response, where the red line is the reference speed and the black line is the system response, and as the figure shows, there is an overshoot of 7%, a peak time of 0.2 s, and a time response of 0.3 s, which is confirmed by the experiment part, as shown in (Fig. 4). For the fuzzy logic controller, the simulation response (Fig. 3) is the same as the experiment response (Fig. 5), where the system gets a good response in 0.2 s without overshoot or peak time. When the fuzzy logic controller is used, there is a lack of overshoot, lack of peak time, and less rise time, as shown in the simulation and experiment responses. This increases its efficiency and makes it suitable for a wide range of applications.

4 Conclusion

This work confirms that the fuzzy logic controller gives good results than PID controller, where the results shown that Fuzzy logic controller eliminate the overshoot and the peak time however minimize the rise time of the system. Thus, it gives a good performance due to improving these constrained in industrial domain.

Thanks to Quanser QNET 2.0 DC motor laboratory, the test of DC motor will be easier and the analyses of different controller will be done without the need of electronic cards and their programing. Therefore, the visualization of the responses won't require the measure instruments owing to the virtual instruments of this laboratory.

References

1. Choi, J.: Modelling of DC motors: control systems lectures, pp. 1–15. Department of Mechanical Engineering, University of British Columbia (2008)

2. Shirazul, I., Farhad, I.B., Atif, I.: Stability analysis of a three-phase converter controlled DC motor drive. In: Third International Conference on Advanced Computing & Communication Technologies, 978–0–7695–4941–5/12 $26.00 © 2012 IEEE (2013)
3. Burns, R.S.: Advanced Control Engineering. Jordan Hill, Oxford (2001)
4. Engle, B.J., Watkins, J.M.: A Software platform for implementing digital control experiments on the QUANSER DC motor control trainer. In: Proceeding IEEE International Conference Control Applications CCA, pp. 510–515 (2008)
5. Quanser, N.I.: Quanser Engineering Trainer for NI-ELVIS. QNET Practical Control Guide. Quanser Inc., Canada (2009)
6. Brito Palma, L., Vieira Coito, F., Gomes Borracha, A. Francisco Martins, J.: A platform to support remote automation and control laboratories. In: 1st Experiment@ International Conference, Nov 17–18, Lisboa-Portugal (2011)
7. Nagrath, I.J., Gopal, M.: Control systems engineering, 5th edn. Delhi, India (2010)
8. Aisha, J., Sadi, M., Syed, O.: Controlling speed of dc motor with fuzzy controller in comparison with ANFIS controller. Intell. Control. Autom. **6**, 64–74 (2015)
9. Ravinder, K., Vineet, G.: High performance fuzzy adaptive control for D.C. motor international archive of applied sciences and technology IAAST; vol. 3, Society of Education, India, pp. 1–10 (September 2012)
10. Attila, S., et al.: Fuzzy-logic controller for smart drivers. In: 9th International Conference on Information technology and Quantitative Management, Procedia Computer Science, vol. 214, pp. 1396–1403 (2022)
11. Bature, A.A., et al.: Design and real time implementation of fuzzy controller for DC motor position control. Int. J. Scien. Technol. Res. **2**(11), 254–256 (2013)
12. Ramadan, E.A., El-bardini, M., Fkirin, M.A.: Design and FPGA-implementation of an improved adaptive fuzzy logic controller for DC motor speed control. Ain Shams Eng. J. 2090–4479 (2014)
13. Suradkar, R.P., Thosar, A.G.: Enhancing the performance of DC motor speed control using fuzzy logic. Int. J. Eng. Res. Technol. IJERT 103–110 (2012)

Unleashing Collective Intelligence for Innovation: A Literature Review

Ghita Ibrahimi(✉), Wijdane Merioumi, and Bouchra Benchekroun

Faculty of Law Economics and Social Sciences, Sidi Mohammed Ben Abdellah University, 28810 Fez, Morocco
ghita.ibrahimi@usmba.ac.ma

Abstract. In recent years, "Collective Intelligence" and "Innovation" has gained popularity in the academic world. Recognizing this growing interest, our study aims to contribute to the field of business and management research by conducting a comprehensive literature review on the role of collective intelligence in fostering innovation. Through our analysis, we identified that collective intelligence is a multidimensional concept, often manifested under crowdsourcing. Our findings show that collective intelligence plays a key role in driving various forms of innovation (e.g. disruptive innovation, social innovation) in both private and public sectors. In addition, our review underlined the mediating role of technologies in innovation processes. However, it appears that the reviewed papers did not explore this connection in sufficient depth. Hence, we concluded by acknowledging the limitations of our study and proposing several avenues for future research to further investigate this intriguing relationship.

Keywords: Collective intelligence · Innovation · Crowdsourcing

1 Introduction

Innovation is the twentieth-century engine. It's the driving force behind globalization, growth, and the ongoing evolution of human civilization. As for businesses, the proliferation of information and communication technologies, the changing environment, and fierce competition have made innovation a survival tool rather than an option. It is in demand not only from businesses but also from public organizations and various industries that are keen to unlock new opportunities, overcome challenges, enhance competitiveness, and stay ahead of the curve. Therefore, organizations constantly seek ways to promote innovative potential, this entails incorporating collective intelligence as a problem-solving tool through exchange and collaboration, involving both internal and external parties [1].

The concept of "Collective Intelligence" (CI) has attracted academics from different fields for many years [2, 3]. It traces its origins to evolutionary processes in numerous living forms, including humans and social animals. It emerged in response to efforts to achieve collective wisdom through the development of new coordination, collaboration, and knowledge-sharing tools. Furthermore, research on collective intelligence in human

Y. Farhaoui et al. (Eds.): ICAISE 2023, LNNS 837, pp. 172–179, 2024.
https://doi.org/10.1007/978-3-031-48465-0_23

groups has shown its importance as a key component in group learning [4]. This is due to its ability to empower collaboration, foster a culture of inclusivity and diversity, as well as promote adaptation via experimentation and continuous improvement to generate new ideas and possibilities. These solutions are not limited to organizational issues, but may also address major global challenges such as climate change, poverty reduction, recessions, or healthcare improvements.

Against this background, Collective Intelligence looks to be tricky to describe. Especially, when it comes to narrowing the importance of collective intelligence in fostering innovation. Accordingly, in this paper, we conduct a literature review that examines the current body of scholarly literature on the association between these two concepts in Business, management, and accounting area. It aims to analyze the published works that shed light on collective intelligence and innovation. As well as why collective intelligence plays a key role in driving innovation. Hence, the paper is organized as follows. We first describe the methodology used to conduct this literature review. Next, we present our key findings. Then, we analyze our results. Finally, we conclude our research and present future research avenues.

2 Research Methodology

To pursue the aim of this study, the authors used Scopus as one of the most common Databases used to conduct a literature review. This choice was based on the advantages that Scopus offers given its wide breadth, its meticulous indexing process guarantees covering only high-quality journals, its sophisticated search features, and its powerful analysis tools.

To ensure the integrity and relevance of the articles, we conducted a two-phase selection process. In the 1st selection phase, the authors conducted a first investigation on Scopus based on the titles and abstracts with the terms 'Collective Intelligence' and 'Innovation'. The following search query resulted in 396 papers: *TITLE-ABS-KEY (collective-intelligence, AND innovation).*

To refine the findings, the authors limited the search by field and type of document. It resulted in 71 articles and conference papers in Business, management and accounting. The corresponding search query is as follows: *TITLE-ABS-KEY (collective-intelligence, AND innovation) AND (LIMIT-TO (SUBJAREA, "BUSI")) AND (LIMIT-TO (DOCTYPE, "ar") OR LIMIT-TO (DOCTYPE, "cp")).*

In the 2nd selection phase, we first examined and selected the most relevant papers related to collective intelligence in innovation based on the following primary inclusion criteria (Table 1):

Table 1. Inclusion criteria

Criteria ID	Inclusion criteria
C1	The paper discusses the theoretical basis of innovation
C2	The paper discusses the theoretical basis of collective intelligence

(*continued*)

Table 1. (*continued*)

Criteria ID	Inclusion criteria
C3	The paper addresses a specific type of innovation
C4	The paper examines crowdsourcing as an enhanced type of CI
C5	The paper discusses the role of collective intelligence in innovation
C6	The paper describes the architecture/frameworks/study case including CI

We noticed that some articles were duplicated. Therefore, they have manually removed them to avoid redundancy. Moreover, we selected papers that covered a particular type of innovation, namely open innovation. We also added crowdsourcing since firms are moving beyond standard collaborations. As a result, we found 22 studies that meet the inclusion criteria.

Afterward, we wrapped up this phase by applying the following quality assessment criteria to the primary articles selected:

- QC1: The paper provides a clear research objective.
- QC2: The paper proposes a new framework for an existing CI system.
- QC3: The paper proposes a clearly defined architecture, framework, or design.
- QC4: The paper compares a new framework against an established one.
- QC5: The paper explores the role, importance, and behavior of individuals.
- QC6: The paper proposes solutions to innovation issues using collective intelligence.

The quality evaluation is based on an independent assessment by the authors using the criteria listed above. Each paper's criteria were scored based on this paper's author's responses. Papers with a score of 3 or higher were selected for data synthesis and presented in the Table 2.

Table 2. Comparison of selected papers

Reference	QC1	QC2	QC3	QC4	QC5	QC6	Total
[3]	1	0	1	0	1	0	3
[4]	1	0	0	1	1	1	4
[5]	1	0	0	0	1	1	3
[6]	1	1	1	0	1	1	5
[7]	1	0	0	0	1	1	3
[8]	1	0	0	0	1	1	3
[9]	1	0	0	0	1	1	3
[10]	1	1	1	0	1	1	5
[11]	1	0	1	0	1	0	3
[12]	1	0	0	0	1	1	3
[13]	1	0	0	0	1	1	3

(*continued*)

Table 2. (*continued*)

Reference	QC1	QC2	QC3	QC4	QC5	QC6	Total
[14]	1	1	1	1	1	1	6
[15]	1	1	1	0	1	1	5
[16]	1	1	1	0	1	1	5
[17]	1	1	1	1	1	1	6
[18]	1	0	0	0	1	1	3
[19]	1	0	0	0	1	1	3
[20]	1	0	0	0	1	1	3
[21]	1	1	1	0	1	1	5
[22]	1	1	1	0	1	1	5
[23]	1	0	0	0	1	1	3
[24]	1	0	0	0	1	1	3

3 Results

The papers cover a wide range of topics that we summarized below:

Several authors discussed the role of collective intelligence in enhancing knowledge management, including Boder (2006) that proposes a practical model for knowledge management integrating collective intelligence. Next, Karakas and Kavas (2009) introduce a model embracing nonlinear approaches to knowledge and action. Afterward, Boulesnane and Bouzidi (2013) proposed a conceptual model that promotes knowledge management and collective intelligence in an organization's management. Finally, Taylor et al. (2014) proposed a methodology based on a bilayer social wiki network to facilitate open knowledge accumulation in manufacturing process innovation.

Furthermore, the selected papers explored the synergies between open innovation and collective intelligence, namely Karakas and Kavas (2009) incorporated the global brain metaphor to highlight the importance of collective intelligence and open innovation in nurturing creativity. Martínez-Torres (2014) explores open innovation within online communities as a collective intelligence assessment technique, and analyzes the behavior of community members using social network analysis. Then, Celis and García (2019) discussed how public organizations innovate under the lens of open innovation by harnessing collective intelligence. Eventually, Attalah et al. (2023) tackled the role of collective intelligence tools with an emphasize on Hackathons in open innovation and the overall innovation process. Whereas, some authors focused on the public sector including Papadopoulos et al. (2012) who advocate the shift towards open innovation practices such as collective intelligence and participation in voluntary communities, particularly in economies facing crises. In addition to Mergel and Desouza (2013) who investigated the implementation of open innovation and collective intelligence in the public sector using the Challenge.gov platform to source innovative ideas and solutions from citizens.

Besides, several papers discussed the potential of crowdsourcing in maximizing collective intelligence effectiveness, such as Fähling et al. (2011) who discussed the integration of the consumers throughout the whole innovation process by exploiting the collective intelligence phenomenon. Similarily, Majchrzak and Malhotra (2013) discussed the role of crowdsourcing and co-creation in innovation. They emphasized on the crowd's size and expertise to gather diverse innovative ideas. Sharma et al. (2014) discuss the use of Web 2.0 technologies and collective intelligence for capturing user-generated value. Likewise, Beretta et al. (2018) highlight the role of moderators in managing web-enabled ideation systems to harness both internal and external parties' collective intelligence. Unlike Chiu et al. (2014) who considered collective intelligence as a type of crowdsourcing and explore its applications in managerial decision-making and problem solving. Finally, Cappa et al. (2019) defined conditions that allow firms to harvest value from the crowd and identified the benefits of open innovation on the firm's performance.

Other authors like Elia et al. (2020) and Erbguth et al. (2022) focused on the role of open innovation and collaborative systems in promoting a well-orchestrated collective intelligence to solve sustainable development challenges. Lastly, Chandra and Yang (2012) highlight the importance of collective intelligence in disruptive innovation.

4 Analysis

The 22 papers covered different aspects of innovation and collective intelligence. They investigated the theoretical foundations of each concept and provided architecture, frameworks, or case studies that include both. Additionally, these papers can provide practitioners and scholars with key insights into the dynamics, applications, and importance of collective intelligence in innovation processes. Figure 1 illustrates the selected studies' keyword cloud.

Fig. 1. Keyword cloud of the selected studies

However, numerous observations have been noticed. Firstly, an assessment of previous studies demonstrates the importance of collective intelligence in generating new knowledge and fostering innovation [5]. It enriches innovation by tapping into groups' uniqueness, diversity, and expertise to gather innovative ideas and solutions [6, 7]. It also fosters critical thinking skills by promoting collaboration, knowledge sharing, and encouraging perspective differences, all of which are necessary components of innovation [8, 9]. Furthermore, specific types of innovation and their association with collective intelligence were also investigated in these studies. One of the most frequently discussed types was open innovation [9, 10, 19, 11–18]. Combining them allows for a more holistic approach to innovation by encouraging creativity and knowledge co-creation [6, 9, 17, 20, 21]. It also pushes firms to go beyond internal collaborations by engaging with third parties to access a broader range of knowledge and expertise, spark creativity, and bring disruptive innovations. Similarly, some papers focused on entrepreneurship and disruptive innovation, suggesting ways to manage innovation to survive and navigate challenging environments [21] (Fig. 2).

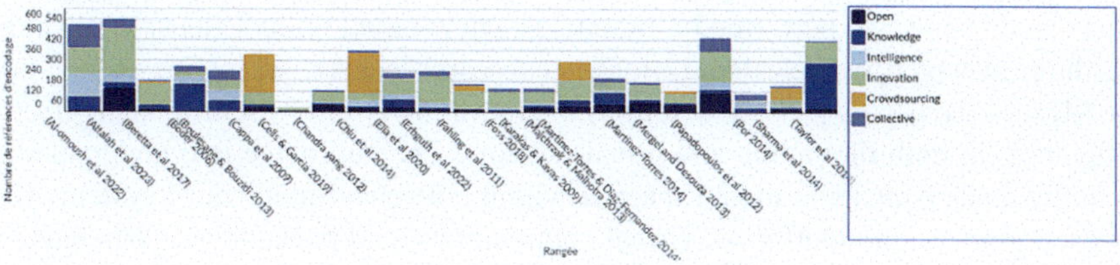

Fig. 2. Matrix crossover graph

Secondly, we observed that various terminologies were used to describe collective intelligence. The main one was crowdsourcing as an enhanced form of collective intelligence [6, 14–16, 22, 23]. The papers studied the potential of crowdsourcing to influence each stage of the innovation process by dipping into a wide network of talent and ideas [22]. We also noticed that crowdsourcing may occur both through traditional methods such as Hackathons, and digital technologies including online platforms, and virtual communities [14, 18, 24, 25]. This acknowledges the mediating role of information technology, digital technologies, and information systems in decision-making, innovation processes, and the entrepreneurial ecosystem [7, 15, 18, 21–23, 26] (Fig. 3).

Mot	Longueur	Nombre	Pourcentage pondéré (%)
innovation	10	1879	1,42
knowledge	9	897	0,68
open	4	733	0,55
crowdsourcing	13	721	0,54
collective	10	612	0,46
intelligence	12	560	0,42

Fig. 3. Word frequency chart

Thirdly, we noticed that collective intelligence can be leveraged not only in the private sector but also in driving public affairs. According to several studies, public organizations should ensure collaboration between policymakers and citizens [11, 17]. This enables them to create new solutions that fulfill the requirements of citizens, improve the performance of public institutions, and deal with challenging issues in society [11, 17]. Along the same lines, collective intelligence has the potential to reach sustainable development aims [8, 18]. As its applications extend into fostering social innovation by providing insightful solutions for social concerns, especially in times of crises such as the Greek debt crisis, and the COVID-19 crisis [27].

Last but not least, after a deep analysis of the described studies we noticed that most of them don't focus on developing new frameworks, providing technical details, and making comparisons with existing frameworks. Additionally, they do not explore the link between innovation and collective intelligence properly. However, they provide insightful data about individuals and their interaction within collective intelligence systems to provide innovative solutions.

5 Conclusion

In this paper, we assessed the potential of collective intelligence in promoting innovation. To this end, we performed a literature review based on an extensive screening procedure, in which we selected 22 relevant research papers. According to our findings, Most of the papers propose solutions to innovation issues by leveraging collective intelligence, in both the public and private sectors. They all delve into understanding how individuals contribute to and interact within collective intelligence systems (e.g. crowdsourcing, online platforms, virtual communities, and hackathons). Whereas, we noticed that they do not generally focus on proposing new frameworks nor compare them to established ones. This implies that the focus of these articles lies elsewhere. Thus, future studies must focus on developing new frameworks that only involve collective intelligence and innovation components. Secondly, we noticed that these studies do not delve deeper into the relationship between the two key concepts. Thus, future research must undertake more in-depth studies that tackle their key aspects. Thirdly, another common feature in the evaluated papers is the lack of comparisons between existing and new frameworks. This suggests that the author's primary goal is not assessing or comparing other approaches but rather focusing on their unique perspectives. In addition, one of the principal limitations stems from our preliminary search. As we only relied on Scopus, future research should include other Databases to cover more journals. In summary, businesses that embrace the potential of collective intelligence can break old barriers and unlock limitless innovation possibilities. This paradigm shift may embark them on a revolutionary path shaped by shared value, creativity, adaptability, success, and unlimited growth.

References

1. Lee, J.Y., Jin, C.H.: How collective intelligence fosters incremental innovation. J. Open Innov. Technol. Mark Complex [Internet]. **5**(3), 53 (2019)
2. Riedl, C., Blohm, I., Leimeister, J.M., Krcmar, H.: Rating scales for collective intelligence in innovation communities: why quick and easy decision making does not get IT right. In: ICIS 2010 Procceding—Thirty First International Conference Information System (2010)
3. Suran, S., Pattanaik, V., Draheim, D.: Frameworks for collective intelligence: a systematic literature review. ACM Comput. Surv. **53**(1), 1–36 (2020)
4. Gurumurthy, K.: Driving the Economy through Innovation and Entrepreneurship. Springer, pp. 886 (2013)
5. Boder, A.: Collective intelligence: a keystone in knowledge management. J. Knowl. Manag. **10**(1), 81–93 (2006)
6. Majchrzak, A., Malhotra, A.: Towards an information systems perspective and research agenda on crowdsourcing for innovation. J. Strateg. Inf. Syst. **22**(4), 257–268 (2013)

7. Foss, R.A.: A self-organizing system for innovation in large organizations. Syst. Res. Behav. Sci. **35**(3), 324–340 (2018)
8. Erbguth, J., Schörling, M., Birt, N., Bongers, S., Sulzberger, P., Morin, J.H.: Co-creating innovation for sustainability. Grup Interaktion Organ Zeitschrift fur Angew Organ. **53**(1), 83–97 (2022)
9. Karakas, F., Kavas, M.: Service-learning 2.0 for the twenty-first century: Towards a holistic model for global social positive change. Int. J. Organ. Anal. **17**(1), 40–59 (2009)
10. Papadopoulos, T., Stamati, T., Nikolaidou, M., Anagnostopoulos, D.: Technological forecasting and social change from open source to open innovation practices : a case in the Greek context in light of the debt crisis. Technol. Forecast. Soc. Chang. [Internet] (2012)
11. Mergel, I., Desouza, K.C.: Implementing open innovation in the public sector: the case of challenge.gov. Public Adm. Rev. **73**(6), 882–890 (2013)
12. Martínez-Torres, M.R.: Analysis of open innovation communities from the perspective of social network analysis. Technol. Anal. Strateg. Manag. **26**(4), 435–451 (2014)
13. Taylor, P., Wang, G., Tian, X., Geng, J., Guo, B.: A knowledge accumulation approach based on bilayer social wiki network for computer-aided process innovation. 37–41 (2014)
14. Sharma, R.S., Soe, K.M.Y., Balasubramaniam, D.: Case studies on the exploitation of crowdsourcing with Web 2.0 functionalities. Int. J. Electron. Bus. **11**(4), 384–408 (2014)
15. Chiu, C, Liang T, Turban E. What Can Crowdsourcing Do for Decision Support? Decis Support Syst [Internet]. 2014
16. Cappa, F., Oriani, R., Pinelli, M., De Massis, A.: When does crowdsourcing benefit firm stock market performance? Res. Policy [Internet]. **48**(9), 103825 (2019). https://doi.org/10.1016/j.respol.2019.103825
17. Celis, J., García, E.C.: Madrid Escucha Y Experimenta Distrito Dos Experiencias De Colaboraci6n Y Experimentaci6n Social. Rev Gesti6n Pública. **VIII**(2), 179–210 (2019)
18. Elia, G., Margherita, A., Passiante, G.: Digital entrepreneurship ecosystem: how digital technologies and collective intelligence are reshaping the entrepreneurial process. Technol. Forecast. Soc. Change [Internet]. **150**, 119791 (2020)
19. Attalah, I., Nylund, P.A., Brem, A.: Who captures value from hackathons ? innovation contests with collective intelligence tools bridging creativity and coupled open innovation. 1–15 (2023)
20. Al-Omoush, K.S., Sim6n-Moya, V., Sendra-García, J.: The impact of social capital and collaborative knowledge creation on e-business proactiveness and organizational agility in responding to the COVID-19 crisis. J. Innov. Knowl. **5**(4), 279–288 (2020)
21. Chandra, Y., Yang, S.J.S.: Managing disruptive innovation: entrepreneurial strategies and tournaments for corporate longevity. J. Gen. Manag. **37**(2) (2012)
22. Fähling, J., Blohm, I., Krcmar, H., Leimeister, J.M., Fischer, J.: Accelerating customer integration into innovation processes using Pico Jobs. Int. J. Technol. Mark. **6**(2), 130–147 (2011)
23. Beretta, M., Björk, J., Magnusson, M.: Moderating ideation in web-enabled ideation systems. J. Prod. Innov. Manag. **35**(3), 389–409 (2018)
24. Attalah, I., Nylund, P.A., Brem, A.: Who captures value from hackathons? innovation contests with collective intelligence tools bridging creativity and coupled open innovation. Creat. Innov. Manag. 266–80 (2023)
25. Pór, G.: Augmenting the collective intelligence of the ecosystem of systems communities: introduction to the design of the CI enhancement lab (CIEL). Syst. Res. Behav. Sci. **31**(5), 595–605 (2014)
26. Boulesnane, S., Bouzidi, L.: The mediating role of information technology in the decision-making context. J. Enterp. Inf. Manag. **26**(4), 387–399 (2013)
27. Al-Omoush, K.S., Ribeiro-Navarrete, S., Lassala, C., Skare, M.: Networking and knowledge creation: social capital and collaborative innovation in responding to the COVID-19 crisis. J. Innov. Knowl. **7**(2), 100181 (2022)

Which Data Quality Model for Recommender Systems?

Meriem Hassani Saissi(✉) and Ahmed Zellou

SPM Research Team, ENSIAS, Mohammed V University in Rabat, Rabat, Morocco
{meriem_hassanisaissi,ahmed.zellou}@um5.ac.ma

Abstract. Although data quality has been acknowledged as a significant issue in a variety of information systems research studies, it has received little attention in recommender systems. Data quality is crucial for the performance and effectiveness of recommender systems. Recommender systems rely on historical data to make predictions and recommendations, and the accuracy of these recommendations heavily depends on the quality of the data used. However, the reliability of these recommendations are substantially impacted by the quality of the data used. To evaluate and enhance the performance of recommender systems, it is crucial to comprehend the dimension of data quality. This paper addresses the gap by conducting a comprehensive literature assessment on data quality dimensions and models in the context of recommender systems. It draws attention to the various dimension of data quality, looks at the data models and offers suggestion models for assessing and enhancing data quality. This paper lays the groundwork for future studies and advancements in data quality for recommender systems, which will ultimately result in recommendations for users that are more accurate and trustworthy.

Keywords: Data quality · Data quality dimensions · Data quality models · Recommender systems

1 Introduction

Recommender systems have become an integral part of our digital lives, helping us discover new products, services, and content that align with our interests and preferences [1].These systems leverage vast amounts of data to make accurate predictions and generate personalized recommendations. However, the effectiveness and reliability of these recommendations heavily rely on the quality of the underlying data. The paper is divided into four section, in the first one, the paper explore the main aspects of data quality dimension, including accuracy, completeness, timeliness and consistency. The implications of each dimension of the recommender systems are unique and have a significant impact on their performance, as well as how users perceive the recommendations. Additionally, the paper focuses on exploring different models for data quality that can be used to improve recommender systems. These models, including the Six Sigma Model (SSDQ), Total Data Quality Management (TDQM) Model, and The Data

Y. Farhaoui et al. (Eds.): ICAISE 2023, LNNS 837, pp. 180–185, 2024.
https://doi.org/10.1007/978-3-031-48465-0_24

Management Model (DAMA), offer structured approaches to managing and enhancing data quality. The rest of this paper analyzes the strengths and appropriateness of each model within the context of recommender systems.

2 Data Quality

There is currently no consensus on a single definition of data quality that can be applied to all data domains. The quality of data should be defined in terms of fit for a particular purpose. As, Wang and Strong [2] defined data quality as 'fitness for use' by data consumers i.e. the ability of a data collection to meet user's needs [3], and it can be defined as "the level of utility (or value) perceived by users when using specific data in a specific context" [4, 5]. The purpose of this paper is to present the findings of experts on data quality dimensions and to identify the most important data quality dimensions according to the purpose of recommender systems.

3 Data Quality Dimension

The data quality literature offers a complete taxonomy of data quality dimensions [4, 6, 7]; however, there are many differences in the definitions of most dimensions due to the contextual nature of quality. Wand and Wang [7] provide a classification of six main quality dimensions; Redman [8]; Jack et al. [9]; Bovey et al. [10]; and Naumann [10]. By analyzing these classifications, a set of basic data quality dimensions can be defined, including accuracy, completeness, consistency, and timeliness, which is the focus of most authors Catarci and Scannapieco. However, there is no consensus on which set of dimensions defines the quality of data, and what exactly each dimension means. In Wang and Strong's proposal, the data quality dimension is selected by surveying data consumers, and as a result, nearly 179 data quality dimensions from the user's perspective are surveyed. From them, the author selected 15 different dimensions and grouped them into four different categories, such as intrinsic, accessibility, contextual, representational and accessibility.

The proposal of Bovee and Srivastava considered the data quality dimension from the perspective of data consumers and developed a conceptual model consisting of four attributes, namely: Accessibility, Interpretability, Relevance, and Integrity (Fig. 1).

- Accessibility: access to information we may find useful.
- Interpretability: Understanding information and deriving meaning from it.
- Relevance: Find it applicable to the domain and context of interest.
- Integrity: Trust that it is free from defects. The last attribute, Integrity, is further subdivided into four sub-attributes: Accuracy, Completeness, Consistency, and Existence [10].

In this paper, we consider the Data Quality dimensions presented in [7] and for each considered Data Quality dimension, we present its definition and its impact on Recommender System (Table 1).

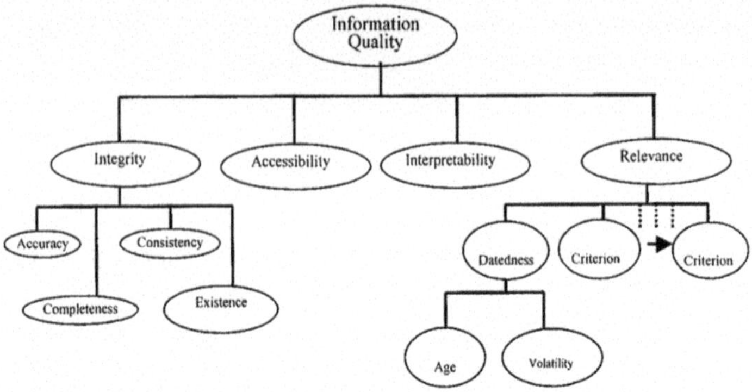

Fig. 1. Data quality model proposed by Bovee and Srivastava [10].

Table 1. Dimensions and their impact on recommender system.

Dimension	Impact in recommender system
Accuracy	Accuracy makes sure that the system's recommendations closely match the preferences of the users, which boosts user engagement and happiness Importance: High accuracy is essential for recommender systems since poor recommendations might make users angry and make them lose faith in the system
Completeness	The system may collect a thorough perspective of user preferences and item features when the data is complete, allowing for more precise and customized recommendations Importance: The system's capacity to offer pertinent and varied recommendations may be constrained by missing data or substantial gaps in user preferences or item information
Timeliness	The system can react to changing user interests by using the data's timeliness to represent the most recent user preferences and item information Importance: Outdated data may produce recommendations that do not reflect the tastes of the current user base, decreasing the system's relevance and efficacy
Consistency	Data consistency guarantees that there are no inconsistencies or contradicting statements, which reduces confusion and increases the accuracy of suggestions Importance: Inconsistent data might result in recommendations that are in conflict or cause user confusion, compromising the recommender system's credibility and trust

4 Data Quality Models

To evaluate and control the quality of data, a number of data quality models have been presented. These models offer frameworks and instructions for assessing and enhancing the quality of data in several fields [11]. While there are not specific data quality models tailored exclusively for recommender systems, the following general data quality models can be effectively applied to assess and enhance data quality in recommender systems. For a recommender system, a variety of data quality models can be useful. The three typical models are Total Data Quality Management (TDQM) [12], Six Sigma Data Quality (SSDQ) [12] and The Data Management Model (DAMA).

Although these models offer a framework for evaluating the data quality in recommender systems, it is crucial to take into account the particular requirements and traits of the domain. These models can be modified and customized to meet the particular requirements of recommender systems; to improve data quality evaluation and personalization of recommendations.

5 Comparison and Discussion

When comparing data quality dimensions in the context of recommender systems, it is important to consider how each dimension affects the performance and effectiveness of the recommendation process. In [13] the authors of this paper focuses on the data quality dimension completeness of item content data and investigates its impact on the prediction accuracy of recommender systems, and presents a theoretical model based on the literature and derives ten hypotheses. The question now, how those models can benefit a recommender system? In order to address this inquiry, we will conduct a quick comparison of the most salient characteristics (Table 2).

Table 2. Key elements of data quality models

Model	Key elements
Six sigma (SSDQ)	-Statistical analysis to assess and enhance data quality is one of the Six Sigma Model's key elements -Establishing performance objectives and defining quantifiable quality criteria -Using statistical methods like process capability analysis and control charts
The total data quality management (TDQM)	-Planning and strategy creation for data quality -Control and improvement techniques for data quality -Incorporating data quality procedures across the entire data lifecycle -Proactive supervision and preservation of data quality
The data Management (DAMA)	-Determining the dimensions and measurements of data quality is one of the DAMA Model's key components -Data profiling and evaluation to find problems with data quality -Creating guidelines and criteria for data quality

The decision between these models is influenced by various elements, including the organizational context and the particular needs, objectives, and resources of the recommender system. It is crucial to understand that there is not a single solution that works for all situations because many models each have advantages and disadvantages. However, the Total Data Quality Management (TDQM) paradigm is one that is well known and useful for recommender systems. It takes into account attributes like accuracy, completeness, consistency, integrity, timeliness, uniqueness, and accessibility. A comprehensive strategy for maintaining data quality throughout the data lifecycle is provided by the Total Data Quality Management (TDQM) model. It places a strong emphasis on incorporating data quality procedures into every facet of data management, including data collection, processing, analysis, and maintenance. Organizations can set up proactive procedures for data quality assessment, measurement, planning, control, and continuous improvement by adopting the Total Data Quality Management (TDQM) paradigm.

6 Conclusion and Future Works

Having high-quality data is crucial for recommender system success. This article investigated the different data quality models and features that are pertinent to these systems. During our discussion, we delved into crucial dimensions of data quality. These dimensions encompass accuracy, completeness, timeliness and consistency. It is imperative that these aspects of data quality are taken into account during the development and upkeep of recommender systems. In addition, we have analyzed various models of data quality that are suitable for recommender systems. Among these models are the Six Sigma (SSDQ) Model, the Total Data Quality Management (TDQM) Model, and the Data Management (DAMA) Model. Each of these models presents a methodical strategy for managing data quality with a focus on elements such as data accuracy, integration, continuous improvement, and general data management practices. In future research, it would be valuable constructing solid assessment metrics with a focus on evaluating the quality of the data in the context of recommender systems and we will propose another model for collaborative filtering in recommender systems, we can also examine how different user contexts or types of recommendation may alter the needs for data quality. The data quality required for movie suggestions, for example, can be different from the data quality required for healthcare recommendations.

References

1. Idrissi, N., Zellou, A.: A systematic literature review of sparsity issues in recommender systems. Soc. Netw. Anal. Min. **10**(1), 15 (2020)
2. Wang, R.Y., Strong, D.M.: Beyond Accuracy: What Data Quality Means to Data Con-sumers. J. Manag. Inf. Syst. **12**(4), 5–33 (1996). https://doi.org/10.1080/07421222.1996.1151809
3. Liebchen, G.A., Shepperd, M.: Data sets and data quality in software engineering. In: Proceedings of the 4th International Workshop on Predictor models in Softwar Engineering—PROMISE, p. 39 (2008). https://doi.org/10.1145/1370788.1370799
4. Strong, D.M., Lee, Y.W., Wang, R.Y.: Data quality in context. Commun. ACM **40**(5), 103–110 (1997)

5. Liebchen, G.A., Twala, B.: Assessing the quality and cleaning of a software project dataset: an experience report. In: Proceedings of Evaluation and Assessment in Software Engineering (EASE 2006), pp. 1 (2006)
6. Batini, C., Scannapieco, M.: Data and information quality—dimensions, principles and techniques springer (Data-Centric Systems and Applications) (2016)
7. Wand, Y., Wang, R.Y.: Anchoring data quality dimensions in ontological foundations. Commun. ACM **39**, 86–95 (1996)
8. Redman, T.C.: Data quality for the Information Age. Artech House (1996)
9. Jarke, M.: Fundamentals of Data Warehouses. Springer Verlag (2003)
10. Bovee, M., Srivastava, R.P., Mak, B.: A conceptual framework and belief-function approach to assessing overall information quality. Int. J. Intell. Syst. **18**(1), 51–74 (2003). https://doi.org/10.1002/int.10074
11. Mimouni, L., Zellou, A., Idri, A.: MDQM: Mediation data quality model aligned data quality model for mediation systems (2018)
12. Linstedt, D., Olschimke, M.: The data vault 2.0 methodology. Data Vault **2.0**, 33–88 (2016)
13. Heinrich, B., Hopf, M., Lohninger, D., Schiller, A., Szubartowicz, M.: Data quality in recommender systems: the impact of completeness of item content data on prediction accuracy of recommender systems. Electron. Mark. **31**, 389–409 (2021)

Logistics Blockchain: Value-Creating Technology for the Port of Casablanca's Container Terminal

Fouguig Nada(✉)

Faculty of Legal, Economic and Social Sciences Ain Chock, University of Hassan II, Casablanca, Morocco
nada.fouguig@gmail.com

Abstract. The intensity of trade has profoundly transformed port logistics and global maritime traffic. Asian and European ports are competing fiercely with their technical and technological lead. New technologies – the Internet of Things (IoT), Industry 4.0, artificial intelligence (AI) and blockchain technology - are steadily granting a competitive edge for port logistics chains while playing a major role in this global trade battle. Morocco is well aware of the opportunities offered by digital technologies applied to port logistics, and of the intense competitiveness between the world's ports in attracting trade flows. The country has developed a network of modern port infrastructures that meets international standards. The national port agency (NPA), owner of the port regulations, has adopted a strategy making digitalization and smart technologies such as Logistics 4.0, Blockchain and the Cloud, its battle tools to improve port logistics, integrate its partners into a common digital ecosystem and further enhance the performance of its container port terminals (CPTs) and all the partners present in the ecosystem. To understand this trend and its consequences, we conducted a study on the digitalization of NPA to survey the improvement of Casablanca port terminal performance indicators and on the future contribution of blockchain technology to the container terminal port logistics and all its actors. Our paper aims to present the main findings of the study.

Keywords: Blockchain technology · Logistics 4.0 · Port community · Performance · Port logistics · Stakeholders

1 Introduction

1.1 The Study Context

The dawn of new technologies, combined with the intensity of international trade, is causing major shifts in port logistics. Nowadays, all the ports face the same challenge: how to be more attractive while staying competitive. "Smarter Ports" have then emerged. Based on the digitalization strategy, the port processes have been reengineered to align to intelligent logistics 4.0 [1]. This includes smart network systems with supply chain stakeholders interconnected in a decentralized way. The rush for this technology has already begun using Blockchain, a technology that is being encouraged by logistics 4.0 among other technologies.

Y. Farhaoui et al. (Eds.): ICAISE 2023, LNNS 837, pp. 186–192, 2024.
https://doi.org/10.1007/978-3-031-48465-0_25

In 2017, several Blockchain technology integration projects [2] are starting to be tested in European ports[1] and Asian ports.[2]

Morocco is no exception to the dynamics of international technological change with its strategic maritime position and 14 ports dedicated to international trade. The National Port Authority is aware of this new situation and has launched a digital strategy for ports in 2011. It has made digitalization a fundamental pillar to improve the performance of maritime terminals, increase the attractiveness of national ports and keep pace with global dynamics in the field of new technologies.

1.2 Interest and Challenge of the Study

The interest of the study carried out on the impact of the digitalization strategy on the performance indicators of the processes of Casablanca port terminals and the benefits offered by recent technologies aims to assess the findings of the study with regard of the new opportunities offered by the blockchain technology. The fundamental challenge of this paper is to address the following question:

- Would Blockchain technology contribute to supporting the digitalization efforts of the ANP aiming to strengthen the performance of the port process and satisfy its stakeholders?

Two questions arise from this: has the digitalization strategy of the NPA led to the desired results? Is it not currently the time to move towards Blockchain, which is recognized as a value-creating technology by and for port stakeholders?

The paper consists of two parts: the first demonstrates the findings following the study of the digitalization strategy of the National Ports Authority on the performance of the port container terminal process and discusses the appreciation of the stakeholders of the port community. The second, illustrates the potential of applying the blockchain technology in ports logistics and the benefits that will procure to the actors of the port community.

1.3 Approach and Research Methodology

To carry out this study, we relied on the documentation of the National Port Authority,[3] and on the data of our thesis.

The data analysis allowed us to make trend curves of CPT's process performance indicators. Based on these trends, we have created interview guides[4] for different populations in the port community that are part of the end-to-end process[5] from the upstream to downstream.[6]

[1] Rotterdam project, Antwerp project.

[2] Japanese consortium, Korean consortium, Singaporean project, Hong Kong consortium Global Shipping Business Network (GSBN) and Abu Dhabi project.

[3] The monthly reports for the years 2017 to 2022, the annual reports of the same years.

[4] The interviews are either direct and semi-directive or by telephone interviews.

[5] The target population that was the subject of the study is the stakeholders who operate in the port of Casablanca as mentioned in Table 1.

[6] From the arrival of the vessel to the exit of the container from the storage yard.

The interview guides are meant to measure the degree of satisfaction of these populations with regards to the performance indicators of the container port terminal and the digitalization strategy. Furthermore, it was a mean to collect their opinion on the opportunities offered by the new Blockchain technology as a source of value creation.

Our theoretical approach allowed us to understand that the digitalization strategy of the NPA based on a centralized interrelational model has limits in terms of achieving the desired results. Hence, we opted to present the potential contributions of the Blockchain as a new decentralized interrelational model that could integrate all the port community stakeholders and lead to their satisfaction (Table 1).

Table 1. The description of the sample.

Stakeholders of port community	Number of interviews	Response rate (%)
Port container terminal of Casablanca	2	100
Shipowners	2	80
Freight forwarders	10	30
NAP	1	100
Custom authority	1	100

2 Findings of the Digitization of the NPA on the Performance of the Port Container Terminals (PCTs) Process

The NPA has put the digitalization at the heart of its strategy to meet the growing expectations of port stakeholders. To draw relevant conclusions, it is necessary to evaluate the impact of digitalization on the performance indicators [3] of the port process.

Our study conducted in the PCTs of Casablanca, shows that three indicators have improved; one indicator has shown a slight progress; but three other important indicators have not shown the expected development.

- **The performance indicator "Vessel arrival notification processing times"** has improved steadily and significantly. The processing time for vessel arrival notices has been reduced by 62.3%, resulting in a considerable time saving of 16.5 min per arrival notice; it benefited the three port stakeholders - the NPA, the shipping operators and the terminal handling operator in which they expressed their satisfaction with the trend of the curve of the indicator studied and the significant gains.
- **The performance indicator "Average annual manifest[7] filing time"**, has also improved. They have essentially contributed to reducing the average annual manifest filing time from an average of 29 h before the arrival of the ship in 2016 to 74 h

[7] The manifest is a document sent by the vessel captain to the government authorities in the exporting and importing countries. It includes information about the goods contents, the exporter, the importer, and who is the transporting the cargo.

in 2022. This good performance gives plenty of time to the NPA, the customs and the port terminals to make arrangements and easily prepare the reception of the ship. The three actors have also indicated their satisfaction with these results.

- **The indicator "Average quarterly time between the single customs declaration for goods and container release order"** performed well. Its improvement allowed a time saving of 2.25 days for obtaining the container release order over a period of 5 years. The actors involved in this process also expressed their satisfaction with the time gain on this indicator.
- **The indicator "Average quarterly time, in days, between the inspection of containers and the creation of the single customs declaration for goods"** has certainly registered an improving trend. Significant time savings have been achieved in the sixteen quarters of the years 2017 to 2020. However, the trend began to deteriorate over the next eight quarters and tends to reverse the gains made in the previous years. The handling operator, the customs administration and the inspection organisms seem to be running out of steam. They did not have a steady and a better impact on the performance indicator studied in this point, which showed through their dissatisfaction with the result obtained.
- **The "Time taken to validate the manifest by the Customs and Indirect Tax Administration" indicator** recorded a sharp increase from 117 min in 2016 to 334 min in 2018. The expected time saving did not succeed. The trend towards an increase in this indicator proved to be a serious hurdle for the continuity of the port process. This delay has an impact on the rest of the operations chain, more precisely the delay of the ship's Port call and the increase in its waiting time in the harbor. Stakeholders were dissatisfied with the delay shown by this indicator.
- **The "Waiting time in the harbor for ships" indicator** is an accurate performance indicator. Its reduction increases the satisfaction of shipowners and consequently improves port attractiveness. Despite its importance, the data concerning it are not provided by the port authority. Thus, we believe that since the delay in the previous indicators will automatically affects the waiting time in harbor, its trend would be the same, reason why most stakeholders are disappointed with this result.
- **The indicator "Average quarterly containers stay duration"**, the result from all the performance indicators of the port process, recorded three trends: firstly, an upward trend of the 8 quarters of the years 2017 and 2018, followed by a downward trend of four quarters, due to the Covid pandemic in 2020. Thus, it ends with a new upward trend during the 8 quarters of the years 2021 and 2022. The result of these three trends is moving toward a stability of this indicator around 8 days of stay on average quarterly per container. The expected improvement did not yield the expected results which shows the dissatisfaction of the port community.

Overall, the digitization of the ANP could not reach the desired results. The time savings achieved at the level of the four indicators are whittled by the negative trend (upward) of the remaining three indicators of the container shipping process. Hence, it affected the satisfaction of the stakeholders. Given the analysis above, and taking into account the rush of international ports toward new technologies, isn't it opportune, for the Moroccan ports, to evolve towards the Blockchain recognized to be a technology that creates value by and for the port stakeholders?

3 Blockchain Logistics: An Opportunity to Achieve Desired Goals in PTC

We have retained the following definition: "A blockchain is a register, a wide database which has the particularity of being shared in the same time with all its actors, all of whom are also holders of this register, who also all have the ability to access it and register data, according to specific rules set by a computer protocol very well secured thanks to cryptography" [4].

There are four main points to retain from the blockchain definition: It is a register, a large database for storing and transmitting data [5]. It has its very specific users and who are its holders equally. It has the particularity of being shared simultaneously by its users. It offers its holders the ability to inscribe information and carry out transactions, without middleman, by abiding to specific rules ensuring security through a computer protocol and cryptography.

The blockchain is a fairly advanced technology. It offers an opportunity to seize, in the various fields of human activity, to create more value by strengthening the will of organizations to achieve their goals lies in performance and competitiveness. Therefore, it is leveraged in several other sectors as well: financial, economic, commercial, logistics, etc. Recently, it has made its way to the world of port logistics. Since 2018, several ports have tested it and are currently working to adopt it. Several projects are currently in progress. Some experts believe that it is an invaluable source of value creation for interconnected and integrated activities in value chains of container port terminals [6]. It offers them, at least, six great advantages.

Building Trust and the Cooperation Between Actors. In the port sector, it is crucial that the actors build relationships based on trust in order to be able to cooperate and help each other seriously. The blockchain offers the opportunity to establish relationships of trust through the process of cryptography [7], the consensus protocol [8], decentralized network management and smart contracts [9]. The entrance to the network is monitored by a cryptographic process. All users must be spotted and correctly identified before allowing them to enter the network. To ensure this condition, each user has an identifier or a key (a code) that enables him to access the common network. The identification of the user is done by a cryptographic process which authorizes or refuses his access to the system. Adding new data to the system is subject to the consensus protocol of all "nodes" [10]. This protocol guarantees the security and reliability of transactions. The added data is decrypted and authenticated by "miners" or by "validator nodes", once validated, they are added to the register in the form of a block. The network's decentralized management system protects registry data from being tampered. As soon as a new block is added to the chain, a copy is sent, respecting the transmission chronology, to all the "nodes" of the network. Each node keeps its copy. It becomes difficult to corrupt or remove the block. The terms of agreements between users are guaranteed by Smart Contracts. The Smart Contract is used to negotiate, execute and uphold the terms of a legal agreement without going through the trouble and slow rules of regular contracts.

Data Security. Securing data in the port sector is very crucial. Errors at this level can be disastrous. Using the process of cryptography, the users of the data are identified. Through the consensus protocol between all the nodes, the new added information is authenticated, validated and stored in the register. The data stored in a decentralized network the data is preserved against any attempt of falsification or destruction. The Smart Contract, in addition to defining the rules of an agreement between several parties, freezes the rules in the blockchain by ensuring a "notarization" of the contractual process.

Real-Time Monitoring of Information and Transactions [11]. Access to real-time information allows port actors to ensure simultaneous monitoring of the stages of the logistics process from the arrival of the ship in the harbor, through its docking, loading and unloading, to control and exit. The implementation of Blockchain technology offers this possibility. Indeed, once the user is authorized to access the network, he sends the data to a peer (a computer user). The operation runs, simultaneously, on a distributed Peer-to-Peer (P2P) network [12]. Each network user represents a node connected to several peers. It communicates directly from one computer to another and works at the same time as a client and a server. Any user can then access the information stored in the data simultaneously, equally between users and at the real time of its introduction into the network. Hence it will certainly contribute to improving the performance of port indicators.

Fast Data Processing [13]. The blockchain encourages decentralization and autonomy logistics processes management, which ensures a fast and efficient processing of data, hence, reducing delays.

Traceability and Reduction of Time and Costs. Container control procedures in the port are very long. The containers must be pointed, controlled and checked manually which is time-consuming. The transaction is materialized by a heavy exchange of documents which generates maritime freight costs and expenses relating to commercial documentation. The implementation of Blockchain technology in the supply chain of the port terminal helps reduce costs and speed up procedures. The dematerialization of the paper documents will help speed up the procedures [14], and will lead to a reduction in communication time, delays, costs and ensure the security of the data.

The Extension of the Network. Blockchain plays a crucial role in stimulating the expansion and interconnection of networks, which creates an added value in the port logistics sector. It allows actors to connect with each other within a decentralized network without the interference of a centralized authority. Also, many blockchain networks, from around the word, could interact easily with each other, which allows foreign ports to collaborate with each other more efficiently. This approach reduces latency and improves performance. Interconnecting also optimizes security, as direct connections are more secure.

4 Conclusion

Since the digitalization strategy of the ANP in Morocco could not achieve the wanted results, we have explored the latest technologies advanced by blockchain and we tried to leverage it in our study. The aim of this paper is to assess current digitalization status of the port of Casablanca and the potential of blockchain technology to improve the port logistics and the emergence of a smarter port.

We conclude that blockchain could create value, enhance the efficiency of current logistics processes and improve the performance indicators to satisfy stakeholder expectations.

References

1. Paksoy, T., Koçhan, Ç., Samar Ali, S.: Logistics 4.0 Digital Transformation of Supply Chain Management. CRC Press, Taylor & Francis Group (2021)
2. Tsiulin, S., Hegner Reinau, K., Hilmola, O., Goryaev, N., Karam, A.: Blockchain-based applications in shipping and port management: a literature review towards defining key conceptual frameworks. Rev. Int. Bus. Strategy (2020)
3. National Ports Authority Annual Reports, from 2017 to 2022
4. Bercy Infos : «Qu'est-ce qu'une chaîne de blocs (blockchain)?», le 12/04/2022, innovation et data, Ministère de l'économie des finances et de la souveraineté industrielle et numérique
5. Verny, J., Oulmakki, O., Cabo, X., Roussel, D.: Blockchain & supply chain: towards an innovative supply chain design. Dans Projectics **2**(n 26) (2020)
6. De la Raudière L. and Mis J.M., "Report of the mission on the uses of blockchains and other register certification technologies", December 2018 National Assembly
7. Bercy Infos: What is a blockchain? 04/12/2022, Innovation and data. Ministry of the Economy, Finance and Industrial and Digital Sovereignty
8. Laforet, L.: Supply Chain and Blockchain: Innovative Technology Adoption Drivers and Strategic Alignment: Case Studies. Phd Thesis, Publicly defended on November 9, 2022
9. Parma, B.: Blockchain Consensus Mechanisms: A Primer for Supervisors. International Monetary Fund, Washington, DC (2022)
10. Hauri, R.: Contributions of Smart Contracts to Blockchains and How to Create a New Cryptocurrency. Secretariat, Geneva School of Management (HEG-GE), Management IT, September 29, 2017
11. Yang, X., Ning, Z., Li, J., Wenjing L., Thomas H.Y.: Distributed Consensus Protocols and Algorithms
12. Mehboub Khan, K., Arshad, J., Iqbal, W., Sidrah, A., Zaib, H.: Blockchain-Enabled Real-Time SLA Monitoring for Cloud-Hosted Services. Springer (2021)
13. Eisenbarth, J., Cholez, T., Perrin, O.: A Comprehensive Study of the Bitcoin P2P Network. University of Lorraine, CNRS, Inria, France
14. Fu, J., Mixue, X., Yongzhong, H., Xueming, S., Chao, Y.: Data processing scheme based on blockchain. EURASIP J. Wirel. Commun. Netw. **2020**(239)
15. Elommal, N., Manita, R.: How blockchain innovation could affect the audit profession: a qualitative study. J. Innov. Econ. Manage. **1**, 37 (2022)

Using Artificial Intelligence (AI) Methods on the Internet of Vehicles (IoV): Overview and Future Opportunities

Adnan El Ahmadi[1(✉)], Otman Abdoun[2], and El Khatir Haimoudi[1]

[1] Polydisciplinary Faculty, Abdelmalek Essaadi University, Larache, Morocco
adnan.elahmadi@etu.uae.ac.ma

[2] Computer Science Department, Faculty of Sciences, Abdelmalek Essaadi University, Tétouan, Morocco

Abstract. Recent developments in communication technologies, intelligent transportation systems, and computer systems have created many opportunities for intelligent traffic solutions that improve efficiency, ease, and safety. Researchers have recently used AI techniques in many application fields because it can improve data-driven networks. Moreover, The IoV reduces a lot of data from various sources that can be used to choose the best routes, improve driver awareness and passenger comfort and safety, optimize road time, and avoid breakdowns. In this study, we evaluate AI methods from various IoV research projects including their advantages and limitation. In the light of these limitations, we discuss open challenges and future research perspectives related to IoV networks using AI methods.

Keywords: Artificial Intelligence (AI) · Internet of things (IoT) · Internet of vehicles (IoV) · Quality of Service (QoS)

1 Introduction

By 2025, there will be more than 75 billion Internet of Things (IoT)-connected devices in use [1]. Among these devices, vehicles will contribute with a significant number. The IoV is a vital component of the smart city concept which is included to the applications of IoT. IoV represents a dynamic mobile network that connects heterogenous networks using Vehicle-to-Vehicle (V2V), Vehicle-to-humans (V2H), Vehicle-to-Personal Device (V2P), Vehicle-to-Network (V2N), Vehicle-to-Sensors (V2S), Vehicle-to-Infrastructure (V2I) and Vehicle-to-Everything (V2X). Figure 1 [2] shows some types of communication in IoV.

The classical Vehicular Ad-hoc Networks (VANETs) are transforming into the IoV [3]. IoV becomes a wider network that includes entities such as persons, things, and other types of heterogeneous networks, in contrast to VANETs, which only include vehicles [4]. IoV incorporates recent communication technologies namely LTE, 5G, to create a dependable communication. Connection between IoV and its infrastructure is handled by

Y. Farhaoui et al. (Eds.): ICAISE 2023, LNNS 837, pp. 193–201, 2024.
https://doi.org/10.1007/978-3-031-48465-0_26

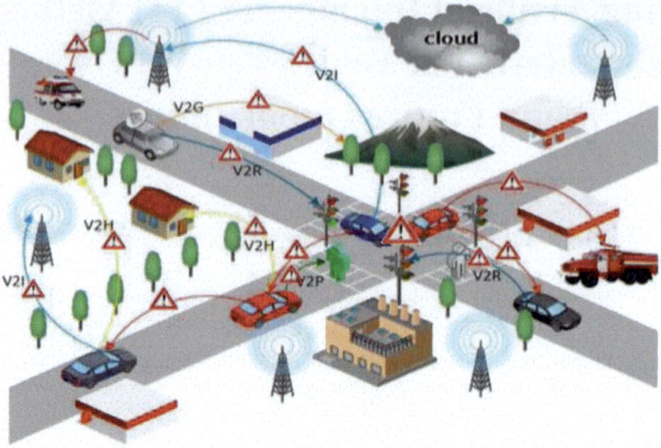

Fig. 1. V2X communications in the IoV [2].

ad-hoc networks, while communication with its backbone network is handled through cellular networks. IoV allows to gather and share data on the environment, vehicles, and the state of the roads [5]. Recently, VANETs haven't been popular since they can't provide sustainable applications and services to the network users in bigger areas. IoV treats a vehicle as a smart-entity with sensors and computing power [3]. IoV integrates users, vehicles, things, and other heterogeneous networks to increase transport system efficiency, security, and safety [6].

In this study, we present a survey about the IoV, we investigate numerous AI approaches and evaluate their applicability as well as limits in resolving various challenges, and we highlight future research areas that need to be investigated further to fully integrate AI with IoV. The following are some of the primary contributions made by this paper:

- We investigated a variety of AI methods, and we emphasized both their advantages and their limitations in resolving a variety of IoV challenges.
- We discuss future research opportunities that need to be investigated further to enable the full integration of AI with IoV.

The remainder of this work is structured as follows. Section 2 presents a brief overview about the IoV networks and different AI algorithms. Section 3 introduces Advantages and Limitations of Integrating AI Methods in the IoV. Section 4 is consecrated for discussing Perspectives and Future directives.

2 IoV and AI Methods Overview

2.1 IoV Architecture

IoV implies the deployment of a reliable architecture that can deal with heterogeneity, scalability, and other requirements of the IoV networks. The main architecture of IoV may be divided into three distinct layers.

- **Perception/Client Layer**: This layer is made up of sensors that are responsible for the collection of data, as well as information on the environment, location, driving patterns, and many other things [7]. The information that has been collected from the sensors is analyzed and communicated to determine the decision to take.
- **Network Layer**: It allows vehicular nodes and other networks and entities to communicate. This layer handles almost all IoV communications.
- **Application Layer**: It manages user interaction, storage, decision making based on data analysis, as well as entertainment and convenience applications such as in-car entertainment, traffic information, and many more.

2.2 IoV Applications

IoV offers a wide range of applications, namely: safety, transport efficiency, and logistics.

2.2.1 Safety

Safety implies that vehicles may automatically transmit real-time collision information, including location data to emergency teams to make the right decision [8].

2.2.2 Transport Efficiency

IoV may be used to achieve transport efficiency applications such as route guiding and optimization and green light efficiency [9]. This will improve traffic management and reduce travel time, fuel consumption, and pollution due to fewer traffic jams.

2.2.3 Logistics

Because it relies so heavily on road transportation, the logistics industry hopes to benefit from IoV. The delivery of goods in a timely and effective manner requires the utilization of road transportation [9]. The application of IoV in logistics may result in several positive outcomes.

2.3 IoV Challenges

Because of the inherent nature of IoV, it faces substantial challenges as compared to other wireless networks. Table 1 summarize the challenges encountered by the IoV paradigm.

2.4 AI Methods and Algorithms: Background

2.4.1 Machine Learning Methods

Supervised Approaches

- *K-Nearest Neighbor (K-NN)*. Classifies data or device characteristics in terms of malicious activity based on the nominated nearest neighbor's votes.
- *Support Vector Machine (SVM)*. SVM creates splitting hyperplane among various class data to classify the given samples.

Table 1. Summary of IoV challenges.

Challenge	Description
Security	As in every network, security is a severe concern and assumes prime importance due to the open nature of IoV [10]
Heterogeneity	IoV encompasses diversity of devices and networks running various protocols leading to heterogeneity, thus hampering interoperability [11]
Scalability	Connected cars are increasing day by day. The unpredictable scale of the network hampers the deploy-ability of the IoV paradigm [11]
Mobility	High mobility causes frequent network disconnections due to low interaction time between vehicles
Reliability	Stringent reliability requirement is a prime concern in IoV as it supports delay sensitive applications [11]

- *Decision Tree (DT).* It is a predictive model which uses a decision tree to observe and reach in the conclusion.
- *Random Forest.* Tree based Supervised ensemble learning model which construct a multitude of DT to predict the output.

Unsupervised Approaches

- *K-Means.* In this algorithm, data is clustered. Initially, k-centroids are randomly chosen. Nodes are then clustered around centroids. The algorithm is responsible for recalculating centroids using a cluster's node average.
- *Principal Component Analysis (PCA)* is the process of computing the principal components and using them to perform a change of basis on the data, sometimes using only the first few principal components and ignoring the rest.

2.4.2 Deep Learning Methods

Supervised Approaches

- *Convolutional Neural Network (CNN).* Reduces the connection between layers and combines convolutional layer with pooling layer to deteriorate training complexity.
- *Recurrent Neural Network (RNN).* Works on graph-like structure to detect malicious data in time-series based threats

Unsupervised Approaches

- *A restricted Boltzmann machine (RBM)* is a generative stochastic artificial neural network that can learn a probability distribution over its set of inputs.

Semi-supervised Approaches

- *Generative Adversarial Networks* (GANs) are a model architecture for training a generative model, and it is most common to use deep learning models in this architecture.

2.5 Swarm Intelligence (SI)

- *Particle Swarm Optimization (PSO).* The PSO is a global optimization method that may be used to solve problems with single-point or n-dimensional surface solutions. It ensures that each particle chooses one of its best previous places or advances to a better one using its speed [12].
- *Ant Colony Optimization (ACO).* This algorithm finds near-optimal graph solutions by choosing the shortest route.

3 Advantages and Limitations of Integrating AI Methods in the IoV

3.1 Advantages and Limitations

The use of Machine learning algorithms increases QoS in IoV applications and solve several encountered problems. Deep Learning has earned a lot of attention for its merits, and it's employed in many fields to improve earlier approaches. We list in Table 2 the advantages and Limitations of Integrating AI Methods (ML, DL, SI) in IoV networks.

3.2 Research Using AI Methods and Algorithms in the IoV

Recently, people all over the world have begun to rely on innovative modes of transportation, which make driving more comfortable, simple, and secure, and which also help to prevent accidents and the deaths of people. This can be achieved using AI methods. The Table 3. Represents a summary of some research using the AI methods and Algorithms in the IoV.

4 Future Opportunities

Among the unresolved issues and challenges that present potential avenues for the development of effective and safe solutions are the following:

- **Security and Privacy**. IoV stores a great amount of private information pertaining to vehicle owners: videos and voices of drivers, images, locations, and other data. Therefore, the information on the cognitive IoV network is very vulnerable and requires additional protection. Encryption should be used when transmitting data to protect vehicular devices with low intelligence and limited computer resources, from attackers. Consequently, it's important to develop a safe and automated future driving scenario using the AI-based methods.
- **Automation**. IoV contains a significant number of heterogeneous vehicles, resulting in a system structure that is highly complex. In a such environment, researchers should consider the high-performance computing. As a result, the issue of how to achieve automatic coordination and administration across networks of various sorts is also complex. With the advent of innovative communication technologies, a substantial amount of research is required to investigate the issues of task automation.
- **Energy Awareness**. Batteries are likely going to take over as the primary energy storage option for self-driving cars in the future. While IoV is always progressing, energy-awareness must be achieved using intelligent methods.

Table 2. Advantages and limitations of integrating AI methods in the IoV

AI Algorithm or Method		Advantages	Limitations	Application
Machine learning algorithms and methods	KNN	– High accuracy for node classification	– Complex – Processing time is long	– Detect crashes or incidents – Maintain stable clusters for IoV cluster-based techniques
	SVM	– Processes big data with few input variables	– Hard to comprehend and perceive SVM-models – Challenge of Optimized kernel choice	– Optimize clustering of vehicular nodes – Detect/Prevent attacks
	DT	– Simple – Transparent	– Requires big memory – Complex when several DTs are necessary	– Enhances route selection – Traffic signal automation – Identifying abnormalities of harmful vehicles
	RF	– Handles over-fitting – Requires a few input values	– Complex when dealing with big data – Long processing time – More security challenges	– Predict road congestion – Detect attacks
	K-means	– Processing anonymous data	– Less efficient for treat detection – Fix number of clusters	– Detection of road jam – Securing communication between vehicular nodes
	PCA	– Useful for dimensionality reduction of data	– Requires implementation of other ML methods	– Identify driving risks – Predict Denial of Service (DoS) attacks

(*continued*)

Table 2. (*continued*)

AI Algorithm or Method		Advantages	Limitations	Application
Deep learning algorithms and methods	CNN	– Scalable – High efficiency – Less complex	– High computation capabilities – Difficult to deploy on resource-constrained nodes	– Analysis of multimedia data – Accident prediction based on video data
	RNN	– Performant with discrete data	– Overloading	– Improving mobility – Sharing intelligence with edge, fog, and cloud – Anticipation of risks
	RBM	– Feedback techniques	– Training time is too long – Difficult to apply on highway roads	– Detecting traffic density – Detect/Prevent intrusions – Identify road status
	GAN	– Pre-defined number of iterations	– Complex – Not producing discrete data while training	– Reduce delay – Determine path of a vehicle
Swarm intelligence algorithms and methods	PSO	– Does not need hypotheses about optimizing the problem	– Requires the cooperation of several vehicles – More security challenges – Vehicle identification requires to be enforced	– Optimize QoS of Routing protocols – Deal with traffic congestion
	ACO	– Gradient information not required for problem optimization	– High processing time	– Detect network abnormalities

- **Performance**. In harsh weather or challenging road conditions, the observation accuracy of radar or cameras of intra-vehicle sensors may decrease. In the case that sensors with high accuracy and more performance are to be invented in response to these unique circumstances, customers could not afford the appropriate cost. AI technology presents us with potential solutions; however, the system performance issue cannot be handled immediately. Future IoV research will focus on how to link environment collaborative perception technology in physical space with flow data mining and prediction technology in network space using AI algorithms.

Table 3. Research using AI methods and algorithms in the IoV.

Year	Reference	Simulator/Dataset	Main contributions	Evaluation parameters	AI methods
2019	[13]	OMNET++	– Anomaly detection – Network optimization – Classification of the vehicular network data	– Detection Rate – Accuracy – False Positive Rate	– Support Vector Machine (SVM)
2023	[14]	SUMO/NS2	– Improving QoS – Clustering-based network topology	– Packet Delivery Ratio – Packet Dropped Ratio – Overhead – Throughput – Latency	– Particle swarm optimization (PSO)
2020	[15]	–	– Design of data-driven IDS	– Accuracy - Precision – Recall – False Alarm – F-Score	– Convolutional Neural Network (CNN)
2023	[16]	VeReMi	– Implementation of False BSM Detection Scheme (FBDS)	– Precision – Recall – Accuracy	– Random Forest (RF)
2019	[17]	SUMO/NS2	– Implementing CACOIOV algorithm to increase stability	– Packet Delivery Ratio – Packet Dropped Ratio – Throughput – Latency	– Ant Colony Optimization
2022	[18]	MATLAB	– Enhancing the security (Blockchain-based) – Enhancing efficiency (AI-based)	– Average Risk Score – Throughput – Latency – Assign accuracy rate	– Generative Adversarial Networks (GANs)

5 Conclusion

In this paper, we have provided a thorough analysis of the AI methods that are able to enhance the performance of IoV applications. The ML, DL, and SI subfields of AI may collaborate to produce optimum solutions that overcome the limits of ML, DL, and SI separately. We have examined how some of the advantages of AI approaches may be used in the IoV context despite the constraints associated with some AI techniques.

References

1. Statista: IoT devices installed base worldwide 2015–2025. https://www.statista.com/statistics/471264/iot-number-of-connected-devices-worldwide/. Accessed 14 Jun 2022
2. Azzaoui, N., Korichi, A., Brik, B., Fekair, M.E.: Towards optimal dissemination of emergency messages in internet of vehicles: a dynamic clustering-based approach. Electronics **10**(8), (2021). https://doi.org/10.3390/electronics10080979
3. Verma, H.K., Sharma, K.P.: Evolution of VANETS to IoV: applications and challenges. Tehnički Glasnik **15**(1), 143–149 (2021)
4. Mchergui, A., Moulahi, T., Zeadally, S.: Survey on artificial intelligence (AI) techniques for vehicular ad-hoc networks (VANETs). Veh. Commun. 100403 (2021)
5. Benalia, E., Bitam, S., Mellouk, A.: Data dissemination for Internet of vehicle based on 5G communications: a survey. Trans. Emerg. Telecommun. Technol. **31**(5), e3881 (2020)
6. El Ahmadi, A., Abdoun, O., Haimoudi, E.K.: A comprehensive study of integrating AI-based security techniques on the internet of things. In: International Conference on Advanced Intelligent Systems for Sustainable Development. Springer, pp. 348–358 (2022)
7. Ang, L.-M., Seng, K.P., Ijemaru, G.K., Zungeru, A.M.: Deployment of IoV for smart cities: applications, architecture, and challenges. IEEE Access **7**, 6473–6492 (2018)
8. Mahmood, A., Siddiqui, S.A., Sheng, Q.Z., Zhang, W.E., Suzuki, H., Ni, W.: Trust on wheels: towards secure and resource efficient IoV networks. Computing **104**(6), 1337–1358 (2022)
9. Dhanare, R., Nagwanshi, K.K., Varma, S.: A study to enhance the route optimization algorithm for the internet of vehicle. Wirel. Commun. Mob. Comput. **2022** (2022)
10. Al-Shareeda, M.A., Manickam, S.: Security Methods in Internet of Vehicles. arXiv preprint arXiv:2207.05269 (2022)
11. Seth, I., et al.: A taxonomy and analysis on internet of vehicles: architectures, protocols, and challenges. Wirel. Commun. Mob. Comput. **2022**, 9232784 (2022). https://doi.org/10.1155/2022/9232784
12. Shami, T.M., El-Saleh, A.A., Alswaitti, M., Al-Tashi, Q., Summakieh, M.A., Mirjalili, S.: Particle swarm optimization: a comprehensive survey. IEEE Access (2022)
13. Garg, S., Kaur, K., Kaddoum, G., Gagnon, F., Kumar, N., Han, Z.: Sec-IoV: a multi-stage anomaly detection scheme for internet of vehicles. In: Proceedings of the ACM MobiHoc Workshop on Pervasive Systems in the IoT Era, pp. 37–42 (2019)
14. Kayarga, T., Ananda, K.S.: Improving QoS in internet of vehicles integrating swarm intelligence guided topology adaptive routing and service differentiated flow control. Int. J. Adv. Comput. Sci. Appl. **14**(4), (2023)
15. Nie, L., Ning, Z., Wang, X., Hu, X., Cheng, J., Li, Y.: Data-driven intrusion detection for intelligent internet of vehicles: a deep convolutional neural network-based method. IEEE Trans Netw Sci Eng **7**(4), 2219–2230 (2020)
16. Anyanwu, G.O., Nwakanma, C.I., Lee, J.M., Kim, D.-S.: Novel hyper-tuned ensemble random forest algorithm for the detection of false basic safety messages in internet of vehicles. ICT Express **9**(1), 122–129 (2023)
17. Ebadinezhad, S., Dereboylu, Z., Ever, E.: Clustering-based modified ant colony optimizer for internet of vehicles (CACOIOV). Sustainability **11**(9), 2624 (2019)
18. Priscila, S.S., et al.: Risk-based access control mechanism for internet of vehicles using artificial intelligence. Security Commun. Netw. **2022** (2022)

A Novel for Seizure Prediction Using Artificial Intelligent and Electroencephalography

Ola Marwan Assim(✉) and Ahlam Fadhil Mahmood

University of Mosul, Mosul, Iraq
ola.marwan@uomosul.edu.iq

Abstract. Seizure prediction is critical in effectively managing epilepsy, a chronic neurological disorder characterized by sudden abnormal brain activity. This study proposes a Long Short-Term Memory (LSTM) neural network deep learning model for seizure prediction using non-invasive scalp Electroencephalography (EEG) recordings. The LSTM model is well-suited for time series analysis, making it an ideal candidate for detecting patterns in EEG data. The model is designed to distinguish between ictal and interictal states, accurately predicting the onset of seizures with high sensitivity and minimal false alarms. The results demonstrate the effectiveness of the proposed model, achieving accuracy rates ranging from 99.07 to 99.95% and a sensitivity of 1, indicating that all seizure states are correctly predicted. The model's low false positive and false negative rates validate its potential for early and accurate seizure prediction. Future works are presented at the end of this paper to enhance the model's performance and translate it into practical clinical applications. The proposed LSTM model holds promising prospects for timely seizure warnings and improving the quality of life for patients with epilepsy.

Keywords: Seizure · Prediction · Long short term memory · Electroencephalography

1 Introduction

Epilepsy is a chronic neurological disorder that occurs suddenly and abnormally in the brain, with a recurrence characteristic. The cause of the disorder is related to abnormal activity of brain neurons [1]. When afflicted by this abrupt neurological disorder, the patient experiences a disruption in normal brain function, leading to various abnormal reactions such as fainting, physical imbalance, convulsions, muscle contractions, and loss of consciousness [2]. Epilepsy patients face significant consequences as seizures can profoundly affect all aspects of their lives, even posing life-threatening risks [3]. Hence, the early prediction of epilepsy becomes an essential prerequisite for effective seizure management. The significance of early detection lies in providing patients with timely awareness of potential dangers, enabling them to take preventive measures to control seizures and avoid potentially life-threatening situations during such episodes [4–6]. Early prediction holds immense importance not only for patients and their families but

Y. Farhaoui et al. (Eds.): ICAISE 2023, LNNS 837, pp. 202–209, 2024.
https://doi.org/10.1007/978-3-031-48465-0_27

also for healthcare professionals. Several screening techniques for epilepsy exist, such as Electroencephalography (EEG) [7]. EEG allows for continuous monitoring of electrical brain activity, effectively capturing hidden features of neurological disorders. EEG stands out as a convenient and cost-effective option. The seizure process is typically divided into four states: the ictal state, preictal state, postictal state, and interictal state, as shown in Fig. 1. The key to developing a seizure prediction system that can forecast upcoming seizures is effectively separating the preictal periods from the interictal periods. The preictal period, which varies in duration across studies, refers to the period preceding a seizure. The interictal period encompasses the segments of the signal that are neither preictal nor ictal [8]. Identifying the pre-ictal state as early and reliably before seizure onset is crucial to improving seizure prediction accuracy.

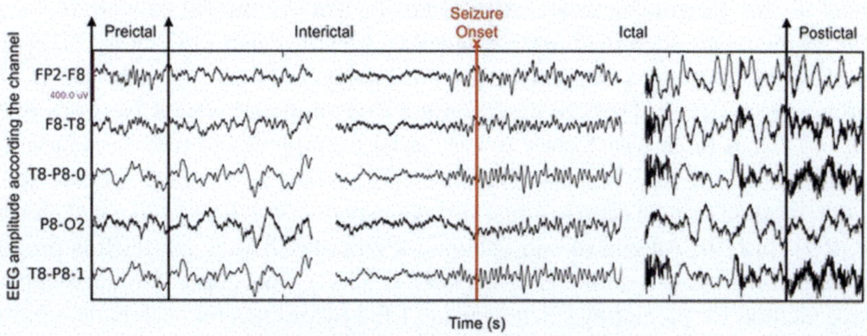

Fig. 1. Brain states for epileptic patients.

Over the past few decades, there has been a proliferation of algorithms dedicated to predicting seizures. These algorithms aim to anticipate seizures by analyzing preictal changes [9]. Deep learning is One of the subfields of artificial intelligence. In contrast, Deep learning has surpassed traditional machine learning methods by excelling in automatic representation learning, significantly reducing the need for manual feature engineering [10, 11]. Despite some research progress in epilepsy prediction, to predict a seizure must consider the uniqueness of each person's epilepsy and the significant variability in seizure patterns. What may work effectively for one person may not yield the same results for another [12]. The primary purpose of epilepsy prediction is to allow patients enough time to take preventive measures or prepare for impending seizures to control them or avoid accidents, the main contributions of our research are as follows:

- Development of a two-layer LSTM model optimized for time series analysis, particularly in the context of seizure prediction for each patient individually.
- Effective classification of ictal and interictal states is present and resulting in accurate and early seizure predictions.
- Robust performance across different patients makes the model's generalizability and potential for clinical applications.

The paper is organized as follows: Sect. 2 presents previous research, Sect. 3 shows methodology the results and insights gained from the experiments presented in Sect. 4, Sect. 5 presents a discussion, and Sect. 6 offers final remarks and prospects.

2 Related Works

In seizure prediction, researchers have leveraged well-established classification algorithms and evaluation metrics. Seizure prediction is often approached as a binary classification task, where the objective is to distinguish between pre-seizure and non-seizure states. Classification models are trained using input data to predict whether a seizure will occur within a specific time window. In recent years, the ascent and widespread adoption of deep learning techniques have yielded promising outcomes in automatically extracting features from time series data. In [13], Convolutional neural networks were used to extract spatial features, and recurrent neural networks predict seizures early time. Paper [14] presented Long-term recurrent convolutional network (LRCN) is proposed for predicting epileptic seizures. EEG time series are converted into 2D images for multichannel fusion. Deep features are extracted using a convolutional network block, and preictal segments are identified using a block of LSTM. Paper [15] showed EEG segments of various durations evaluated using Single-layer and two-layer LSTM models. The proposed models in [16] are based on the Convolutional Neural Network (CNN) model. In [17], A three-transformer tower model is employed to fuse and classify the extracted features of EEG signals. The study [18] proposes a patient-specific seizure prediction method using a deep residual shrinkage network (DRSN) and gated recurrent unit (GRU). In [19], An end-to-end epileptic seizure prediction approach is proposed based on the long short-term memory network (LSTM). This paper introduces a highly effective method for predicting seizures using EEG recordings for each patient individually. It specifies the period before seizures for each patient to take the necessary to reduce the risks. The pre-seizure period is segmented into several time windows, and deep learning techniques are employed to classify these windows, resulting in accurate seizure prediction. Sensitivity is a commonly used indicator when evaluating seizure prediction algorithms' performance. Sensitivity represents the precise prediction of seizures divided by the total number of seizures recorded [20]. Additional performance indices related to specificity include time in warning, which denotes the fraction of time the system makes positive predictions and false positive error rate [21, 22].

3 Methodology

3.1 Dataset

In this study, a neonatal EEG dataset was used. It includes 79 raw EDF files capturing newborn EEG recordings and three annotation files in CSV formats. This comprehensive collection is a valuable resource for studying brain activity and exploring possible neurological conditions in newborns. Some patients diagnosed with epileptic seizures, according to the first specialist, were selected, and the model was applied separately. Patients 1, 15, 19, 25, 38, 41, 50, and 66 with an EEG record length of 6993, 6898, 9006, 6709, 6095, 9684, 9850 and 11,350 s were selected. These lengths provide valuable insights into the duration of EEG recordings for specific patients, which is vital for further analysis and research in neonatal EEG and seizure prediction.

3.2 The Proposed Model

The proposed model to predict seizures is an LSTM (Long Short-Term Memory) neural network designed specifically for processing sequential data and well-suited for tasks involving time series analysis and other sequential data domains.

The model architecture comprises three main layers: the first LSTM layer consists of 64 units (neurons), and The second LSTM layer includes 32 units. It processes the output sequences from the first LSTM layer. The final Dense layer has a single unit, representing the model's output. It performs a regression task, predicting a continuous value. This layer contains 33 trainable parameters, Fig. 2 present the architecture of the model with the details about the training process. This LSTM model aims to capture complex patterns and dependencies present in sequential data. It's a relatively small model with a moderate number of parameters.

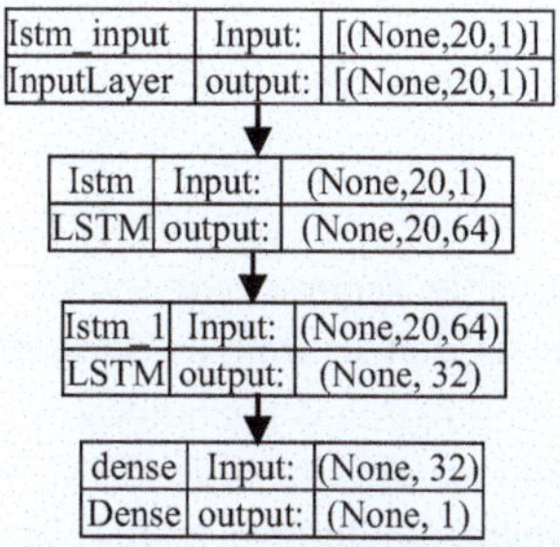

Optimizer	Adam
Learning Rate	0.001
Loss Function	MSE
Number of Epochs	50
Batch Size	32
Total Trainable Parameters	29,345
MSE: Mean Square Error	

Fig. 2. The proposed model architecture.

4 Results

The performance of the proposed model: Train RMSE (Root Mean Squared Error) score, Test score, accuracy (Acc.), sensitivity (Sen.), False Positive Rate (FpR), False Negative Rate (FnR), and Time/epoch in seconds for predicting seizures at different time intervals before the seizure occurs is shown in Table 1. The model's performance is consistently strong, achieving high accuracy and sensitivity for all patients at this time interval. The test scores (RMSE) remain low, indicating accurate predictions close to the actual seizure occurrence.

Pre-seizure period second, the model shows slightly reduced performance, particularly in sensitivity, as the model is not capturing the pre-seizure patterns effectively. The accuracy remains high, but the sensitivity is 0 for all patients, indicating that the model is not correctly predicting the pre-seizure state. As is evident in the results, the period before the seizure that can be expected differs from one patient to another, for patients (1, 15, 25, 41) it was predicted 28 min before the onset. Patient 19 can predict his seizures 27 min before their occurrence. Patient 32's seizures can be predicted 23 min in advance, while Patient 51 seizures can be predicted 31 min in advance.

Table 1. The results before 20 s of a seizure occurring.

P. No.	No. of samples	Train		Test		Acc		Sen		FpR		FnR		Time/epoch	
		R1	R2	R1	R2	R1	R2	R1	R2	R1	R2	R1	R2	R1	R2
P1	6933	0.08	0.18	0.10	0.12	0.990	0.976	1	0	0	0	0	1	38	6
P15	6899	0.08	0.18	0.07	0.12	0.995	0.986	1	0	0	0	0	1	37	5
P19	9007	0.05	0.05	0.03	0.03	0.999	0.997	1	0	0	0	0	1	52	6
P25	6710	0.07	0.07	0.04	0.05	0.998	0.992	1	0	0	0	0	1	38	5
P38	6096	0.08	0.18	0.08	0.12	0.984	0.983	1	0	0	0	0	1	33	15
P41	9685	0.10	0.10	0.08	0.09	0.993	0.986	1	0	0	0	0	1	74	8
P50	9851	0.05	0.05	0.05	0.05	0.998	0.991	1	0	0	0	0	1	57	7
P66	11350	0.02	0.04	0.02	0.03	0.999	0.999	1	0	0	0	0	1	61	12

Results before 20 s of a seizure occurring, R2: results of pre-seizure occurring

5 Discussion

The obtained results from various studies reflect the performance of different seizure prediction and detection methods. [13] achieved a notable accuracy of 99.6% while maintaining a low false alarm rate of 0.004 h^{-1}, and the prediction time for seizures was reported to be around 1 h. Conversely, Reference [14] demonstrated a balanced accuracy of 93.40% accompanied by a sensitivity of 91.88% and specificity of 86.13%, with a corresponding false positive rate of 0.04 F P/h. The aggregated outcomes from [15] showcased consistently high performance, with an average accuracy of 98.14%, sensitivity of 98.51%, and specificity of 97.78%. Similarly, Reference [16] reported a solid accuracy rate of 95%. Additionally, [17] exhibited a sensitivity of 92.1%, emphasizing the method's capability to correctly identify positive instances. Furthermore, [18] indicated a sensitivity of 90.54% and an AUC of 0.88, suggesting its effectiveness in distinguishing between positive and negative cases. The corresponding false prediction rate was noted as 0.11/h. In the context of mean sensitivity, Reference [19] reported an average of 91.76%, with an associated false prediction rate of 0.29/h. In the proposed approach, as outlined, the accuracy values demonstrated variability across different patients, ranging from 99.07 to 99.95%. Impressively, a sensitivity of 1 indicated accurate predictions for all ictal states. Equally noteworthy, there were no false alarms indicated by a false positive rate of 0, and no missed ictal states as denoted by a false negative rate of 0. The period before seizure occurrences was estimated to be between 23 and 32 s. Collectively, these results underline the advancements in seizure prediction and detection techniques, showcasing substantial accuracy rates and sensitivities across various studies. However, the proposed approach stands out for its personalized accuracy rates and the ability to accurately predict imminent seizures, minimizing false alarms and missed events. Figure 3 presents the comparison between previous studies and the proposed model in the performance metrics.

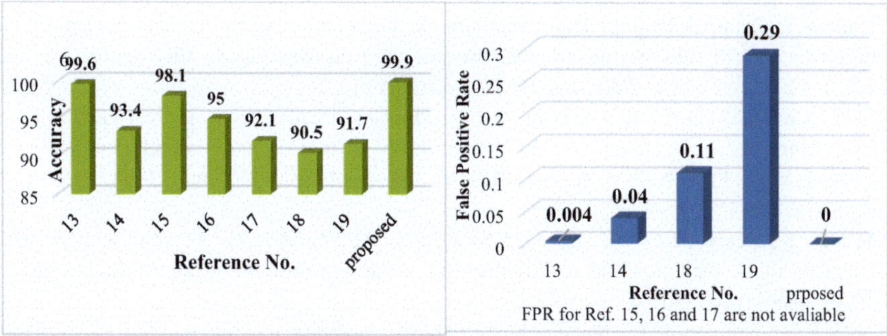

Fig. 3. Comparison of performance metrics in the proposed model and previous studies.

6 Conclusion

This study presents a novel Long Short-Term Memory (LSTM) neural network model designed for seizure prediction using non-invasive scalp EEG recordings. The model has shown remarkable accuracy in distinguishing between ictal and interictal states, allowing for effective seizure prediction.

The LSTM architecture has established its proficiency in the reliable detection of seizure occurrences by demonstrating a notable combination of high accuracy, sensitivity, and low rates of false positives and false negatives. Despite the promising results obtained with the proposed LSTM model, several avenues for further refinement and exploration are identified:

1. Dataset Expansion: Extending the dataset to include a broader range of patients would help to improve the model's generalizability and real-world utility.
2. Multimodal Data Integration: The incorporation of supplementary data modalities, such as clinical insights or other physiological signals, has the potential to improve the model's precision and provide additional perspectives, thereby advancing the comprehensiveness of seizure prediction.
3. Real-time Deployment: Integrating our model into real-time seizure prediction systems, which allows for continuous monitoring and timely alerts to patients and caregivers, is valuable for optimizing seizure management.

References

1. World Health Organization: Epilepsy: A Public Health Imperative. Available online: https://www.who.int/mental_health/neurology/epilepsy/report_2019/en/ (2019)
2. Siuly, S., Li, Y.: Discriminating the brain activities for brain–computer interface applications through the optimal allocation-based approach. Neural Comput. Appl. **26**, 799–811 (2015). https://doi.org/10.1007/s00521-014-1753-3
3. Vaurio, L., Karantzoulis, S., Barr, W.B.: The impact of epilepsy on quality of life. In: Changes in the Brain: Impact on Daily Life, pp. 167–187 (2017). https://doi.org/10.1007/978-0-387-98188-8_8

4. Kapoor, B., Nagpal, B., Jain, P.K., Abraham, A., Gabralla, L.A.: Epileptic seizure prediction based on hybrid seek optimization tuned ensemble classifier using EEG signals. Sensors **23**(1), 423 (2022). https://doi.org/10.3390/s23010423
5. Schroeder, G.M., Diehl, B., Chowdhury, F.A., Duncan, J.S., de Tisi, J., Trevelyan, A.J., Wang, Y.: Seizure pathways change on circadian and slower timescales in individual patients with focal epilepsy. Proc. Natl. Acad. Sci. **117**(20), 11048–11058 (2020). https://doi.org/10.1073/pnas.1922084117
6. Mlinar, S., Petek, D., Cotič, Ž., Mencin Čeplak, M., Zaletel, M.: Persons with epilepsy: between social inclusion and marginalisation. Behav. Neurol. **2016** (2016). https://doi.org/10.1155/2016/2018509
7. Assi, E.B., Nguyen, D.K., Rihana, S., Sawan, M.: Towards accurate prediction of epileptic seizures: a review. Biomed. Signal Process. Control **34**, 144–157 (2017). https://doi.org/10.1016/j.bspc.2017.02.001
8. Rogowski, Z., Gath, I., Bental, E.: On the prediction of epileptic seizures. Biol. Cybern. **42**(1), 9–15 (1981). https://doi.org/10.1007/BF00335153
9. Alzubaidi, L., Zhang, J., Humaidi, A.J., Al-Dujaili, A., Duan, Y., Al-Shamma, O., Farhan, L.: Review of deep learning: Concepts, CNN architectures, challenges, applications, future directions. J. Big Data **8**, 1–74 (2021). https://doi.org/10.1186/s40537-021-00444-8
10. Nafea, M.S., Ismail, Z.H.: Supervised machine learning and deep learning techniques for epileptic seizure recognition using EEG signals—a systematic literature review. Bioengineering **9**(12), 781 (2022). https://doi.org/10.3390/bioengineering9120781
11. Sarker, I.H.: Deep learning: a comprehensive overview on techniques, taxonomy, applications and research directions. SN Comput. Sci. **2**(6), 1–20 (2021). https://doi.org/10.1007/s42979-021-00815-1
12. Bandarabadi, M., Rasekhi, J., Teixeira, C.A., Karami, M.R., Dourado, A.: On the proper selection of preictal period for seizure prediction. Epilepsy Behav. **46**, 158–166 (2015). https://doi.org/10.1016/j.yebeh.2015.03.010
13. Daoud, H., Bayoumi, M.A.: Efficient epileptic seizure prediction based on deep learning. IEEE Trans. Biomed. Circ. Syst. **13**(5), 804–813 (2019)
14. Wei, X., Zhou, L., Zhang, Z., Chen, Z., Zhou, Y.: Early prediction of epileptic seizures using a long-term recurrent convolutional network. J. Neurosci. Methods **327**, 108395 (2019). https://doi.org/10.1016/j.jneumeth.2019.108395
15. Singh, K., Malhotra, J.: Two-layer LSTM network-based prediction of epileptic seizures using EEG spectral features. Complex Intell. Syst. **8**(3), 2405–2418 (2022). https://doi.org/10.1007/s40747-021-00627-z
16. Ouichka, O., Echtioui, A., Hamam, H.: Deep learning models for predicting epileptic seizures using iEEG signals. Electronics **11**(4), 605 (2022). https://doi.org/10.3390/electronics11040605
17. Yan, J., Li, J., Xu, H., Yu, Y., Xu, T.: Seizure prediction based on transformer using scalp electroencephalogram. Appl. Sci. **12**(9), 4158 (2022). https://doi.org/10.3390/app12094158
18. Xu, X., Zhang, Y., Zhang, R., Xu, T.: Patient-specific method for predicting epileptic seizures based on DRSN-GRU. Biomed. Signal Process. Control **81**, 104449 (2023). https://doi.org/10.1016/j.bspc.2022.104449
19. Wu, X., Yang, Z., Zhang, T., Zhang, L., Qiao, L.: An end-to-end seizure prediction approach using long short-term memory network. Front. Hum. Neurosci. **17**, 1187794 (2023). https://doi.org/10.3389/fnhum.2023.1187794
20. Medvedovsky, M., Taulu, S., Gaily, E., Metsähonkala, E.L., Mäkelä, J.P., Ekstein, D., Paetau, R.: Sensitivity and specificity of seizure-onset zone estimation by ictal magnetoencephalography. Epilepsia **53**(9), 1649–1657 (2012). https://doi.org/10.1111/j.1528-1167.2012.03574.x

21. Ren, Z., Han, X., Wang, B.: The performance evaluation of the state-of-the-art EEG-based seizure prediction models. Front. Neurol. **13**, 1016224 (2022). https://doi.org/10.3389/fneur.2022.1016224
22. Leal, A., Curty, J., Lopes, F., Pinto, M. F., Oliveira, A., Sales, F., Teixeira, C.A.: Unsupervised eeg preictal interval identification in patients with drug-resistant epilepsy. Sci. Rep. **13**(1), 784 (2023). https://doi.org/10.1038/s41598-022-23902-6

Artificial Intelligence in Dentistry: What We Need to Know?

Rachid Ait Addi[1(✉)], Abdelhafid Benksim[1,2], and Mohamed Cherkaoui[1]

[1] Department of Biology, Laboratory of Human Ecology, School of Sciences Semlalia, Cadi Ayyad University, Marrakech, Morocco
dr.rachid.aitaddi@gmail.com, cherkaoui@uca.ac.ma
[2] High Institute of Nursing and Technical Health, Marrakech, Morocco

Abstract. Although dated back to 1950, artificial Intelligence (AI) has not become a practical tool until two decades ago. In fact, AI is the capacity of machines to do tasks that normally require human intelligence. AI applications have been started to provide convenience to people's lives due to the rapid development of big data computational power, as well as AI algorithm. Furthermore, AI has been used in every dental specialties. Most of the applications of AI in dentistry are in diagnosis based on X-ray or visual images, whereas other functions are not as operative as image-based functions mainly due to data availability issues, data uniformity and computing power for processing 3D data. AI machine learning (ML) patterns assimilate from human expertise whereas Evidence-based dentistry (EBD) is the high standard for the decision-making of dentists. Thus, ML can be used as a new precious implement to aid dental executives in manifold phases of work. It is a necessity that institutions integrate AI into their theoretical and practical training programs without forgetting the continuous training of former dentists.

Keywords: Artificial Intelligence (AI) · Machine learning (ML) · Deep learning (DP) · Dentistry · Diagnostic · Convolutional neural networks (CNN)

1 Introduction

This Artificial Intelligence (AI) is developing fast in all sectors. It may assimilate human expertise and do tasks that required human intelligence. It can be defined by the theory and development of computer systems capable of executing tasks that need human understanding, such as seeing perception, talk identification, resolution, and translation [1]. Also, it a machine's ability to express its own intelligence by solving problems based on data. Machine learning (ML) uses algorithms to anticipate outcomes from a set of data. The aim is to facilitate it for machines without human contribution to study from data and fix problems (Fig. 1) [2]. Artificial intelligence has been employed in every domain such as industry, medicine, dentistry, research, portable display, hospital monitoring, automatic and non-Human assistants.

AI may be often used as a practical implement helping dentists to minimize their work time. In addition of diagnosing utilizing data feed directly, AI is able to acquire

Y. Farhaoui et al. (Eds.): ICAISE 2023, LNNS 837, pp. 210–216, 2024.
https://doi.org/10.1007/978-3-031-48465-0_28

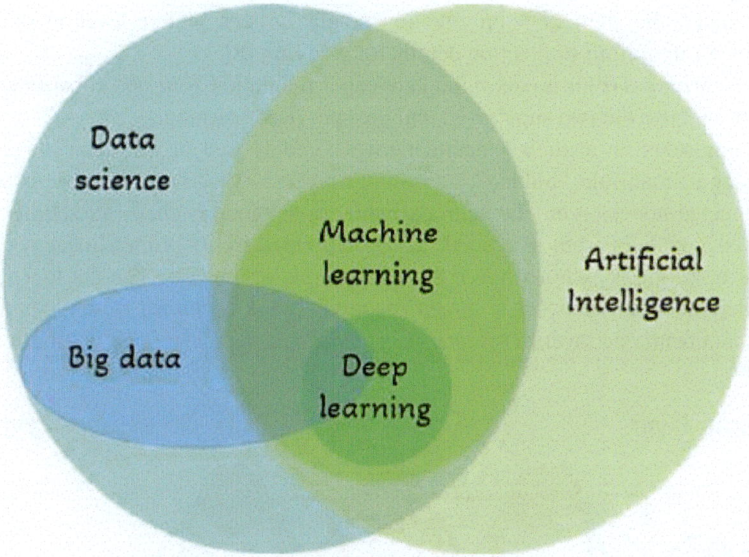

Fig. 1 Key elements of artificial intelligence systems [2]

a knowledge from several information origins to make a diagnostic further on human capabilities.

2 Classifications of AI

AI may be sorted as weak AI and strong AI. Weak AI, utilizes an application skilled to fix unique or precise functions.

Now, the most utilized AI is weak AI. For example, of AI in strengthening studying we can cite AlphaGo, and talk operating we have Google translation, and Amazon chat robot [3].

Strong AI calls attention to the competence and cleverness of AI equaling that of humans. It possesses its proper understanding and behavior whose supplessness is comparable to humans [4].

Therefore, till then no strong AI applications are available. In Addition, ML is classified as supervised, semi-supervised and unsupervised learning.

Supervised learning utilizes labelled inputs for learning to supervise the algorithm. The algorithm studies from the labelled input, releases and recognizes the shared characteristics of the labelled input to take auguries about unlabeled input [5].

At variance, unsupervised learning, performs automatically to discover the different characteristics of unlabeled data [6]. Semi-supervised learning reposes in mid of supervised and no-supervised learning, which employs a little size of labelled input jointly with a big size of unlabeled data during training [7]. Lately, a novel process named weakly supervised learning has been progressively common in the AI domain to reduce labelling expenses.

Especially, the item division function solely utilizes picture-level marks as an alternative of item limit or position details for studying [8].

Deep learning (DP) is now a significant experimentation zone and constitutes a part of ML. It may use the two supervised and unsupervised learning.

DP represents an artificial "neural network" composed of a base of three nodal layers—input, manifold "hidden", and output layers. Every layer is made of several interconnected nodes (synthetic neurons) while every node is characterized by weight and biased threshold from m crucial factors, provided by its proper linear regression model. The weight is assigned when there is an input of the node. Similar to a decision tree model, the neural network as a feedforward network is defined from the process of passing data from one layer to the next (Fig. 2) [9].

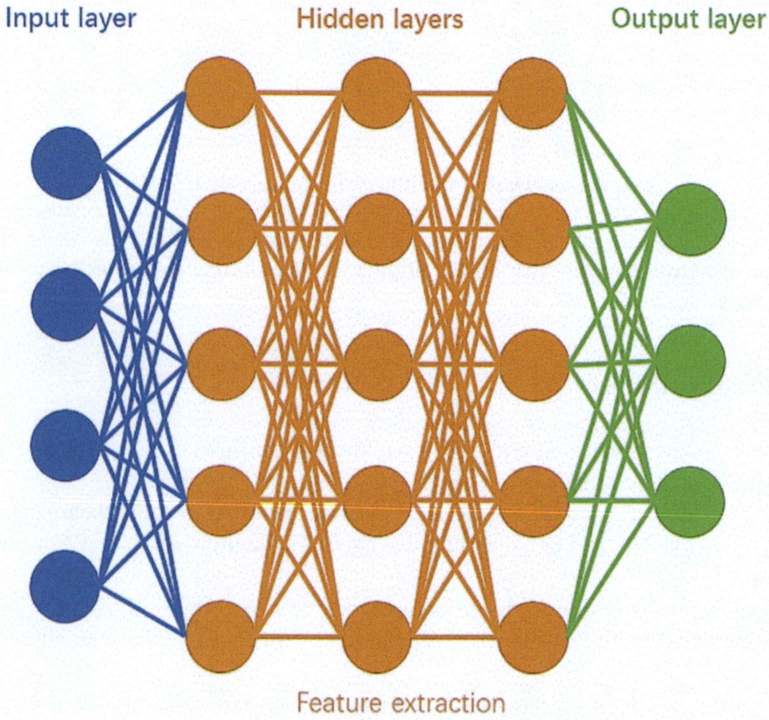

Fig. 2 Schematic diagram of deep learning [9]

A deep neural connection may bring characteristics from the input, without human intervention. Neural networks (NN) are the mainstays of deep learning algorithms. In fact, there are different variants of NN, the most important sorts of neural networks are artificial neural networks (ANN), convolutional neural networks (CNN) and generative adversarial networks (GAN).

2.1 Artificial Neural Networks (ANN)

ANN is composed of a set of neurons and layers, a group of three layers corresponds to an elemental pattern for deep learning. Only forward direction is allowed to inputs. Input neurons bring out characteristics of input data from the input layer and dispatch data to hidden layers, and the data traverse all the hidden layers consecutively. At the end, the output layers expose and summarize the results. From previous layers, all the hidden layers in ANN may weigh the data and perform adjustments to send data to the next layer. Each hidden layer acts as an input and output layer, allowing the ANN to understand more complex features [10].

2.2 Convolutional Neural Networks (CNN)

CNN is a sort of deep learning pattern mostly utilized for picture identification and production. The presence in CNN of convolution layers, the pooling layer and the fully connected layer in the hidden layers is the principal difference between CNN and ANN. Utilizing convolution kernels, characteristic maps of input data were produced by convolution layers. The input picture is bended by the kernels. Because of the weight sharing convolution, the intricacy of pictures is decreased. The pooling layer is mostly continued by every set of convolution layers, which decreases the size of characteristic maps for more characteristic taking out. The fully connected layer is utilized succeeding the convolution layer and pooling layer. The fully connected layer links to every activated neurons in the previous layer and converts the 2D characteristic maps into 1D.

1D characteristic maps are then coupled with nodes of groups for categorization [11, 12].

Finally, image recognition is showing greater leverage and preciseness in CNN compared to ANN due to the use of the functional hidden layers.

2.3 Generative Adversarial Networks (GAN)

GAN is a sort of deep learning algorithm conceived by Goodfellow from the input data, this unsupervised learning method automatically discover and generates new data with alike characteristics or models in comparison with the input data [13]. 2 neural networks: a generator and a discriminator. The principal aim for the generator is to produce input which doesn't allow to the discriminator to identify if the input is produced by the generator or from the initial input data. The essential goal for the discriminator is to differentiate between the output produced by the generator and the initial input data as much as possible. The two GAN networks ameliorate themselves and supplement each other. Furthermore, GAN has been spread fast after its creation. They are mostly used to picture-to-picture movement and creating credible pictures of items, environments, and individuals [14, 15].

A new 3D-GAN structure was created founded on a conventional GAN connection [16]. It generates 3D items from a specified 3D spot by joining new discoveries in GAN and dimensional convolutional networks. Different from a traditional GAN network, it is competent to create items in 3D automatically or from 2D images. It provides a larger spectrum of feasible utilizations in 3D input operating juxtaposed with its 2D shape.

3 AI in Dentistry

3.1 AI in Operative Dentistry

Dentists identify dental decays by ocular and manual investigation or by X-ray assessment but detecting early-stage lesions is difficult when profound fissures, close interproximal joining. In fact, several damages are detected uniquely in the late phases of dental decay, which conduct to supplementary sophisticated treatment. Furthermore, most of diagnosis belong to dentists' experience despite of the wide use of dental radiography and explorer in dental caries diagnosis.

Each pixel has a degree of gray in two-dimensional X-Ray which represents the object density. An AI algorithm may assimilate the model and provide auguries to several dental lesions from this concept.

In fact, several studies performed a CNN algorithm dental caries detection on periapical x-rays and intraoral images [16, 17].

Others found that AI in proximal caries detection was further productive and cheaper than dentists.

3.2 AI in Periodontics

Periodontitis is one of the most prevalent troubles. It is a charge for billions of people and, if not well fixed, may conduct to tooth mobility or loss. It is well known that Prompt discovery and care are required to avert acute periodontitis. In clinical practice, periodontal illness determination is based on assessing pocket probing profundity and gingival regress.

Researchers used AI in diagnostic and periodontal disease classification.

Others researchers utilized CNN in the discovery of periodontal bone damage on panoramic radiographs.

3.3 AI in Orthodontics

Orthodontic treatment organization is generally found on the experience and priority of the orthodontists. In fact, orthodontists spend a great effort to identify malocclusion, due to the multitude of changeable that must be examined in the cephalometric investigation, which makes difficult to establish the treatment program and anticipate the result.

Moreover, treatment planning and prediction of treatment results, such as simulating the changes in the appearance of pre- and post-treatment facial photographs are the most applications of AI in orthodontics. Actually, thanks of AI, the orthodontic treatment outcome, the skeletal class, and the anatomic landmarks in lateral x-rays may be examined.

A study performed an algorithm to diagnose if there is a requirement for treatment by orthodontics on the base of orthodontics-related data.

3.4 AI in Oral and Maxillofacial Pathology

By utilizing x-rays, pictures from microscope and ultrasonography AI may be utilized for tumor and cancer identification by CNN algorithms [9, 10].

AI is used to handle cleft lip and palate in risk augury [11].

Further, with intrabuccal visual pictures and using a CNN model, it was possible to spot buccal latent malignant troubles and oral squamous cell carcinoma (OSCC).

Also, optical Coherence Tomography (OCT) has been utilized in the recognition of benign and malignant lesions in the buccal mucosa in addition to intrabuccal visual pictures. In addition, a study has used ANN and Support Vector Machine (SVM) patterns to identify neoplastic buccal lesions [13].

In other study, researchers were able to mechanically identify oral squamous cell carcinoma using a CNN algorithm from confocal laser endomicroscopy pictures [12].

Finally, a study has used a CNN algorithm to recognize and determine ameloblastoma and keratocystic odontogenic tumor (KCOT) [14].

3.5 AI in Prosthodontics

AI is mostly used in prosthodontics to perform the restoration design.

CAD/CAM has digitalized the design work in profit-oriented yields, like CEREC, 3Shape, etc. Some studies demonstrated novel methods founded on 2D-GAN patterns creating a crown by studying shape technicians' designs. Transformed from 3D mouth models, the forming input was 2D depth maps. Other study utilized 3D data directly generating crown using a 3DDCGAN network [7, 8].

In addition, associating AI and CAD/CAM or 3D/4D printing could bring a high effectiveness [16]. Also, in debonding prediction and shade matching of restorations, AI may be an unavoidable support [15].

3.6 AI in Endodontics

Using properties of periapical radiolucency, AI algorithms may identify periapical disease. Also, radiolucencies can be recognized on periapical on panoramic radiographs with deep learning algorithm model.

A study utilizing AI system identified 142 out of 153 periapical lesions with a detection accuracy rate of 92.8%. In addition, utilizing artificial neural connections the detection of cystic lesions has been done. Furthermore, a separation of granuloma from periapical cysts using CBCT images was performed and three-dimensional teeth segmentation using the CNN method was demonstrated.

AI can assimilate further on that human competence. Also, the growth of computer tech is vital to promote the AI development.

4 Conclusion

A multiple of AI systems are being developed for diverse dental disciplines and have produced encouraging results which predicts a bright future for AI in dentistry.

Thereafter, it is now a necessity that institutions integrate AI into their theoretical and practical training programs without forgetting the continuous training of former dentists.

References

1. Smith, T.F., Waterman, M.S.: Identification of common molecular subsequences. J. Mol. Biol. **147**, 195–197 (1981). https://doi.org/10.1016/0022-2836(81)90087-5
2. May, P., Ehrlich, H.-C., Steinke, T.: ZIB structure prediction pipeline: composing a complex biological workflow through web services. In: Nagel, W.E., Walter, W.V., Lehner, W. (eds.) Euro-Par 2006. LNCS, vol. 4128. Springer, Heidelberg, pp 1148–1158 (2006). https://doi.org/10.1007/11823285_121
3. Foster, I., Kesselman, C.: The Grid: Blueprint for a New Computing Infrastructure. Morgan Kaufmann, San Francisco (1999)
4. Czajkowski, K., Fitzgerald, S., Foster, I., Kesselman, C.: Grid information services for distributed resource sharing. In: 10th IEEE International Symposium on High Performance Distributed Computing, pp. 181–184. IEEE Press, New York (2001). https://doi.org/10.1109/HPDC.2001.945188
5. Foster, I., Kesselman, C., Nick, J., Tuecke, S.: The physiology of the grid: an open grid services architecture for distributed systems integration. Technical Report, Global Grid Forum (2002)
6. National Center for Biotechnology Information. http://www.ncbi.nlm.nih.gov
7. Farhaoui, Y.: Design and implementation of an intrusion prevention system. Int. J. Netw. Secur. **19**(5), 675–683 (2017). https://doi.org/10.6633/IJNS.201709.19(5).04
8. Farhaoui, Y.: Big Data Mining and Analytics **6**(3), I–II (2023). https://doi.org/10.26599/BDMA.2022.9020045
9. Farhaoui, Y.: Intrusion prevention system inspired immune systems. Indonesian J. Electr. Eng. Comput. Sci. **2**(1), 168–179 (2016)
10. Farhaoui, Y.: Big data analytics applied for control systems. Lect. Notes Netw. Syst. **25**, 408–415 (2018). https://doi.org/10.1007/978-3-319-69137-4_36
11. Farhaoui, Y.: Big Data Mining and Analytics **5**(4), I–II (2022). https://doi.org/10.26599/BDMA.2022.9020004
12. Alaoui, S.S., Farhaoui, Y.: Hate speech detection using text mining and machine learning. Int. J. Decis. Support Syst. Technol. **14**(1), 80 (2022). https://doi.org/10.4018/IJDSST.286680
13. Alaoui, S.S., Farhaoui, Y.: Data openness for efficient e-governance in the age of big data. Int. J. Cloud Comput. **10**(5–6), 522–532 (2021). https://doi.org/10.1504/IJCC.2021.120391
14. El Mouatasim, A., Farhaoui, Y.: Nesterov step reduced gradient algorithm for convex programming problems. Lect. Notes Netw. Syst. **81**, 140–148 (2020). https://doi.org/10.1007/978-3-030-23672-4_11
15. Sossi Alaoui, S., Farhaoui, Y.: A comparative study of the four well-known classification algorithms in data mining. Lect. Notes Netw. Syst. **25**, 362–373 (2018). https://doi.org/10.1007/978-3-319-69137-4_32
16. Farhaoui, Y.: Securing a local area network by IDPS open source. Procedia Comput. Sci. **110**, 416–421 (2017). https://doi.org/10.1016/j.procs.2017.06.106
17. Kuhnisch, J., Meyer, O., Hesenius, M., Hickel, R., Gruhn, V.: Caries detection on intraoral images using artificial intelligence. J. Dent. Res. **101** (2021)

Predicting Ejection Fractions from Echocardiogram Videos Using Deep Learning

Donya Hassan(✉) and Ali Obied

College of Computer Science and Information Technology, Al-Qadisiyah University, Al Diwaniyah, Iraq
{Com21.post3,ali.obied}@qu.edu.iq

Abstract. Echocardiography is a widely used imaging modality for assessing cardiac function, with ejection fraction (EF) being a critical metric for diagnosing heart conditions. Analyzing video echocardiograms poses special problems because of the dynamic Data type and the requirement to capture temporal relationships. This study proposes 3D Convolutional Neural Networks (3DCNN) and (2DCNN with LSTM) working parallel to analyze Echonet's dynamic dataset. The primary objective is to accurately estimate the heart's ejection fraction from input video echocardiograms. The proposed model employs 3DCNNs to capture spatial patterns across different frames and LSTM layers to model temporal dependencies. By combining the strengths of both architectures, the model aims to extract informative representations and capture the dynamic changes in the heart's structure and function. The experimental results on Echonet's dynamic dataset demonstrate the effectiveness of the parallel 3DCNNand 2DCNN + LSTM model for ejection fraction estimation. The model achieves competitive performance with RMSE of 1.1, MAE of 3.2, and R2 of 0.80, indicating its potential for accurate and reliable evaluation of cardiac function from video echocardiograms. In conclusion, the proposed model offers a promising approach for estimating ejection fraction from video echocardiograms. Its ability to capture spatial and temporal information improves accuracy and provides a valuable tool for diagnosing and monitoring heart conditions. Also, the proposed model outperforms all other recent models that use the same dataset.

Keywords: Echocardiography · Ejection fraction · 3DCNN · LSTM · EchoNet dynamic dataset

1 Introduction

The main factor contributing to illness and early mortality globally is cardiovascular disease. Accounting for 43.8% of yearly fatalities, especially in people with low socioeconomic levels. Effective risk factor management and therapy depend on early diagnosis. CT, cardiac magnetic resonance imaging (MRI), and echocardiography are some tests that can be used to examine the heart. The least expensive option is the

Y. Farhaoui et al. (Eds.): ICAISE 2023, LNNS 837, pp. 217–225, 2024.
https://doi.org/10.1007/978-3-031-48465-0_29

echocardiogram [1]; Echocardiogram combines rapid image acquisition with the lack of ionizing radiation and is a widely accessible diagnostic tool for assessing heart structure and function. A crucial step in determining the end-diastolic (ED) and systolic (ES) phases of the echo cine sequence is quantifying the size and function of the heart chambers [2]. Deep learning and neural networks are two cutting-edge technologies that have significantly increased the efficacy of echocardiography. As a result, Faster and more precise methods are now available for standard section identification of cardiac anatomical features, automatic detection and segmentation of cardiac systems, cardiac functional evaluation, and auxiliary illness diagnosis [3].

Intelligent systems are quickly growing due to computer technology; artificial intelligence-based frameworks are rapidly revolutionizing the healthcare sector. These intelligent systems provide a promising, supplemental diagnostic technique for front-line clinical doctors and surgeons using robust models based on deep learning and machine learning for early disease diagnosis [4].

Medical practitioners often rely on computer-aided medical image analysis to help them make accurate clinical diagnoses and select the best course of treatment. Convolutional neural networks (CNNs) are currently the method of choice for analyzing medical images. Additionally, 3D deep learning techniques are becoming increasingly popular in examining medical images because of the rapid development of three-dimensional (3D) imaging equipment and the accessibility of superior hardware and software support for processing massive amounts of data [5].

The remaining parts of the document are arranged as follows: Sect. 2 contains background (deep learning, the 3DCNN, and LSTM). Section 3 explains the proposed approach. In Sect. 4, it is described how to use the implementation tools and results.

2 Background

Cardiovascular imaging technology has advanced over the past few decades into more complex tools that generate complex and detailed data, such as real-time videos of the heart's chambers and valves. By analyzing these data, scientists and cardiologists have developed unique methods to identify and evaluate heart issues. As a result, various techniques have been created to detect heart diseases by looking at myocardial activity. These techniques concentrate on processing images, signals, or, most recently, videos [6].

2.1 Deep Learning

A machine learning algorithm that works with neural networks is called deep learning (DL). Deep neural networks are also called neural networks with a deep structure or more than two hidden layers. Multiple layers of representation are used in representation learning, or DL. In recent years, DL has developed into a well-liked tool that draws scholars from various disciplines [4]. To get better results, it aids in overcoming the limitations of conventional approaches and resolving complex issues. The most recent technology, including graphical processing units (GPUs), boosted DL's capabilities in addition to the software base innovation that contributed to its popularity. Deeper layers enhance the system's performance by learning information from data and simplifying complex structures. As a result, it is a groundbreaking discovery for dealing with issues in fields with high data. Deep neural networks are constructed from numerous hidden layers between input and output and are inspired by how the brain operates [2].

2.1.1 3D-CNN

Convolutional neural networks (CNNs) are specific models that extract features from 2D inputs, such as photographs. You can train learning models using supervised or unsupervised methods. Instead of straightforward 2D frames, video streams are examined in various applications, including surveillance, action detection, and study of the scene. Three-dimensional CNN models using video streams to extract spatial and temporal properties. Thus, 3D CNNs can capture motion, which is, by definition, present in numerous adjacent frames. More complicated 3D CNN architectures with long-term temporal convolutions have recently been intended to record full temporal-scale video representations [8].

The (CNN) can be defined mathematically as:

$$x_k^l = f\left(w_k^l * x^{l-1} + b_k^l\right) \tag{1}$$

$$Y_k^l = pooL\left(x_k^l\right) \tag{2}$$

where $X^L{}_k$ represents the output of the current layer, $W^L{}_k$ and bL k stand for the weights and bias of the k-th filter of the l-th layer, respectively, and X^{L-1} k represents the layer's output before that. The functions f(.) and pool(.) are the activation and pooling functions, and $Y_k{}^L$ is the reduced feature map, respectively [9].

2.1.2 Long Short-Term Memory (LSTM)

Hochreiter and Schmidhuber presented the long short-term memory (LSTM) RNN architecture 1997. The ability of this network to handle long-term dependencies is its defining feature. The LSTM cell, which functions as a memory cell and can recall and forget information, is the primary distinction between it and an RNN. The cell state, which describes the data flow in an LSTM, depends on a set of gates to decide if an input is essential enough to be remembered and whether previously recorded information should still be remembered or forgotten [5]. These gates include input, output, and forget gates. Instead of images, applied videos offer a more challenging input because they comprise frames. The vanishing gradient problem can be solved, and long-term dependency can be eliminated using LSTM [9]. Three gates are fitted in the LSTM device. They are specifically used to choose which data to maintain. The LSTM network employs temporary switching gates used for memory to prevent gradient fading. The initial cell state c(t-1), The current input vector xt, and hidden state h(t-1) are external inputs for the basic LSTM unit. A set of equations can be used to explain the complete calculation as follows:

$$i_t = \sigma\left(w_{x_i}x_t + w_{h_i}h_{t-1} + b_i\right) \tag{3}$$

$$f_t = \sigma\left(w_{x_f}x_t + w_{h_f}h_{t-1} + b_f\right) \tag{4}$$

$$O_t = \sigma\left(w_{x_0}x_t + w_{h_0}h_{t-1} + b_0\right) \tag{5}$$

$$c_t = f_t \cdot c_{t-1} + i_t \cdot \tanh(w_{xc}x_t + w_{hc}h_{t-1} + b_c) \tag{6}$$

$$h_t = O_t \tanh(c_t) \tag{7}$$

The bias vector is represented by b, and the weights are Wx and Wh. Designates the activation actions. As the sigmoid function for gates, the activation function is typically appropriate σ. Tanh symbolizes the nonlinear activation function. The cell's output is then determined using the output gate's result and the Tanh of the cell state [6].

3 Proposed Model

A 3D CNN and a (2D CNN + LSTM) are used in the proposed model and work in parallel. Figure 1 demonstrates the suggested model architecture.

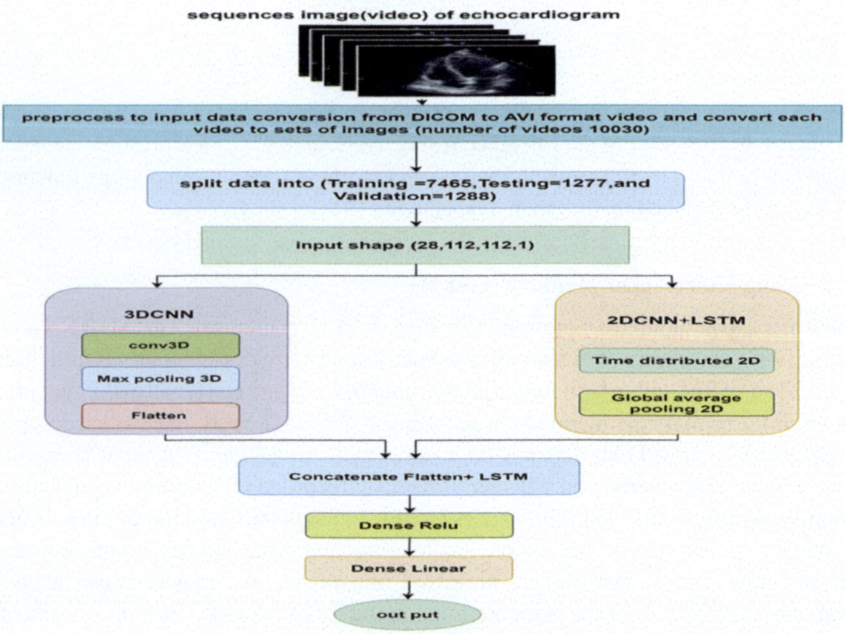

Fig. 1. Displays the suggested model architecture.

The model needs two inputs with the shapes (28, 112, 112, 1) (input_1 and input_2). The outputs of the 2D CNN + LSTM and 3D CNN models are concatenated using concatenate (). An example is adding additional layers, such as a layer with 128 units and ReLU activation that is fully connected. Following the model's creation with the desired inputs and outputs, as shown above.

The left branch represents the 3D CNN model, which uses Input_1 as an input and applies one subsequent 3D convolutional layer before max pooling. After that, the output is flattened.

The (2D CNN + LSTM) model on the right branch applies a time-distributed 2D convolutional layer before global average pooling on each frame using the input Input_2. The LSTM layer is then given the output. The results from both branches are combined in a layer with 128 units and ReLU activation that is fully connected. The output layer is then subjected to a linear activation function, with a single unit representing the predicted output.

3.1 Implementation Details

TensorFlow, the Keras sequence for TensorFlow, and metrics from scikit-learn are among the necessary libraries that are imported. You utilized the implementation specifics of the tested experiments. Modelling and experimentation were conducted on a Windows 10 operating system workstation. The system comprises a cori7_10750H processor, an 8 GB NVIDIA GeForce RTX 2070Ti GPU, and 16 GB of RAM. Python (3.9.13) programming language and Jupyter Notebook software were used to create the algorithms. The algorithms used the PyTorch deep learning module. The network was trained for all trials using an initial learning rate of 0.001 and batch size of 32. There were 80M parameters in the network for 20 epochs (15 h) of training data on the EchoNet dynamic datasets.

4 Experimental Results and Discussion

First provided the datasets and training data in this section. Additionally, contrast the effectiveness of sophisticated deep learning approaches for echocardiography with existing methods to show how well ED and ES frames can be distinguished when evaluating EF. The same data from the test set and training set were used to compare the proposed approach with other methods to demonstrate the efficacy of 3DCNN + (2DCNN + LSTM).

4.1 Data Set

Compiled the 10,030 echocardiography videos that comprise the EchoNet-Dynamic dataset, which covers a range of typical lab imaging acquisition circumstances. All images have measurements, including the left ventricle's expert tracings, LV volume at end-systole and end-diastole, and Ejection Fraction. Videos from patients who received apical-4-chamber echocardiography exams between (2016 and 2018) at Stanford University Hospital are included in the dataset. Each video was downscaled into standardized (112 X 112) pixel movies after being cropped to remove text and material outside the scanning region. The videos are in.avi format [7]. Experienced sonographers used five different ultrasound devices to get the echocardiograms. The video lengths were between (28, 1.002) frames encompassing one to several cardiac cycles, and the frame rates ranged from 18 to 138 fps and had an average frame rate of 50 fps. Videos were randomly divided into (7.465), (1.277), and (1.288) samples to create training, validation, and test sets [10].

4.2 Evaluation Metrics

The mean absolute error (MAE), root mean square error (RMSE), and R-square (R^2) are utilized as the evaluation criteria of the methodologies to assess the forecasting impact of 3DCNN-(2DCNN + LSTM).

The formula for calculating MAE is as follows:

$$MAE = \frac{1}{n}\sum_{i=1}^{n}\left|y_i - \hat{y}_i\right| \tag{8}$$

where the true value is y_i and $\hat{y}_i$ is the predictive value. The prediction is more accurate the lower the MAE number is[8].

The following is the RMSE calculation formula:

$$RMSE = \left(\sqrt{MSE}\right) = \sqrt{\frac{1}{n}\sum_{i-1}^{n}\left(y_i - \hat{y}_i\right)^2} \tag{9}$$

where the true value is y_i and $\hat{y}_i$ is the predictive value. The accuracy of the forecast increases with decreasing RMSE values.

The following is the R^2 calculation formula:

$$R^2 = 1 - \frac{\sum_{i=1}^{n}\left(y_i - \hat{y}_i\right)^2}{\sum_{i=1}^{n}\left(y_i - \overline{y_i}\right)^2} \tag{10}$$

Y_i represents the genuine value, y_i represents the average, and $\hat{y}_i$ represents the forecast value. R^2 value range is 0 to 1. The smaller the difference between the anticipated and actual values, and the greater the forecasting accuracy, the closer the values of MAE and RMSE are to 0. The more accurately the model fits the data, the closer R^2 is to 1 [9].

4.3 The Results

The 3DCNN and (2DCNN + LSTM) fared better than other techniques when compared among different models for segmenting the left ventricle. Depending on the stage, end-diastolic or end-systolic cardiac contraction. Use the same EchoNet dynamic dataset and the (MAE), (RMSE), and R^2 measures.

As shown in Table 1 (in Sect. 4), the proposed model 3DCNN and (2DCNN + LSTM) parallel technique is successful in estimating ejection percent from the video echocardiograms Echonet dynamic dataset and achieved high performance compared to the other, potentially less effective, methods of evaluating EF's performance. To evaluate the models, we used the R2 (Coefficient of Determination), RMSE (Root Mean Squared Error), and MAE (Mean Absolute Error) metrics. The proposed model had an R2 of 0.80, RMSE of (1.1), and MAE of (3.21). This finding outperformed all other publications using the same dataset. Lower MAE and RMSE values indicate higher performance. Displays how the suggested model outperforms other models in terms of efficacy. The shape below displays more results and details about the proposed model.

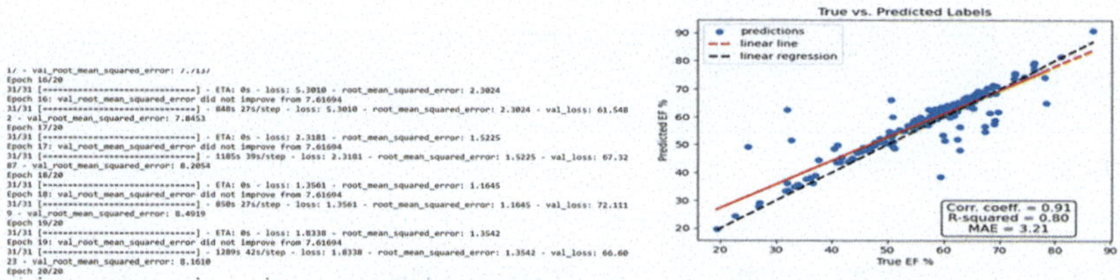

(a) (b)

Fig. 2. The executed as (a) runs a screenshot displaying execution details to RMSE, (b) shows the actual and forecast EF the R2, MAE, and dice coefficient are displayed to demonstrate distributed data. Using 20 epochs to train the proposed model and achieved the (1.1) to RMSE in epoch 18 as shown in the image above and batch size 32, Adam optimizer, and learning rate of 0.001.

In Fig. 2(b), it's clear to notice actual and forecast EF, and draw, the R2, MAE, and dice coefficient are displayed to demonstrate distributed data.

Also, Fig. 3 illustrates how closely the cardiac muscle's actual and predicted values match up. (a) displays two boxes, one for the actual data and the other for the anticipated data. While (b) illustrates how closely the orange line, which represents the expected values of the heart muscle, and the first blue line, which represents the true values of the ejection fraction, are related.

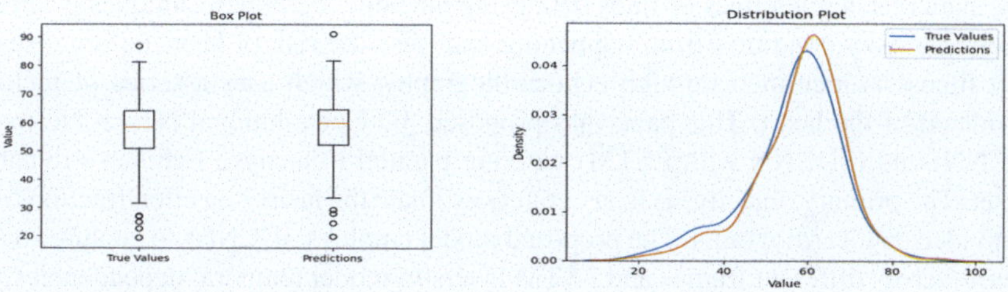

(a) (b)

Fig. 3. The graph displays how closely real values match expected values. (a) The Figure is presented as an of boxes plot (b). The Figure depicts the data's normal distribution for the actual and predicted values.

4.4 Comparison with Other Models

The proposed model outperforms all other recent models that use the same dataset, as shown in Table 1.

Table 1. The RMSE, MAE, and R2 and the techniques employed when comparing papers using the same dataset.

Reference	Model	Data set	MAE	RMSE	R2
[10]	R2 + 1D, MC3, and R3D	Echonet dynamic	4.05	5.32	0.81
[11]	UltraSwin-base	Echonet dynamic	5.59	7.59	0.59
[12]	Ultrasound video transformers (UVT)	Echonet dynamic	5.95	8.38	0.52
[13]	Deep Heartbeat	Echonet dynamic	6.34	8.59	0.50
[14]	(MAEF-Net)	Echonet dynamic	6.29	8.21	0.54
[15]	VGG16 + lstm	Echonet dynamic	8.08	11.98	-
[16]	DeepLabV3	Echonet dynamic	6.55	-	0.61
Proposed model	3DCNN and(2DCNN + LSTM) parallel	Echonet dynamic	3.21	1.1	0.80

5 Conclusion

The number of persons receiving a heart disease diagnosis is increasing daily. A framework that creates guidelines or finds data using machine learning techniques is needed to stop this risky condition from happening and raise the risk of heart failure disease. Heart illness is identified through echocardiography, which uses a series of pictures (or videos) of the heart. This paper has proposed 3D Convolutional Neural Networks (3DCNN) and (2DCNN with LSTM) working parallel to analyze Echonet's dynamic dataset. The primary objective is to accurately estimate the heart's ejection fraction from input video echocardiograms. The proposed model employs 3DCNNs to capture spatial patterns across different frames and LSTM layers to model temporal dependencies. By combining the strengths of both architectures, the model aims to extract informative representations and capture the dynamic changes in the heart's structure and function. The experimental results on Echonet's dynamic dataset demonstrate the effectiveness of the parallel 3DCNNand 2DCNN + LSTM model for ejection fraction estimation. The model achieves competitive performance with RMSE of 1.1, MAE of 3.2, and R2 of 0.80, indicating its potential for accurate and reliable evaluation of cardiac function from video echocardiograms.

Additionally, helping to enhance the network's prediction is the Attention module. The results show that the suggested technique can successfully work on the used datasets. Expanding the datasets and understanding the clinical repercussions of inaccurate classifications. Also, the proposed model outperforms all recent models using the same dataset.

The quality of the model in evaluating EF in the video echocardiogram dataset is demonstrated by the suggested model's achievement of the lowest error rate among the earlier models that operated on the same database.

References

1. Farhad, M., Masud, M.M., Beg, A., Ahmad, A., Ahmed, L., Memon, S.: Cardiac phase detection in echocardiography using convolutional neural networks. Sci. Rep. **13**(1), 1–16 (2023). https://doi.org/10.1038/s41598-023-36047-x
2. Dezaki, F.T., et al.: Cardiac phase detection in echocardiograms with densely gated recurrent neural networks and global extrema loss. IEEE Trans. Med. Imaging **38**(8), 1821–1832 (2019). https://doi.org/10.1109/TMI.2018.2888807
3. Zhou, J., Du, M., Chang, S., Chen, Z.: Artificial intelligence in echocardiography: detection, functional evaluation, and disease diagnosis. Cardiovasc. Ultrasound **19**(1), 1–11 (2021). https://doi.org/10.1186/s12947-021-00261-2
4. Yanik, E., et al.: Deep neural networks for the assessment of surgical skills: a systematic review. J. Def. Model. Simul. **19**(2), 159–171 (2022). https://doi.org/10.1177/15485129211034586
5. Lara Hernandez, K.A., Rienmüller, T., Baumgartner, D., Baumgartner, C.: Deep learning in spatiotemporal cardiac imaging: a review of methodologies and clinical usability. Comput. Biol. Med. **130**, 104200 (2021). https://doi.org/10.1016/j.compbiomed.2020.104200
6. Zhang, W., Li, H., Li, Y., Liu, H., Chen, Y., Ding, X.: Application of deep learning algorithms in geotechnical engineering: a short critical review. Artif. Intell. Rev. **54**(8), (2021). https://doi.org/10.1007/s10462-021-09967-1
7. Ouyang, D., et al.: EchoNet-dynamic: a large new cardiac motion video data resource for medical machine learning. In: 33rd Conference on Neural Information and Processing Systems (NeurIPS 2019), pp. 1–11 (2019)
8. Kumari, D.: Study of Heart Disease Prediction Using CNN Algorithm, vol. 8, no. 7, pp. 792–830 (2021). Available: www.jetir.org
9. Lu, W., Li, J., Li, Y., Sun, A., Wang, J.: A CNN-LSTM-based model to forecast stock prices. Complexity **2020**, (2020). https://doi.org/10.1155/2020/6622927
10. Ouyang, D., et al.: Video-based AI for beat-to-beat assessment of cardiac function. Nature **580**(7802), 252–256 (2020). https://doi.org/10.1038/s41586-020-2145-8
11. Fazry, L., et al.: Hierarchical vision transformers for cardiac ejection fraction estimation. In: IWBIS 2022—7th International Workshop on Big Data Information Security Proceedings, pp. 39–44 (2022). https://doi.org/10.1109/IWBIS56557.2022.9924664
12. Reynaud, H., Vlontzos, A., Hou, B., Beqiri, A., Leeson, P., Kainz, B.: Ultrasound video transformers for cardiac ejection fraction estimation. In: Lecture Notes in Computer Science (including Subseries Lecture Notes in Artificial Intelligence and Lecture Notes in Bioinformatics), vol. 12906. LNCS, pp. 495–505 (2021). https://doi.org/10.1007/978-3-030-87231-1_48
13. Laumer, F., et al.: DeepHeartBeat: latent trajectory learning of cardiac cycles using cardiac ultrasounds. Proc. Mach. Learn. Res. **136**, 194–212 (2020) [Online]. Available: http://proceedings.mlr.press/v136/laumer20a.html
14. Zeng, Y., et al.: MAEF-Net: multi-attention efficient feature fusion network for left ventricular segmentation and quantitative analysis in two-dimensional echocardiography. Ultrasonics **127**, 2023 (2022). https://doi.org/10.1016/j.ultras.2022.106855
15. Blaivas, M., Blaivas, L.: Machine learning algorithm using publicly available echo database for simplified ‘visual estimation’ of left ventricular ejection fraction. World J. Exp. Med. **12**(2), 16–25 (2022). https://doi.org/10.5493/wjem.v12.i2.16
16. Duffy, G., Jain, I., He, B., Ouyang, D.: Interpretable deep learning prediction of 3d assessment of cardiac function. Pac. Symp. Biocomput.Biocomput. **27**, 231–241 (2022). https://doi.org/10.1142/9789811250477_0022

Mechanical Intelligence Techniques for Precision Agriculture: A Case Study with Tomato Disease Detection in Morocco

Bouchra El Jgham[1(✉)], Otman Abdoun[2], and Haimoudi El Khatir[3]

[1] Information Technology and Modeling Systems Research Unit (TIMS), Faculty of Sciences, Abdelmalek Essaadi University, Tetouan, Morocco
eljghambouchra@gmail.com

[2] Information Security, Intelligent Systems and Applications (ISISA), Faculty of Sciences, Abdelmalek Essaadi University, Tetouan, Morocco
o.abdoun@uae.ac.ma

[3] Computer Science Department, Polydisciplinary Faculty, Abdelmalek Essaadi University, Larache, Morocco

Abstract. This paper provides a brief review of the application of machine learning in agriculture. For this purpose, several machine-learning algorithms were considered as SVM, ANN and CNN. Furthermore, a case study for the detection of various diseases of tomato with DNN. The efficiency of the DNN algorithm for the detection of pepper diseases has been demonstrated from the results obtained. This paper can assist researchers in this area in developing an optimal and efficient machine-learning model for various agricultural applications in the future. Within this way, and with highlighting Morocco as a developing State, this study can provide effective support information to decision makers, decision makers, professionals and end-users in introducing new techniques and utilizing follow-up techniques in the agricultural sector through their adoption.

Keywords: Agriculture · Disease detection · DNN

1 Introduction

Agriculture is the oldest and largest economic growth sectors of any country whose backbone is agriculture. As time went on, it became not only the main source of food for the ever-growing population but also played the role of commodity supplier for most industries. They were developing and adapting to new technologies, there was limited or no access to climate data by the farmer himself, the demands of the market-place, the movement of pests such as grasshoppers, or plant animals and diseases. All of this made sure that even if agriculture is an industry of subsistence it is lagging in relation to integration with data analysis and other pertinent techniques, which has an impact on overall performance.

In this paper, a case study is proposed to detect different diseases of tomato leaves in Morocco. Morocco has achieved one of the highest tomato yields, surpassing Spain

Y. Farhaoui et al. (Eds.): ICAISE 2023, LNNS 837, pp. 226–232, 2024.
https://doi.org/10.1007/978-3-031-48465-0_30

and the Netherlands. Tomato production in Morocco reached 1409.44 million kilos, for which its devoted an extension of 15,955 hectares, obtaining a yield per square meter of 8.83 kilos in 2018, According to FAO, this is one of the top 15 tomato-producing countries in the world.

In March this year (2023), Morocco imposed restrictions on the export of tomatoes to curb rising prices locally because the bad weather caused the disease of tomato plants in Morocco, This year's vegetable crops were disturbed, shrinking the availability of fresh vegetable dishes in Europe and rising prices, which contributed to UK inflation rising to 10.4% in February.

Inflation in Morocco brought the central bank's benchmark interest rate up three consecutive times 50 a basis point to 3% last Tuesday and food inflation jumped to 20.1% last month, bringing general inflation to 10.1% at a level not reached since the 1980s.

The primary purpose of this paper is to integrate these two concepts, IoT and AI, to multiply the value of IoT by using all the data generated by smart devices to facilitate learning and intelligence. Any deployment across multiple domains where resource-constrained devices are a major concern.

This document is divided into the following sections. Section 2 deals with weed problems in crops and solutions. We review the methodology in Sect. 3.We demonstrate the proposed method of detecting tomato illnesses in Sect. 4. Results and discussion and analysis are given in Sect. 5. The conclusion is presented in Sect. 6.

2 Weed Problems in Crops and Solutions

Farmers are working hard to protect our environment by making tremendous changes to their farming practices. The major issue is that weeds increase the amount of biological competition compared to existing crops, Examples include overusing fertilizers, water, and manual labor. The handwork of weeding and clearing the land downs more moment. However, robotic applications are capable of performing excellent weed detection. It's a mechanical removal task [1].

Later, with technological progress, herbicides were used to kill weeds. The image treatment is then used for weeding. In this work, the emphasis is on the treatment of images using DNN to detect weeds in forests. Raja et al. [2] have developed a real-time weed management system that uses a robotic machine to detect weeds and remove them using a knife.

Then you have to label every plant, whether it is weeds or crops. Moreover, identifying weed species is important for the application of specific treatments. Lottes et al. [3] suggested a fully convoluted network approach (FCN) for this task; see also Fig. 1 grid architecture overview. A single encoder shares the suggested table of task-specific decoders to detect bar regions and perform pixel segmentation.

Considering the large amount of work on AI weed detection, the aim of this article is to suggest a "DNN" model designed to extract images from the "Tomato" dataset to detect tomatoes at an early stage. The purpose provides insight into how a regional convolutional neural network can help separate weeds (tomatoes) from the original crop, allowing the crop to be nutritious and increase yield.

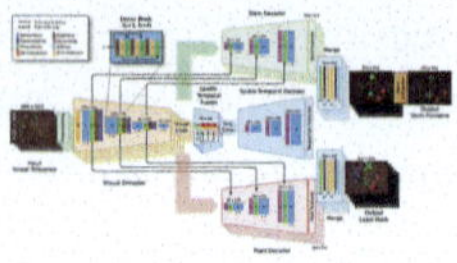

Fig. 1. Architecture of systems for sequential classification of common stocks and plants/weeds

This section presents relevant advanced technologies to date. Next, case studies on the detection of various diseases in pepper leaves are presented. For this, several machine learning algorithms have been considered. Despite the different approaches used, deep learning algorithms have established themselves as one of the most effective techniques. Neural networks also play an important role in task execution. In this section, datasets of four different crops and two weeds are collected. Morphological feature extraction and performance analysis were performed on three classification algorithms, that is to support the vector machine, the network of artificial neurons and the network of convolutional neurons. SVM is a supervised learning model that analyzes an input matrix of morphological features, recognizes patterns and aims to find an optimal hyperplane. The SVM initially uses the kernel function to transform the input data into a space of higher dimensions, and then constructs a linear optimal hyperplane in the transformed space to separate the two classes.

The entry layer of the ANN is supplied with a form vector followed by hidden layers for the localization of the salient characteristics. Estimate or estimate approximately the value of each node using a non-linear feature. The ANN architecture for weed detection.

CNNs use consecutive convolutional layers with a non-linear ReLU function to store the characteristics of an image of a specific dimensionality. Maximum aggregation strata are used for subsampling.

The confusion matrix in Table 1 is used to validate the classification of crops and weeds by various classifiers. This shows that weeds are more easily recognized as crops in SVM and ANN than in CNN. The equivocation may be due to similarities in lighting, background, and patterns between crops and weeds. The SVM and ANN accuracy can be enhanced by the application of more robust feature extraction methods in the recognition system. If either image contains crops and weeds, a region of interest is extracted and passed to the model to predict whether the region of interest is a crop or a weed. See Fig. 2(a) for the captured field image. The hiding of the floor is shown in the Fig. 2(b). With the help of the analysis of the connected components, the number of objects is determined and boxes are drawn, as shown in Fig. 2(c). Each object receives a CNN model and forecasts if it is a weed or a crop.

Performance of graders by SVM, ANN and CNN is analyzed. In comparison to SVM and ANN, CNN has better performance because of its deep learning capacity to learn image-related features. In next sections, we propose new approach using DNN for tomato leaf disease detection in Morocco.

Table 1. SVM /ANN/CNN Confusion Matrix

	SVM		ANN		CNN	
	Crop	Weed	Crop	Weed	Crop	Weed
Crop	125	0	116	9	125	0
Weed	21	104	11	114	4	121

(a) (b) (c)

Fig. 2. Prediction for real-time image

3 Methodology

In our assessment Sankaran et al. [4], we focused on a single plant, the tomato.

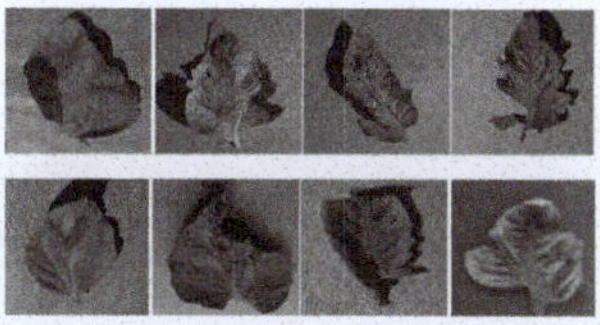

Fig. 3. Pictures of diseased tomato leaves from the vegetal village dataset.

Current advancements in ML and computer vision require distinctive features to generate precise inference. Figure 3 illustrates the distinct types of tomato plants depicted in the pictures used in this publication, which are taken from the Internet.

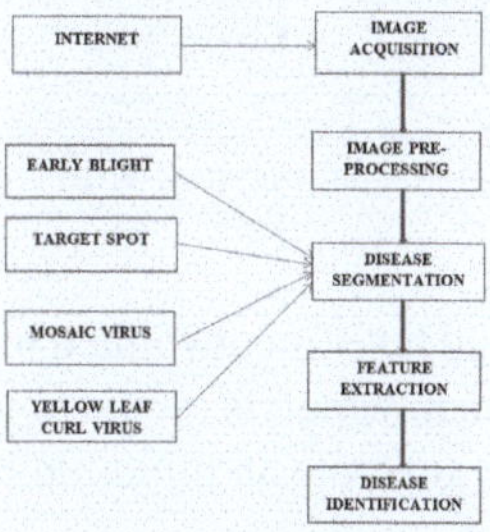

Fig. 4. Flowchart for early detection of tomato seedling diseases.

Image acquisition is the primary step in extracting the characteristics of tomato leaf images, image pre-processing, and the training process, as illustrate in Fig. 4.

3.1 Image Acquisition

Kaggle has amassed a large collection of tomato sample images. Image-based experimental data collection process of tomato took nearly 30 days. The built-up data set is in jpg format with 24-bit color.

3.2 Image Pre-Processing

Different image processing mechanisms are suitable for different levels of accumulated images. The dataset is then divided into two classes, randomly split into training data and test data. Classifiers are used for predicting and classifying images according to built image attributes. This achieves weed detection in tomato images.

3.3 Training Process

The model is formed using the training data set, whereas another part is used for testing purposes - to get results for the proposed work. The results of the proposed work and the speed of learning of the dataset are functions of the learning rate, which indirectly indicates that the optimized model is skipped.

4 Method of Detecting Tomato Illnesses

Using functions extracted earlier, a classifier based on machine or profound learning can categorize diseases in tomato seedlings and an early predictor of the disease will predict the disease. To implement this approach, it is necessary to build a faster DNN model. DNN is a neural network that is advanced and based on a machine learning model, with multiple layers present in both the inlet and outlet layers. The difference between a single neural network and DNN can be seen in Fig. 5.

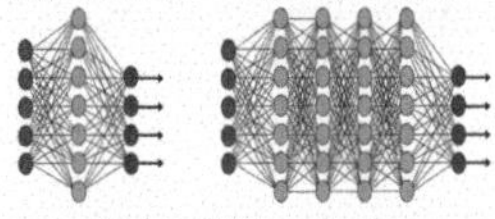

Fig. 5. Simple neuronal network with respect to DNN.

The activated map of the passed layer is convoluted in the convolutional layer through the convolutional (or p ice) channel, which is included with the predisposition and thus fed to the activation capability to generate an initiation card for the next layer. It is used next to the convolutional network.

The mutualisation layer is one of the components of the DNN which gradually reduces the size of the activation card to decrease the number of factors and the time it takes for the neuronal network to compute. Each data card has its own layer that operates separately. In DNN, Max pooling is one of the most frequently used pooling layers. Maximizing activation on a non-overlapping inlet section can reveal the output of a maximum pooling layer [5–7]. The enable function matches a result with a set of inputs. We typically use an activation feature after each convoluted layer. It is a non-linear transfer function used on entry data. The next layer is where the transfer output is sent for entry.

5 Results and Discussion

In this paper, 1227 images of tomato leaves were taken from data from the plant village. Of this number, 227 were photos of healthy tomato leaves and 1000 were photos of sick tomato leaves. Assess similarities or differences among diseases. The histogram of each analyzed image is examined first and compared to a sample of different diseases in Fig. 6. The characteristics of a healthy tomato are compared to those of a medium tomato. There is very little difference in mean values, whereas mean values vary more when examining diseased tomato leaves. Table 2 displays the accuracy of the various machine-learning models in detecting diseases in tomato leaves.

The DNN is faster than the other models. DNN functions better with more hidden layers. The proposed approach to detecting leaf disease is approximately 97% accurate based on the DNN model. In addition, the SVM model is superior to CNN, ANN and random forest.

Table 2. Yield of different graders for the detection of various diseases

Classifiers	Different tomato leaf diseases and accuracy			
	Early blight (%)	Mosaic virus (%)	Target spot (%)	Yellow leaf curl virus (%)
DNN	93.75	91.02	97.59	90.63
ANN	88.58	85.69	87.321	86.61
CNN	82.69	89.58	87.90	87.99
Random forest	79.00	79.69	87.57	88.06
SVM	90.39	89.98	88.58	87.79

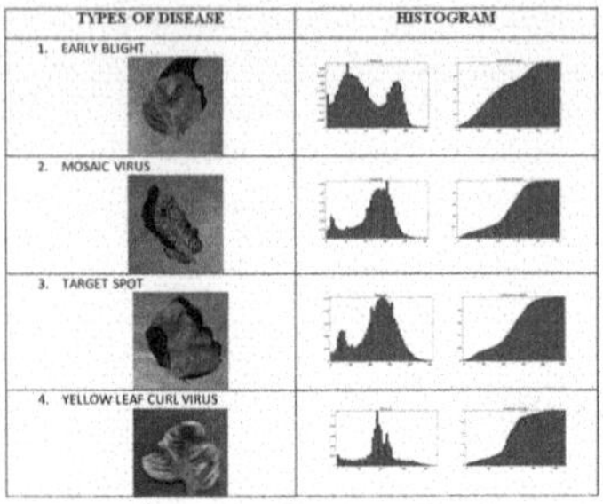

Fig. 6. Histograms for four disease types in tomato leaves.

6 Conclusion

Agriculture is a key component of improving the economy of any country. Weeds slow down cultivation. Previously, weeds have been identified by hand, but this was an extremely costly and lengthy process. Now, weeds are detected in a robotic manner, using automated sprayers and weeding.

This study develops a new DNN for good discrimination of plants and weeds in precision agriculture. The proposed model correctly identified weeds in crops, reducing herbicide use and increasing productivity.

The comparison shows that his newer method performs better on several measures. Moreover, the SVM model outperforms CNN, ANN and random forest. As a future extension, the proposed DNN technique can be used in IoT and smart greenhouses.

References

1. El Jgham, B., Abdoun, O., El Khatir, H.: Review of Weed Detection Methods Based on Machine Learning Models. Springer Science and Business Media LLC (2023)
2. Raja, R., Nguyen, T.T., Slaughter, D.C., Fennimore, S.A.: Real-time robotic weed knife control system for tomato and lettuce based on geometric appearance of plant labels. Bi Syst. Eng. **194**, 152–164 (2020)
3. Lottes, P., Behley, J., Chebrolu, N., Milioto, A., Stachniss, C.: Robust joint stem detection and crop-weed classification using image sequences for plant-specific treatment in precision farming. J. Field Robot. **37**(1), 20–23 (2020)
4. Sankaran, S., Mishra, A., Ehsani, R., Davis, C.: A review of advanced techniques for detecting plant diseases. Comput. Electron. Agric. **72**(1), 1–13 (2010)
5. Wu, G., Shao, X., Guo, Z., Chen, Q., Yuan, W., Shi, X., Xu, Y., Shibasaki, R.: Automatic building segmentation of aerial imagery using multi-constraint fully convolutional networks. Remote Sensing **10**(3), 407 (2018)
6. Mohapatra, S.K., Mohanty, M.N.: Analysis of Diabetes for Indian Ladies Using Deep Neural Network Cognitive Informatics and Soft Computing, pp. 267–279. Springer (2019)
7. Mohapatra, S.K., Srivastava, G., Mohanty, M.N.: Arrhythmia classification using deep neural network. Paper presented at the 2019 International Conference on Applied Machine Learning (ICAML) (2019)

Predict Fires with Machine Learning Algorithms

Adil Korchi[1(✉)], Ahmed Abatal[2], and Fayçal Messaoudi[3]

[1] Faculty of Juridical, Economic and Social Sciences, Chouaib Doukkali University, El Jadida, Morocco
korchi.a@ucd.ac.ma
[2] Faculty of Sciences and Techniques, Hassan Premier University, Settat, Morocco
a.abatal@uhp.ac.ma
[3] National School of Commerce and Management, Sidi Mohamed Ben Abdellah University, Fez, Morocco
faycal.messaoudi@usmba.ac.ma

Abstract. In a previous article, we described the steps involved in creating a Machine Learning project, which are frequently difficult to construct and require the problem to be divided into stages in order to be solved [1]. We were able to pinpoint the top 5 steps we believe are necessary to complete this project. These five processes are: defining the problem, gathering the data, selecting the appropriate algorithms, refining the outcomes, and presenting the results. In this publication, we suggest the use of several methods for assessing categorization models via an algorithm that forecasts whether or not there will be a fire in a specific location. We are aware that it is challenging to find a solution to this issue, particularly when we must forecast the future when there is no fire, fail to anticipate when there is actually one, or fail to predict when there isn't. This study's methodology will demonstrate how to pick the best algorithm and how to evaluate it. The confusion matrix and classification model technique, which provide a 93% accuracy in fire detection, are credited with the experiment's encouraging results.

Keywords: Algorithms · Machine learning · Classification algorithms · Linear regression · Confusion matrix

1 Introduction

Computers can learn without being particularly programmed for it thanks to "Machine Learning" technology [2, 3]. For computers to learn and advance, they need data to analyze and train on. Despite the fact that machine learning has been around for a while, many individuals are still unsure of its precise nature. It is a modern science that uses statistics, data mining, pattern recognition, and predictive analysis to draw conclusions from data patterns.

In this research, we'll use a classification model approach to solve the problem of predicting if a fire would occur anywhere in particular. The confusion matrix, one of the tools employed in this study, determines whether a classification system is effective by assigning a correct class to each line and an anticipated class to each column. The confusion matrix comparing predictions to real values will also be included [4].

Y. Farhaoui et al. (Eds.): ICAISE 2023, LNNS 837, pp. 233–241, 2024.
https://doi.org/10.1007/978-3-031-48465-0_31

2 State of Art

2.1 Machine Learning Algorithms

There are many algorithms in the machine learning industry to suit various demands. Each has unique mathematical and computational features. It might not be simple to understand for a newcomer to the field [5]. Eight of the most fundamental but indisputable machine learning algorithms are: Vector Machine Support, Linear Regression, Logistic Regression, Anomaly Detection, Naïve Bayes, Neural Networks, and K-Means Decision Trees.

2.2 Selecting the Appropriate Algorithm

Data scientists have access to a wide range of algorithm options. The nature of the issue that has to be solved informs this selection in part. For instance, we won't attempt to solve a regression problem using a classification algorithm [6]. However, understanding how to assess any algorithm based on its dataset is essential. An important phase in an algorithm's deployment is a thorough evaluation of its performance [7]. While clustering, regression, and classification are only a handful of the issues that machine learning might potentially solve, there may be a multitude of viable solutions for each problem category. The number of attributes, the amount of data, and other factors can all affect which method is optimal. The machine learning project manager is directed toward the algorithm that appears to be most suited to complete his project using Fig. 1.

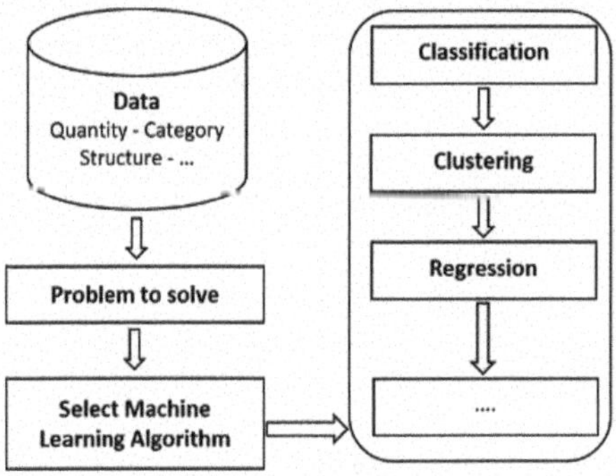

Fig. 1. Help in choosing the appropriate algorithm

2.3 Classification Algorithms

The purpose of classification algorithms is to use the vast amounts of information that have been gathered and stored in databases to help us learn more. By assigning a class to a specific set based on the features of a newly added component, the classification process differs from other operations. This procedure involves two steps.

The first step is building the model, followed by using it. The second step is understanding the term "classification," using the sonar object prediction example to do so. Whether it's a fish or merely a thing, we want to know. Therefore, there are only two ways to group our prediction into one of the two categories. This classification issue [6] exists.

2.4 Model for Classification and Evaluation

In order to illustrate the classification models, we will use labeled data to forecast the class to which an object belongs. The majority of our discussion will center on binary classification, which asks us to determine whether an object is a member of a class. It must be emphasized that we considered the quantity of mistakes as a performance indicator when assessing classification models [7]. However, it's not the only factor! True, not all errors are created equally.

3 Fire Prediction Methodology Using a Classification Model Algorithm

In this section, we will briefly review a few recent works that are pertinent to our topic, after that, we'll show how to evaluate a classification model utilizing a forecasting algorithm who check whether a fire will occur somewhere or not. Setting off a fire alarm when there isn't one is less risky than doing so when a home is on fire. The confusion matrix test [8], which evaluates how effectively a classification system performs, was the tool employed in this experiment. Each column represents an approximation of a class, while each row represents a real class.

3.1 Similar Works

Faroudja [9] examined a variety of methods for forecasting and spotting forest fires. Additionally, he showed how machine learning methods are commonly used to forecast and locate forest fires.

Preeti [10] claimed in his research that forest fire prediction is a crucial aspect of controlling forest fires and that there are numerous fire detection algorithms available with various approaches to fire detection. Using meteorological characteristics like temperature, rain, wind, and humidity, the author suggested a system procedure to solve the problem. He employed the RandomizedSearchCV coefficient with Hyperparameter adjustment. Root mean square error (RMSR) was 0.07, Mean Absolute Error (MAE) was 0.03 and Mean Square Error (MSE) was 0.004.

A thorough analysis of the application of several machine learning algorithms in managing forest fires or wildfires is given by Muhammad Arif [11]. The author has identified some possible areas where new data and technology can aid in making better fire management decisions. He also asserted that despite extensive research into the management and forecast of forest fires and wildfires using various machine learning techniques, we were unable to locate any well-structured data sets that were openly accessible.

3.2 Confusion Matrix for Predicting Fires

Bringing up the problem of predicting fires. How to forecast the future when there is none, how not to forecast when there is one, or how not to forecast. This issue is challenging to resolve! And we'll utilize the confusion matrix to identify the answer (Table 1).

Table 1. Matrix of confusion

		Real class	
		–	+
Predicted class	–	True Negatives	False Negatives
	+	False Positives	True Positives

In this context, "Positive" denotes the classification that a fire belongs to while "Negative" denotes the opposite. A truly "positive" prediction is when we anticipate there will be a fire and it actually occurs. In contrast, if this prediction proves to be incorrect, it is a false positive, etc. They are sometimes referred to as "false-type II" false negatives and "Type-I errors," which are also known as false-positives. With scikit-learn [3], also referred to as sklearn, a free toolset for machine learning for the Python, all we need to do is utilize the "confusion_matrix function" [12].

3.3 Results of the Confusion Matrix and Evaluation Indicators

The clash between the observed and expected classes is materialized by the confusion matrix we apply [3]. The confusion matrix command is used to create interpretable (metric) indicators. The target modality can be specified using a third parameter, which is required for the computation of some indicators. In our scenario, we prioritize spotting fake messages (fire = pos). The command code of confusion matrix that was developed to forecast a fire is excerpted below in Fig. 2.

```
# Confusion matrix
mat                                                              <-
confusionMatrix(data=pred,reference=FireTest$fire,positive="pos")
print(mat)
## Confusion Matrix and Statistics
## ## Reference
## Prediction pos neg
## - 799 72 ## pos 37 471
## ## Accuracy : 0.921
## 95% CI : (0.9054, 0.9347)
## No Information Rate : 0.6062
## P-Value [Acc > NIR] : < 2.2e-16
## ## Kappa : 0.8325
## Mcnemar's Test P-Value : 0.001128
## ## Sensitivity : 0.8674
## Specificity : 0.9557
## Pos Pred Value : 0.9272
## Neg Pred Value : 0.9173
## Prevalence : 0.3938
## Detection Rate : 0.3416
## Detection Prevalence : 0.3684
## Balanced Accuracy : 0.9116
## ## 'Positive' Class : pos ##
```

Fig. 2. Command code of confusion matrix

When compared to the standard display, the matrix is reversed. Online predictions of the courses have been seen as columns. The accuracy (not precision) success rate is 92.1%. It includes the 95% confidence interval. It is uncommon enough to warrant reporting. We also have other indicators, such as the sensitivity, which is equivalent to 471/(471 + 72) = 86.74% and is related with the positive class "fire = pos".

There are various characteristics of the "confusion matrix" object. We use the vector $overall, which has named values, to have access to the global indicators (Fig. 3).

```
# access to global indicators print(mat$overall)
## Accuracy Kappa AccuracyLower AccuracyUpper AccuracyNull
## 9.209572e-01 8.325489e-01 9.054394e-01 9.346511e-01 6.062364e-01
## AccuracyPValue McnemarPValue
## 1.111938e-156 1.127515e-03
```

Fig. 3. Command code of $overall vector

We'll follow these steps to get to the "Accuracy" cell (Fig. 4).

```
#accuracy print(mat$overall["Accuracy"])
## Accuracy
## 0.9209572
```

Fig. 4. Command code to get to the accuracy cell

The #byclass field allows access to indicators by class (Fig. 5):

```
#by class print(mat$byClass)
## Sensitivity Specificity Pos Pred Value
## 0.8674033 0.9557416 0.9271654
## Neg Pred Value Precision Recall
## 0.9173364 0.9271654 0.8674033
## F1 Prevalence Detection Rate
## 0.8962892 0.3937636 0.3415518
## Detection Prevalence Balanced Accuracy
## 0.3683829 0.9115725
```

Fig. 5. Command code to indicators' access by class

After executing each of the confusion matrix's command codes, we obtained a fire detection accuracy value of 471/(471 + 37) = 92.71%. Here, we achieved excellent precision.

3.4 Optimization Criteria and Generated Values

From the confusion matrix that was previously created, we may infer various performance criteria. It is typically preferable to report a percentage of errors rather than the complete number of errors when we don't know how many points are in the test set (5% of errors) [13].

To evaluate the performance of our model in identifying positive class events [14, 15]. Given that detect fire is the positive class, sensitivity quantifies the percentage of actual detect fire events that are correctly anticipated. We refer to the sensitivity as the proportion of correctly identified positives, or true positives. This is the ability of our model to detect every fire. Thus, in order to calculate the "Sensitivity" [16], and the "Precision", we shall utilize the abbreviations TP (True Positive), FN (False Negatives), and FP (False Positives).

$$Sensitivity = \frac{\mathrm{TP}}{TP + FN} \quad (1)$$

By methodically foreseeing "positive," one can quickly have a very good reminder. Although our model is not very useful, we won't miss any fires. Since accuracy is the proportion of correctly anticipated points among positively projected points, we will focus on it. Only in the event of a true fire will our model have the ability to activate an alert.

$$Precision = \frac{\mathrm{TP}}{TP + FP} \quad (2)$$

By making extremely few positive predictions (we are less likely to get them wrong), we can comparably and quickly achieve very high accuracy [4]. We can calculate the "F-measure", which is their harmonic average, to evaluate a recall-precision trade-off.

In conclusion, confusion matrices have the advantage of being simple to use and understand. It enables you too easily to display statistics and data for the examination of model performance and trend detection that could lead to configuration changes. By adding rows and columns, a confusion matrix can also be used to solve classification problems involving three or more classes.

3.5 Measures of Error

We began by calculating the model's predicted error rate while discussing classification model evaluation. This is not the right solution for a regression issue. To be more specific, how precisely can data be used to determine whether a prediction is correct? On the other hand, we typically choose a model closer to true values overall than one that is practically exact at some spots. The picture below (Fig. 6) provides information on forecasts and the actual parameters needed to forecast fire [6].

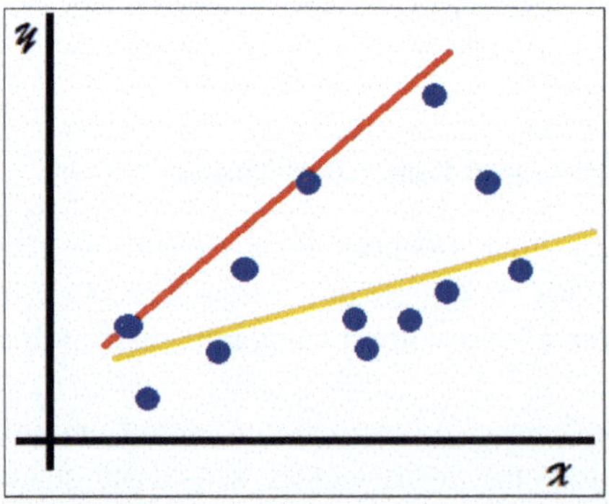

Fig. 6. Location of the closest and exact points of the real values

The orange line more accurately depicts the data and is generally closer to the true values to anticipate than the red line, which produces nearly precise predictions for two of the points.

Let's examine this idea of being "closer to true values". We will figure out and sum the separation between each point xi in the test set's label and the expected value. The sum of the residual squares, abbreviated RSS for Residual Sum of Squares, is the outcome.

$$RSS = \sum_{i=1}^{n} (f(xi) - yi)^2 \tag{3}$$

RSS has the drawback of being more effective with data. Because of this, we will normalize it using the n-point test set. This provides the mean squared error, also known as MSE [17].

$$MSE = \frac{1}{n} \sum_{i=1}^{n} (f(xi) - yi)^2 \tag{4}$$

Take the root of the MSE to return to the unity of y. The result is the Root Mean Squared Error (RMSE).

$$RMSE = \sqrt{\frac{1}{n} \sum_{i=1}^{n} (f(xi) - yi)^2} \tag{5}$$

But when labels can accept values that range over several orders of magnitude, RMSE does not behave very well [18]. Consider making a mistake on a label that is 4 that is 100 units off; the corresponding term in the RMSE is 1002 = 10,000. It's precisely the same as writing 100 units incorrectly on a label that says 8000. To me, a miscalculation of 8100 instead of 8000 seems much larger than a prediction of 104 instead of 4, a wrong of two orders of magnitude.

Before calculating the RMSE, one can take this into consideration by inputting both the true and the projected values into the log. Thus, the RMSLE (Root Mean Squared Log Error) is obtained.

$$RMSLE = \sqrt{\frac{1}{n} \sum_{i=1}^{n} (\log(f(xi) + 1) - \log(yi + 1))^2} \tag{6}$$

Let's go back to the first illustration. The equivalent term in the RMSLE been changed to 1.75 for the forecast of 104 fire detections as opposed to 4. The prediction of 8100 instead of 8000 is now 3.10–5, which is in line with the data. There, we see that the log transit method had indeed worked. But while RMSE doesn't show relative values, we'll utilize the coefficient of determination instead, which provides a highly accurate indication of how well a phenomenon is explained by your model.

4 Conclusion

The confusion matrix is intended to illustrate the different ways that every categorization model could make predictions that are incorrect. It gives a general overview of classifier faults as well as the different kinds of errors that can occur.

In this article, we used metrics to assess the model's performance and classification using the confusion matrix to forecast fire. A confusion matrix can be useful in demonstrating how well our classification model is performing: For instance, it is possible to verify whether the system scheduled both good and bad events correctly or poorly. Each of these procedures provides the basis for the statistical metric for the more general category of matrix misunderstanding. In our case, we use the most widely used, namely the F-measure, sensitivity, specificity, and accuracy. A confusion matrix and class statistics were used in the fire detection problem. It is simple modifiable to handle issues with multinomial classifications.

References

1. Korchi, A., Messaoudi, F., Oughdir, L.: Successful machine learning project. Int. J. Sci. Eng. Res. **10**(9), 1540–1543 (2019)
2. Sharifani, K., Amini, M.: Machine learning and deep learning: a review of methods and applications. World Inf. Technol. Eng. J. **10**(07), 3897–3904 (2023)
3. Heininger, M., Ortner, R.: Predicting packaging sizes using machine learning. In: Operations Research Forum, vol. 3, no. 3, p. 43. Springer International Publishing, Cham (2022)
4. Singal, A.G., Mukherjee, A., Elmunzer, B.J., Higgins, P.D., Lok, A.S., Zhu, J., Waljee, A.K.: Machine learning algorithms outperform conventional regression models in predicting development of hepatocellular carcinoma. Am. J. Gastroenter. **108**(11), 1723 (2013)
5. Abubakar, A.I., Ahmad, I., Omeke, K.G., Ozturk, M., Ozturk, C., Abdel-Salam, A.M., Imran, M.A.: A survey on energy optimization techniques in UAV-based cellular networks: from conventional to machine learning approaches. Drones **7**(3), 214 (2023)
6. Manzali, Y., Elfar, M.: Random forest pruning techniques: a recent review. In: Operations Research Forum, vol. 4, no. 2, pp. 1–14. Springer International Publishing (2023)
7. Rahman, A., Lu, Y., Wang, H.: Performance evaluation of deep learning object detectors for weed detection for cotton. Smart Agric. Technol. **3**, 100126 (2023)
8. Masood, F., Masood, J., Zahir, H., Driss, K., Mehmood, N., Farooq, H.: Novel approach to evaluate classification algorithms and feature selection filter algorithms using medical data. J. Comput. Cogn. Eng. **2**(1), 57–67 (2023)
9. Abid, F.: A survey of machine learning algorithms based forest fires prediction and detection systems. Fire Technol. **57**(2), 559–590 (2021)
10. Preeti, T., Kanakaraddi, S., Beelagi, A., Malagi, S., Sudi, A.: Forest fire prediction using machine learning techniques. In: 2021 International Conference on Intelligent Technologies (CONIT), pp. 1–6. IEEE (2021)
11. Arif, M., et al.: Role of machine learning algorithms in forest fire management: a literature review. J. Robot. Autom **5**, 212–226 (2021)
12. Miranda, F.M., Köhnecke, N., Renard, B.Y.: Hiclass: a python library for local hierarchical classification compatible with scikit-learn. J. Mach. Learn. Res. **24**(29), 1–17 (2023)
13. Li, S., et al.: Short-term electrical load forecasting using hybrid model of manta ray foraging optimization and support vector regression. J. Clean. Prod. **388**, 135856 (2023)

14. Turner, K.E., Sohel, F., Harris, I., Ferguson, M., Thompson, A.: Lambing event detection using deep learning from accelerometer data. Comput. Electron. Agric.. Electron. Agric. **208**, 107787 (2023)
15. Moustafa, S.S., Khodairy, S.S.: Comparison of different predictive models and their effectiveness in sunspot number prediction. Phys. Scr. **98**(4), 045022 (2023)
16. Khan, M.A.R., Afrin, F., Prity, F.S., Ahammad, I., Fatema, S., Prosad, R., Uddin, M.: An effective approach for early liver disease prediction and sensitivity analysis. Iran J. Comput. Sci. 1–19 (2023)
17. Mitra, A., Jain, A., Kishore, A., Kumar, P.: A comparative study of demand forecasting models for a multi-channel retail company: a novel hybrid machine learning approach. In: Operations Research Forum, vol. 3, no. 4, p. 58. Springer International Publishing, Cham (2022)
18. Galante, N., Cotroneo, R., Furci, D., Lodetti, G., Casali, M.B.: Applications of artificial intelligence in forensic sciences: current potential benefits, limitations and perspectives. Int. J. Legal Med. **137**(2), 445–458 (2023)

Identifying ChatGPT-Generated Essays Against Native and Non-native Speakers

Anoual El kah[1(✉)], Ayman Zahir[2], and Imad Zeroual[3]

[1] Polydisciplinary Faculty, Sultan Moulay Slimane University, Beni Mellal, Morocco
elkah.anoual.mri@gmail.com

[2] Engineering Cycle in Computer Engineering, FST Errachidia, Moulay Ismail University of Meknes, Meknes, Morocco

[3] L-STI, T-IDMS, FST Errachidia, Moulay Ismail University of Meknes, Meknes, Morocco

Abstract. Native Language Identification (NLI) consists of applying automatic methods to determine the native language (i.e., L1) of the writer of a given text. In the last few years, the NLI has gained more attention from scholars in several fields, such as authorship profiling, forensic and security, and language teaching. On the other hand, using the ChatGPT as a writing aid, especially by students, raises serious concerns for the academic world about the potential misuse of AI-generated content. Therefore, in this study, we examined the performance of relevant classification models (i.e., Naïve Bayes, Support Vector Machine, and Random Forest) in identifying essays that were written by native and non-native Arabic learners as well as those were totally generated by the ChatGPT. Further, we implemented four language-independent feature extraction techniques, primarily Countvectorizer, TF-IDF, Word2vec, and Glove. The dataset used for training and evaluation includes roughly 800 short essays for each category. Our findings reveal that the SVM-based classifier with TF-IDF achieved the highest accuracy of 91.14%, and the most accurately identified essays are those written by Arabic native speakers.

Keywords: ChatGPT · Native language identification · Text classification · Arabic language

1 Introduction

Since its first appearance as fine-tuning-based representation models, large language models (LLM) (e.g., BERT) are achieving state-of-the-art results for a range of natural language processing tasks [1]. Further, Generative pre-trained transformers (GPT) are currently one of the biggest LLMs, and they can provide a much more accurate representation of the data. The ChatGPT is considered a leading example of GPT-based intelligent agents, which was developed by OpenAI. Based on written prompts, ChatGPT has led to significant advances in digital communication, especially writing articles, scripts, blogs, email templates, letters, stories, and jokes, among others.

Regardless of its many benefits, the rapid adoption of ChatGPT by students has brought many concerns that are not limited to the possibility that a writing sample

Y. Farhaoui et al. (Eds.): ICAISE 2023, LNNS 837, pp. 242–247, 2024.
https://doi.org/10.1007/978-3-031-48465-0_32

can be done or aided by an intelligent agent. The main problem, as cited on its website, "ChatGPT sometimes writes plausible-sounding but incorrect or nonsensical answers"[1]. Simultaneously, its use to write scientific papers is increasing. In a recent study [2], academic reviewers only caught 63% of AI-generated abstracts that were submitted for possible publication. Therefore, the current trend is to detect this AI-generated content through AI-based tools.

In this study, we examined the performance of relevant classification models (i.e., multinomial Naive Bayes, Support Vector Machine, and Random Forest) in identifying essays that were written by native and non-native Arabic learners as well as those were totally generated by the free version of the ChatGPT. Further, we implemented four language-independent feature extraction techniques, primarily Countvectorizer, TF-IDF, Word2vec, and Glove.

In addition to this introduction, the rest of the paper is structured in three other sections. Section 2 describes the keys steps of our methodology, including data collection, text pre-processing, feature extraction, and essays classification. Section 3 summarizes the results achieved using most common performance evaluation metrics. At last, the paper is concluded in Sect. 4.

2 Methodology

To conduct this study, we carried on the following the key steps outlined below and depicted in Fig. 1:

1. Data Collection: We used two primary resources to prepare our dataset. The first one is the Arabic Learner Corpus V2 (ALC) [3], which includes a collection of written essays produced by Arabic and international students in the kingdom of Saoudi Arabia. From the ALC, we extracted 791 essays of Native Arabic Speakers (NAS) and 794 essays of Non-Native Arabic Speakers (NNAS). The second source is the free version of ChatGPT. We executed the corresponding prompts on ChatGPT to generate 790 essays that have similar topics and average size as those written by the students. Due this, the size and topics of the essays will not be considered as fitting criteria by the classifiers implemented. Thus, only the essays' content and structure will be mainly considered.
2. Data Pre-processing: We retrieved and kept only the essay that fulfils the executed prompt, otherwise, we re-executed the prompt until an appropriate essay is generated from the ChatGPT. Then, we performed an automatic cleaning process to remove numbers, dates, punctuation marks, and stop-words.
3. Data Representation and Reduction: We converted the textual content of the essays into the appropriate numerical features, which will be fitted into the classifiers, using four different features extractors. The Countvectorizer which converts the text into to a vector based on the term/token frequency. The TF-IDF is a multiplying result of the Term Frequency in a given document and the logarithm of the Inverse Document Frequency that contain that term. The lexical embedding model Word2vec [4] is a standard method for generating word embedding using a neural network model to

[1] https://openai.com/blog/chatgpt.

learn word associations. Finally, an unsupervised learning algorithm called GloVe [5] (i.e., Global Vectors) that maps words into a meaningful space using semantic similarity.

4. Data Classification: The numerical features extracted are then fitted to three relevant classifiers for training and testing. Of course, there are numerous classifiers that have been implemented in Arabic text classification. However, Naïve Bayesian (NB), Support Vector Machines (SVM), and Random Forest (RF) have shown promising results in predicting target categories even with small training dataset like ours [6]. The NB-based classifier relies on Bayes' Theorem [7]. The SVM-based classifier which has a non-linear formula even if it was originally designed for binary classification [8]. The RF-based classifier is an ensemble model that randomly assembles various decision tree classifiers and determine the final category through unweighted vote [9].

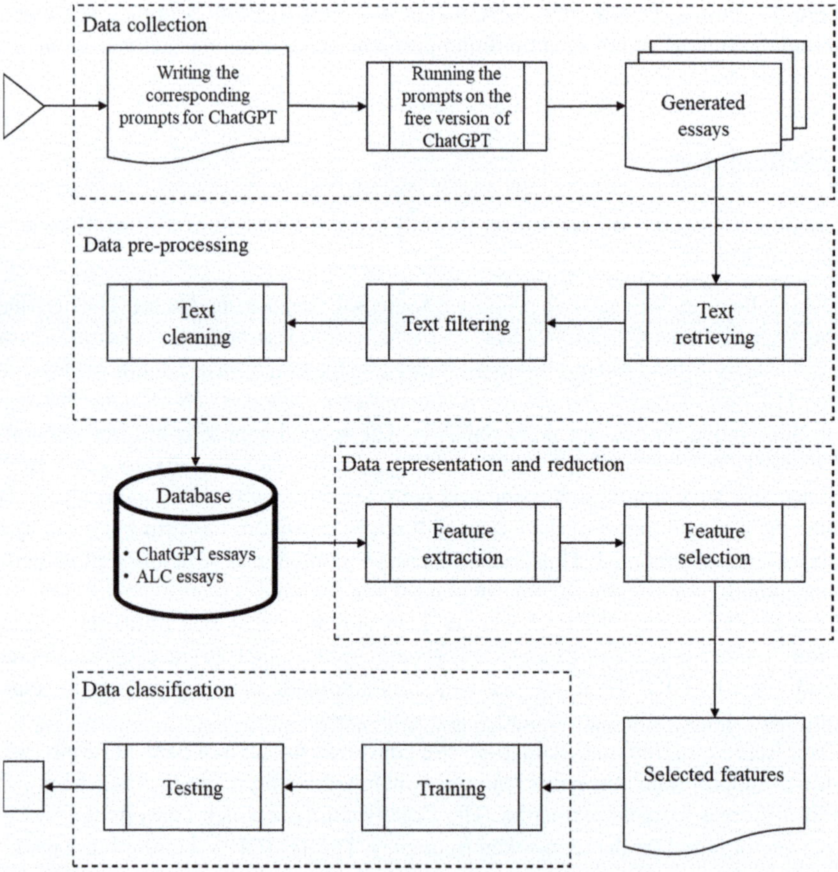

Fig. 1. Flow diagram of the methodology conducted.

3 Results and Discussion

Once the datasets of the three categories (i.e., NAS, NNAS, and ChatGPT) have been ready to be fitted into the classifiers, each dataset is divided into two parts: a training set and a test set. These two sets are manually proportioned to avoid a random distribution of the proportions and to ensure that the two sets are equally distributed among the three categories. Further, we used the ratio 80:20, which means 80% of the data is for training and 20% for testing. Finally, we evaluated the three classifiers NB, SVM, and RF by measuring their performance using commonly metrics: accuracy, precision, recall, and F1-score. Table 1 exhibits the average results reported by these classifiers.

Table 1. Average results reported by the three classifiers.

Classifier	Feature extractor	Accuracy (%)	Recall (%)	Precision (%)	F-score (%)
Multinomial NB	Countvectorizer	85.86	88.64	82.28	85.34
	TF-IDF	89.87	92.38	86.92	89.57
	Word2vec	64.98	65.50	63.29	64.38
	Glove	54.85	54.47	59.07	56.68
SVM	Countvectorizer	90.08	93.18	86.50	89.72
	TF-IDF	91.14	94.52	87.34	90.79
	Word2vec	76.79	79.26	72.57	75.77
	Glove	59.49	59.57	59.07	59.32
Random forest	Countvectorizer	90.30	93.21	86.92	89.96
	TF-IDF	88.40	90.99	85.23	88.02
	Word2vec	61.18	61.88	58.23	60.00
	Glove	58.02	57.98	58.23	58.11

As observed in Table 1, the SVM-based classifier outperformed the other classifiers with an accuracy rate of 90.14% and an F-score of 90.79% when the TF-IDF method was used as the feature extractor. On the other hand, the NB-based classifier has the lowest accuracy rate of 54.85% and an F-score of 56.68% when the Glove model was used for feature extraction. In general, the bag-of-words approach (i.e., Countvectorizer) and the TF-IDF improve the performance of all three classifiers when compared to the word-embedding models Word2vec and Glove. The reason could be that the training dataset used in the current study is not large enough for word-embedding techniques since these latter require a large training dataset consisting of millions or even billions of words to provide an accurate result.

To test whether these classifiers will show an improvement if trained on a larger dataset, we retrained the classifiers using different ratios for the training set while we used the same 10% portion as the test set. Table 2 presents the accuracies obtained by the three classifiers while using the TF-IDF as the feature extraction method.

Table 2. Accuracies achieved by the three classifiers.

Classifier	Training portion				
	20%	40%	60%	80%	90%
Multinomial NB (%)	63.22	64.43	75.71	87.09	89.71
SVM (%)	61.31	66.81	78.65	91.36	91.41
Random Forest (%)	64.36	65.29	78.50	88.66	90.08

According to Table 2, the amount of the training data is still a primary factor that influences the performance of the classifiers implemented. However, compiling such data requires considerable effort and time since it is mostly done manually. Fortunately, the ALC corpus is publicly accessible; also, we planned to make ChatGPT-generated essays freely available for academic research.

After examining the correct and misclassified essays, we observed that the essays samples written in the native language are identified the most accurately among the rest. On the other hand, the non-native writing essays were more likely to be misclassified, and all three classifiers consistently misclassified them as ChatGPT-generated.

4 Conclusion

LLM's high performance in a wide range of natural language processing tasks increases the risk of misuse of AI-generated content. Although many detection tools have been developed to distinguish AI from human-generated content, their effectiveness is still limited, especially when approaching a language other than English (e.g., [10–12]). This study is another contribution that aims to fill this gap by evaluating the performance of widely used classifiers and feature extractors using writing samples from native and non-native Arabic learners as well as their counterparts in size and topic that are fully generated by ChatGPT.

From the results found, we can infer that it is possible to build AI-based models to detect AI-generated content. These results encourage us to continue working on this project, especially for under-resourced languages such as Arabic. Also, we plan to increase the size of the dataset compiled and expand the scope of testing and verifying the performance of other models while considering the detection of AI-generated content at different levels like sentence structure, vocabulary, and colocations, just to name a few.

References

1. Sun, C., Qiu, X., Xu, Y., Huang, X.: How to fine-tune BERT for text classification? In: Chinese Computational Linguistics, pp. 194–206. Springer, Cham (2019). https://doi.org/10.1007/978-3-030-32381-3_16
2. Thorp, H.H.: ChatGPT is fun, but not an author. Science **379**, 313 (2023). https://doi.org/10.1126/science.adg7879

3. Alfaifi, A.Y.G., Atwell, E., Hedaya, I.: Arabic learner corpus (ALC) v2: a new written and spoken corpus of Arabic learners. Proc. Learn. Corpus Stud. Asia World **2014**(2), 77–89 (2014)
4. Mikolov, T., Chen, K., Corrado, G., Dean, J.: Efficient estimation of word representations in vector space. ArXiv Preprint ArXiv13013781 (2013)
5. Pennington, J., Socher, R., Manning, C.D.: Glove: global vectors for word representation. In: Proceedings of the 2014 Conference on Empirical Methods in Natural Language Processing (EMNLP), pp. 1532–1543 (2014)
6. El Kah, A.E., Zeroual, I.: Is Arabic text categorization a solved task? In: 2022 International Conference on Intelligent Systems and Computer Vision (ISCV), pp. 1–6 (2022). https://doi.org/10.1109/ISCV54655.2022.9806076
7. Joyce, J.: Bayes' theorem. In: Zalta, E.N. (ed.) The Stanford Encyclopedia of Philosophy. Metaphysics Research Lab, Stanford University (2021)
8. Vapnik, V., Chervonenkis, A.Y.: A class of algorithms for pattern recognition learning. Avtomat Telemekh. **25**, 937–945 (1964)
9. Breiman, L.: Random forests. Mach. Learn. **45**, 5–32 (2001)
10. Cingillioglu, I.: Detecting AI-generated essays: the ChatGPT challenge. Int. J. Inf. Learn. Technol. **40**, 259–268 (2023)
11. Chaka, C.: Detecting AI content in responses generated by ChatGPT, YouChat, and Chatsonic: the case of five AI content detection tools. J. Appl. Learn. Teach. **6**, (2023)
12. Yan, D., Fauss, M., Hao, J., Cui, W.: Detection of AI-generated essays in writing assessment. Psychol. Test Assess. Model. **65**, 125–144 (2023)

Technology in Education: An Attitudinal Investigation Among Prospective Teachers from a Country of Emerging Economy

Manilyn A. Fernandez[1(✉)], Cathy Cabangcala[1], Emma Fanilag[1,2], Ryan Cabangcala[1], Keir Balasa[1,3], and Ericson O. Alieto[1]

[1] Western Mindanao State University, Zamboanga City, Philippines
{xt202001458,ericson.alieto}@wmsu.edu.ph
[2] San Roque Elementary School, Zamboanga City, Philippines
[3] Jose Rizal Memorial State University, Zamboang Del Norte, Tampilisan, Philippines

Abstract. The integration of technology for teaching and learning created many changes in the academe. This added an expectation for teachers and prospective teachers to be competent in using different technological tools for classroom instruction. To ensure successful technology integration in education, factors such as attitudes toward it must be explored. Given this, studies have been conducted in the Philippines focusing on individuals' attitudes toward technology integration and the factors that influence such attitudes; however, these studies delved only on the views of teachers. Thus, this study aimed to explore the attitude of teacher education students toward technology integration as well as whether factors such as perceived ease of use, IT self-efficacy, IT anxiety, adoption of IT, relative advantage, compatibility, and an institute's support are significantly related to their attitude. This study utilized a quantitative descriptive survey design. A questionnaire with a reliability score of 0.9 from the study of Ramdhony et al. [4] was used, and 100 teacher education students responded in the survey. The results reveal that the respondents display a positive attitude toward technology integration. Apart from IT anxiety, the other six mentioned factors have significant relationships with attitude. The results are further discussed in the study.

Keywords: Attitude · Education · Prospective teachers · Technology integration

1 Introduction

Continuous advancements and developments in technology have greatly influenced people's work, business, entertainment, health, and education [1]. As the rise of technology provides the world with innovations and resources that bring convenience and efficiency in performing everyday tasks, Hue and Ab Jalil [2] claim that individuals are now expected to be well equipped with handling technology. Hence, with the increasing demand for technology-related skills in the workplace, education plays a significant role in preparing students and molding them into quality human resources with the necessary qualities to thrive in today's time, which, according to Hussein [3], makes technology integration in educational institutions a must.

Y. Farhaoui et al. (Eds.): ICAISE 2023, LNNS 837, pp. 248–255, 2024.
https://doi.org/10.1007/978-3-031-48465-0_33

The integration of technology in education is seen as essential by stakeholders of learning institutions in the world [4], including the Philippines, in elevating the quality of teaching and learning. Aytekin and Isiksal-Bostan [5] opine that it could help improve and provide support for educational systems. It is also believed to help achieve objectives and improve the quality of lessons [2] by providing teachers and learners with various learning tools and strategies to explore. This in turn also develops the skills of students. To ensure that learners gain the necessary 21st century skills for their future careers, the role of teachers in achieving success on such aims cannot be denied. Thus, it is of great importance for teachers and aspirants to be knowledgeable and competent in using and effectively integrating technology in the classroom [6]. In line with this, the technology acceptance model (TAM) has been widely used to understand one's tendency to accept or reject technology [4, 7]. The Theory of Reasoned Action [8] is where this model emanated, in which attitude is a main determinant of one's behavior. According to Ajzen [8], Alieto [9], Alieto et al. [10], and Alieto and Rillo [11], attitude greatly affects one's willingness or acceptance toward something. This is supported by the claim of Cullen and Green [12] that having a positive attitude toward technology determines the success of integrating it. Making the cognizance of teachers and teacher aspirants' attitudes imperative in the success of technology integration and the development of students' skills. This entails that to maintain successful technological integration, a positive attitude toward it must also be maintained. Therefore, factors that could possibly influence one's attitude must also be cognized.

In consideration of this, a plethora of studies have explored the attitude of individuals toward the integration of technology among various contexts and various respondents. Some have delved into the attitude of teachers [2, 13–17], while some have considered students [5, 6, 18–20], all of which yielded a positive attitude toward technology integration, and the factors that influence such attitudes, including perceived ease of use (PEU) [4, 21], self-efficacy (SE) [4, 13], anxiety [22], experience with technology [4, 23], relative advantage [24], compatibility, and support of one's institution [4], were explored as extensions of TAM according to Ramdhony et al. [4] and John [21]. However, these studies are from foreign contexts, and only a few studies in the Philippines have explored the attitudes of Filipinos toward technology integration, such as Dela Rosa [25], and the factors that influence their attitudes [26, 27]. These studies, however, are only centered on teachers. It must be noted that teacher aspirants are also expected to be competent and skilled in using technology in the classroom. To address this gap in the literature, this study aims to analyze the attitudes of Filipino teacher education students who have taken the subject of technology for teaching and learning (TTL) and at the same time determine whether the aforementioned factors significantly influence their attitudes. Knowledge on the matter may be used in assessing the success of technology integration in the country, provide insight to better support the use of technology in education, and add knowledge to the literature.

Research Questions

The questions below are what this study seeks to answer:

1. What is the attitude of education students toward technology integration in education?

2. Is the respondents' attitude toward technology integration significantly related to the following factors: (1) perceived ease of use, (2) self-efficacy, (3) anxiety, (4) experiences with technology, (5) relative advantage, (6) compatibility, and (6) the institute's support?

2 Methodology

2.1 Research Design

This study employed a quantitative descriptive survey design. Quantitative design is used when the main goal of a study is to quantify and measure identified variables from gathered data [28], which is aligned with the objective of the present study in measuring and analyzing the attitude of prospective teachers toward the integration of technology in the teaching and learning process and the factors that determine their attitude. Mean, standard deviation, and frequency will also be used for the interpretation of data gathered from a survey questionnaire.

2.2 Respondents

The target respondents of this study are prospective teachers who are being prepared for and exposed to technology integration. Their attitude provides imperative knowledge for the success of effective integration and gaining competence in using technology in the teaching and learning process. Given this, criteria are set for the inclusion of the respondents of this study: (1) the respondent must be currently taking a course under the College of Teacher Education, and (2) they must have already taken at least one Technology for Teaching and Learning (TTL) course by the time this study is conducted. With the presented parameters, 100 randomly selected students (67% are females and 33% are males) studying different courses in the teacher education program responded in the survey.

2.3 Research Instrument

The present study adopted the questionnaire from the study of Ramdhony et al. [4]. It contains 39 items divided into 8 subparts: attitude (5), PEU (3), IT SE (5), IT anxiety (5), experiences (5), relative advantage (5), compatibility (4), and support of the institution (7). A five-point Likert scale with a range of (1) strongly disagree to (5) strongly agree was utilized as the response. The instruments' validity was checked by an expert in the field of English and research for its alignment with the study's purpose. The questionnaire has a reliability result of 0.9, and the Cronbach's alpha scores for each part are as follows: 0.93, 0.73, 0.95, 0.93, 0.92, 0.95, 0.93, and 0.96.

2.4 Data Gathering Procedure

Data were gathered from undergraduate prospective teachers who are currently enrolled in the College of Teacher Education (CTE) at Western Mindanao State University (WMSU). For convenience and safety purposes, the respondents were asked to access a digitized instrument (Google Form) given by the researcher. To do this, the researcher contacted one representative from each course and asked the representatives to disseminate the link with their respective sections. The form also contains the consent form where it was explained that their participation in the survey is voluntary and no other identifying information, except for their gender and course, will be collected.

2.5 Data Analysis Technique

Analysis of the data gathered was performed using the Statistical Package for the Social Sciences (SPSS) version 25. Descriptive statistics such as the mean and standard deviation were utilized to describe the attitude of teacher education students toward the integration of technology in the teaching and learning process. The Pearson product-moment correlation coefficient was used to determine the relationship between the respondents' attitudes and the factors mentioned in this study.

3 Results and Discussion

Prospective Teachers' Attitudes Toward Integrating Technology in Education

The respondents' attitudes toward the use of technology for teaching and learning were analyzed using computed mean scores and standard deviations of the data collected. A scale with equal intervals was utilized for the interpretation of the results.

Table 1. Attitude toward technology integration

	Mean	SD	Interpretation
Attitude	4.12	0.58	Positive attitude

Scale: 1.0 to 1.79—Very Negative Attitude, 1.80 to 2.59—Negative Attitude, 2.60 to 3.39—Neutral Attitude, 3.40 to 4.19—Positive Attitude, and 4.20 to 5.0—Very Positive Attitude. N = 100.

Table 1 presented reveals that teacher education students display an attitude that is favorable to the use of technology in the process of teaching and learning. This means that they view the rise of technology usage for classroom instruction and student learning positively. This result is in concordance with the studies of Giles [6], Hussein [3], Salele and Khan [18], Yapici & Hevedanli [19], and Zhou et al. [20], which also sought knowledge on the attitude of teacher aspirants toward technology integration in education. This could be due to the belief that individuals need to be exposed to and use technology in educational settings to be prepared and able to thrive in today's time [3].

Factors Affecting Teacher Education Students' Attitudes Toward Technology Integration

The Pearson-product moment correlation coefficient was utilized for the analysis of relationships between the teacher education students' attitudes and the different factors. Interpretations were then given based on the p value and r-value results.

Table 2. Correlation between attitude and the different factors

Variables		p value	r-value	Interpretation
Attitude	Perceived Ease of Use	0.000	0.594	Significant/Moderate correlation
	IT Self-efficacy	0.000	0.690	Significant/Moderate correlation
	IT Anxiety	0.525	−0.064	Not Significant
	Experiences (Adoption of IT)	0.000	0.655	Significant/Moderate correlation
	Relative Advantage	0.000	0.655	Significant/Moderate correlation
	Compatibility	0.000	0.636	Significant/Moderate correlation
	Institute's Support	0.000	0.514	Significant/Moderate correlation

Significant at Alpha = level 0.01.

Table 2 reveals that six (6) factors are significantly related to the respondent's attitude in integrating technology in teaching and learning. As Hussein [3] opines, factors under TAM are highly influential on an individual's attitude toward technology use and integration in education. Perceived ease of use (PEU) is one such factor [21, 25]. With the findings presented, a significant relationship between attitudes and PEU is seen. This means that the more the respondents perceive technology to be easy to use, the more positive their attitude toward its integration in education is. The result shows congruence with the studies of David and Aruta [26] and Nueva [27], who focused on Filipino teachers. It must be due to the respondents having enough experience and skills in using certain technologies for teaching and learning that such results are seen.

Attitude and self-efficacy (SE) show a significant and moderate positive relationship, which means that the higher the SE is, the more positive the attitude. This is supported by Compeau and Higgins [29], who claim that self-efficacy influences how an individual views and acts toward something, which in this study is the respondents' attitude. Similar findings were revealed in the studies of Aktas and Karaka [13] and Ramdhony et al. [4]. This positive relationship must be because of the respondents' capabilities in using technology for their learning as well as having the training for common technologies used in the classroom as given in their Technology for Teaching and Learning (TTL) subject.

A correlation between attitude and anxiety was expected; however, there is no occurrence of such a relationship within the data, which indicates that the attitude of Filipino teacher education students is not affected by their perceived anxiousness about using technology. This shows alignment with the studies of Ramdhony et al. [4] and Ventakesh et al. [30], who investigated the influence of anxiety on attitude. Such a result could be due to the respondents having enough experience and skill in using technology.

A significant relationship between the respondents' attitude and their experiences in technology adoption is presented in the Table 2. This finding is the same as the studies of Ramdhony et al. [4] and Scherer et al. [23], where it was revealed that the more experienced an individual is in using technology, the more positive their attitude becomes. This must be because the respondents have been exposed to technology usage and integration both inside and outside the school setting.

Relative advantage is significantly related to teacher aspirants' attitudes toward technology integration in education. This is congruent with the study of Erskine et al. [24], where the same variables were also explored. This finding indicates that the more the respondents view technology as advantageous, the more positive their attitude becomes. This found knowledge could be aligned with what Hussein [3] maintains, that being exposed to classroom instruction integrated with technology usage is beneficial for individuals to progress in today's time.

Compatibility and attitude display a significant relationship as well, with a moderate positive correlation. This reveals that the more compatible the use of technology in teaching and learning is the more positive the respondents' attitude becomes. This is aligned with what was found in the analysis done by Ramdhony et al. [4], where attitude is influenced by how compatible a certain object is to an individual's life. This result might be due to teacher education students often being exposed to technological devices and experiences in most of their classes and even outside the university, as they have their own technological devices at their disposal.

The institute's support for technology integration in education and the respondents' attitude are significantly related, with a moderate correlation found between variables. Ramdhony et al. [4] found the same results in his study, where the support of the institution highly influences the attitude of the students. This is supported by the claim of Butler and Sellbom [31] and Vahed and Cruickshank [32] that the higher the institute's support toward the integration of technology, the more positive the attitude and performance of the students become. Given this result, it must be because the teacher aspirants receive the needed technological support from their university.

4 Conclusion and Recommendation

The integration of technology in educational settings has made great changes to how teachers teach, including instructional materials and resources, assessments, and content, which in turn affect how learners learn. The modern time set the standard for teachers and teacher aspirants to be equipped with skills relating to technology use. Given this, factors that might affect their intention to integrate technology and gain competence in using it in the classroom must be investigated. One of these factors is attitude. Thus, this study explored the attitudes of aspiring teachers toward the utilization of technology for teaching and learning, as the topic mostly delved into the views of teachers in the Philippines. It was revealed that the respondents perceive the integration of technology in the process of teaching and learning as favorable. This means that the teacher education students view the use of technological resources for their learning and as something that they accepted fully and are willing to learn and use in their teaching career. This supports the theory presented by Ajzen [31] and the claims of Alieto [7] Alieto et al. [10], and Alieto and Rillo [8] that attitude is a determiner of an individual's willingness to do a certain task, which in this study is the integration of technology in education. It also tried to determine whether certain factors influence the attitude that the teacher aspirants display. The findings show that except for IT anxiety, the other factors were significantly related to and influenced the attitude of the respondents. A strong positive correlation with attitude was found among factors such as self-efficacy, adoption of IT, relative advantage, and compatibility, while a moderate positive correlation was found with perceived ease of use and the institute's support.

This implies that educational institutions must continue to provide opportunities and resources for learners to train and improve their skills in using common technology used for teaching and learning, as PEU, SE, experiences with technology, relative advantage, and support of the institution are highly influential on the attitude of learners toward technology use in educational settings. With regard to compatibility, educational planners and institutions must also consider integrating technology that is similar to commonly used platforms by their students, such as social media, which, according to Ramdhony et al. [4], is valued by students of modern times. The findings are also helpful for curriculum planners in assessing the perception of learners toward the success of technology integration in the curriculum. However, this study is limited to teacher education students only. Other students from different college programs should be investigated for further analysis considering their different characteristics.

References

1. Gulbahar, Y., Guven, I.: A survey on ICT usage and the perceptions of social studies teachers in Turkey. Educ. Technol. Soc. **11**, 37–51 (2008)
2. Hue, L.T., Ab Jalil, H.: Attitudes towards ICT integration into curriculum and usage among university lecturers in Vietnam. Int. J. Instr. **6**, 53–66 (2013)
3. Hussein, Z.: Leading to intention: the role of attitude in relation to technology acceptance model in e-learning. Procedia Comput. Sci. **105**, 159–164 (2017). https://doi.org/10.1016/j.procs.2017.01.196
4. Ramdhony, D., Mooneeapen, O., Dooshila, M., Kokil, K.: A study of university students' attitude towards integration of information technology in higher education in Mauritius. Higher Educ. Q. 1–16 (2020). https://doi.org/10.1111/hequ.12288
5. Aytekin, E., Isiksal-Bostan, M.: Middle school students' attitudes towards the use of technology in mathematics lessons: does gender make a difference? Int. J. Math. Educ. Sci. Technol. 1–21 (2018). https://doi.org/10.1080/0020739X.2018.1535097
6. Giles, M.: The influence of paired grouping on teacher candidates' attitude towards technology use and integration. Technol. Pedagogy Educ. **28**, 363–380 (2019). https://doi.org/10.1080/1475939X.2019.1621772
7. Ventakesh, V., Davis, F.: A theoretical extension of the technology acceptance model: four longitudinal field studies. Manage. Sci. 46:186–204. https://doi.org/10.1287/mnsc.46.2.186.11926
8. Ajzen, I.: The theory of planned behavior. Organ. Behav. Hum. Decis. Process. **50**, 179–211 (1991). https://doi.org/10.1016/0749-5978(91)90020-T
9. Alieto, E.O.: Language shift from English to mother tongue: exploring language attitude and willingness to teach among pre-service teachers. TESOL Int. J. **13**, 134–146 (2018)
10. Alieto, E.O., Ricohermoso, C., Abequibel, B.: An investigation on digital and print reading attitudes: samples from Filipino preservice teachers from a non-metropolitan-based university. Asian EFL J. **27**, 278–311 (2020)
11. Alieto, E.O., Rillo, R.: Language attitudes of English language teachers (ELTS) towards Philippine English. J. Hum. Soc. Sci. **13**, 84–110 (2018)
12. Cullen, T.A., Green, B.A.: Preservice teachers' beliefs, attitudes, and motivation about technology integration. J. Educ. Comput. Res. **45**, 29–47 (2011)

13. Aktas, N., Karaka, F.: The relationship between Turkish high school administrators' technology leadership self-efficacies and their attitudes and competencies towards technology use in education. Participatory Educ. Res. **9**, 430–448 (2022). https://doi.org/10.17275/per.22.122.9.5
14. Albirini, A.: Teachers' attitudes toward information and communication technologies: the case of Syrian EFL teachers. Comput. Educ. **47**, 373–398 (2006)
15. Hong, J.E.: Social studies teachers' views of ICT Integration. RIGEO **6**, 32–48 (2015)
16. Mustafina, A.: Teachers' attitudes toward technology integration in a Kazakhstani secondary school. Int. J. Res. Educ. Sci. (IJRES) **2**, 322–332 (2016)
17. Tou, N.X., Kee, Y.H., Koh, K.T., Camire, M., Chow, J.Y.: Singapore teachers' attitudes towards the use of information and communication technologies in physical education. Eur. Phys. Educ. Rev. 1–14 (2019). https://doi.org/10.1177/1356336X19869734
18. Salele, N., Khan, M.S.H.: Engineering trainee-teachers' attitudes toward technology use in pedagogical practices: extending computer attitude scale (CAS). SAGE Open 1–14 (2022). https://doi.org/10.1177/21582440221102436
19. Yapici, I., Hevedanli, M.: Pre-service Biology teachers' attitudes towards ICT using in Biology teaching. Procedia Soc. Behav. Sci. 64:633–638 (2012)
20. Zhou, Q., Zhao, Y., Hu, J., Liu, Y., Xing, L.: Pre-service chemistry teachers' attitude toward ICT in Xi'an. Procedia Soc. Behav. Sci. 9:1407–1414 (2010)
21. John, S.P.: The integration of information technology in higher education: a study of faculty's attitude towards IT adoption in the teaching process. Contaduría Adm. **60**, 230–252 (2015). https://doi.org/10.1016/j.cya.2015.08.004
22. Hackbarth, G., Grover, V., Yi, M.Y.: Computer playfulness and anxiety: positive and negative mediators of the system experience effect on perceived ease of use. Inf. Manage. **40**, 221–232 (2003)
23. Scherer, R., Siddiq, F., Tondeur, J.: The technology acceptance model (TAM): a meta-analytic structural equation modeling approach to explaining teachers' adoption of digital technology in education. Comput. Educ. **128**, 13–35 (2019). https://doi.org/10.1016/j.compedu.2018.09.009
24. Erskine, M.A., Khojah, M., McDaniel, A.E.: Location selection using heat maps: relative advantage, task-technology fit, and decision-making performance. Comput. Hum. Behav. **101**, 151–162 (2019). https://doi.org/10.1016/j.chb.2019.07.014
25. Dela Rosa, J.P.O.: Experiences, perceptions and attitudes on ICT integration: a case study among novice and experienced language teachers in the Philippines. Int. J. Educ. Dev. Using Inf. Commun. Technol. **12**, 37–57 (2016)
26. David, A., Aruta, J.J.B.: Modeling Filipino teachers' intention to use technology: a MIMIC approach. Educ. Media Int. **59**, 62–79 (2022)
27. Nueva, M.G.C.: Filipino teachers' attitude towards technology—its determinants and association with technology integration practice. Asia Pac. Soc. Sci. Rev. **19**, 167–184 (2019)
28. Creswell, J.: Research design: qualitative, quantitative and mixed methods approaches (2nd ed.). Sage Publications, Thousand Oaks, CA. Williams, C.: Research methods. J. Bus. Econ. Res. **5**(3), 65–71 (2007)
29. Compeau, D.R., Higgins, C.A.: Computer self-efficacy: development of a measure and initial test. MIS Q. **191**, 89–211 (1995). https://doi.org/10.2307/249688
30. Venkatesh, V., Morris, M.G., Davis, G.B., Davis, F.D.: User acceptance of information technology: toward a unified view. MIS Q. **27**, 425–478 (2003). https://doi.org/10.2307/30036540
31. Butler, D.L., Sellbom, M.: Barriers to adopting technology. Educ. Q. **2**, 22–28 (2001)
32. Vahed, A., Cruickshank, G.: Integrating academic support to develop undergraduate research in dental technology: a case study in a South African University of Technology. Innov. Educ. Teach. Int. **55**, 566–574 (2018)

Comparative Analysis of Pre-trained CNN Models for Image Classification of Emergency Vehicles

Ali Omari Alaoui[1](✉), Ahmad El Allaoui[1], Omaima El Bahi[2], Yousef Farhaoui[1], Mohamed Rida Fethi[1], and Othmane Farhaoui[1]

[1] L-STI, T-IDMS, FST Errachidia/Moulay Ismail University, Meknes, Morocco
{al.omarialaoui,m.fethi,ot.farhaoui}@edu.umi.ac.ma,
a.elallaoui@umi.ac.ma, y.farhaoui@fste.umi.ac.ma

[2] L-MAIS, T-MIS, FST Errachidia/Moulay Ismail University, Meknes, Morocco
omaima.elbahi@edu.umi.ac.ma

Abstract. This paper presents a comprehensive study on image classification, focusing specifically on the evaluation of six pre-trained CNN models in the context of emergency vehicle classification. The models examined in this research are VGG19, VGG16, MobileNetV3Large, MobileNetV3Small, MobileNetV2, and MobileNetV1. We followed a systematic research methodology involving dataset preparation, model architecture modification, layer operation restriction, and model compilation optimization. Through extensive experimentation, the performance of each model is analyzed, considering factors such as accuracy, loss and training time. The findings shed light on the strengths and limitations of each model, emphasizing the importance of selecting an appropriate pre-trained CNN model for image classification tasks. Overall, this article provides a comprehensive overview of image classification, highlighting the crucial role of pre-trained CNN models in achieving accurate results for emergency vehicle classification.

Keywords: Pre-trained CNN models · Image classification · Computer vision · Emergency vehicles

1 Introduction

Image classification is a fundamental task in computer vision with a wide range of applications, including object recognition, autonomous driving, medical imaging, and surveillance systems. The ability to automatically categorize and understand the content of images has revolutionized many industries and has become a crucial component of various real-world systems. Deep learning models, particularly Convolutional Neural Networks (CNNs) [1], have emerged as the state-of-the-art approach for image classification tasks. These models can automatically learn hierarchical representations from raw pixel data, enabling them to capture intricate patterns and features that are essential for accurate classification. Pre-trained CNN architectures, trained on massive image datasets

Y. Farhaoui et al. (Eds.): ICAISE 2023, LNNS 837, pp. 256–262, 2024.
https://doi.org/10.1007/978-3-031-48465-0_34

like ImageNet, have further accelerated the progress in image classification. By leveraging transfer learning [3], these pre-trained models can be fine-tuned on specific datasets, saving training time and achieving impressive performance. The objective of this article is to compare the performance of different pre-trained CNN models, including VGG19, VGG16, MobileNetV3Large, MobileNetV3Small, MobileNetV2, and MobileNetV1 for image classification. By conducting a comprehensive analysis, we aim to provide insights into the strengths and limitations of each model and guide researchers and practitioners in choosing the most suitable pre-trained CNN model for their specific image classification tasks.

2 Related Works

Deep convolutional neural networks (CNNs) have revolutionized the field of image recognition, achieving remarkable performance in large-scale visual recognition challenges. The concept of Very Deep Convolutional Networks (VGG) was introduced by Simonyan and Zisserman [4], demonstrating the effectiveness of deep architectures with small convolutional filters. Building upon this, Szegedy et al. [5] developed the VGG16 and VGG19 models, further improving accuracy by increasing network depth. Howard et al. [6] introduced MobileNet, efficient CNN architectures designed specifically for mobile vision applications. Tan and Le [7] proposed EfficientNet, a model that achieved state-of-the-art performance by carefully balancing network depth, width, and resolution. To address the degradation problem in deep networks, He et al. [8] introduced Residual networks (ResNet), which incorporated skip connections. Sandler et al. [9] presented MobileNetV2, which introduced inverted residuals and linear bottlenecks for efficient feature extraction. These works, along with techniques such as large minibatch stochastic gradient descent (SGD) training by Goyal et al. [10] and the ImageNet Large Scale Visual Recognition Challenge [2] have significantly advanced the field of image classification using deep CNNs.

3 Methodology

In our comprehensive study, we conducted an extensive analysis to compare the performance of various pre-trained CNN models for image classification. To facilitate this evaluation, we carefully curated a dataset comprising 384 diverse images from different classes. Prior to training, we applied a crucial preprocessing step, resizing all images to a uniform size of (224, 224).To obtain the necessary image information, we parsed XML files with image annotations using the xml.etree.ElementTree library [11]. This allowed us to efficiently extract image paths, class labels, and bounding box coordinates. By leveraging the power of Google Colab's GPUs, we harnessed their capabilities to expedite the training process.

The dataset was split into a standard 80% training set and a 20% validation set. This partitioning enabled us to train our models on a substantial amount of data while reserving a separate subset for unbiased performance assessment and validation.

To customize the pre-trained models for our specific task, we loaded them without their top classification layers. This strategic choice enabled us to leverage the convolutional base of the models, responsible for extracting meaningful features from images. By excluding the top layers, which were originally designed for specific classification tasks, we could adapt the models to our own image classification problem.

To adapt the models further, we added custom classification layers atop the base models. These additional layers were responsible for learning the class probabilities based on the extracted features.

Our customized architecture included a global average pooling layer to reduce spatial dimensions, followed by a fully connected layer with a ReLU activation function[12] as shown in the Eq. (1), a dense layer with a SoftMax [13] activation function (2) was appended to produce the final class probabilities.

$$ReLU(x) = \max(0, x) \tag{1}$$

$$SoftMax(x)_i = \frac{e^{x_i}}{\sum_{j=1}^{N} e^{x_j}} \tag{2}$$

In order to expedite training, we made the decision to freeze certain layers in the models. By freezing these layers, we ensured that their weights remained unchanged during training, preserving the valuable pre-trained representations they had learned. For our experiments, we chose to freeze all layers except for the last four, striking a balance between leveraging pre-trained knowledge and allowing adaptation to our specific classification tasks.

To compile the models, we utilized the Adam optimizer with a learning rate of 0.0001, a proven effective optimizer for deep learning models. For the loss function, we employed sparse categorical cross-entropy, specifically suited for multi-class classification problems.

4 Results and Discussion

Here are the results of our study on pre-trained CNN models for image classification. We evaluated each model's performance using accuracy and training time metrics and visualized their learning progress. VGG19 model obtained 87.01% accuracy in 80.58 s training time, while VGG16 achieved 89.61% accuracy in 69 s of training as shown in Fig. 1. MobileNetV3Large achieved 58.44% accuracy in 36.99 s whereas MobileNetV3Small get 38.96% accuracy in 33.59 s training as shown in the Fig. 2. MobileNetV2 had 90.91% accuracy in 34.30 s of training while **MobileNetV1** reached 92.21% accuracy in just 32.47 s Fig. 3.

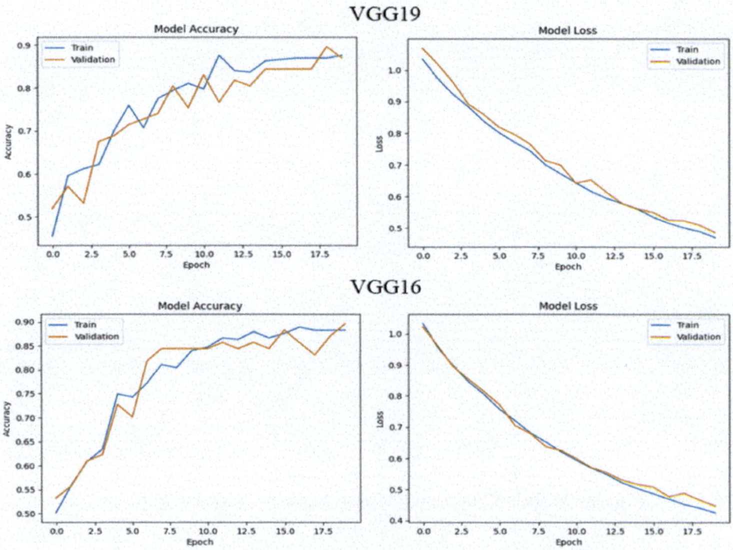

Fig. 1. VGG19 and VGG16 accuracy and loss.

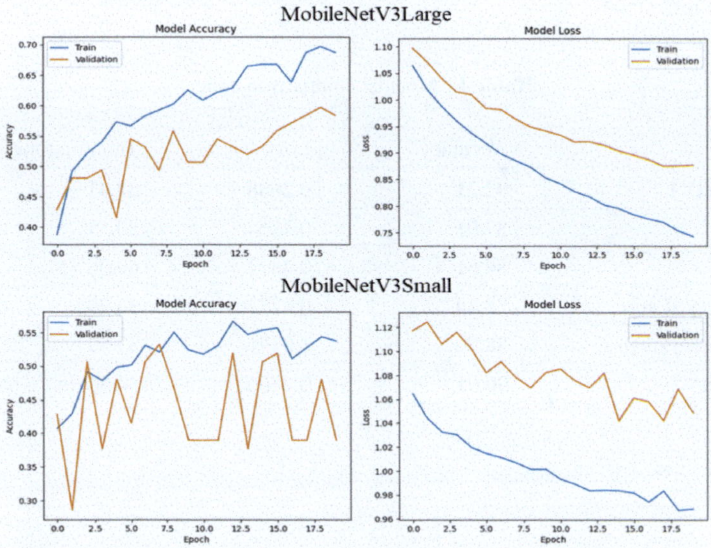

Fig. 2. MobileNetV3Large and MobileNetV3Small accuracy and loss.

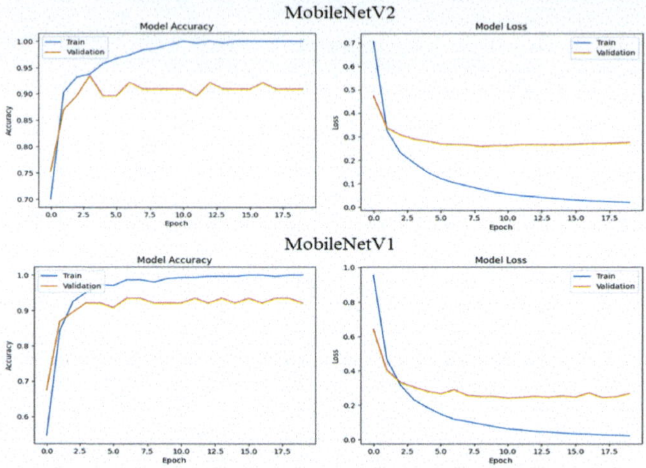

Fig. 3. MobileNetV2 and MobileNetV1 accuracy and loss.

MobileNetV1 performed best according to Table1, with interesting data displayed in Fig. 4. Loss is depicted on the right axis, while Accuracy (%) and training time (s) are on the left.

Table 1. Models' comparison table.

Model	Accuracy	Loss	Training time
MobileNetV1	**92.21**	**0.2688**	**32.47**
VGG19	87.01	0.4864	80.58
VGG16	89.61	0.4472	69.00
MobileNetV3Large	58.44	0.8772	36.99
MobileNetV3Small	38.96	1.0491	33.59
MobileNetV2	90.91	0.2766	34.30

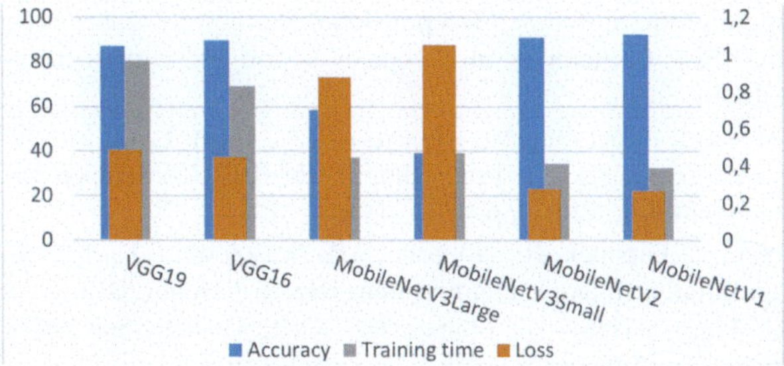

Fig. 4. Comparison models in term of accuracy, loss, and training time

MobileNetV1 is the most accurate model with a complex architecture and short training time.

5 Conclusion

Our study compared CNN models for image classification and found that MobileNetV1 had the highest accuracy. VGG16 and MobileNetV3Small also performed well with shorter training times. Each model has strengths and limitations - VGG16 and VGG19 are more complex but require more computational resources, while MobileNet strikes a balance between accuracy and efficiency. Picking a proper pre-trained CNN model is vital for precise and efficient image classification. Consider architecture, computation requirements, and application specifications. Knowing the strengths and limitations of each model helps in informed decision-making for state-of-the-art classification results. The article highlights leveraging pre-trained CNN models and selecting the ideal one for image classification to boost progress in various fields.

References

1. Li, Z., Liu, F., Yang, W., Peng, S., Zhou, J.: A survey of convolutional neural networks: analysis, applications, and prospects. IEEE Trans. Neural Netw. Learn. Syst. **33**, 6999–7019 (2022). https://doi.org/10.1109/tnnls.2021.3084827
2. Russakovsky, O., Deng, J., Su, H., Fei-Fei, L.: ImageNet large scale visual recognition challenge. Int. J. Comput. Vis. **115**(3), 211–252 (2015)
3. Zoph, B.: Learning Transferable Architectures for Scalable Image Recognition. https://arxiv.org/abs/1707.07012
4. Simonyan, K., Zisserman, A.: Very Deep Convolutional Networks for Large-Scale Image Recognition (2014)
5. Szegedy, C., Liu, W., Jia, Y., Sermanet, P., Reed, S.M., Anguelov, D., Erhan, D., Vanhoucke, V., Rabinovich, A.: Going deeper with convolutions. In: Proceedings of the IEEE Conference on Computer Vision and Pattern Recognition (CVPR), pp. 1–9 (2015)
6. Howard, A.G., Zhu, M., Chen, B., Kalenichenko, D., Adam, H.: MobileNets: Efficient Convolutional Neural Networks for Mobile Vision Applications (2017)
7. Tan, M., Le, Q.-V.: EfficientNet: rethinking model scaling for convolutional neural networks. In: International Conference on Machine Learning, pp. 6105–6114 (2019)

8. He, K., Zhang, X., Ren, S., Sun, J.: Deep residual learning for image recognition. In: Proceedings of the IEEE Conference on Computer Vision and Pattern Recognition (CVPR), pp. 770–778 (2016)
9. Sandler, M., Howard, A.W., Zhu, M., Zhmoginov, A., Chen, L.-C.: MobileNetV2: inverted residuals and linear bottlenecks. In: Proceedings of the IEEE Conference on Computer Vision and Pattern Recognition (CVPR), pp. 4510–4520 (2018)
10. Goyal, P., Dollár, P., Girshick, R., Noordhuis, P., Wesolowski, L., Kyrola, A., Zheng, A.: Accurate, Large Minibatch SGD: Training ImageNet in 1 Hour (2017)
11. 19.7. xml.etree. ElementTree—The ElementTree XML API—Python 2.7.18 documentation. https://docs.python.org/2/library/xml.etree.elementtree.html
12. Bai, Y.: RELU-function and derived function review. SHS Web Conf. **144**, 02006 (2022). https://doi.org/10.1051/shsconf/202214402006
13. Pearce, T.: Understanding Softmax Confidence and Uncertainty. https://arxiv.org/abs/2106.04972

Classification of Depression, Anxiety, and Quality of Life in Diabetic Patients with Machine Learning: Systematic Review

Hind Bourkhime[1,2,3](✉), Noura Qarmiche[4,5], Nassiba Bahra[2,3,5], Mohammed Omari[1,2], Imad Chakri[1,2], Mohamed Berraho[2,3,5], Nabil Tachfouti[2,5], Samira E. L. Fakir[2,5], and Nada Otmani[1,2]

[1] Medical Informatics and Data Science Unit, Laboratory of Epidemiology, Clinical Research and Community Health, Faculty of Medicine and Pharmacy of Fez, Sidi Mohamed Ben Abdellah University, Fez, Morocco
hind.bourkhime@usmba.ac.ma

[2] Diagnostic Center, Hassan II University Hospital, Fez, Morocco

[3] Medical and Pharmaceutical Sciences and Translational Research, Laboratory of Epidemiology and Health Sciences Research, Faculty of Medicine and Pharmacy of Fez, Sidi Mohamed Ben Abdellah University, Fez, Morocco

[4] Laboratory of Artificial Intelligence, Data Science and Emerging Systems, National School of Applied Sciences Fez, Sidi Mohamed Ben Abdellah University, Fez, Morocco

[5] Department of Epidemiology, Clinical Research and Community Health, Faculty of Medicine and Pharmacy of Fez, Sidi Mohamed Ben Abdellah University, Fez, Morocco

Abstract. *Background*: Diabetes is a chronic and costly disease that is emerging in low- and middle-income countries. Current research suggests that diabetes may be associated with depression, anxiety, and/or impaired quality of life. Therefore, assessment of these psychological, mental, physical, and social aspects can significantly improve overall diabetes care services. Recently, studies have been conducted on the classification of these components using new machine learning (ML) techniques. *Objective*: To summarize existing findings on machine learning models classifying diabetics with depression, or anxiety, or quality of life. *Methods*: Systematic review of original research between January 2010 and May 2023. A search covered three databases on (Scopus), (Web of Science), and (PubMed). *Results*: From 126 search results, 4 articles were selected after using the eligibility criteria: 3 machine learning models classifying subjects with versus without depression, and 1 model classifying subjects with versus without pain (Pain as a quality of life dimension). *Conclusion*: Almost all reported machine learning models have shown optimal performance results, although they need to be standardized and comparable. Furthermore, it is strongly recommended that they be improved and implemented in research and clinical practice.

Keywords: Diabetes · Machine learning · Classification · Depression · Anxiety · Quality of life

Y. Farhaoui et al. (Eds.): ICAISE 2023, LNNS 837, pp. 263–270, 2024.
https://doi.org/10.1007/978-3-031-48465-0_35

1 Introduction

Diabetes is a chronic disease characterized by long-term management and the potential for various comorbidities and severe complications. These complications include blindness, kidney failure, myocardial infarction, strokes, and lower limb amputation…, all of which significantly impact the quality of life and psychological well-being of individuals [1–3]. Thus, there is an imperative need to develop precise tools capable of comprehensively assessing the mental, physical, and social health of patients with diabetes, considering its global prevalence as a major public health challenge [4].

While several validated scales and studies exist in this area, the majority of literature predominantly revolves around conventional statistical and assessment methodologies [1–4]. However, recent advancements in technology, particularly within the domain of machine learning (ML), offer promising opportunities for decision support.

ML, a discipline within computer science, involves training computers to learn and make informed decisions without explicit programming. It entails constructing algorithms that continuously refine their models to enhance predictive capabilities [5].

In recent years, there has been a growing interest in leveraging ML techniques to identify high-risk populations for various diseases [5]. Consequently, this paper aims to provide a comprehensive synthesis of existing research on ML models specifically tailored to classify individuals with diabetes based on their susceptibility to depression, anxiety, or deteriorating quality of life.

2 Methods

A systematic review adhering to the recommendations of the Preferred Reporting Items for Systematic Reviews and Meta-Analyses guidelines (PRISMA) statement 2020 (Page et al., 2021) [4] was conducted to identify studies reporting the development or attempted validation of a ML model to classify diabetics with depression, or anxiety, or deteriorate quality of life.

2.1 Information Sources

We performed an exhaustive electronic literature search of English-language studies from PubMed, Scopus, and Web of Science from 01/01/2010 to 18/05/2023.

Selected studies for review were manually searched to determine additional potentially pertinent studies in the reference lists.

2.2 Search Strategy

The search strategy was based on the following equation:

- "Machine learning" AND" Diabetes" AND "Classification" AND ("Depression" OR "Anxiety" OR "Quality of life").

2.3 Eligibility Criteria

Articles were included based on passing all the selection criteria:

- Published as a primary research paper in a peer-reviewed journal and Conference papers only;
- Full paper available in English;
- Described either the development and validation, or a proposal, of a ML model to assess depression, or anxiety, or quality of life in patients with diabetes.
- The source population must be diagnosed with balanced, unbalanced or complicated diabetes involving either diabetic nephropathy, diabetic peripheral neuropathy, diabetic retinopathy or diabetes-related foot ulcers…

2.4 Data Extraction

Data extraction was based on the characteristics of each model.

One table regarding the general characteristics of the study, country, number of participants, year of publication, algorithms applied and their characteristics, model prediction parameters, model performance, including accuracy, discrimination, sensitivity, and specificity rates, area under the receiver operating characteristic curve (AUROC), area under the precision/recall curve (AUPRC).

2.5 Data Synthesis and Analysis

We summarize the data in narrative form, following the structure provided by the features summarized in a table.

3 Results

A systematic review adhering to the recommendations of the Preferred Reporting Items for Systematic Reviews and Meta-Analyses guidelines (PRISMA) statement 2020 (Page et al. 2021) [4] was conducted to identify studies reporting the development or attempted validation of a ML model to classify diabetics with depression, or anxiety, or deteriorate quality of life.

3.1 Search Results

The literature search identified 126 studies. Of these, 45 duplicates were removed. Then 77 articles were excluded because they did not meet the criteria. Only 4 were included after full-text analysis.

3 papers discussed ML models using depression data. And another one paper included the quality of life data (especially pain).

The PRISMA flow chart for the systematic review process is shown in Fig. 1.

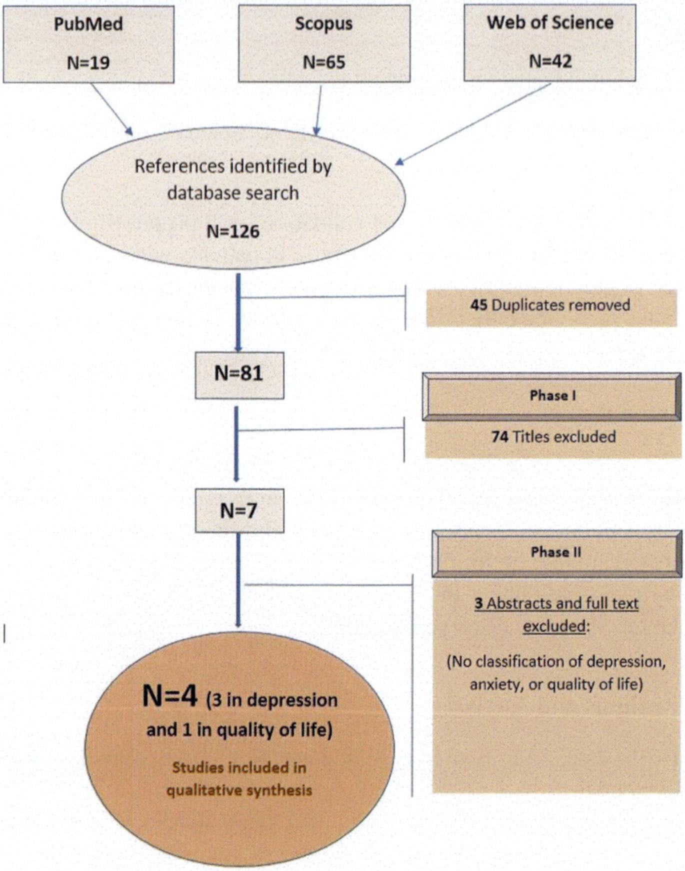

Fig. 1. Flow diagram of process of systematic literature search in accordance with PRISMA guidelines.

3.2 Model Development and Validation

The ML algorithms used to classify depression in diabetic patients were especially supervised such Sarda and al who compared 5 models which are Supporting Vector Machine (SVM) supported by Radial Basis Function (RBF), Decision tree (DT), Random Forest (RF), Adaptive Boosting (AdaBoost), Extreme Gradient Boosting (XGBoost) [6], or M. Khalil and al who compared 4 models which are the K-Mean, F-Cmean, Probabilistic Neural Network (PNN) and SVM [7]. Also a 2-level Poisson Regression was cited by Jin in a paper [8].

Baskozos and al tried to classify quality of life (Pain) in diabetic patients by using 3 ML models which are random forest (RF), Adaptive Regression Splines (ARS), and Naive Bayes (NB) in one study [9]. Table 1 shows the details of the development of these models.

3.3 Study Populations

The 2 studies [6, 7] on depression data used a cross-sectional design to train the models. The third depression's study [8] used a cohort design.

Table 1. Development details of ML models.

First author, Year, Reference	Country	Source of data	Study design	Main Variables
Sarda [6]	India	A diabetes clinic situated in Aurangabad, a city in the state of Maharashtra, India	Cross-sectional (N = 47)	Depression data – PHQ-9 scores – Sociodemographic factors – Level of control over the existing diabetes condition – Activity, mobility, sleep, and communication data
Baskozos [9]	United Kingdom	– DOLORisk Imperial College London, PINS—University of Oxford – DOLORisk Technion – Israel Institute of Technology	Cross-sectional cohorts (N = 1230)	Quality of Life data – EQ5D score – Sociodemographic factors – Lifestyle (smoking, alcohol consumption) – Personality and psychology traits (anxiety, depression, personality traits) – Biochemical (HbA1c) – Clinical variables (BMI, hospital stay and trauma at young age)
Khalil [7]	Australia	Black Lion General Specialized Hospital, Addis Ababa, Ethiopia	Cross-sectional	Depression data – Psychosocial factors data: negative life event, poor social support, fear of diabetic complication, health care cost, doing physical activity – Sociodemographic factors – Clinical factors
Jin [8]	USA	Diabetes-Depression Care-management Adoption Trial (DCAT)	Cohort (N = 4265)	Depression data – PHQ-9 scores – Sociodemographic factors – Depression history

(*continued*)

Table 1. (*continued*)

First author, Year, Reference	ML algorithms	Validation, Calibration, Feature selection, Preprocessing	Performance
Sarda [6]	SVM, DT, RF, AdaBoost, XGBoost	Cross-validation	Best Accuracy (XGBoost) = 81.1% Best sensitivity (XGBoost) = 75% Best specificity (SVM + RBF) = 92.7%
Baskozos [9]	Random Forest, Adaptive Regression Splines; Naive Bayes	10-fold cross-validation, Calibration, Feature selection	Best Balanced accuracy was around 67% Best AUPRC (ARS) was around 77%
Khalil [7]	SVM, K-MEAN, F-CMEAN, PNN		Accuracy for: SVM = 96.87%; FCmean = 95.45%; PNN = 93.75%; Kmean = 87.87%
Jin [8]	2-level Poisson regression (Generaliz-ed multilevel regression)	Randomly selected patients, FS	Areas under the receiver operating characteristic (AUROC) curve about 0.9

The study [9] on quality of life (Pain) data used 2 large cross-sectional cohorts' studies to train the models.

3.4 Features Selection

2 models [6, 8] on depression used Patient Health Questionnaire PHQ-9 score data in addition to the sociodemographic information, the level of diabetes control, or variables derived from the activity, mobility, sleep, and communication data.

Another model [7] on depression used particularly psychosocial factors data such as negative life event, lack of social support, fear of diabetic complication, health care cost, doing physical activity, in addition to sociodemographic and clinical factors.

The left model [9] on quality of life (Pain) used especially the EuroQol-5 Dimension (EQ5D) questionnaire in addition to demographics (age, gender), lifestyle (smoking, alcohol consumption), personality and psychology traits (anxiety, depression, personality traits), biochemical (HbA1c) and clinical variables (BMI, hospital stay and trauma at young age).

3.5 Models Discrimination's Metrics

The accuracy was reported in 3 studies models where values ranged from 0.64 [9] to 0.96 [7]. One reported an area under the receiver operating characteristic curve (AUROC) about 0.9 [8]. And one reported the area under the precision/recall curve (AUPRC) with values around 0.77 [9].

Sensitivity was reported just in one paper where the values can reach 0.75 [6]. Also, specificity was reported just in one study where the values can attain 0.92 [6].

4 Discussion

This systematic review presents a comprehensive examination of machine learning (ML) models used for classifying depression, anxiety, and quality of life in diabetic patients. To the best of our knowledge, this is the first review of its kind, highlighting the limited research conducted in this area.

Diabetes is a well-known condition that profoundly affects the physical, social, and mental well-being of patients [1–3]. However, despite the acknowledged impact of diabetes on multiple dimensions of health, research on ML techniques in this specific context is surprisingly limited. This review brings attention to this crucial knowledge gap and serves as a call to action for further exploration.

By meticulously examining the existing literature, this review identifies four classification models that assess the quality of life and presence of depression and anxiety in diabetic patients. Notably, the study by Raid M. Khalil in 2017 [7] stands out, reporting an impressive discrimination rate of 0.96 using a support vector machine (SVM) model. These findings underscore the potential of ML in enhancing the understanding and management of the complex interplay between diabetes, mental health, and overall well-being.

Given the heterogeneity of the identified models and the nascent stage of research in this area, a qualitative analysis was deemed appropriate for this review. This approach allows for a nuanced interpretation of the results and encourages further exploration of this underdeveloped field.

The first and most valuable element of this work is its focus on three crucial aspects: depression, anxiety and quality of life in diabetic patients. Whereas previous studies have focused mainly on depression, and given only limited attention to the dimensions of anxiety and quality of life, this review broadens the scope and emphasizes the need for a holistic approach. By encompassing these interconnected factors, a more holistic understanding of the challenges faced by diabetic patients can be achieved.

Furthermore, this review highlights the urgent need for standardization and validation of ML algorithms specifically designed to assess depression, anxiety, and quality of life in diabetic patients. The integration of these validated models into clinical practice holds the potential to revolutionize the care provided to this vulnerable population. The development of robust and accurate models can empower healthcare professionals to offer targeted interventions, resulting in improved outcomes and enhanced overall well-being.

One striking finding of this review is the absence of African-developed models in the existing literature. This observation underscores the significance of this research in the Moroccan context, where a high prevalence of diabetes and limited availability of endocrinologists and general practitioners pose significant challenges to delivering quality care [10, 11]. By emphasizing the importance of building African-specific models, particularly a Moroccan one, this review paves the way for locally relevant and culturally sensitive solutions tailored to the unique needs of these populations.

5 Conclusion

This systematic review not only fills a critical gap in the current literature but also highlights the originality and novelty of this work. By elucidating the role of ML in classifying depression, anxiety, and quality of life in diabetic patients, this review establishes a foundation for future research and innovation. It emphasizes the pressing need for standardized and validated ML algorithms, underscores the importance of integrating these models into clinical practice, and advocates for the development of African-specific models to address the specific challenges faced in the region. Ultimately, this review has the potential to drive transformative improvements in the care and well-being of diabetic patients worldwide.

References

1. Holt, R.I.G., de Groot, M., Golden, S.H.: Diabetes and depression. Curr. Diab. Rep. **14**(6), 491 (2014)
2. Smith, K.J., Béland, M., Clyde, M., Gariépy, G., Pagé, V., Badawi, G., Rabasa-Lhoret, R., Schmitz, N.: Association of diabetes with anxiety: a systematic review and meta-analysis. J. Psychosomatic Res. **74**(2), 89–99 (2013)
3. Rubin, R.R., Peyrot, M.: Quality of life and diabetes. Diab. Metab. Res. Rev. **15**(3), 205–218 (1999)
4. World Health Organization. Diabetes [Internet]. [cité 15 oct 2022]. Disponible sur: https://www.who.int/news-room/fact-sheets/detail/diabetes
5. DeepAI. Machine Learning [Internet]. DeepAI. 2019 [cité 4 sept 2021]. Disponible sur: https://deepai.org/machine-learning-glossary-and-terms/machine-learning
6. Sarda, A., Munuswamy, S., Sarda, S., Subramanian, V.: Using passive smartphone sensing for improved risk stratification of patients with depression and diabetes: cross-sectional observational study. JMIR Mhealth Uhealth **7**(1), e11041 (2019)
7. Khalil, R.M., Al-Jumaily, A.: Machine learning based prediction of depression among type 2 diabetic patients. In: 2017 12th International Conference on Intelligent Systems and Knowledge Engineering (ISKE) [Internet]. IEEE, Nanjing, [cité 5 Oct 2022]. pp. 1–5 (2017). Disponible sur: http://ieeexplore.ieee.org/document/8258766/
8. Jin, H., Wu, S., Vidyanti, I., Di Capua, P., Wu, B.: Predicting Depression among patients with diabetes using longitudinal data: a multilevel regression model. Methods Inf. Med. **54**(06), 553–559 (2015)
9. Baskozos, G., Themistocleous, A.C., Hebert, H.L., Pascal, M.M.V., John, J., Callaghan, B.C., Laycock, H., Granovsky, Y., Crombez, G., Yarnitsky, D., Rice, A.S.C., Smith, B.H., Bennett, D.L.H.: Classification of painful or painless diabetic peripheral neuropathy and identification of the most powerful predictors using machine learning models in large cross-sectional cohorts. BMC Med. Inform. Decis. Mak. **22**, 144 (2022)
10. Chetoui, A., Kaoutar, K., Kardoudi, A., Boutahar, K., Chigr, F.: Epidemiology of diabetes in Morocco: review of data, analysis and perspectives. Int. J. Sci. Eng. Res. **9**, 1310–1316 (2018)
11. World Health Organization. Stratégie de coopération les pays de l'OMS (2017–2021): Maroc [Internet]. [cité 13 oct 2022]. Disponible sur: https://apps.who.int/iris/bitstream/handle/10665/272538/ccsbrief-mar-fre.pdf?sequence=1&isAllowed=y

Network Intrusion System Detection Using Machine and Deep Learning Models: A Comparative Study

Asmaa Benchama[1](✉), Rajae Bensoltane[1], and Khalid Zebbara[2]

[1] Faculty of Science Ibn Zohr, University, Agadir, Morocco
{a.benchama,r.bensoltane}@uiz.ac.ma
[2] Faculty of Science AM, Ibn Zohr University, Agadir, Morocco
k.zebbara@uiz.ac.ma

Abstract. As Internet technology advances, the exponential growth of network security risks is becoming increasingly apparent. Safeguarding the network poses a significant challenge in the realm of network security. Numerous security measures have been put in place to detect and pinpoint any malicious activities occurring within the network. Among these mechanisms, the Intrusion Detection System (IDS) stands out as a widely utilized mechanism to reduce the impact of these risks. In this paper, a comparative study of different machine learning classifiers namely decision tree, random forest, SVM, KNN and different deep learning models including CNN, GRU and LSTM is performed to select the best approach for intrusion detection. Our approach consists of three main steps: pre-processing, feature extraction and selection, and classification and validation. The experiments are conducted using two large datasets namely KDDCUP99 and Hogzilla datasets in both binary and multi-class classification fashion. The evaluation results show the superiority of deep learning methods over machine learning classifiers on both datasets.

Keywords: Cybersecurity · KDDCUP99 · Hogzilla dataset · binary-classification · Multiclassification · IDS

1 Introduction

The security of computer networks is facing serious difficulties as a result of the sophistication and frequency of cyberattacks, which have grown in recent years. It has been shown that conventional security measures and rule-based intrusion detection systems (IDS) fall short in recognizing and addressing these changing threats. The use of ML techniques in intrusion detection has drawn a lot of attention in order to solve this problem.

By utilizing the strength of automated pattern recognition and anomaly detection, ML offers a potential strategy to improve the detection capabilities of IDS. ML algorithms can recognize complicated attack patterns and tell them apart from genuine network

Y. Farhaoui et al. (Eds.): ICAISE 2023, LNNS 837, pp. 271–280, 2024.
https://doi.org/10.1007/978-3-031-48465-0_36

traffic by learning from prior data. This makes it possible to create proactive and flexible defense systems that can quickly identify and stop different kinds of invasions.

Therefore, the aim of this paper is to provide a comparative study to evaluate the performance of ML and DL techniques in the intrusion detection task. The experiments are conducted on two different datasets namely KDDCUP99 and Hogzilla in both binary and multi-class classification manners.

The remainder of this paper is structured as follows: Sect. 2 discussed related work to intrusion detection. Methodology and materials are provided in Sect. 3. Section 4 provides a detailed description of the experimental setup. In Sect. 5, we present and analyze the results obtained from our experiments. Finally, in Sect. 6, we draw conclusions based on the findings of our study and offer suggestions for future research directions.

2 Related Works

Recent studies on intrusion detection have adopted two main approaches: machine learning-based approach and deep learning-based approach.

Li and Yi [1] used the K-Nearest Neighbor (KNN) classifier for detecting intrusions in wireless sensor networks. The system takes advantage of the characteristics of wireless sensor networks and employs the KNN algorithm to effectively classify and detect potential intrusions. It achieved an accuracy of 87%, indicating its ability to correctly classify instances as either normal or intrusive. Shapoorifard and Shamsinejad [2] combined the KNN classifier with K-MEANS clustering to enhance intrusion detection. The proposed system achieved an accuracy score of 92%.

Farnaaz and Jabbar [3] utilized Random Forest for anomaly detection. The Random Forest algorithm is employed to construct an ensemble of decision trees, where each tree is trained on a different subset of the data and features. The trees' predictions are combined through voting to determine the final classification decision. The proposed system achieved an accuracy rate of 92%.A recent paper of (An Analysis of Intrusion Detection Classification using Supervised Machine Learning Algorithms on NSL-KDD Dataset) an overview of various supervised machine learning techniques, namely SVM, Naive Bayes, KNN, Random Forest, Logistic Regression, and Decision tree, applied to the NSL-KDD dataset. The experiments were conducted for both multi-class and binary classification tasks. The evaluation results revealed that Random Forest exhibited the highest accuracy score among all classifiers, achieving 98.70% accuracy for binary classification and 98.46% accuracy for multi-class classification.

For deep learning-based models, Staudemeyer [4] emphasize the significance of modeling network traffic data as a time series to enhance IDS performance. They train LSTM models on the KDD Cup dataset using both full and minimal feature sets, achieving a maximum accuracy of 93.82% after 1000 epochs. Yin and Zhu [5] examined the performance of RNN for handling intrusion detection task. The model is trained on the NSL-KDD dataset, binary and multi-class classification are performed. The performance of RNN based IDS is far superior in both classification when compared to other traditional approaches and the author claims that RNN based IDS has a strong modeling capability for IDS. In a comparative study [6], the effectiveness of CNN (Convolutional Neural Network) and CNN-RNN (Recurrent Neural Network) models is explored. Different models such as CNN, CNN-LSTM, CNN-GRU, and CNN-RNN are trained on the KDD Cup dataset, with the CNN-based model outperforming hybrid CNN-RNN models. In Shone and Ngoc [7], a novel NIDS (Network Intrusion Detection System) based on a stacked non symmetric deep autoencoder (NDAE) is introduced. The model is trained on both the KDD Cup 99 and NSLKDD benchmark datasets and compared to a DBN-based model. The experimental analysis reveals that the NDAE approach improves accuracy by up to 5% while reducing training time by 98.8% compared to the DBN approach. Furthermore, a highly scalable deep learning framework for intrusion detection at both the network and host levels is proposed by Vinayakumar and Alazab [8]. Multiple ML and DNN models are trained on datasets including KDD Cup, NSLKDD, WSN-DS, UNSW-NB15, CICIDS 2017, ADFA-LD, and ADFA-WD, and their performances are compared.

3 Methodology and Materials

Our approach has three main steps namely preprocessing, feature extraction and selection process, and the classification process [9]. The input datasets suffer from the lack of data distribution, resulting in a class imbalance problem.

The evaluation of datasets plays a crucial role in analyzing high-dimensional imbalance problems through three main aspects: preprocessing the datasets, evaluating the metrics, and employing K-fold Cross-Validation.

This research uses K-fold cross validation for performance evaluation, for each fold, the training data and target labels are evaluated with the corresponding test data and target labels.

The K-fold cross validation for $k = 1,2,3\ldots\ldots K$, The fitted function denoted by:

$$\hat{y}^{-k}(x)$$

Computed with kth part of the data removed.

$$\mathrm{CV}(\hat{y}) = \frac{1}{N}\sum_{i=1}^{N} L(x_i, \hat{y}^{k(i)}(x_i))$$

[1]

This comparative study is conducted based on various DL and ML IDS models in both binary and multi-class classification fashion.

3.1 IDS Machine Learning Approaches

ML enables a machine to learn automatically from a set of data and to perform improvements based on previous experience. ML is categorized into different types based on learning style and the way they work with new data. Three main learning techniques: supervised learning, unsupervised learning and semi-supervised learning [10]. In the

[1] K-fold cross validation formula at https://stats.stackexchange.com/questions/17431/a-mathematical-formula-for-k-fold-cross-validation-prediction-error

case of supervised learning, the machine is trained on the basis of labeled data in order to make predictions. Supervised learning algorithms are used for classification and regression problems. Examples include support vector machines, the random forest algorithm [11] and the linear regression algorithm. On the other hand, in unsupervised learning, the machine is trained with unlabeled data to make predictions for unknown cases. Examples include K-means clustering, association rules and principal component analysis. Semi-supervised learning, which includes the two techniques mentioned above, is trained using both labeled and unlabeled data, semi-supervised SVM, spectral graph transducer, Gaussian field approach are examples. ML techniques are among the solutions for the efficient development of IDS with better detection and a low false alarm rate [12]. ML Techniques used for our evaluation experimentations are Decision Tree, KNN [12], Random forest and SVM.

3.2 IDS Deep Learning Approaches

DL is a subfield of ML that focuses on training artificial neural networks with multiple layers, hence the term "deep." It involves building and training complex models known as deep neural networks to automatically learn hierarchical representations of data.

DL algorithms leverage these multi-layered neural networks to learn patterns and features directly from raw input data. The networks consist of interconnected nodes or "neurons," organized into layers. Each layer progressively extracts more abstract and high-level features from the input data, allowing the network to understand complex relationships and make predictions or classifications.

The DL techniques used in our experiments are: CNN, LSTM, and GRU. CNNs are specialized for processing grid-like data such as images, using convolutional layers to detect spatial patterns and features. LSTM networks are a type of recurrent neural network designed for sequential data, equipped with memory cells and gating mechanisms to capture long-range dependencies. GRUs are also used for sequential data, employing gating mechanisms for efficient information flow; however, they combine hidden and cell states to simplify the architecture, which can be advantageous in scenarios where computational efficiency is crucial.

3.3 Hybrid Approaches

Mixed approaches can be subdivided into two categories: the combination of statistical and artificial learning, and the combination of mathematical methods and artificial learning. One advantage of hybrid models is that they combine the strengths of various models to provide a robust modeling framework.

The architecture of the CNN-LSTM model typically consists of two main components: the CNN module and the LSTM module. The CNN module is responsible for extracting spatial and local features from network traffic data. It applies a series of convolutional and pooling layers on the input sequence to capture patterns and features at different scales [13]. The convolutional layers use filters to convolve over the input data and learn spatially localized features, while the pooling layers down sample the feature maps to retain the most important information. The LSTM module is designed to capture temporal dependencies and sequence patterns in the extracted features. It takes the output of the CNN module as input and processes it through LSTM layers. This module captures the temporal relationships and dependencies within the data, allowing the model to understand and classify sequences of features.

Another hybrid model is CNN-GRU. Similar to CNN-LSTM, CNNs process spatial features from input data. However, instead of using LSTMs, GRU is used in the subsequent layers. GRUs, like LSTMs, are designed to capture sequential dependencies, but they have a simpler architecture by combining the hidden state and memory cell into a single state.

In this study, we explore the aforementioned hybrid models (i.e., CNN-LSTM and CNN-GRU) in both binary and multi-class classification settings.

4 Experiments

4.1 Datasets

In this study, two main intrusion detection datasets are used to evaluate the different ML and DL methods namely KDDCUP99 and HOGZILLA datasets.

KDDCUP99. The kddcup99[2] dataset was created in 1999 for a ML competition [14]. The aim of the competition was to correctly classify network connections into 5 categories: normal, denial of service (DoS), network probe, remote to local (R2L), user to root (U2R). Each connection has 41 characteristics that enable the classifier to correctly predict its class. These features are statistics calculated from listening to a simulated local network of the U.S. Air Force in 1998: connection duration, protocol type, percentage of connections to the same service, etc.

Hogzilla dataset. The Hogzilla[3] dataset is a combination of network records from the CTU-13 botnet and the ISCX 2012 IDS datasets.

Each stream contains 192 behavioral features, which were tagged by the Snort IDS, the nDPI library and the original dataset.

The result was the assembly of data from datasets containing behavioral information on botnets, found in the CTU-13 dataset, and on normal traffic found in the ISCX 2012 IDS dataset [15, 16].

4.2 Preprocessing Data

The preprocessing steps involved in this study include data cleaning, handling missing values, feature selection, and dimension reduction. It is important to note that the dataset contains a significant number of redundant records, which may have an impact on the performance of our models.

For the kddcup 99 dataset, multi-classification concerns the attribution of column names for dataset who doesn't have it, in addition attack labels were changed to their respective attack classes 'DOS','R2L','Probe','U2R' as shown below.

['apache2', 'back', 'land', 'neptune', 'mailbomb', 'pod', 'processtable', 'smurf', 'teardrop', 'udpstorm', 'worm'] ➔ DOS.

[2] http://kdd.ics.uci.edu/databases/kddcup99/kddcup99.html

[3] https://ids-hogzilla.org/dataset/

['ftp_write', 'guess_passwd', 'httptunnel', 'imap', 'multihop', 'named', 'phf', 'sendmail', 'snmpgetattack', 'snmpguess', 'spy', 'warezclient', 'warezmaster', 'xlock', 'xsnoop'] ➔ R2L.

['ipsweep', 'mscan', 'nmap', 'portsweep', 'saint', 'satan'] ➔ Probe.

['buffer_overflow', 'loadmodule', 'perl', 'ps', 'rootkit', 'sqlattack', 'xterm'] ➔ U2R (Table 1).

Table 1. Distribution of attack classes for the kddcup 99 dataset.

Class attack	Number of instances
Dos	321458
Probe	97277
R2L	4107
U2R	1126
Normal	52

Concerning the binary classification of kddcup99 dataset, we proceed to the creation of a dataframe with binary labels with an encoding (0,1) for the two binary labels (abnormal, normal).

For the hogzilla dataset, the pre-processing was carried out on the original data; this pre-processing mainly concerned the change of attack labels to their respective attack class: 'acceptable', 'unrated' and 'unsafe' (Table 2).

Table 2. Distribution of attack classes for Hogzilla dataset.

Type of flow	Number of flows
Acceptable	2523
Safe	106
Fun	10
Unrated	5647
Unsafe	4546
Total of flows	12832

The attribution of attack labels to their respective attack class is made as follows:

['Acceptable', 'Safe'] ➔ Acceptable;

['Unrated', 'Fun'] ➔ Unrated;

['Unsafe'] ➔ Unsafe.

The main steps are: the selection of numeric attributes columns from data using one hot encoding, Standard scaler for normalizing on the selected numeric attributes and finally producing a label encoding (0,1,2) for the multi-class labels ('acceptable', 'unrated', 'unsafe').

The attack labels are classified into two categories 'normal' and 'abnormal' for the binary classification, after encoding labels of the dataset with binary labels and label encoded column.

4.3 Evaluation Metrics

In order to evaluate the performance of each technique, Precision, Recall and F1-score are used. Moreover, because of the imbalanced distribution of the evaluated datasets, the macro score of these metrics is computed.

5 Results

5.1 Binary Classification

The results of the binary classification on the Hogzilla dataset are illustrated in Table 3. For the ML classifiers, SVM obtained the lowest results among other classifiers by achieving a macro F1-score of 73.17%, whereas the best results were achieved by KNN (F1-score = 96.06%).

For DL models, hybrid CNN-LSTM method demonstrated the highest performance by obtaining a macro F1-score of 98.81%, demonstrating the ability of hybrid methods to leverage the strengths of multiple models.

On the other hand, the results on the KDD dataset are shown in Table 4. For ML classifiers, Random Forest has the least performance by achieving a macro F1-score of 83.60% while KNN also obtained the best F1-score of 97.97%.

The comparison between DL methods reveals that LSTM and GRU methods had the better performance by obtaining closely matched values respectively 99.13% and 99.12%.

For both datasets, DL methods had the better performance compared to ML methods. Additionally, KNN performed well on both datasets. However, this classifier required very high computation costs compared to other ML algorithms.

Table 3. The average of evaluation metrics on Hogzilla dataset

		Machine learning			Deep learning				
Metrics	SVM (%)	RF (%)	KNN (%)	DT (%)	LSTM (%)	GRU (%)	CNN (%)	CNN−GRU (%)	CNN−LSTM (%)
Precision	71.46	93.11	97.90	93.28	98.40	98.46	98.51	98.64	98.73%
Recall	76.52	96.25	94.46	95.72	97.20	97.43	97.09	97.82	98.90
F1−score	73.17	94.56	96.06	94.43	97.79	97.93	97.78	98.23	98.81

Table 4. The average of evaluation metrics on KDD99

	Machine learning				Deep learning				
Metrics	SVM (%)	RF (%)	KNN (%)	DT (%)	LSTM (%)	GRU (%)	CNN (%)	CNN−GRU (%)	CNN − LSTM (%)
Precision	96.41	93.53	98.64	95.69	98.78	98.76	98.39	98.06	98.37
Recall	95.22	78.69	97.31	98.19	99.50	99.49	98.98	99.28	99.27
F1−score	9.80%	83.60	97.97	96.87	99.13	99.12	98.68	98.66	98.81

5.2 Multi-class Classification

The results of our simulation for the multi-class classification on Hogzilla dataset are shown in Table 5. KNN method gives better results among other ML classifiers and performed competitively with CNN by obtaining a macro F1-score of 96.85%. For DL methods, the best performance was obtained by CNN-LSTM model with a macro F1-score of 98.67%.

The results of kddcup 99 are presented in Table 6. DT classifier achieved better results among other ML methods with a macro F1-score of 78.07%. For DL, The CNN-LSTM model demonstrated superior performance; achieving a macro F1-score of 89.59%.This proves the effectiveness of employing hybrid architectures for handling the intrusion detection task.

Table 5. The average of evaluation metrics on Hogzilla dataset

	Machine learning				Deep learning				
Metrics	SVM (%)	RF (%)	KNN (%)	DT (%)	LSTM (%)	GRU (%)	CNN (%)	CNN−GRU (%)	CNN−LSTM (%)
Precision	96.98	93.99	97.37	96.42	97.89	97.43	97.17	98.16	98.59
Recall	95.63	95.08	96.35	93.55	97.05	96.64	96.69	97.97	98.76
F1−score	96.30	94.45	96.85	94.73	97.45	97.03	96.93	98.06	98.67

Table 6. The average of evaluation metrics on KDD99

	Machine learning				Deep learning				
Metrics	SVM (%)	RF (%)	KNN (%)	DT (%)	LSTM (%)	GRU (%)	CNN (%)	CNN−GRU (%)	CNN−LSTM (%)
Precision	74.28	77.94	76.37	87.20	91.80	91.23	90.95	89.20	92.79
Recall	69.65	60.43	75.83	75.62	85.98	86.21	86.14	87.75	87.5
F1−score	71.73	61.62	73.86	78.07	88.12	87.93	88.04	88.37	89.59

6 Conclusion

In this paper, a comparative study in intrusion network detection was conducted using four different machine learning models (Decision Tree, Random Forest, K Nearest Neighbor, SVM), DL techniques (LSTM, GRU,CNN), and hybrid models (CNN-LSTM and CNN-GRU) on two datasets: Hogzilla and kddcup99.

The experimental results demonstrate that deep learning methods are more efficient in handling intrusion detection task on both kddcup 99 and hogzilla datasets compared to ML classifiers. Feature selection and the quality of the datasets are major factors in improving the efficiency of attack prediction models. For the kddcup99, the low level of rates in ML methods on multi-class classification is due to the huge number of replicated records which eventually may lead learning algorithms to be partial on multiple classes. This proves that the quality of the dataset eventually affects the efficiency of a Network Intrusion Detection System.

For future research directions, we aim to optimize the feature selection process and explore hybrid combinations of ML and deep learning models to enhance the efficiency of detection rates across multiple datasets.

References

1. Li, W. et al.: A new intrusion detection system based on KNN classification algorithm in wireless sensor network. J. Electri. Comput. Eng. (2014)
2. Shapoorifard, H., Shamsinejad, P.: Intrusion detection using a novel hybrid method incorporating an improved KNN. Int. J. Comput. Appl. **173**(1), 5–9 (2017)
3. Farnaaz, N., Jabbar, M.: Random forest modeling for network intrusion detection system. Proc. Comput. Sci. **89**, 213–217 (2016)
4. Staudemeyer, R.C.: Applying long short-term memory recurrent neural networks to intrusion detection. South Afri. Comput. J. **56**(1), 136–154 (2015)
5. Yin, C., et al.: A deep learning approach for intrusion detection using recurrent neural networks. IEEE Access **5**, 21954–21961 (2017)
6. Vinayakumar, R., et al.: Deep learning approach for intelligent intrusion detection system. IEEE Access **7**, 41525–41550 (2019)
7. Shone, N., et al.: A deep learning approach to network intrusion detection. IEEE Trans. Emerg. Topics in Computat. Intell. **2**(1), 41–50 (2018)

8. Vinayakumar, R., Soman, K., Poornachandran, P.: Applying convolutional neural network for network intrusion detection. In: 2017 International Conference on Advances in Computing, Communications and Informatics (ICACCI), IEEE (2017)
9. Herve Nkiama, S.Z.M.S., Saidu, M: A subset feature elimination mechanism for intrusion detection system. (IJACSA) Int. J. Adv. Comput. Sci. Appl. **7**(4) (2016)
10. Lantz, B.: Machine learning with R. 2nd ed. Birmingham, Packt Publishing (2015)
11. Zulkernine, Z.A.: Network intrusion detection using random forests. Citeseer (2005)
12. Kilian, Q., Weinberger, L.K.S.: Distance metric learning for large margin nearest neighbor classification. J. Mach. Learn. Res. **10,** 207–244 (2009)
13. Yi, T. et al.: Review on the application of deep learning in network attack detection. J. Netw. Comput. Appl. **212**, 103580 (2023)
14. Mining, K.D.a.D.: In: The Fifth International Conference on Knowledge Discovery and Data Mining. http://kdd.ics.uci.edu/databases/kddcup99/kddcup99.html
15. Resende, P.A.A., Drummond, A.C.: An active labeling approach for behavioral-based intrusion detection systems. Comput. Secur. (2020)
16. Resende, P.A.A., Drummond, A.C.: HTTP and contact-based features for Botnet detection. Secur. Privacy (2018)

AI-Assisted English Language Learning and Teaching in a Developing Country: An Investigation of ESl Student's Beliefs and Challenges

Gemma Amante-Nochefranca[1], Olga Orbase-Sandal[1], Ericson Olario Alieto[2](✉), Izar Usman Laput[2], Salman Ebod Albani[3], Rochelle Irene Lucas[4], and Manuel Tanpoco[4]

[1] Western Mindanao State University, Zamboanga City, Philippines
[2] RUPID Center, Western Mindanao State University, Zamboanga City, Philippines
ericson.alieto@wmsu.edu.ph
[3] Mindanao State University—Sulu, Sulu, Philippines
[4] Dela Salle University, Taft Avenue, Manila, Philippines

Abstract. This empirical study addresses the beliefs and challenges of English as a Second Language (ESL) students in a developing country in relation to AI-assisted English language learning. The research utilizes a descriptive-quantitative design to answer four key research questions. Firstly, it explores the challenges faced by ESL students when learning English with AI assistance in a developing country. Secondly, it examines the beliefs of ESL students in this context regarding AI-assisted English language learning. Additionally, the study investigates potential gender differences in beliefs about using AI for English language learning. Finally, it explores variations in challenges faced when using AI-assisted English language learning across different gender. An adopted research questionnaire was employed to gather data from senior high school students. The analysis revealed some significant gender differences in the beliefs and challenges faced by male and female respondents when learning English with AI assistance.

Keywords: Artificial intelligence · English language learning · Gender

1 Introduction

AI is a transformative technology with the potential to revolutionize multiple aspects of our lives. AI, a term introduced by John McCarthy in 1955, refers to the field of creating intelligent machines [1]. Intelligent machines possess the ability to mimic human intelligence and execute tasks that traditionally necessitate human cognition, such as speech recognition, language processing, and visual interpretation. AI has become a valuable tool in our modern world, finding numerous valuable applications in our daily lives [1]. AI is widely used in language learning. AI is transforming ESL teaching through AI-Assisted Language Learning (AI-ALL). AI language tools like Elsa and others use

Y. Farhaoui et al. (Eds.): ICAISE 2023, LNNS 837, pp. 281–289, 2024.
https://doi.org/10.1007/978-3-031-48465-0_37

advanced NLP systems to enhance learning outcomes through personalized and engaging methods [2]. AI-powered chatbots have been shown to enhance learning and improve understanding of course material when compared to traditional methods [3–5].

Previous research has predominantly examined the integration of AI in language learning within developed nations like Malaysia, China, and Korea. Limited research has investigated the integration of AI in language learning in developing countries such as the Philippines. Countries with limited internet connectivity and restricted access to devices may encounter difficulties in adopting AI-assisted language learning [5]. Learners in developing countries face socioeconomic challenges that hinder effective language learning. This study aims to explore the beliefs and challenges faced by ESL students in the Philippines when using AI-assisted English language learning, filling a gap in the current literature. This research aims to contribute to the existing literature by exploring AI-assisted English language learning in a developing country context, an area that has not been extensively studied before.

To achieve these objectives, this study aims to answer the following research questions:

1. What beliefs do ESL students in a developing country have about AI-assisted English language learning?
2. What challenges are faced by ESL students in a developing country when learning English with AI assistance?
3. Do male respondents significantly differ in their beliefs and challenges regarding the use of AI-assisted English language learning compared to their female counterparts?

This study contributes to understanding the perceptions and experiences of ESL students in the Philippines using AI-assisted English language learning, thereby filling a gap in the existing literature. The findings will provide valuable insights that can inform future practice and policy decisions related to AI integration in language learning, especially in developing-country contexts. Overall, the integration of AI in language learning holds immense potential for enhancing the English language proficiency of students. However, it is essential to consider the beliefs, challenges, and contextual factors specific to developing countries to ensure equitable and effective implementation of AI-assisted language learning initiatives [6–8].

2 Methodology

The study used a descriptive research design to explore and analyze people's behavior and experiences [9, 10]. This study explored ESL students' beliefs and challenges in using AI-assisted English language learning in a developing country using a survey method. A link that directed the respondent to Google form was given after the school's focal person approved the permission for data collection. The respondents were given 30 min to complete the survey. This link was available for two weeks and was deactivated after. The study was comprised of 200 senior high school students in Zamboanga Peninsula, enrolled in SY 2022–2023, primarily STEM, ABM, GAS, and ICT. The instrument used was an adapted tool used by [2]. The tool was validated using Cronbach Alpha of 0.791 for students' perception and 0.754 for students' problem.

The responses were coded to facilitate the analysis of the data gathered through the research instruments. For the demographic profile, gender (1 = male, 2 = female). For responses in the in beliefs and challenges in AI-powered learning tools the responses were coded as, 1 for strongly disagree, 2 for disagree, 3 for neutral, 4 for agree and 5 for strongly agree. Further, the study focuses on beliefs and challenges of AI-powered tools used in learning English, descriptive statistics, particularly mean and standard deviation, were used.

3 Results and Discussion

AI-assisted English language learning holds significant promise for ESL students in developing countries. Analyzing the beliefs of these students provides valuable insights into their perceptions, highlighting both the positive aspects and areas that require improvement (Table 1).

On the positive side, ESL students in developing countries have a positive perception of AI-based language learning tools, as evidenced by their high level of trust in the content provided (Mean = 3.865). They also appreciate the time-saving advantages of using such tools (Mean = 3.924, Rank = 2), recognizing the benefits of personalized learning paths and instant feedback. These findings suggest that AI-powered language learning has the potential to significantly enhance the language skills of ESL students in developing nations.

However, there are areas that require attention. One such area is the perceived usability of AI-based applications (Mean = 3.800, Rank = 9). ESL students generally find these applications user-friendly, but there is room for improvement in terms of enhancing their ease of use and user experience. Affordability is another concern, with the belief that few AI applications for English language learning are cost-effective (Mean = 3.697, Rank: 10). This suggests that students perceive a lack of affordable options, limiting their access to AI-assisted language learning tools.

On the other hand, ESL students appreciate the flexibility offered by AI-powered applications (Mean = 3.918, rank: 3). They recognize the opportunity to learn English anywhere, at any time, which aligns with the needs of students with busy schedules or limited access to educational resources. Furthermore, they acknowledge the potential benefits of AI applications available on smartphones for learning English (Mean = 3.827, Rank = 7). This belief highlights the student's awareness of the availability and potential advantages of mobile AI applications, although there may be scope to expand their understanding of the range of beneficial options.

In general, ESL students in a developing country hold positive beliefs about AI-assisted English language learning, particularly regarding the reliability of the tasks and the time-saving benefits. However, there are areas that require attention, such as improving the usability of AI-based applications and addressing the affordability concerns. The students highly value the flexibility and the potential benefits of AI applications available on smartphones. By understanding these high points and low points, educators, policymakers, and developers can work towards enhancing the positive aspects and addressing the challenges faced by ESL students, ultimately improving the effectiveness and accessibility of AI-assisted English language learning in developing countries (Table 2).

Table 1. Student's beliefs on AI-assisted English language learning

No.	Beliefs	Mean	SD	Rank
1	AI-powered applications allow me to access reliable English language learning tasks	3.865	0.826	6
2	AI-based applications are user-friendly and easy to use	3.800	0.877	9
3	The benefit of utilizing AI-powered applications for English language learning is that it saves time	3.924	0.783	2
4	Various English language skills can be learned using AI applications	3.811	0.939	8
5	Few AI applications are cost-effective for English language learning	3.697	0.798	10
6	Applications driven by artificial intelligence provide students that opportunity to learn English anywhere, at any time	3.918	0.914	3
7	There are a few AI applications available on smartphones that are beneficial for learning English	3.827	0.809	7
8	AI application via good internet connectivity provides an instant response to us at anywhere and anvtime	4.021	0.853	1
9	Apps with AI technologies will improve English language learning more engaging	3.914	0.816	4
10	Using AI-based apps facilitates interactive English language learning activities	3.887	0.775	5

When examining the challenges faced by ESL students in a developing country when learning English with AI assistance, it is important to consider their perceptions and level of concern. Based on the mean ratings provided, it appears that only two items stand out as challenges, while the others are viewed more neutrally or with less concern by the students.

The most significant challenge identified by ESL students is poor internet connectivity with a mean rating of 3.935. This challenge holds substantial concern for students as limited access to stable internet connections significantly hampers their inter-action with

Table 2. Student's challenges on AI-assited English language learning

No.	Challenges	Mean	SD	Rank
1	Poor internet connectivity leads to less interaction with AI applications	3.935	1.111	1
2	AI applications are being used by individuals for non-academic purposes	3.151	0.994	8
3	English language learning is not supported by the characteristics of AI apps	2.795	0.995	9 1
4	Smartphones do not have good AI apps for language learning	2.649	1.053	**10**
5	Battery becomes a problem during interaction with AI applications	3.297	1.044	**6**
6	Data cost is very high when using any Al-powered applications	3.410	1.028	4
7	Less familiarity among students with utilizing AI applications for English language instructions	3.443	0.871	3
8	To receive an immediate answer, slow internet speed becomes a bigger issue	3.843	1.012	2
9	Al-powered apps are quite expensive	3.31S	1.027	5
10	There is currently a shortage of English language based on AI applications	3.216	0.889	7

AI applications. It is crucial to address this challenge by improving internet infrastructure to ensure equitable access to AI-assisted language learning tools.

Another notable challenge is the need for immediate answers but facing slow internet speed with a mean rating of 3.843. ESL students perceive this as a challenge, as slow internet speeds can hinder their ability to receive timely responses from AI applications. Addressing this challenge involves finding solutions to improve internet speed or developing AI applications that can function effectively even with slower connections. Regarding the remaining items, ESL students express more neutral attitudes or lesser concern. They may perceive them as less significant in terms of posing challenges for their English language learning with AI assistance. This neutral perception suggests that students may not consider these items as major obstacles or may not have encountered significant issues in these areas (Table 3).

The perceptions of male and female respondents were compared to see if there were any statistically significant variations at the .05 level. The results revealed that there were no significant differences in beliefs between genders for any of the items in the scale. Although there were some slight variations in mean ratings, these differences were likely due to chance rather than meaningful distinctions. Despite the lack of significant differences, both males and females generally held positive views towards AI-assisted

Table 3. Beliefs of ESl students on AI-assisted English language learning significantly vary across gender

No.	Beliefs	Gender	Mean	SD	Sig.
1	AI-powered applications allow me to access reliable English language learning tasks	Male	3.717	0.885	0.091
		Female	3.936	0.791	
2.	AI-based applications are user-friendly and easv to use	Male	3.667	0.951	0.153
		Female	3.864	0.836	
3.	The benefit of utilizing Al-powered applications for English language learning is that it saves time	Male	3.850	0.777	0.373
		Female	3.960	0.787	
4.	Various English language skills can be learned using AI applications	Male	3.700	0.869	0.267
		Female	3.864	0.970	
5.	Few AI applications are cost-effective for English language learning	Male	3.550	0.8522	0.082
		Female	3.768	0.763	
6	Applications driven by artificial intelligence provide students that opportunity to learn English anywhere, at any time	Male	3.717	0.922	
		Female	4.016	0.898	0.037
7	There are a few AI applications available on smartphones that are beneficial for learning English	Male	3.767	0.745	0.4S4
		Female	3.856	0.840	
8	AI application via good internet connectivity provides an instant response to us at anywhere and anvtime	Male	3.833	0.924	0.037
		Female	4.112	0.805	
9	Apps with AI technologies will improve English language learning more engaging	Male	3.767	0.767	0.090
		Female	3.984	0.833	
10	Using AI-based apps facilitates interactive English language learning activities	Male	3.733	0.686	0.063
		Female	3.960	0.807	

English language learning. They believe that AI-powered applications can provide reliable language learning tasks, are user-friendly, and can save time. It's worth noting that women tended to have slightly higher mean ratings compared to men, indicating a slightly stronger belief in the effectiveness and benefits of AI-assisted language learning. However, these differences were not substantial enough to draw definitive conclusions (Table 4).

While no significant variations were found, it's important to consider the overall positive attitudes towards AI-assisted language learning among both males and females. Both genders recognized the value of AI-powered applications in facilitating language learning tasks, improving accessibility, and enhancing engagement. The absence of significant differences suggests a shared belief in the potential of AI tech-nologies for language learning.

Table 4. Challenges of ESL Students on AJ-assisted english language learning significantly vary across gender

No.	Beliefs	Gender	Mean	SD	Sig.
1	Poor internet connectivity leads to less interaction with AI applications	Male	3.917	1.062	0.876
		Female	3.944	1.138	
2	AI applications are being used by individuals for non-academic purposes	Male	3.233	1.095	0.438
		Female	3.112	0.944	
3	English language learning is not supported by the characteristics of AI apps	Male	3.017	1.033	0.035
		Female	2.688	0.962	
4	Smartphones do not have good AI apps for language learning	Male	3.017	0.100	0.001
		Female	2.472	1.036	
5	Battery becomes a problem during interaction with AI applications	Male	3.433	1.031	0.221
		Female	3.232	1.04 S	
6	Data cost is very high when using any Al-powered applications	Male	3.633	0.974	0.041
		Female	3.304	1.041	
7	Less familiarity among students with utilizing AI applications for English language instructions	Male	3.433	0.91 1	0.666
		Female	3.424	0.854	
8	To receive an immediate answer, slow internet speed becomes a bigger issue	Male	3.817	0.911	0.805
		Female	3.856	1.060	
9	Al-powered apps are quite expensive	Male	3.500	1.033	0.097
		Female	3.232	1.017	
10	There is currentlv a shortage of English language based on AI applications	Male	3.350	0.917	0.157
		Female	3.152	0.871	

To investigate variations in beliefs based on gender, the perceptions of male and female respondents regarding the challenges faced when learning English with AI assistance were examined. The data provided in the table includes mean ratings and significance levels for each challenge, enabling the determination of statistically significant differences in beliefs between males and females.

The analysis revealed significant variations in beliefs between male and female respondents for three challenges at a 0.05 level of significance.

Firstly, gender variations were observed in the perception of characteristics that hinder AI-based English language learning ($p = 0.035$). While both genders expressed neutrality, females tended to view AI apps as more supportive of language learning compared to males. Further research is needed to determine the factors underlying these differences. Developing AI apps that cater to the distinct needs and learning styles of both male and female ESL students can enhance the effectiveness of language learning.

Secondly, in relation to "The lack of good AI apps for language learning on smartphones," significant gender differences were found ($p = 0.001$).

Thirdly, significant gender differences were also observed in t males expressed a neutral stance, while females tended to believe that there are good AI-based language learning apps on smartphones. These variations may be due to differences in smartphone usage patterns between male and female respondents. Developing high-quality and user-friendly AI apps that address the preferences and constraints faced by both genders can ensure equal access and enhance the overall language learning experience. The challenge of "High data costs when using AI-powered applications" ($p = 0.041$). Males agreed that data costs are high when using AI-powered applications, while females expressed neutrality on this issue. This finding aligns with the previous observation that females perceive some AI apps for English language learning to be good, while males maintain a neutral stance. Males may consider data costs high due to not yet fully recognizing the value of AI apps in English language learning. Addressing this challenge involves exploring cost-effective options and optimizing data usage to make AI-assisted language learning more affordable for both genders.

On the other hand, no significant gender differences were observed for the challenges of poor internet connectivity, battery problems, and familiarity with using AI applications, the cost of AI-powered apps, and the lack of English language-based AI applications. This implies that both male and female respondents have similar perceptions and experiences regarding these challenges. The analysis indicates that poor internet connectivity is the most critical issue, leading to reduced interaction in AI applications and slower response times. Both genders expressed similar views on this matter.

4 Conclusion

The use of AI-assisted English language learning has shown a positive impact on ESL students in developing countries. These students place trust in AI-powered language learning apps due to the time-saving benefits and personalized learning features they offer. However, certain challenges such as inadequate internet connectivity and limited affordability hinder widespread access to these tools. To overcome these obstacles, it is imperative to improve internet infrastructure and optimize costs associated with these apps. While AI-based language learning apps are generally user-friendly, there is room for enhancements to further improve their effectiveness. Gender differences were found to be insignificant in most aspects, except for beliefs related to AI app characteristics and data costs. Nonetheless, both male and female ESL students appreciate the flexibility of learning English at their own pace and convenience through AI-powered smartphone apps. However, they also express the need for additional information and support. Overall, AI-assisted English language learning shows promise in enhancing accessibility and engagement for ESL students, regardless of gender, in developing countries.

References

1. McCarthy, J.: Reminiscences on the history of time sharing. IEEE Annals of the History of Comput. **14**, 19–24 (1995)
2. Moulieswaran, N., Prasantha, K.: Google assisted language learning (GAALL): ESL learner's perception and problem towards AI-powered google assistant-assisted english language learning. **11**, 122–130 (2023). https://doi.org/10.11114/smc.v11i4.5977

3. Ali, Z.: Artificial Intelligence (AI): a review of its uses language teachings and learning. Mater. Sci. Eng. **179**, 1–6 (2020). https://doi.org/10.1088/1757-899X/769/1/012043
4. Keerthiwansha, N.W.B.S.: Artificial intelligence education (AIEd) in english as a second language (ESL) Ccassroom in Sri Lanka. Int. J. Concept. Comput. Inform. Technol. **6**, 31–36 (2018)
5. Neo, M.: The MERLIN project: malaysian student's acceptance of an AI Chatbot in their learning process. Turkish Online J. Distance Educ. **23**, 31–48 (2022)
6. Burch, Z.A., Mohammed, S.: Exploring faculty perceptions about classroom technology integration and acceptance: a literature review. Int. J. Res. Educ. Sci. (IJRES) **5**, 722–729 (2019)
7. Wang, C., Lan, Y., Tseng, W., Lin, Y., Gupta, K.: On the effects of 3D virtual worlds in language learning—a meta-analysis. Comput. Assist. Lang. Learn. **33**, 891–915 (2019). https://doi.org/10.1080/09588221.2019.1598444
8. Daniel, J.: Education and the AI revolution from reactive to proactive policy. Educat. Sci. **9**, 235 (2019)
9. vanden Boom, G., Paas, F., van Merriënboer, J.: Reflection prompts and tutor feedback in a web-based learning environment effects on student's self- regulated learning competence. Comput. Human Behav. **20**, 551–567 (2004). https://doi.org/10.1016/j.chb.2003.10.001
10. Nasaji, S.: Effectiveness of cognitive- behavioral intervention on coping responses and cognition emotion regulation strategies. J. Behav. Sci.Behav. Sci. **4**, 35–43 (2010)

Face and Eye Detection Using Skin Color and Viola-Jones Detector

Hicham Zaaraoui[1](✉), Samir El Kaddouhi[2], and Mustapha Abarkan[1]

[1] LSI, Polydisciplinary Faculty of Taza, Sidi Mohamed Ben Abdellah University of Fès, Fès, Morocco
{hicham.zaaraoui,mustapha.abarkan}@usmba.ac.ma

[2] IEVIA Team, IMAGE Laboratory, Ecole Normale Supérieure, Moulay Ismail University of Meknès, Meknès, Morocco
s.elkaddouhi@umi.ac.ma

Abstract. In this paper, we proposed a robust face and eye detection method based on skin color and the Viola-Jones detector. First the images are segmented into skin region and non-skin region. Then, the faces are detected using a Viola-Jones face detector applied only to the skin regions. Then, in the detected faces we applied a Viol-Jones eye detector on the non-skin regions to localize the position of the eyes. The results obtained show that our method is robust and provides superior performance compared to other recently published methods.

Keywords: Face detection · Eye detection · Viola-Jones detector · Skin color

1 Introduction

Face and eye detection is a very interesting field of research that allows to verify the presence of faces and eyes in an image and to locate their positions. It has attracted a lot of interest in recent years from the computer vision community. This is primarily because a variety of applications such human-computer interface systems, face recognition, robotics, video surveillance, facial expression recognition, age estimate, gaze tracking systems, and driver fatigue monitoring systems have emerged [1, 2].

Recently, several techniques can be used to detect faces and eyes: The template matching [3, 4], the Haar-like features [5–9], the neural networks [10–13], the support vector machine [9, 13–15], the local binary pattern [16, 17], the skin color [18–20], the corners points [21, 22].

The majority of face and eye detection methods are based on the concept of image scanning. The latter consists of scanning the image with windows of different sizes, then classifying them as face (eye) or non-face (non-eye). This can take a long time because the detection involves a lot of testing and can result in a lot of false positives. In order to avoid the scanning step, we have proposed to first determine regions likely to be faces or eyes and then to determine among these regions those which are faces or eyes. Therefore, this article presents a face and eye detection method based on skin color and Viola-Jones detector. It has two algorithms. In the first, we used a simple and effective

Y. Farhaoui et al. (Eds.): ICAISE 2023, LNNS 837, pp. 290–296, 2024.
https://doi.org/10.1007/978-3-031-48465-0_38

method for face detection based on the Viola-Jones detector and skin color. The second algorithm consists in determining the non-skin regions, then applying the Viola-Jones eye detector on these regions to detect the eyes.

Figure 1 below illustrates the different steps of our face and eye detection method.

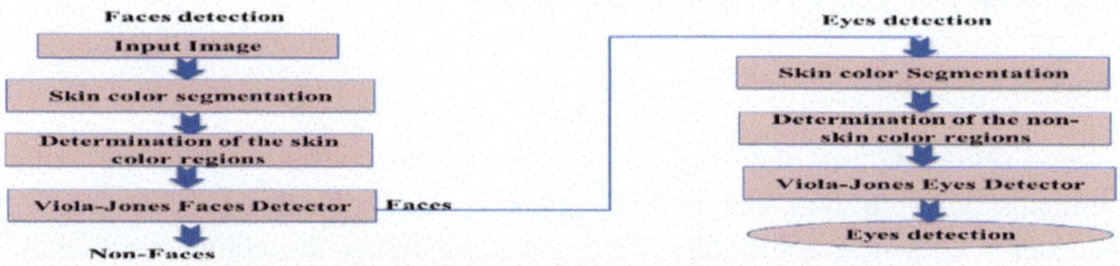

Fig. 1. Diagram of our face and eyes detection method.

The paper is structured as follows: the second section lists some previous works. The third and fourth parts discuss our methodology as well as the results and their interpretations. The fifth part presents the conclusion and the work's perspectives.

2 Related Works

Several face and eyes detection approaches have been developed in recent decades, including:

The method for detecting eyes in [9] is based on discriminatory Haar features (DHFs) and a novel, effective support vector machine (eSVM). The DHFs are obtained by applying a discriminating feature extraction (DFE) method to the 2D Haar wavelet transform. The eSVM significantly improves the computational efficiency upon the conventional SVM for eye detection without sacrificing the generalization performance.

Yu et al. [14] present a method for detecting eyes that is based on the support vector machine and the grayscale variance filter. To keep fewer eye candidate locations, the variance filter is employed to remove the majority of images from non-eye regions. Then, the SVM classifier determines the precise location of the areas of both eyes.

Bhatta et al. [19] suggest a method that involves three processes for extracting face features (the eyes, nose, and mouth). First, the Sobel edge detector and the edge intensity values are used to detect the faces. After that, in the detected faces, the skin regions are extracted in the YCbCr space. Then, the left eye is located in the left part of the skin color image considering that the point with minimum intensity value is as position of the left eye.

Chen et al. [20] provide a method of face detection based on a model of skin color that eliminates the interference of non-skin-color regions by using an enhanced "reference white" method. Then they apply a Gaussian model in YCbCr space to segment skin regions. After that they verify certain conditions the hair and the face surface to remove regions of the skin that are not faces.

El Kaddouhi et al. [21], propose a robust eye detection method based on the Viola and Jones method and corner points. Firstly, faces are detected by a system composed of two detectors of Viola-Jones (one for the frontal faces and the other for the profile faces). Secondly, the Shi-Tomasi detector (to detect corner points) and K-means are used to determine eye candidate regions. Thirdly, the localization of eyes is achieved by matching of these regions with an eye template.

3 The Steps of the Proposed Method

3.1 Faces Detection

Skin region segmentation. Segmentation of skin regions consists in building a rule that discriminates between skin and non-skin pixels. Generally, this process takes place in three steps: choosing a suitable color space, modeling the color of the skin, and classifying the pixels into skin and non-skin pixels.

To model skin color, we used a set of images from the Brazilian FEI database FEI [23] containing 2800 images of 200 people taken under different conditions (lighting, pose, etc.). From these images we have defined the rules that determine the regions of the skin in three normalized rgb color spaces, HSV and YCbCr. These rules are shown in Table 1.

Table 1. Skin region segmentation rules

Spaces	**Rules**	
Normalized RGB	**A**	$0.45 \leq r \leq 0.7$
	B	$0.23 \leq g \leq 0.55$
YCbCr	**C**	$77 \leq Cr \leq 127$
	D	$133 \leq Cb \leq 173$
HSV	**E**	$0 \leq H \leq 0.2$
	F	$0.2 \leq S \leq 0.7$

Application of the Viola-Jones Face Detector to skin regions. Face detection by the Viola-Jones method [5] takes place in two stages: The training stage which is based on the AdaBoost algorithm and the detection stage.

In the training step a set of rectangular features called Haar features, and a set of face images are used. The AdaBoost algorithm selects a set of features and then trains them using the training images to train a set of weak classifiers. These classifiers are divided into groups, and each group of classifiers is combined to form a strong classifier. The strong classifiers obtained are grouped in a cascade structure to form the detector of Viola-Jones. This step is shown in Fig. 2.

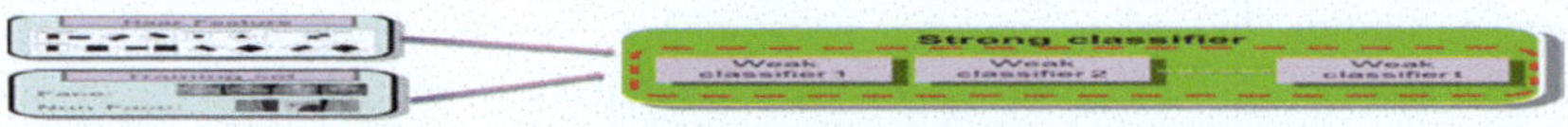

Fig. 2. Training of classifiers

In the detection step each region of the skin is presented to the cascades of classifiers, and if it passes through all strong classifiers it is detected as face and if it is rejected by only one strong classifier it is detected as non-face (see Fig. 3).

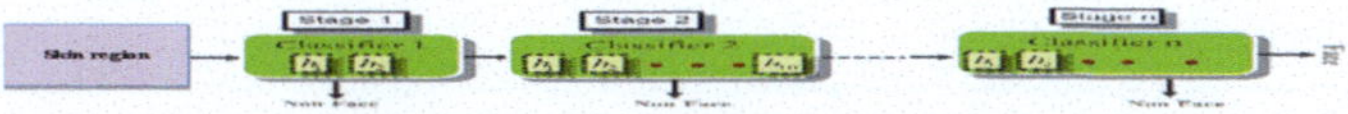

Fig. 3. Cascade of strong classifiers

3.2 Eyes Detection

Determination of non-skin regions in face images. The segmentation of face images makes it possible to obtain two types of regions: skin areas, and non-skin areas. Each of these is likely to be an eye. Thus, the location of these regions makes it possible to determine areas likely to be eyes. Detected faces are segmented into skin and non-skin regions using rules presented in Table 1.

Application of the Viola-Jones eye detector on non-skin regions. Eye detection consists of presenting each non-skin region to a Viola-Jones eye detector to classify it as eye or non-eye. This detector contains a set of strong classifiers mounted in cascade (see Fig. 3). Strong classifiers are trained by the AdaBoost algorithm by combining weak classifiers obtained from Haar features (see Fig. 2). The weak classifiers of the detector are formed, from a set of training images representing eyes.

4 Experimental Results

To evaluate the performance of our proposed method, we used the FEI database [23], a set of personal images and other taken from the Internet. The evaluation is based on two indicators: The true detection rate (TDR), and the false detection rate (FDR) defined by:

$$\mathbf{TDR}(\%) = \frac{\text{number of faces detected}}{\text{total number of faces}} \times 100$$

$$\mathbf{FDR}(\%) = \frac{\text{number of false faces detected}}{\text{total number of detections}} \times 100$$

4.1 Simulation

Performance analysis. We tested our method on the FEI database available online [23]. It contains a set of images of 200 people taken in 14 different situations. we tested our method on 2800 images, each image contains a single face. The results obtained are presented in Table 2. This table contains the number of images, the number of faces, the number of eyes, the number of faces detected, the number of eyes detected, the number of false faces, the number of false eyes, and the detection rates (**TDR**, **FDR**) for total, frontal, profile, dark, with glasses and facial expressions.

Table 2. Face and eye detection results obtained using our approach

Characteristics of the database FEI				Face and eye detection with our method							
Images	Number of images	Number of faces	Number of eyes	Number of correct detection				Number of false detection			
				Faces(%)		Eyes(%)		Faces(%)		Eyes(%)	
Frontal	1400	1400	2800	1400	100	2757	98.46	33	2.35	33	2.87
Profil	800	800	1171	783	97.87	1103	94.19	25	3.12	115	8.94
Dark lighting	400	381	762	363	95.27	685	89.98	5	1.31	23	3.01
Occultation (glasses)	66	66	115	63	95.45	107	93.04	3	4.54	12	9.44
Facial expression	200	200	400	200	100	391	97.75	4	2	15	3.61
Total	**2800**	**2781**	**5133**	**2746**	**98.74**	**4936**	**96.16**	**67**	**2.40**	**186**	**3.62**

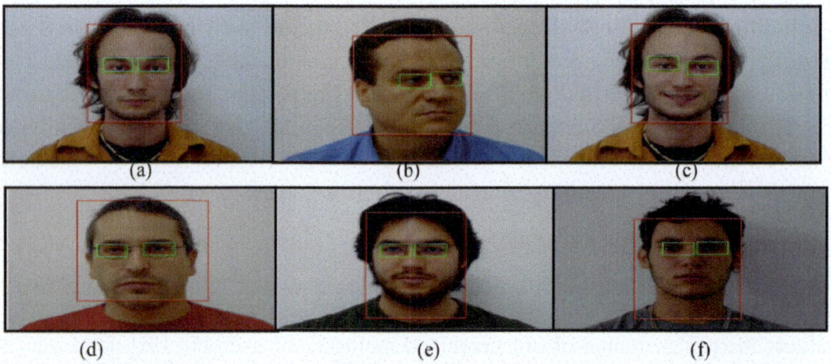

Fig. 4. Face and eye detection in the presence of different constraints. a: frontal face, b: profile face, c: Image with low light, d: face with glasses, e: face with beard, f: face with smile

Figure 4 shows sample results that illustrate face and eye detection.

Comparison. To show the effectiveness of our method, we compared our results with those recently published in the field and using the same database (FEI). The methods chosen for face detection are those presented in [12, 20, 22]. And for eye detection are those shown in [14, 19, 21]. The results obtained are presented in Table 3.

4.2 Complex Images

To study the efficiency of our detection technique on complex images. We thus applied it to a database of 50 images containing 168 faces and 336 eyes. These images are characterized by complex backgrounds, are taken in different lighting conditions and contain several faces of different sizes. Table 4 presents the rates of true and false detection of faces and eyes obtained for complex images.

Table 3. Face and eye detection results obtained using our approach

		True detection rate (%)
Faces detection	**Our method**	**98.74**
	Robin et al. [12]	92.78
	Chen et al. [20]	96.04
	El Kaddouhi et al. [22]	98.02
Eyes detection	**Our method**	**96.16**
	Mingxin et al. [14]	95.2
	Bhatta et al. [19]	94.5
	El Kaddouhi et al. [21]	98.29

Table 4. Face and eye detection results obtained

Total number		Number of true detections		Number of false detections	
Faces	Eyes	Faces	Eyes	Faces	Eyes
168	336	157 (93.45%)	310 (92.26%)	13 (7.73%)	31 (9.09%)

Figure 5 below presents an example of results that illustrates the detection of faces and eyes in the presence of the various factors that disturb the detection process.

Fig. 5. Example of face and eye detection results in complex images

5 Conclusion

In this article, we presented a face and eye detection method based on skin color and the Viola-Jones detector. We detailed its different stages, then we presented the results obtained on the FEI database, and on a set of personal images and others taken on the Internet. We also compared these results with those of other recently published methods. The test and comparison results are very satisfactory. They show that our approach is robust against facial expressions, in the presence of occlusions and against complex backgrounds.

References

1. Al-Rahayfeh, A., Faezipour, M.: Eye tracking and head movement detection: a state-of-art survey. IEEE J. Transl. Eng. Heal. Med. **1**, 11–22 (2013)
2. Song, F., Tan, X., Chen, S., Zhou, Z.H.: A literature survey on robust and efficient eye localization in real-life scenarios. Pattern Recognit. **46**, 3157–3173 (2013)
3. Nanaa, K., Rizon, M., Abd Rahman, M.N., Almejrad, A., Abd Aziz, A.Z., Bahri Mohamed, S.: Eye detection using composite cross-correlation. Am. J. Appl. Sci. **10**, 1448–1456 (2013)
4. Ge, S., Yang, R., He, Y., Xie, K., Zhu, H., Chen, S.: Learning multi-channel correlation filter bank for eye localization. Neurocomputing **173**, 418–424 (2016)
5. Paul Viola, M.J.J.: Robust real-time face detection. Int. J. Comput. Vis. **57**, 137–154 (2004)
6. Ghazali, K.H., Jadin, M.S., Jie, M., Xiao, R.: Novel automatic eye detection and tracking algorithm. Opt. Lasers Eng. **67**, 49–56 (2015)
7. Jian, M., Lam, K.M., Dong, J.: Facial-feature detection and localization based on a hierarchical scheme. Inf. Sci. (Ny) **262**, 1–14 (2014)
8. Ancheta, R.A., Reyes, F.C., Caliwag, J.A., Castillo, R.E.: FED security: implementation of computer vision thru face and eye detection. Int. J. Mach. Learn. Comput. **8**, 619–624 (2018)
9. Chen, S., Liu, C.: Eye detection using discriminatory Haar features and a new efficient SVM. Image Vis. Comput. **33**, 68–77 (2015)
10. Rusek, K., Guzik, P.: Two-stage neural network regression of eye location in face images. Multimed. Tools Appl. **75**, 10617–10630 (2016)
11. Xiang, Y., Yang, H., Hu, R., Hsu, C.Y.: Comparison of the deep learning methods applied on human eye detection. In: Proceedings 2021 IEEE International Conference Power Electronics Computer Applications ICPECA 2021, pp. 314–318. (2021)
12. Robin, M.H., Ur Rahman, M.M., Taief, A.M., Nahar Eity, Q.: Improvement of face and eye detection performance by using multi-task cascaded convolutional networks. In: 2020 IEEE Region 10 Symposium TENSYMP 2020. pp. 977–980. (2020)
13. Yu, M., et al.: An eye detection method based on convolutional neural networks and support vector machines. Intell. Data Anal. **22**, 345–362 (2018)
14. Yu, M., Lin, Y., Wang, X.: An efficient hybrid eye detection method. Turkish J. Electr. Eng. Comput. Sci. **24**, 1586–1603 (2016)
15. Kim, H., Jo, J., Toh, K.A., Kim, J.: Eye detection in a facial image under pose variation based on multi-scale iris shape feature. Image Vis. Comput. **57**, 147–164 (2017)
16. Choi, I., Kim, D.: A variety of local structure patterns and their hybridization for accurate eye detection. Pattern Recognit. **61**, 417–432 (2017)
17. Choi, I., Kim, D.: Generalized binary pattern for eye detection. IEEE Signal Process. Lett. **20**, 343–346 (2013)
18. Vijayalaxmi, B., Sekaran, K., Neelima, N., Chandana, P., Meqdad, M.N., Kadry, S.: Implementation of face and eye detection on DM6437 board using simulink model. Bull. Electr. Eng. Informatics. **9**, 785–791 (2020)
19. Bhatta, L.K., Rana, D.: Facial feature extraction of color image using gray scale intensity value. Int. J. Eng. Res. Technol. (2015)
20. Chang, Z.C.C.L.F., Han, X.: Fast face detection algorithm based on improved skin-color model. Arab J Sci Eng. 629–635 (2013)
21. El Kaddouhi, S., Saaidi, A., Abarkan, M.: Eye detection based on the Viola-Jones method and corners points. Multimed. Tools Appl. **76**, 23077–23097 (2017)
22. El Kaddouhi, S., Saaidi, A., Abarkan, M.: A new robust face detection method based on corner points. Int. J. Softw. Eng. its Appl. **8**, 25–40 (2014)
23. http://fei.edu.br/~cet/facedatabase.html

Intrusion Detection System, a New Approach to R2L and U2R Attack Classification

Chadia El Asry[1](✉), Samira Douzi[2], and Bouabid El Ouahidi[1]

[1] IPSS, Faculty of Sciences, Mohammed V University, Rabat, Morocco
chadia.elasry@um5r.ac.ma

[2] FMPR, Mohammed V University, Rabat, Morocco

Abstract. Intrusion detection systems have played and will continue to play an essential role in detecting network attacks and anomalies. Numerous authors have investigated the use of neural networks to accomplish this objective. Most existing models proposed in the literature struggle to detect various attack types, especially User-to-Root (U2R) and Remote-to-Local (R2L) attacks. The current models appear to be less accurate at detecting these two types of attacks. Consequently, we propose in this paper a detection technique based on feature selection and Long-Short-Term-Memory (LSTM) to address the challenges. Using the Shap values feature selection method, each class of attacks' NSLKDD attacks dataset features are reduced. Then, with the reduced feature set, LSTM is performed for record classification. The NSL-KDD dataset is utilized to train and assess the performance of our model. In terms of accuracy, precision, recall, and F-score values, it outperforms LSTM with all features and other state-of-the-art models. In addition, our model provides a more accurate classification of R2L (Remote-to-Local) and U2R (User-to-Root) attacks.

Keywords: Intrusion detection systems · Deep learning · Shap values · LSTM · Feature selection

1 Introduction

In general, intrusion detection systems are divided into three categories [1]:

IDSs use the behavioral approach (those that seek to detect anomalies), the scenario approach (those that seek to detect signatures), and the specification approach.

In general, behavioral analysis has two phases: the learning phase and the detection phase. During the learning phase, the system learns to identify normal behavior, and during the detection phase, abnormal behavior [2] is identified. In scenario analysis, there is no learning phase. They are equipped with a foundation of predefined attack scenarios that they must recognize (signature). The specification approach is a method for constructing the normal profile without the use of learning algorithms [3].

The development of detection IDSs that employ machine learning and deep learning has consumed a substantial amount of time and resources [4, 5].

In this paper, we propose employing a Shap values method to detect the essential characteristics and Long-Short-Term Memory for network intrusion detection. To train

Y. Farhaoui et al. (Eds.): ICAISE 2023, LNNS 837, pp. 297–304, 2024.
https://doi.org/10.1007/978-3-031-48465-0_39

and evaluate the performance of our model, we utilize the NSL-KDD dataset. In the following ways, this work contributes significantly:

i. Propose a model that can recognize and classify R2L (Remote-to-Local) and U2R (User-to-Root) attacks.
ii. Comparing our classification model to the LSTM (Long-Short-Term-Memory) model using all dataset features.
iii. Our proposed model outperformed a large number of other recently proposed methods.

The document is divided into sections, beginning with an introduction. Section 2 presents methods and materials. Section 3 contains the suggested model and the experimental analysis. The paper is concluded in Sect. 4 with the conclusion.

2 Related Works

Numerous researchers advocate combining intrusion detection and machine learning (ML) technology to identify network threats through the development of effective models. The authors of [6] use PCA (Principal Component Analysis) to extract features and multiple deep learning classification models. A comparison of the model's performances revealed that DNN had the highest accuracy across all datasets used in this investigation (NSL-KDD, UNSW-NB15, and CSE-CIC-IDS2018 datasets). The authors of [7] proposed a bidirectional LSTM deep learning approach to detect various attack types, particularly U2R and R2L attacks; the model achieves a higher detection accuracy for the attack types than the conventional LSTM. The authors of [8] proposed an RNN-based classifier for IDS. They investigated six unique optimizers for LSTM–RNN. Among the six optimizers, Nadam had the best attack detection performance. The authors of [9] present a convolutional neural network based detection method that converts traffic data to images, thereby eliminating the need for manual feature construction in a separate study. The authors of [10] proposed an IDSbased LSTM and attention mechanism with the NSLKDD dataset containing five types of attacks; the model achieves high performance, but U2R attacks are frequently misclassified as normal.

3 Background

In this section, we will first discuss the concept behind LSTM architecture. The architecture of Shap values is then defined. We describe in greater detail the NSL-KDD dataset used to train and validate our model.

3.1 Long Short-Term Memory (LSTM)

Long-Short-Term-Memory (LSTM) [11] was developed by Hochreiter and Schmidhuber in 1995. Long Short-Term Memory (LSTM) is a Recurrent Neural Network (RNN) architecture employed in deep learning to represent time series data [12]. In contrast to conventional feedforward neural networks, LSTM features feedback connections between hidden units that are connected at discrete time steps, enabling the learning and prediction of long-term sequence dependencies based on the sequence of previous data. LSTMs were developed to address the issue of vanishing and exploding gradients observed during conventional RNN training.

3.2 Feature Selection: Shap Values

The selection of features for an intrusion detection system is a crucial step due to the complexity and noise of network data under certain conditions. Consequently, feature selection methods are required to reduce the dimensionality of a dataset [13]. In the present research, Shap Value has been used to select subset characteristics from four sets of data. The Shap value has selected the most essential characteristics. The Shap value method is described in detail as follows:

SHAP, which stands for Shapley Additive exPlanations, was created by Lundberg and Lee (2017) [14] as a method for explaining the output of any machine learning model. It links game theory with local explanations [14, 15]. Shapley values represent model predictions in the form of binary linear combinations indicating the presence or absence of each covariate.

3.3 NSLKDD Data Set

The NSL-KDD Data set [16] was created by McHugh [17]; it contains 4 898 431 instances, 42 features, and four primary attack types. Normal and attack are the two labels present in the dataset. The NSL KDD dataset is frequently used by numerous researchers to train and evaluate their proposed approach in the intrusion detection domain; consequently, we will employ it in our work.

4 Proposed Approach and the Experimental Results

The proposed intrusion detection system is implemented with Python, a Core i5 processor, and 4 GB of RAM. Several evaluation metrics, including Accuracy, Recall, Precision, and F1-score, have been utilized to test the proposed method.

The work proposed in Fig. 1 contains numerous steps, beginning with Dataset preprocessing and ending with Classification.

The proposed approach comprises four steps: data preprocessing, data splitting, feature selection, and classification. Initially, data cleaning is carried out to enhance data quality. Next, the features are standardized by scaling them within a specific range, and non-numeric data is converted into numeric data. In the second stage, the data is partitioned into subsets based on the four types of attacks, with each subset containing one type of attack. The third step involves the implementation of SHAP-Values to identify the most significant features. Lastly, the selected features serve as input for LSTM in the classification process. A detailed explanation of this systematic investigation is provided below.

4.1 Datasets Preprocessing

Several preprocessing steps have been applied to the dataset, which are detailed below:

- We categorized the attacks in the NSL-KDD dataset into four groups and assigned numerical values to each attack type. 0, 1, 2, 3, and 4 correspond to Normal, DoS, Probe, R2L, and U2L, respectively.
- The pandas.factory (function) maps symbolic-valued features such as protocol, service, and flag to numeric-valued attributes.
- The Standard Scaler rescaling algorithm is employed to standardize the features.
- The dataset is subdivided into four subsets based on the four types of attack (DoS-set, Probe-set, R2L-set, and U2L-set), with each subset containing a single type of attack.

4.2 Feature Selection: Shap Value

Each subset is analyzed with Shap Value method to determine which characteristics are crucial for each type of attack. Ten features for each subset have relatively the highest SHAP values as shown in Fig. 2.

The collection of features selected for each subset is merged and considered relevant for the entire NSLKDD dataset. Table 1 presents the reduced characteristics for each subset.

After combining the features selected for each attack subset, we arrive at a total of 25 relevant features that will be applied to the entire dataset to improve the classification of the two attack types R2L and U2R.

4.3 Experimental Results and Discussion

Tables 2 and 3 contain the results of LSTM with all features of NSL KDD and LSTM with 25 relevant features selected from the four subsets of attacks, respectively.

Table 1. List of selected features for each subset attack.

Subset	Selected features
DoS-set	'count', 'dst bytes', 'logged in', 'dst host srv serror rate', 'same srv rate', 'rerror rate', 'dst host serror rate', 'flag', 'wrong fragment', 'num compromised'
Probe-se	'src bytes', 'service', 'dst bytes', 'logged in', 'same srv rate', 'dst host rerror rate', 'dst host diff srv rate', 'dst host same src port rate', 'dst host srv count', 'flag'
R2L-set	'service', 'dst host same src port rate', 'hot', 'is guest login', 'count', 'num failed logins', 'duration', 'src bytes', 'dst host srv diff host rate', 'srv count'
U2R-set	'root shell', 'dst host srv count', 'num file creations', 'src bytes', 'num compromised', 'dst bytes', 'service', 'same srv rate', 'dst host count', 'duration'

Table 3 displays the performance of the LSTM algorithm with all NSLKDD dataset features. R2L and U2R have very low-performance metrics compared to other attacks. The outcomes of the proposed model are presented in Table 3. It is observed that the

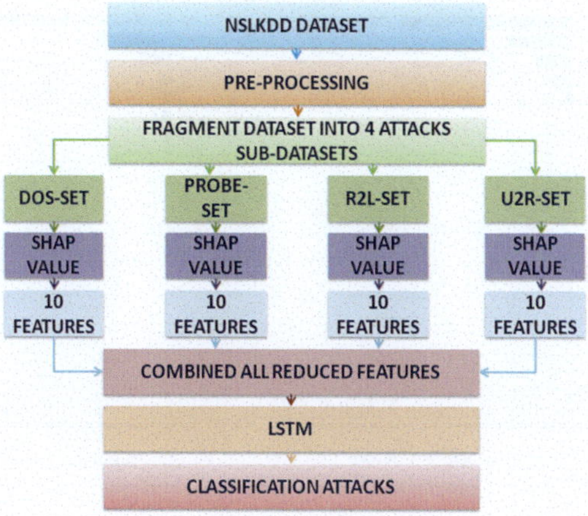

Fig. 1. Proposed architecture

Table 2. Classification performance of each attack with all features

Attacks	Precision	Recall	f1-score
Normal	99.72	99.79	99.76
DoS	99.92	99.97	99.95
Probe	99.85	99.45	99.65
U2R	75.36	92.04	82.87
R2L	66.67	33.33	44.44

Table 3. Classification performance of each attack of our proposed method

Attacks	Precision	Recall	f1-score
Normal	99.54	99.77	99.66
DoS	99.93	99.97	99.95
Probe	99.75	98.99	99.37
U2R	**78.95**	**92.92**	**85.37**
R2L	**83.33**	**83.33**	**83.33**

Bold indicates the best results of the proposed model

proposed model outperformed the LSTM in classifying R2L and U2R attacks across all features.

Figure 3 compares the proposed model's accuracy, precision, recall, and F-measure performance to that of LSTM with all features. The proposed model's outperformed the LSTM in terms of all metrics of evaluation.

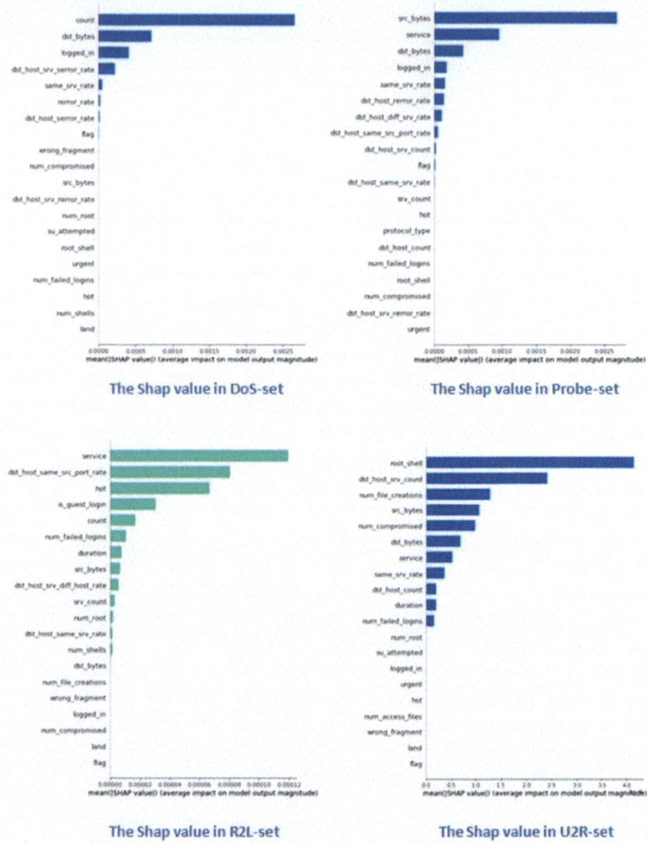

Fig. 2. Shap value for each subset

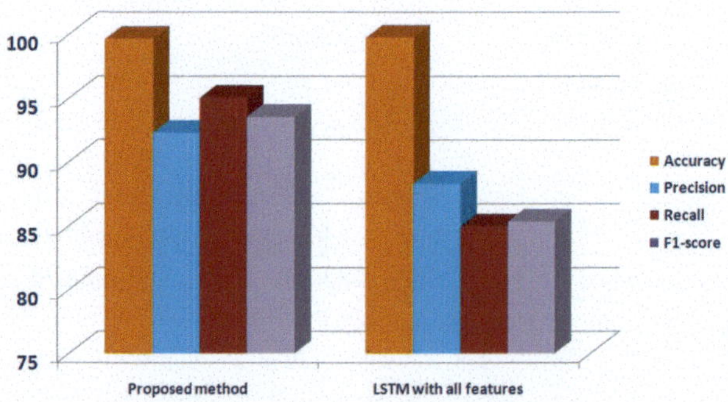

Fig. 3. Performance accuracy of proposed model and LSTM with all features.

Table 4 show the comparison of the performance of our technique to other research in this field.

Table 4. Comparison with our approach and some of IDS techniques.

Approach In	Attacks	Precision %	Recall%	F1-score%
[7]	U2R	37.99	43.50	40.56
	R2L	97.97	70.04	81.69
[7]	U2R	62.42	49.00	54.90
	R2L	98.97	73.60	84.42
[18]	U2R	–	24.00	–
	R2L	–	65.76	–
Proposed model	U2R	**78.95**	**92.92**	**85.37**
	R2L	**83.33**	**83.33**	**83.33**

Bold indicates the best results of the proposed model

In conclusion, according to the above table, it is determined that the proposed model can detect various types of attacks with greater metrics of performance than other existing systems.

5 Conclusion

Using the feature selection method and neural networks, we propose a modular architecture for network intrusion detection systems in this paper. The modularity of our architectures makes it possible to eliminate the components that hinder the IDS's performance, particularly in detecting low-frequency attacks such as U2R and R2L. Experimental results indicate that classification for all four attack categories is improved relative to the U2R and R2L categories. Our strategy performed significantly better than the other strategies. It exhibited 78.95% classification precision, 92.92% recall, and 85.37 percent F1-score for U2R, and 83.33% classification precision, 83.33% recall, and 83.33% F1-score for R2L.

References

1. Shaker, A., Gore, S.: Importance of intrusion detection system. Int. J. Scient. Eng. Res. Janvier (2011)
2. Me, L., Alanou, V.: Detection d'intrusion dans un systeme informatique : methodes et outils. TSI **15**(4), 429–450 (1996)
3. Hiet, G.: Detection of instructions parameterized by the security policy thanks to the collaborative control of information flows within the operating system and applications: implementation under Linux for java program's Universite de Rennes, December (2008)
4. Liu, H., Lang, B.: Machine learning and deep learning methods for intrusion detection systems: a survey. Appl. Sci. **9**(20) (2019). https://doi.org/10.3390/app9204396
5. Xin, Y., et al.: Machine learning and deep learning methods for cybersecurity. IEEE Access **6**, 35365–35381 (2018)

6. Amaizu, G.C., Nwakanma, C.I., Lee, J.M., Kim, D.S.: Investigating network intrusion detection datasets using machine learning. In: 2020 International Conference on Information and Communication Technology Convergence (ICTC), October, pp. 1325–1328. (2020)
7. Imrana, Y., Xiang, Y., Ali, L., Abdul-Rauf, Z.: A bidirectional LSTM deep learning approach for intrusion detection. Expert Syst. Appl. **185**, 115524 (2021). https://doi.org/10.1016/j.eswa.2021.115524
8. Le, T., Kim, J., Kim, H.: An effective intrusion detection classifier using long short-term memory with gradient descent optimization. In: 2017 International Conference on Platform Technology and Service (PlatCon), Busan, pp. 1–6. IEEE (2017)
9. Wang, W., Zhu, M., Zeng, X., Ye, X., Sheng, Y.: Malware traffic classification using convolutional neural network for representation learning. In: Proceedings of the 2017 International Conference on Information Networking (ICOIN), Da Nang, Vietnam, 11–13 January 2017, IEEE, Manhattan, NY, USA, pp. 712–717. (2017)
10. Fatimaezzahra, L., Samira, D., Douzi, K., Badr, H.: IDS-attention: an efficient algorithm for intrusion detection systems using attention mechanism. J. Big Data. **8** (2021). https://doi.org/10.1186/s40537-021-00544-5
11. Hochreiter, S., Schmidhuber, J.: Long short-term memory. Neural Comput. **9**(8), 1735–1780 (1997)
12. Ibtissam, B., Samira, D., Bouabid, O.: Credit card fraud detection model based on LSTM recurrent neural networks. J. Adv. Inform. Technol. **12**, 113–118 (2021). https://doi.org/10.12720/jait.12.2.113-118
13. Rafeef, F., Rafeef, A.-S.: New Approach for Classification R2L and U2R Attacks in Intrusion Detection System (2018)
14. Lundberg, S.M., Lee, S.-I.: A unified approach to interpreting model predictions. In: Advances in Neural Information Processing Systems, vol. 30, pp. 4768–4777. Curran Associates, Inc. (2017)
15. Magesh, P.R., Myloth, R.D., Tom, R.J.: An explainable machine learning model for early detection of Parkinson's disease using LIME on DaTSCAN imagery. Comput. Biol. Med. **126**, 104041 (2020)
16. NSL-KDD dataset.: [online] Available: http://nsl.cs.unb.ca/nsl-kdd/
17. lakhina, S., Joseph, S., Verma, B.: Feature reduction using principal component analysis for effective anomaly–based intrusion detection on NSL-KDD. Int. J. Eng. Sci. Technol. **2**(6), 1790–1799 (2010)
18. Fu, Y., Du, Y., Cao, Z., Li, Q., Xiang, W.: A deep learning model for network intrusion detection with imbalanced data. Electronics **11**, 898 (2022). https://doi.org/10.3390/electronics11060898

An Enhanced Internet of Medical Things Data Communication Based on Blockchain and Cryptography for Smart Healthcare Applications

Joseph Bamidele Awotunde[1](✉), Yousef Farhaoui[2], Agbotiname Lucky Imoize[3], Sakinat Oluwabukonla Folorunso[4], and Abidemi Emmanuel Adeniyi[5]

[1] Department of Computer Science, Faculty of Information and Communication Sciences, University of Ilorin, Ilorin 240003, Kwara State, Nigeria
awotunde.jb@unilorin.edu.ng

[2] STI Laboratory, T-IDMS, Department of Computer Science, Faculty of Sciences and Techniques Errachidia, Moulay Ismail University of Meknes, Meknes, Morocco
y.farhaoui@fste.umi.ac.ma

[3] Department of Electrical and Electronics Engineering, Faculty of Engineering, University of Lagos, Akoka, Lagos 100213, Nigeria
aimoize@unilag.edu.ng

[4] Department of Mathematical Sciences, Olabisi Onabanjo University, Ago-Iwoye, Ogun State, Nigeria
sakinat.folorunso@oouagoiwoye.edu.ng

[5] Department of Computer Science, College of Computing and Communication Studies, Bowen University, Iwo, Nigeria
abidemi.adeniyi@bowen.edu.ng

Abstract. This paper proposes a novel data communication model for the Internet of Medical Things (IoMT) powered smart healthcare applications based on blockchain and cryptography. The proposed model enhances the data communication security of IoMT applications through the distributed ledger technology of blockchain and encryption algorithms of cryptography. The model is designed to provide a secure and reliable channel for data transmission and storage between IoMT nodes and healthcare stakeholders. Moreover, the model is capable of detecting data manipulation and unauthorized access attempts. In addition, the messaging protocol of the proposed model is capable of reducing the communication overhead of the IoMT network. Finally, the proposed model is tested using a scenario of smart healthcare applications for the elderly, and the results show the effectiveness and reliability of the model for IoMT-powered smart healthcare applications.

Keywords: Blockchain technology · Internet of medical things · Artificial intelligence · Cryptography · Smart healthcare systems

Y. Farhaoui et al. (Eds.): ICAISE 2023, LNNS 837, pp. 305–313, 2024.
https://doi.org/10.1007/978-3-031-48465-0_40

1 Introduction

Intensive care is one of the many services offered by healthcare with various other services like emergency care, and outpatient care [1]. Treatment remains a primary concern in medical services for society welfare, and various methods have been employed for this same purpose [2]. Additional resources are made available by healthcare systems to provide a secure environment for patients [3]. Access to comprehensive, top-notch medical treatment is essential to promote and uphold the integrity of all patient's wellness, preventative medicine, management, and rehabilitation [4]. A crucial feature of handling information in the healthcare institution is developing a system that unifies and oversees data access across all departments [5, 6]. Health data management improves patient care, enhances decision-making, and protects the privacy and confidentiality of doctors [7]. Patient information is collected in health records in a meaningful and private manner. The Internet of Medical Things (IoMT) is a subset of IoT that gathers and analyzes data for testing and monitoring in the medical and healthcare industries [8, 9]. IoT devices can support safe living for patients, while implants are medical devices that replace or enhance biological structures [10].

Healthcare data collecting aids in developing comprehensive patient perspectives, personalizing procedures, advancing treatment modalities, enhancing doctor-patient relationships, and enhancing overall health [11]. Smart healthcare systems can now provide more complicated real-time services because of IoMT's fast development [12]. E-health systems require extensive health data for critical patient diagnoses and data transfer to new hospitals or clinics. Blockchain innovation has been investigated for its capacity to maintain privacy and confidentiality in decentralized systems on mobile devices, in the fog, and the cloud. It is a secure decentralized transaction platform that protects user information and has been proposed as a conceptual underpinning for Fog and mobile Cloud computing [13, 14].

To disseminate EHR among several cloud-based parties, authors in [15] propose an access control method based on smart contracts. A blockchain-based attribute based decentralized file access control system was proposed by Zhang et al. [16]. A secure blockchain for EHR sharing on a mobile cloud is suggested along with that. Nguyen et al. [17] also suggested employing smart contracts as a reliable access control method. A decentralized access control paradigm for IPFS was presented by the authors in [18] and is based on attribute-based encryption technology. The work in [19, 34] presented a distributed IoT system access control technique based on smart contracts. The author put forth a capability-based access control solution for IoT devices that is decentralized and enabled by blockchain [20, 32]. The author in [21, 33] introduced the blockchain-based access control mechanism for high scalability in IoT critical systems. The studies already mentioned are mostly concerned with access control techniques. The costs associated with storage and delay, however, have not yet been covered.

Blockchain technology revolutionizes various fields, including healthcare and supply chain, by providing decentralized, persistent, anonymous, and audible transactions [35]. Its key advantages include decentralization, persistence, anonymity, and audibility, reducing costs, and improving efficiency. The goal of this study is to use cryptography and blockchain innovation to guarantee the confidentiality and privacy of healthcare data without the involvement of third parties [36]. Hospitals and clinics may now access particular patient data thanks to blockchain, which speeds up data collection and protection against fraud and loss. The following are the contributions of this study:

i. The proposed model applies a public key encryption algorithm to secure the IoMT-based cloud storage before data sharing within a group of physicians. This helps in data encryption, standardization, and digital signature.
ii. designing a secure IoMT-based framework for healthcare data management and accessibility.
iii. utilizing blockchain technology and cryptography together to provide context-aware security, confidentiality, and decentralized access control.

2 Materials and Methods

To secure the IoMT-based system on the cloud, this paper suggests an intelligently secured smart healthcare monitoring system that makes use of blockchain technology and cryptography. The proposed system will get around the issues and challenges the IoMT-based system encountered by utilizing a range of devices and sensors. The suggested system enables the interconnection of a variety of wearable medical devices to track a person's health. Instead of being stored in a single central database, the blockchain approach was used to encrypt the private and sensitive patient EHR and health data collected by wearable sensors [29, 30].

Only approved users, including caregivers, medical professionals, pharmaceutical firms, and brokers for healthcare insurance, among others, have access to the data kept in the cloud database. Patient's data and information can only be accessible by doctors or other caregivers with the patient's consent because the patient will be informed that the knowledge and records will be accessible by a third party. Only with the patient's permission can a doctor or caregiver access such information and data. The complete system is interconnected via a wireless gateway, and a Wearable Sensors Network (WSN) was used to collect the data. Figure 1 depicts the framework of the proposed Blockchain enabled with a cryptography model for a secured IoMT-based system [28, 31].

2.1 The Five (5) Entities that Made up the Framework and Their Respective Roles Are Listed Below

IoMT-based Sensors and Actuators Devices: The IoMT-based sensors and actuators devices play a crucial role in modern healthcare by enabling remote monitoring, real-time data analysis, and improved patient care [22, 26]. IoMT sensors are devices that collect data from the physical world, such as vital signs, environmental conditions, or patient activities [23, 27]. These sensors can be integrated into wearable devices, implanted in the body, or placed in the patient's environment [24, 25].

Actuators are devices that perform actions based on the data collected by sensors or through remote commands. In the context of IoMT, actuators can trigger actions like medication reminders, adjusting medical equipment settings, or even delivering treatment. Every device in our suggested framework has a distinct embedded identity (ID_{dev}). Additionally, it is extremely difficult to use device-embedded identification to manage system resources in large organizations. Each node registers with the busines's edge systems as a result, receiving a unique set of identities, comprising service identity (SI_{div}) and device identification (ID_{div}).

The Clinical Data/Medical records: Clinical data refers to information collected during a patient's medical care and includes various types of data such as patient demographics, medical history, diagnoses, treatments, laboratory results, imaging studies, and more. Medical records are the organized documentation of this clinical data. The shift from paper-based medical records to electronic health records (EHRs) has transformed the way medical data is managed and accessed. EHRs are digital versions of medical records that offer numerous advantages, including faster data retrieval, improved data sharing between healthcare providers, and enhanced patient engagement through patient portals.

Encryption/Decryption and Standardization: Encryption and decryption are processes used to secure and protect sensitive information by converting it into a format that is unreadable without the proper decryption key. Standardization in this context refers to the establishment of consistent and widely accepted methods, algorithms, and protocols for encryption and decryption, ensuring interoperability and security across different systems and applications.

Blockchain layer: Blockchain technology provides privacy-preserved data exchange networks for easy access to patient data through smart contracts. In the suggested approach, business logic for making decisions at the cloud computing layer was written using smart contacts. Smart contracts fall into two categories: those that automate business logic and those that construct service agreements between different parties. The people who are involved in the services service providers like doctors and insurance agents and service consumers like patients are the ones who create and advertise the service agreements. The adaptive smart contract method is shown in Fig. 2. The smart contract building module asks the two parties for their terms (T), conditions (C), and agreements (A), which then turns into the smart contract. Then, using the service provider wallet account, two parties sign and publish a smart contract. Additionally, the patient service identity $\left(SI_{pat}\right)$ and public identity $\left(DI_{pat}\right)$ are linked with the smart contract address using hybrid computing to automate services.

End Users: Patients use the system via smart gadgets, while medical professionals interact with it via a smartphone or web interface. This comprises the patients, medical professionals, nurses, pharmacists, druggists, clinicians, insurance companies, and academics.

3 Experimental Results

The Hyper-ledger Calliper system is a versatile tool for the blockchain network, compatible with various Hyper-ledger architectures like Fabric, Composer, Sawtooth, and Iroha. It implements Crypto Hash for secure encryption and decryption, ensuring lightweight and secure mechanisms. The proposed model uses Crypto Hash for secure, lightweight encryption, utilizing the Calliper tool for verification and implementation. The system parameters include encryption and decryption times, latency, throughput, and computational cost.

The proposed model throughput parameter is compared with PoW, Crypto Hash, and PoW-CHA. Figure 1 displays the throughput of the four models. The node's sizes have an impact on transaction delay. Figure 2 demonstrates how the model throughput increases over time, which suggests that the number and frequency of nodes have an impact on the proposed Blockchain technology throughput.

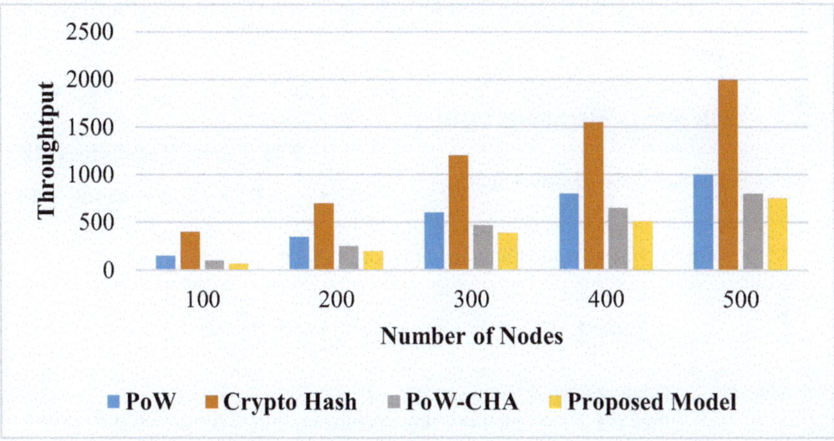

Fig. 1. The proposed model throughput

The proposed model was compared with the Crypto-Hash to evaluate its performance. We calculated the average time needed for the suggested model to upload and download each file operation with various file sizes to assess the network file operation latency. The experiment's findings show that our proposed approach outperforms the Crypto-Hash at handling several simultaneous requests for operations on files. This study clearly shows that the approach, as shown in Fig. 2, has resulted in lower latency for the relevant file operations than the Crypto-Hash.

Tables 1 and 2 display, for the encryption and decryption modules that were implemented in Python, respectively, the memory usage by each line of code.

3.1 The Comparison of Both Encryption and Decryption Performance

Table 1 displays IoMT-based patient data encryption performance metrics, including computing time, memory, processor consumption, and power consumption for 500kb plaintext data. The patient-encrypted information includes things like age, blood type, illness, diseases identified, and laboratory test results.

Table 2 displays IoMT-based patient data decryption performance metrics, revealing the algorithm outperforms conventional algorithms across all metrics.

Blockchain and cryptography play crucial roles in securing IoMT-based systems, which involve medical devices, sensors, and other healthcare-related devices connected to the internet. These technologies provide data integrity, confidentiality, and trust in such sensitive and critical environments. Incorporating blockchain and cryptography

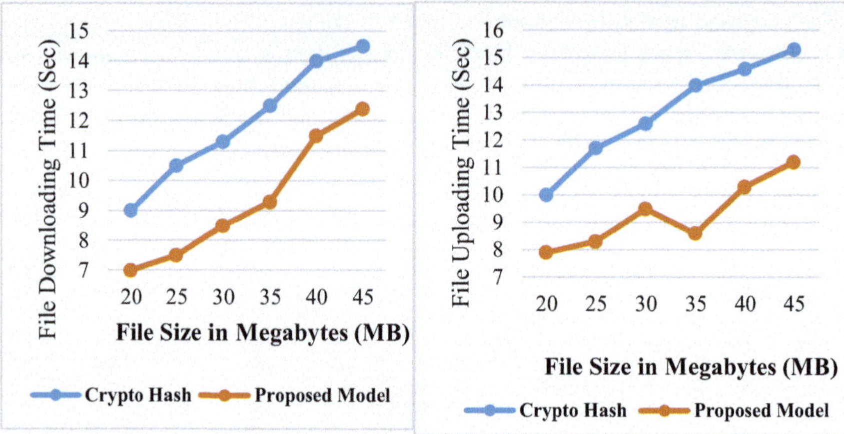

Fig. 2. File Downloading and uploading of the proposed model

Table 1. Cryptography encryption performance

Data size	Algorithm	Computation time (s)	Computing memory (kb)	Processor consumption (%)	Battery consumption (w)	Accuracy (%)
500 kb	AES	0.025	2.37	0	1.56E-05	94
	RSA	5.487	2.08	0.7	0.004141	73
	AESRSA	5.502	4.17	0.7	0.005091	70
	Proposed model	0.301	1.25	0.3	1.16E-07	99

Table 2. Cryptography decryption performance

Data size	Cryptography	Computation time (s)	Computing memory (kb)	Processor consumption (%)	Battery consumption (w)	Accuracy (%)
500kb	AES	0.022	2.32	0	1.50E-05	93
	RSA	5.48	2.03	0.6	0.00411	75
	AESRSA	5.480	4.03	0.6	0.00502	73
	Proposed Model	0.245	2.04	0.3	1.24E-07	98.8

into IoMT systems provides a shared, tamper-resistant ledger for critical healthcare data and ensures that communication between devices and parties remains confidential and secure. The various end users of such frameworks benefit from improved data security, privacy, and trust in the IoMT ecosystem.

4 Conclusion

In this work, a blockchain-enabled by cryptography was utilized to deal with the problems that decentralized IoMT-based smart healthcare systems have with latency, data security, confidentiality, anonymity, and accountability. Additionally, it demonstrates how the blockchain, hash function, smart contract, and digital signature may be used to provide architecture-level solutions to the problems highlighted. System-level traceability is made possible by blockchain-based tamper-proof public ledgers. The proposed cryptographic techniques ensure the security and confidentiality of medical data. Smart contracts automate core healthcare services and medical emergency notifications. An infrastructure for digital contracts to be created by multiple healthcare industry partners is also provided by the proposed framework. The suggested model accomplished the logical analysis's required functions, such as minimizing data transfer latency under urgent conditions. Future work will focus on AI/ML-based models for driven critical patient monitoring, addressing pandemic requirements, and enhancing healthcare services capability and QoS.

References

1. Awotunde, J.B., Misra, S., Ajagbe, S.A., Ayo, F.E., Gurjar, R.: An IoT machine learning model-based real-time diagnostic and monitoring system. In: Machine Intelligence Techniques for Data Analysis and Signal Processing: Proceedings of the 4th International Conference MISP 2022, Volume 1, May, pp. 789–799. Singapore, Springer Nature Singapore (2023)
2. Albahri, A.S., Duhaim, A.M., Fadhel, M.A., Alnoor, A., Baqer, N.S., Alzubaidi, L., ... Deveci, M.: A systematic review of trustworthy and explainable artificial intelligence in healthcare: assessment of quality, bias risk, and data fusion. Inform. Fusion
3. Onasanya, A., Elshakankiri, M.: Smart integrated IoT healthcare system for cancer care. Wireless Netw. **27**, 4297–4312 (2021)
4. Qobbi, Y., Abid, A., Jarjar, M., El Kaddouhi, S., Jarjar, A., Benazzi, A.: An image encryption algorithm based on substitution and diffusion chaotic boxes. In: The International Conference on Artificial Intelligence and Smart Environment, November, pp. 184–190. Cham, Springer International Publishing (2022)
5. Li, Z., Liang, F., Hu, H.: Blockchain-based and value-driven enterprise data governance: a collaborative framework. Sustainability **15**(11), 8578 (2023)
6. Abikoye, O.C., Oladipupo, E.T., Imoize, A.L., Awotunde, J.B., Lee, C.C., Li, C.T.: Securing critical user information over the internet of medical things platforms using a hybrid cryptography scheme. Future Internet **15**(3), 99 (2023)
7. Kumar, P.M., Gandhi, U.D.: Enhanced DTLS with CoAP-based authentication scheme for the internet of things in healthcare applications. J. Supercomput. **76**, 3963–3983 (2020)

8. Douiba, M., Benkirane, S., Guezzaz, A., Azrour, M.: A collaborative fog-based healthcare intrusion detection security using blockchain and machine learning. In: The International Conference on Artificial Intelligence and Smart Environment, November, pp. 1–6. Cham, Springer International Publishing (2022)
9. Gaber, T., Awotunde, J.B., Folorunso, S.O., Ajagbe, S.A., Eldesouky, E.: Industrial internet of things intrusion detection method using machine learning and optimization techniques. In: Wireless Communications and Mobile Computing (2023)
10. Awotunde, J.B., Folorunsho, O., Mustapha, I.O., Olusanya, O.O., Akanbi, M.B., Abiodun, K.M.: An enhanced internet of things enabled type-2 fuzzy logic for healthcare system applications. In: Recent trends on type-2 fuzzy logic systems: theory, methodology, and applications, pp. 133–151. Springer International Publishing, Cham (2023)
11. Abbas, A., Alroobaea, R., Krichen, M., Rubaiee, S., Vimal, S., Almansour, F.M.: Blockchain-assisted secured data management framework for health information analysis based on internet of medical things. Personal and Ubiquitous Comput. 1–14 (2021)
12. Ajagbe, S.A., Awotunde, J.B., Adesina, A.O., Achimugu, P., Kumar, T.A.: Internet of medical things (IoMT): applications, challenges, and prospects in a data-driven technology. In: Intelligent Healthcare: Infrastructure, Algorithms and Management, pp. 299–319. (2022)
13. Jarjar, M., Abid, A., Qobbi, Y., El Kaddouhi, S., Benazzi, A., Jarjar, A.: An image encryption scheme based on DNA sequence operations and chaotic system. In: The International Conference on Artificial Intelligence and Smart Environment, November, pp. 191–198. Cham, Springer International Publishing (2022)
14. Annane, B., Alti, A., Lakehal, A.: Blockchain-based context-aware CP-ABE schema for internet of medical things security. Array **14**, 100150 (2022)
15. Xia, Q.I., Sifah, E.B., Asamoah, K.O., Gao, J., Du, X., Guizani, M.: MeDShare: trustless medical data sharing among cloud service providers via blockchain. IEEE Access **5**, 14757–14767 (2017)
16. Zhang, Y., He, D., Choo, K.K.R.: BaDS: blockchain-based architecture for data sharing with ABS and CP-ABE in IoT. Wirel. Commun. Mob. Comput. **2018**, 1–9 (2018)
17. Nguyen, D.C., Pathirana, P.N., Ding, M., Seneviratne, A.: Blockchain for secure EHRs sharing of mobile cloud-based e-health systems. IEEE Access **7**, 66792–66806 (2019)
18. Wang, S., Zhang, Y., Zhang, Y.: A blockchain-based framework for data sharing with fine-grained access control in decentralized storage systems. IEEE Access **6**, 38437–38450 (2018)
19. Zhang, Y., Kasahara, S., Shen, Y., Jiang, X., Wan, J.: Smart contract-based access control for the internet of things. IEEE Internet Things J. **6**(2), 1594–1605 (2018)
20. Xu, R., Chen, Y., Blasch, E., Chen, G.: Blendcac: a blockchain-enabled decentralized capability-based access control for IoT. In: 2018 IEEE International Conference on Internet of Things (iThings) and IEEE Green Computing and communications (GreenCom) and IEEE cyber, physical and social computing (CPSCom) and IEEE Smart Data (SmartData), July, pp. 1027–1034. IEEE (2018)
21. Novo, O.: Blockchain meets IoT: an architecture for scalable access management in IoT. IEEE Internet Things J. **5**(2), 1184–1195 (2018)
22. Awotunde, J.B., Misra, S., Pham, Q.T.: A secure framework for internet of medical things security based system using lightweight cryptography enabled blockchain. In: International Conference on Future Data and Security Engineering, November, pp. 258–272. Singapore, Springer Nature Singapore (2022)
23. Punugoti, R., Vyas, N., Siddiqui, A.T., Basit, A.: The convergence of cutting-edge technologies: leveraging AI and edge computing to transform the internet of medical things (IoMT). In: 2023 4th International Conference on Electronics and Sustainable Communication Systems (ICESC), July, pp. 600–606. IEEE (2023)

24. Ogundokun, R.O., Awotunde, J.B., Adeniyi, E.A., Misra, S.: Application of the internet of things (IoT) to fight the COVID-19 pandemic. In: Intelligent internet of things for healthcare and industry, pp. 83–103. Springer International Publishing, Cham (2022)
25. Farhaoui, Y.: Design and implementation of an intrusion prevention system. Int. J. Netw. Secur. **19**(5), 675–683 (2017). https://doi.org/10.6633/IJNS.201709.19(5).04
26. Farhaoui, Y. et al.: In: Big Data Mining and Analytics, vol. 6(3), pp. I–II. (2023). https://doi.org/10.26599/BDMA.2022.9020045
27. Farhaoui, Y.: Intrusion prevention system inspired immune systems. Indones. J. Electri. Eng. Comput. Sci. **2**(1), 168–179 (2016)
28. Farhaoui, Y.: Big data analytics applied for control systems. Lecture Notes in Netw. Syst. **25**, 408–415 (2018). https://doi.org/10.1007/978-3-319-69137-4_36
29. Farhaoui, Y. et al.: In: Big Data Mining and Analytics, vol. 5(4), pp. I–II. (2022). https://doi.org/10.26599/BDMA.2022.9020004
30. Alaoui, S.S., Farhaoui, Y.: Hate speech detection using text mining and machine learning. Int. J. Decis. Support Syst. Technol. **14**(1), 80 (2022). https://doi.org/10.4018/IJDSST.286680
31. Alaoui, S.S., Farhaoui, Y.: Data openness for efficient e-governance in the age of big data. Int. J. Cloud Comput. **10**(5–6), 522–532 (2021). https://doi.org/10.1504/IJCC.2021.120391
32. El Mouatasim, A., Farhaoui, Y.: Nesterov step reduced gradient algorithm for convex programming problems. Lecture Notes in Netw. Syst. **81**, 140–148 (2020). https://doi.org/10.1007/978-3-030-23672-4_11
33. Tarik, A., Farhaoui, Y.: Recommender system for orientation student. Lecture Notes in Netw. Syst. **81**, 367–370 (2020). https://doi.org/10.1007/978-3-030-23672-4_27
34. Sossi Alaoui, S., Farhaoui, Y.: A comparative study of the four well-known classification algorithms in data mining. Lecture Notes in Netw. Syst. **25**, 362–373 (2018). https://doi.org/10.1007/978-3-319-69137-4_32
35. Farhaoui, Y.: Teaching computer sciences in Morocco:an overview. IT Professional, vol. 19(4), pp. 12–15, 8012307 (2017). https://doi.org/10.1109/MITP.2017.3051325
36. Farhaoui, Y.: Securing a local area network by IDPS open source. Proc. Comput. Sci. **110**, 416–421 (2017). https://doi.org/10.1016/j.procs.2017.06.106

Prediction of Student's Academic Performance Using Learning Analytics

Sakinat Oluwabukonla Folorunso[1], Yousef Farhaoui[2], Iyanu Pelumi Adigun[1], Agbotiname Lucky Imoize[3], and Joseph Bamidele Awotunde[4](✉)

[1] Department of Mathematical Sciences, Olabisi Onabanjo University, Ago-Iwoye, Ogun State, Nigeria
sakinat.folorunso@oouagoiwoye.edu.ng

[2] STI Laboratory, T-IDMS, Department of Computer Science, Faculty of Sciences and Techniques Errachidia, Moulay Ismail University of Meknes, Meknes, Morocco
y.farhaoui@fste.umi.ac.ma

[3] Department of Electrical and Electronics Engineering, Faculty of Engineering, University of Lagos, Akoka, Lagos 100213, Nigeria
aimoize@unilag.edu.ng

[4] Department of Computer Science, Faculty of Information and Communication Sciences, University of Ilorin, Ilorin 240003, Kwara State, Nigeria
awotunde.jb@unilorin.edu.ng

Abstract. Academic performance is the assessment of knowledge gained by students for a particular period. Therefore, predicting student's academic performance is a vital part of quality assurance in higher learning. As such, their prediction becomes pertinent to higher institutions as it can be used to monitor student's progress and forestall the risk of students derailing from their academic paths. This study proposes a hybrid of linear regression and a k-means clustering model to predict student's academic performance. (The gain of these models is that useful and interesting patterns can be discovered from the data provided)? The proposed hybrid model has been tested on the Covenant University, Nigeria, student dataset. The main features of the dataset include department, gender, Secondary School Grade Point Average (SGPA), Cumulative Grade Point Average (CGPA) for each academic session, and final CGPA. The linear regression model results showed improved performance with a coefficient of determination (R2), Mean Squared Error (MSE), and F-statistic value of 0.9842, 0.00077, and 10448.24, respectively. In contrast, the Clustering model showed good measures with between-cluster error, within-group error, and variance of 462.27, 197.12 and 0.70, respectively. This new hybrid approach displays expressively improved performance. The analysis of the prediction performance indicates that the proposed ensemble scheme performs well. At the same time, the percentage of correctly classified instances is increased as new performance attributes are added during the academic year. Additionally, the investigation was carried out on how soon predictions could be made to offer a timely intervention and enhance students' performance.

Keywords: Academic performance · Prediction · Learning analytics · Regression · Clustering

Y. Farhaoui et al. (Eds.): ICAISE 2023, LNNS 837, pp. 314–325, 2024.
https://doi.org/10.1007/978-3-031-48465-0_41

1 Introduction

The academic performance of students is a vital aspect of higher institutions of learning which makes its early prediction very important [1]. The performance of a student at the point of entry (first year) has a great influence on the student's overall performance at the point of graduation. This often makes some people refer to the first year of undergraduate students as a "make or break" year [2]. Early prediction of student's academic performance will help to set the tone for improvement and ensure quality assurance. Predicting students' performance will also assist academic institutions in planning and making decisions before the students reach their final year [3]. The application of data mining techniques is one of the most popular means of evaluating and analyzing student's academic performance. This form of evaluation is often carried out based on certain factors. The work [4, 20] identified some of these factors, which include Cumulative Grade Point Average (CGPA), demographic data (age, gender, family background, and disability), psychometric factors (students' interest in the course of study, support from family, behavioral study pattern), extracurricular activities, high school background, and social interaction network. According to [5, 21], CGPA appeared to be to most influential factor in determining how students will survive during their studies. There are limited studies on the Educational Data Mining (EDM) approach for predicting students' performance in Nigeria. Therefore, this study proposes a hybrid of regression and clustering models to analyze and predict students' academic performance from data collected from one Nigerian Private University [6, 22].

EDM is an emerging subfield of data mining dealing with the development of methods used to analyze student's academic behavior and predicts their performance [7, 23]. Predicting student's academic performance has become an essential and challenging topic in the educational field. It is regarded as one of the most interesting and well-studied aspects of EDM. Classification and regression are familiar machine learning methods successfully applied to the educational field to detect student's failure or dropout rates [8, 24]. Regression is used in the case of a continuous output variable. A popular regression model used with learning analytics is Linear Regression (LR) [9, 25].

This study models a regression algorithm to predict student performance for each of the student's levels. A clustering model was built for the proper grouping of students. Also, a regression model is proposed to predict students' performance in the five clusters.

2 Related Work

Several studies have been conducted on using data mining to extract meaningful information to help predict students' performance. Lau et al. [10] applied Artificial Neural Network (ANN) model to predict the student's CGPA using the data about their socio-economic background and entrance examination results of undergraduate students from a Chinese University.

Adekitan and Salau [11] proposed a predictive analysis approach to predict the final Cumulative Grade Point Average (CGPA) of engineering students in a Nigerian University based on variables such as the program of study, the year of entry, and the Grade Point Average (GPA), for the first three years of study on Konstanz Information Miner

(KNIME) model. Six (6) data mining models: Probabilistic Neural Network (PNN) based on the DDA (Dynamic Decay Adjustment), Random Forest (RF), Decision Tree (DT), Naïve Bayes (NB), Tree Ensemble Predictor and Logistic Regression (LR) models were evaluated on 1841 students from 2002 to 2014, across seven engineering departments based on student's grades classified as first class ($CGPA > 4.50$); second class upper division ($3.50 \leq CGPA \leq 4.49$); second class lower division ($2.50 \leq CGPA \leq 3.49$) and third class ($1.50 \leq CGPA \leq 2.49$).

The results showed that LR outperformed other models with a classification accuracy of 89.15%, while the pure quadratic regression model obtained $R^2 = 0.975$. Oladokun et al. [12, 26] identified several factors influencing students' performance. These include ordinary-level subject grades, age on admission, parental background, type and location of secondary school attended, and gender. They developed a predictive model based on Multilayer Perceptron (MLP) using data from five graduation years in an Engineering Department at the University of Ibadan, Nigeria. Their result showed that the ANN model could predict performance with an accuracy of 74%. However, the model only considered records gathered for a single department and did not consider other departments in the University. Ramesh et al. [13] developed a model based on some input variables obtained through a questionnaire. The authors aimed to identify weak students early enough and find a means of assisting such students. MLP algorithm was identified as the best-suited algorithm to predict students' performance with an accuracy of 72%. Guo et al. [14] used a deep learning approach to develop a classification model for predicting student's performance.

Devasia et al. [3, 27] developed a web-based application that used the Naïve Bayes technique to extract meaningful information. The system takes students' academic history, including class tests, seminars, and assignments, as input and predicts their performance on a semester basis. Their authors conducted an experiment was conducted on 700 students in a university located at Mysuru in the School of Arts and Sciences. Students' performance was rated as poor, average, good, and excellent. The result showed that those with "good" performance had the highest percentage. Asif et al. [15, 28] used data mining techniques to study the performance of undergraduate students at the end of a four-year study program. The students were classified as either low achieving (with low marks) or high achieving (with high marks). Their results showed that low-achieving groups of students could be given timely warnings while high-achieving students could be exposed to opportunities that benefit them. Yang et al. [16, 29] aimed to predict students' performance early, thereby providing them with timely assistance.

3 Materials and Methods

In this study, there are five steps involved in the process of predicting student's academic performance. These include data collection, pre-processing, visualization, model training, and performance evaluation. These steps are described in Fig. 1.

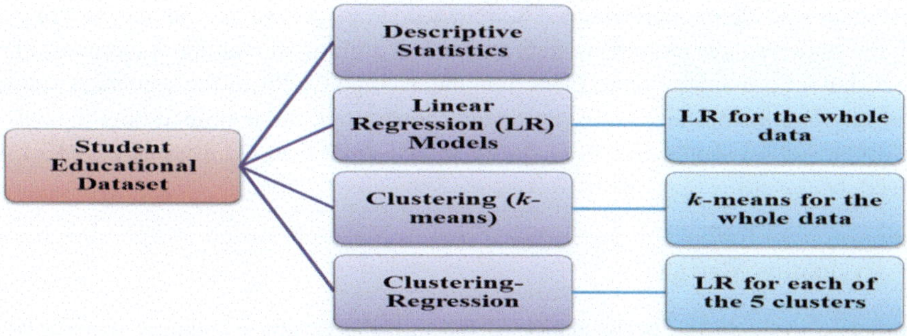

Fig. 1. Model of the academic performance

3.1 Statistical Analysis

Table 2 shows the descriptive statistics of the collected dataset. Statistical analysis performed in this study includes descriptive statistics and Pearson correlation coefficients, shown in Table 1. Pearson correlation coefficients are calculated to measure the linear relationships of all CGPAs across all levels. For example, a student with SGPA with the least standard deviation value of 0.6162. All skewness lies between -0.5 and 0.5, indicating approximate symmetric distributions. With positive skewness of SGPA (0.1500), most students in secondary school scored less than the average SGPA of 3.1196. In comparison, the academic performance shows that the majority scored more than the average in 100L (3.6361), 200L (3.3217), 300L (3.4186), and 400L (3.5325). This conclusion is also reflected in the overall performance (−0.25 CGPA) of the student's academic performance. All the excess kurtoses, CGPA (−0.59), SGPA (−0.57), 100L CGPA (−0.37), 200L CGPA (−0.61), 300L CGPA (−0.53) and 400L CGPA (−0.28), showed low kurtosis (Platykurtic). This shows the presence of less serious outliers or extreme values in

Table 1. Descriptive statistics

Statistics	CGPA	SGPA	CGPA100	CGPA200	CGPA300	CGPA400
Count	3046	3046	3046	3046	3046	3046
Mean	3.4948	3.1196	3.6361	3.3217	3.4186	3.5325
Standard deviation	0.6916	0.6162	0.6793	0.7825	0.8585	0.8022
Minimum	1.5200	1.4600	1.5700	1.1700	0.6300	0.0000
First quartile	3.0000	2.6600	3.1800	2.7600	2.8100	3.0000
Median	3.5600	3.0600	3.6900	3.3400	3.5100	3.6200
Third quartile	4.0100	3.5700	4.1500	3.9200	4.1000	4.1500
Maximum	4.9900	4.9300	5.0000	5.0000	5.0000	5.0000
Skewness	−0.2500	0.1500	−0.3700	−0.1300	−0.4000	−0.4700
Excess Kurtosis	−0.5900	−0.5700	−0.3700	−0.6100	−0.5300	−0.2800

the student's academic performance both in secondary school and on campus. In general, the students have more consistent and non-disturbing (no outlier or extreme value) scores, but with majority scored less than the average CGPA in the secondary school. However, the students have more consistent, coordinated, and non-disturbing (no outlier or extreme value) scores and the majority scored more than the average CGPA in the University.

3.2 Evaluation Metrics

In order to evaluate the performance of the models, certain evaluation metrics were used. R-Squared, MSE, and F-statistic metrics were used to evaluate the performance of the adapted linear regression model [31]. On the other hand, Analysis of Variance (ANOVA), including p-value, Sum Squared Error (SSE), and F-statistics, was used to evaluate the clustering model. The evaluation metrics used in this study are described in Table 2.

Table 2. Evaluation Metrics

Metric	Formula	Remark
R-Squared $\left(R^2\right)$	$1 - \frac{\sum_{i=1}^{n}(y_i - \hat{y}_i)^2}{\sum_{i=1}^{n} y_i^2}$	R^2 measure is used to determine the goodness of fit of a regression model. R^2 values range between 0 and 1, where values nearer to 0 signify a poor fit while values nearer to 1 signify a perfect fit
SSE (Sum Squared Error)	$SSE = \sum_{i-1}^{n}(y_i - \hat{y}_i)$ y_i—the observed value $\hat{y}$—the value estimated by the regression line	SSE measures the difference between actual data and the values predicted by an estimation model. So, a small SSE value indicates a tight fit of the model to the data
MSE (Mean Squared Error)	$MSE = \frac{1}{n}\sum_{i=1}^{n}(y_i - \tilde{y}_i)^2$	MSE expresses the closeness of a set of points to a regression line. So, it measures the quality of an estimator, and the smaller the value, the better
StdErr (Standard Error)	$Standard\ Deviation \sigma = \sqrt{\frac{\sum_{i=1}^{n}(x_i - \overline{x})^2}{n-1}}$ Variance $= \sigma^2$ Standard error $\sigma_{\overline{x}} = \frac{\sigma}{\sqrt{n}}$ Where $\overline{x} = the sample's mean$ n = the sample size	StdErr measures the accuracy with which a sample represents a population

(*continued*)

Table 2. (*continued*)

Metric	Formula	Remark
p-value (significance)	$Z = \frac{\bar{x}-\mu_o}{\frac{\sigma}{\sqrt{n}}}$	The P-value is the probability, under the null hypothesis, about the unknown distribution of the random variable for the variate to be observed as a value equal to or more extreme than the value observed
Analysis of variance (ANOVA)	$F = \frac{MST}{MSE}$ F = ANOVA Coefficient MST = Mean sum of squares due to treatment MSE = Mean sum of squares due to error	ANOVA is a collection of statistical models and their associated estimation procedures (such as the "variation" among and between groups) used to analyze the differences among group means in a sample
SSE	$\sum_{i=1}^{n} (x_i - \bar{x})^2$	SSE is the fraction of the between-group sum of squares to the model degrees of freedom. The between-group sum of squares is a measure of the variation between cluster means
F-Statistics	$\frac{\sigma 1^2}{\sigma 2^2}$	F-statistic is the fraction of the between-group variance to the total variance. The higher the F-statistic value, the better the corresponding variable for distinguishing between clusters
t-value	$t = \frac{\bar{x}-\mu}{\frac{S}{\sqrt{n}}}$	The t-value is the calculated difference represented in units of standard error. The greater the magnitude of t, the greater the evidence against the null hypothesis
The between-group sum of squares	$SS_B = \sum N_A (\overline{X}_A - \overline{X}_G)^2$	This measure quantifies the distance between clusters as a sum of squared distances between each cluster's center mean value, weighted by the number of data points assigned to the cluster and the center of the data set. So, a large value indicates a better separation between clusters
The within-group sum of squares	$SS_w = \sum (X_i - \overline{X}_A)^2$	This measure quantifies the cohesion of clusters as a sum of squared distances between the center of each cluster and the individual points in the cluster. So, a small value indicates better cohesion and a better model

3.3 Clustering Model

Clustering analysis is an unsupervised learning that divides data points into groups/clusters where data points in each cluster are more alike to one another than they are to data points in other clusters. A cluster is a gathering of similar data points [17]. Figure 2 presents the k-means algorithm for clustering; the algorithm's goal is to minimize the objective function (squared error function) given by Eq. (2).

$$J(V) = \sum_{i=1}^{c} \sum_{j=1}^{c_i} \left(\|x_i - v_j\| \right)^2 \tag{1}$$

where $\|x_i - v_j\|$ = Euclidean distance between x_i *and* v_j, c_i is the number of data points in ith cluster. 'c' is the number of cluster centers.

Input: Set of data points $X = \{x_1, x_2, x_3 \cdots x_n\}$

Output: Set of cluster centers $V = \{v_1, v_2, v_3 \cdots v_c\}$

Clustering Algorithm: *k*-means

i. Randomly select 'c' cluster centres.
ii. Compute the distance between each data point and cluster centres.
iii. Allocate the data point with the minimum distance to the cluster centre.
iv. Re-compute the new cluster centre with: $v_i = (1/c_i) \sum_{j=1}^{c_i} x_i$
v. where, 'c_i' represents the number of data points in i^{th} cluster.
vi. Re-compute the distance between each data point and newly obtained cluster centres.
vii. If no data point was reallocated then stop, else repeat from step iii.

Fig. 2. The k-means algorithm for clustering

4 Results and Discussions

The experiments were performed with the following three major stages with Tableau Desktop. For the first stage, regression analysis was performed on all variables individually (CGPA100, CGPA200, CGPA300, CGPA400, SGPA) and then on all dataset features. This is necessary to predict a student's performance at any level. For the second stage, clustering analysis is performed to group students with similar performances. The third stage simply built a regression model on all individual clusters. The Linear Regression (LR) modeled all instances (3046) in the observations, and there were no filtered observations. The model's degree of freedom is 4 while the residual degrees of freedom (DF) is 3042 with p-value (significance): < 0.0001.

4.1 Stage 1: Regression Analysis Modelling

In the first stage of experiments, a linear trend model is computed for CGPA given CGPA100, CGPA200, CGPA300, CGPA400, and SGPA. The LR model may be significant at $p \leq 0.05$, , p-value (significance): < 0.0001. The final CGPA is the dependent variable, while SGPA, CGPA100, CGPA200, CGPA300, and CGPA400 are made independent variables. Also, a regression model was built for each program code (Prog Code) to predict students' performances.

Tableau Desktop Software from the above-mentioned model was used in this study. The regression results were obtained for the individual trend lines for each variable (SGPA, CGPA100, CGPA200, CGPA300, and CGPA400) and information about each trend Line in the views. Tables 3 and 4 also depict the most statistically significant and list coefficient statistics for individual trend Lines. Each row describes each coefficient in each trend Line model. For example, a linear model with an intercept requires two rows for each trend Line. Each line's p-value and the Degree of Freedom (DF) span all the coefficient rows in the Line column. The DF column shows the residual degrees of freedom available during the estimation of each line. The Standard Error (StdErr) column measures the spread of the sampling distribution of the coefficient estimate. This error shrinks as the quality and quantity of the information used in the estimate grows. The t-value column tests the null hypothesis that the true value of the coefficient is zero. The p-value describes the probability of observing a t-value that large or larger in magnitude if the true value of the coefficient is zero [18]. All the variables contribute significantly to the rejection of the null hypothesis. The regression line for SGPA scores is shown in (3):

$$CGPA = 0.3384 * SGPA + 1.9371 \tag{3}$$

For every unit increment in the SGPA of students, there will be an increase of 0.3384 in the Final CGPA, provided all the other variables are constant. From the result obtained, all the models are statistically highly significant. For Table 3, the p-value of the line is the same for the term CGPA. The p-values of the last column mark the p-values of all the coefficients in the ANOVA model.

Table 3 shows the evaluation of individual trend lines. The evaluation result obtained in Table 6 shows all the linear trend lines models have a good fit: CGPA100 (0.6278), CGPA200 (0.8293), CGPA300 (0.8219), and CGPA400 (0.7858) except SGPA (0.1443). CGPA200, with an R2 value of approximately 83%, indicated that 83% of variations in this dependent variable Y are explained by the independent variables present in the linear model. The Standard Error (StdErr) measures the accuracy with which a sample represents a population. So, the CGPA200 Linear model has the lowest value of (0.3233) and obtained the optimum measure values. In general, all are relatively low values that indicate the extent of variation found around any particular estimate of the mean.

4.2 Linear Regression Model

The overall regression model was built on WEKA [19] platform, a machine-learning suite. The result of the evaluation of the model is illustrated in Table 4.

Table 3. Evaluation of Individual Linear Trend lines for all CGPA

[1]Model Plot	R^2	MSE	SSE	StdErr
SGPA	0.1443	0.3250	989.253	0.5701
CGPA100	0.6278	0.1718	522.858	0.4145
CGPA200	0.8293	0.1045	318.186	0.3233
CGPA300	0.8219	0.1314	399.827	0.3624
CGPA400	0.7858	0.1379	419.736	0.3714

Table 4. Evaluation result of the linear regression model

DF	R2	MSE	F-statistic
3027	0.9842	0.00077	10448.24

4.3 Stage 2: Clustering Modelling

The clustering modeling and analysis were also performed in Tableau Desktop software [18]. The variables serving as input to the clustering were CGPA, CGPA100, CGPA200, CGPA300, CGPA400, and SGPA, with CGPA as the clustering class. All variable values were normalized. Five (5) clusters were created based on the classes of degrees: First class ($CGPA > 4.50$); second class upper division ($3.50 \leq CGPA \leq 4.49$); second class lower division ($2.50 \leq CGPA \leq 3.49$); third class ($1.50 \leq CGPA \leq 2.49$) and Pass ($0.00 \leq CGPA \leq 1.49$). Table 5 describes the result of the clustering analysis. Column 1 lists the 5 clusters. Column 2 describes the number of items in individual clusters in the clustering. The centers column shows the mean values within each cluster. The between-group sum of squares (BSSE) is the error between different clusters, while the Within-group sum of squares (WSSE) is the error within different clusters. The sum of the squares error column totals the Within- and Between- square errors. The variance of the model, which is the fraction of the between-group sum of squares and the total sum of squares, is 0.7. This value indicates a better model as it is closer to 1.

For example, in Table 5, for Cluster 1, 529 students were grouped with variables CGPA, CGPA100, CGPA200, CGPA300, CGPA400, and SGPA with mean values of 3.1564, 3.6668, 2.9329, 2.8521, 3.1135 and 3.3878 respectively. For all variables, Cluster 5 had the highest mean value. The errors between- and within groups of clusters are 462.27 and 197.12, respectively, the variance is 0.70 while the standard error (SE) is 659.39.

[1] The model may be significant at $p <= 0.05$. The factor Clusters may be significant at $p <= 0.05$.

Table 5. k-means Clustering

Clusters	Number of Items	Centers					
		CGPA	CGPA100	CGPA200	CGPA300	CGPA400	SGPA
Cluster 1	529	3.1564	3.6668	2.9329	2.8521	3.1135	3.3878
Cluster 2	892	3.8563	3.8895	3.6924	3.8589	3.9158	3.1465
Cluster 3	514	2.4224	2.7687	2.2371	2.2011	2.3975	2.7796
Cluster 4	545	3.2777	3.1723	3.062	3.3391	3.4494	2.5589
Cluster 5	566	4.4243	4.4424	4.3359	4.4363	4.4308	3.6755

4.4 Regression Analysis of Individual Clusters

The Linear Regression modeled all instances (3046) in the observations, with no filtered observations. The model degree of freedom is 10 while the residual degree of freedom (DF) is 3036 with p-value (significance): <0.0001. The linear trend lines for CGPA for the five clusters are presented in Table 6.

Table 6. Linear Trend lines equation for CGPA given the 5 different clusters

[2]Model plot	R2	MSE	SSE	StdErr
SGPA	0.4007	0.2282	692.7470	0.4777
CGPA100	0.7046	0.1367	414.9590	0.3697
CGPA200	0.8367	0.1003	304.3720	0.3166
CGPA300	0.8383	0.1196	362.9770	0.3458
CGPA400	0.7948	0.1325	402.1350	0.3639

5 Conclusion

Recently, great importance has been placed on prediction of student's performance prediction. Innumerable noteworthy studies and research have demonstrated that learning analytics can be an efficient tool for higher institutions to estimate students' academic success with sufficient accuracy. This study paper presented linear regression, clustering, and clustering regression models to predict student academic performance. The Clustering technique was demonstrated with a k-means clustering algorithm where $k = 5$. Then, a linear regression model was also built on the five different clusters. The regression model was measured based on R2, Mean Squared Error (MSE), Standard

[2] The model may be significant at $p <= 0.05$. The factor Clusters may be significant at $p <= 0.05$.

Error (StdErr), and F-Statistics metrics. For individual regression models for each of the variable GPAs at all levels were also built. The results revealed that the SGPA had the lowest R2 and highest MSE and StdErr values and vice versa for CGPA200. It could be concluded that often, the CGPA200 determines the final CGPA of students. Also, the k-means clustering model with 5 clusters was good, with a variance value of 0.70. The linear models built for each cluster also indicate CGPA200. Then, this Clustering-Regression algorithm serves as a good yardstick to monitor the headway of students' academic performance in higher institutions. It also enhances the decision-making by academic planners to monitor the candidates' performance in each academic.

References

1. Abiodun, K.M., Adeniyi, E.A., Aremu, D.R., Awotunde, J.B., Ogbuji, E.: Predicting students performance in examination using supervised data mining techniques. In: Informatics and Intelligent Applications: First International Conference, ICIIA (2021), Ota, Nigeria, November 25–27, 2021, Revised Selected Papers, pp. 63–77. Cham, Springer International Publishing (2021)
2. Ogundokun, R.O., Daniyal, M., Misra, S., Awotunde, J.B.: Student's perspective on online teaching in higher institutions during COVID-19 pandemic'. Int. J. Networking Virtual Organ. **25**(3–4), 308–332 (2021)
3. Devasia, T., Vinushree, T.P., Vinayak, H.: Prediction of student's performance using educational data mining. In: 2016 International Conference on Data Mining and Advanced Computing (SAPIENCE) (2016)
4. Shahiri, A.M., Husain, W.: A review on predicting student's performance using data mining techniques. Proc. Comput. Sci. **72**, 414–422 (2015)
5. Christian, T.M., Ayub, M.: Exploration of classification using NBTree for predicting student's performance. In: 2014 International Conference on Data and Software Engineering (ICODSE) (2014)
6. John, T.M., et al.: The role of gender on academic performance in STEM-related disciplines: data from a tertiary institution. Data Brief **18**, 360–374 (2018)
7. Romero, C., Ventura, S.: Educational data mining: a review of the state of the art. IEEE Trans. Syst. Man and Cybernet. Part C (Applications and Reviews) **40**(6), 601–618 (2010)
8. Silva, C., Fonseca, J.: Educational data mining: a literature review. In: Rea, Á. (ed.) Europe and MENA Cooperation, Advances in Information and Communication Technologies, pp. 87–94. Springer International Publishing (2016)
9. Oladipo, I.D., Awotunde, J.B., AbdulRaheem, M., Ige, O.O., Balogun, G.B., Tomori, A.R., Taofeek-Ibrahim, F.A.: An improved course recommendation system based on historical grade data using logistic regression. In: Applied Informatics: Fourth International Conference, ICAI 2021, Buenos Aires, Argentina, October 28–30, 2021, Proceedings, vol. 4. pp. 207–221. Springer International Publishing (2021)
10. Lau, E.T., Sun, L., Yang, Q.: Modelling, prediction and classification of student academic performance using artificial neural networks. SN Appl. Sci. **1**, 982 (2019). https://doi.org/10.1007/s42452-019-0884-7
11. Adekitan, A.I., Salau, O.: The impact of engineering student's performance in the first three years on their graduation result using educational data mining. Heliyon **5**, e01250 (2019)
12. Oladokun, V.O., Adebanjo, A.T., Charles-Owaba, O.E.: Predicting student's academic performance using artificial neural network: a case study of an engineering course (2008)
13. Ramesh, V. Parkavi, P., Ramar, K.: Predicting student performance: a statistical and data mining approach. Int. J. Comput. Appl. **63**(8) (2013)

14. Guo, B., Zhang, R., Xu, G., Shi, C., Yang, L.: Predicting student's performance in educational data mining. In: 2015 International Symposium on Educational Technology (ISET) (2015)
15. Asif, R., Merceron, A., Ali, S.A., Haider, N.G.: Analyzing undergraduate student's performance using educational data mining. Comput. Educ. 113, 177–194 (2017)
16. Yang, S.J., Lu, O.H., Huang, A.Y., Huang, J.C., Ogata, H., Lin, A.J.: Predicting student's academic performance using multiple linear regression and principal component analysis. J. Inform. Process. **26**, 170–176 (2018)
17. Govindasamy, K., Velmurugan, T.: Analysis of student academic performance using clustering techniques. Int. J. Pure and Appl. Mathem. **119**(15), 309–323 (2018)
18. Tableau.: Tableau desktop and web authoring help (2018). [Online]. Available https://onlinehelp.tableau.com/current/pro/desktop/en-us/help.html#default.html. (Accessed 17 Dec 2019)
19. Bouckaert, R.R., Frank, E., Hall, M.A., Holmes, B., Reutemann, P., Witten, I.A.: WEKA—experiences with a java open-source project. J. Mach. Learn. Res. **11**, 2533–2541 (2010)
20. Farhaoui, Y.: Design and implementation of an intrusion prevention system. Int. J. Netw. Secur. **19**(5), 675–683 (2017). https://doi.org/10.6633/IJNS.201709.19(5).04
21. Farhaoui, Y., et al.: In: Big Data Mining and Analytics, vol. 6(3), pp. 1–2. (2023). https://doi.org/10.26599/BDMA.2022.9020045
22. Farhaoui, Y.: Intrusion prevention system inspired immune systems. Indonesian J. Electri. Eng. Comput. Sci. **2**(1), 168–179 (2016)
23. Farhaoui, Y.: Big data analytics applied for control systems. Lecture Notes in Netw. Syst. **25**, 408–415 (2018). https://doi.org/10.1007/978-3-319-69137-4_36
24. Farhaoui, Y. et al.: In: Big Data Mining and Analytics, vol. 5(4), pp. 1–2. (2022). https://doi.org/10.26599/BDMA.2022.9020004
25. Alaoui, S.S., Farhaoui, Y.: Hate speech detection using text mining and machine learning. Int. J. Decis. Support Syst. Technol. **14**(1), 80 (2022). https://doi.org/10.4018/IJDSST.286680
26. Alaoui, S.S., Farhaoui, Y.: Data openness for efficient e-governance in the age of big data. Int. J. Cloud Comput. **10**(5–6), 522–532 (2021). https://doi.org/10.1504/IJCC.2021.120391
27. El Mouatasim, A., Farhaoui, Y.: Nesterov step reduced gradient algorithm for convex programming problems. Lecture Notes in Netw. Syst. **81**, 140–148 (2020). https://doi.org/10.1007/978-3-030-23672-4_11
28. Tarik, A., Farhaoui, Y."Recommender System for Orientation Student" Lecture Notes in Networks and Systems, 2020, 81, pp. 367–370.https://doi.org/10.1007/978-3-030-23672-4_27
29. Sossi Alaoui, S., Farhaoui, Y.: A comparative study of the four well-known classification algorithms in data mining. Lecture Notes in Netw. Syst. **25**, 362–373 (2018). https://doi.org/10.1007/978-3-319-69137-4_32
30. Farhaoui, Y.: Teaching computer sciences in Morocco: an overview. IT Professional **19**(4), 12–15, 8012307 (2017). https://doi.org/10.1109/MITP.2017.3051325
31. Farhaoui, Y.: Securing a local area network by IDPS open source. Proc. Comput. Sci. **110**, 416–421 (2017). https://doi.org/10.1016/j.procs.2017.06.106

Comparative Study for Predicting Melanoma Skin Cancer Using Linear Discriminant Analysis (LDA) and Classification Algorithms

Abidemi Emmanuel Adeniyi[1], Joyce Busola Ayoola[2], Yousef Farhaoui[3], Joseph Bamidele Awotunde[4](✉), Agbotiname Lucky Imoize[5], Gbenga Rasheed Jimoh[2], and Devine F. Chollom[4]

[1] Department of Computer Science, College of Computing and Communication Studies, Bowen University, Iwo, Nigeria
abidemi.adeniyi@bowen.edu.ng

[2] Department of Computer Science, Landmark University, Omu-Aran, Kwara State, Nigeria
ayoola.joyce@lmu.edu.ng, jimoh_rasheed@unilorin.edu.ng

[3] STI Laboratory, T-IDMS, Department of Computer Science, Faculty of Sciences and Techniques Errachidia, Moulay Ismail University of Meknes, Meknes, Morocco
y.farhaoui@fste.umi.ac.ma

[4] Department of Computer Science, Faculty of Information and Communication Sciences, University of Ilorin, Ilorin 240003, Kwara State, Nigeria
awotunde.jb@unilorin.edu.ng, chollom.devine@lmu.edu.ng

[5] Department of Electrical and Electronics Engineering, Faculty of Engineering, University of Lagos, Akoka, Lagos 100213, Nigeria
aimoize@unilag.edu.ng

Abstract. Early detection and prediction of melanoma can help control large-scale outbreaks and reduce the transmission of epidemics in rapid response to serious public health events. Melanoma is one of the most dangerous types of skin cancer in the world which is responsible for the greatest number of skin cancer-related deaths. As predicted by the World Health Organization (WHO), the number of cases of diagnosed melanoma skin cancer will rise by 18% from 287,723 to 340,271 in 2025, and the number of deaths caused by melanoma will increase by 20%. WHO also predicts there will be a 20% increase in death from 60,712 in 2018 to 72,886 in 2025 and will reach 105,904, a 74% increase by 2040. However, proffering efficient methods for the early prediction can prevent further spread and melanoma-related deaths, the use of classification algorithms is one of such efficient methods for melanoma prediction. Various Machine Learning algorithms have been used in prediction of diseases which include Support Vector Machines (SVM), K-Nearest Neighbour (KNN), Naïve Bayesian, Convolutional Neural Networks (CNN) and other classification algorithms. Therefore, this study used Linear Discriminant Analysis (LDA) for feature extraction, SVM and KNN as classification algorithms for the prediction of melanoma skin cancer. The result obtained shows that both algorithm perform excellently well when optimized with LDA with an accuracy of 89% and 86% respectively. The study proposed that other algorithms/optimization techniques can be introduced to improve the robustness of the prediction.

Y. Farhaoui et al. (Eds.): ICAISE 2023, LNNS 837, pp. 326–338, 2024.
https://doi.org/10.1007/978-3-031-48465-0_42

Keywords: Melanoma · Skin cancer · Classification algorithms · Machine learning · Feature extraction

1 Introduction

In more recent times, skin cancer has become a major concern worldwide, particularly melanoma cancer which is considered to be the deadliest form of skin cancer. The cases of melanoma have increased by almost 50% to over 287,000, this estimates melanoma-related deaths to more than 60,000 per year [11]. In 2012, 67,753 persons in the United States were diagnosed with melanoma, among which 9251 persons were recorded to have died from the skin disease [2, 19]. The incidence rates of melanoma skin cancer increased by 44% from 2008 up until 2018 with deaths increasing by 32% [11]. According to Perera et al. [14], melanoma also referred to as malignant melanoma, is a tumor gotten from melanocytes that were once healthy but changed and grew out of control. Melanoma can be found in animals such as cats and dogs but it occurs very rarely [18]. In humans, the skin is the most common area of the body affected by melanoma. Men are more likely to develop melanoma skin cancer than women and are more likely to die from melanoma than women. However, early detection of this cancer increases the survival chances of a patient [14].

Identifying people at high risk is also very important for the development of preventive methods. Although sun exposure is the most significant risk factor of the evolution of melanoma, other factors have been associated with an increased risk of developing melanoma such as having a history of melanoma in the family, exposure to the sun, having a light-skinned complexion and being exposed to UV light at tanning salons [15]. The most significant risk factor of melanoma is related to the magnitude of UV radiation exposure which means that proper protection from the rays of the sun plays a very important role in preventing melanoma, consistent use of sunscreen serves as a protective measure against penetration of sun rays [3]. In countries such as Australia, 74% of the population makes use of sunscreen and it has been observed that the incidence rate of melanoma has significantly reduced [8]. The average age at which people get diagnosed with melanoma is at 65 years Melanoma in women is more common in younger people and melanoma in men is more common in older people [12].

In detecting melanoma, the signs of warning to look for in moles include itching, changes in color, and bleeding of the mole. Within 6 weeks, melanoma is capable of putting a life at risk because it grows quickly and spread to the other parts of the body including the vital organs. It can be present in any area, even those which have not been in the sun. Melanoma is serious and can be fatal and also cause death if left untreated. Some of the symptoms that dermatologists consider for detecting melanoma skin cancer include the asymmetry of the lesion shape, different color variations, specific diameters, and irregular borders [13].

Therefore, the use of machine learning techniques and algorithms in classifying the lesion from images can help the dermatologist to achieve higher performance [7, 13].

2 Material and Method

The system developed uses supervised machine learning algorithms, Support Vector Machine and K-Nearest Neighbour to make classifications of skin lesion images, in order to identify melanoma. Datasets are obtained which are then trained and classified passed on into the model [5]. The dataset is pre-processed by using Linear Discriminant Analysis that filters the data. The characteristics of the data set can be categorized using Support Vector Machine and K-Nearest Neighbour algorithms, and the results are compared with precision, accuracy, specificity, f1 score and computational time [7].

2.1 The Stages Below Provide the Methodological Details:

1. Use the Linear Discriminant Analysis as a feature extraction technique.
2. Use the Classification algorithms, Support Vector Machine and K-Neighbour Neighbour [20, 21].
3. Analyse the results in accuracy, f1 score, precision, f1 score, sensitivity and computational time.
4. Compare the results of the classification algorithms.

The flow diagram for the proposed system is shown in Fig. 1.

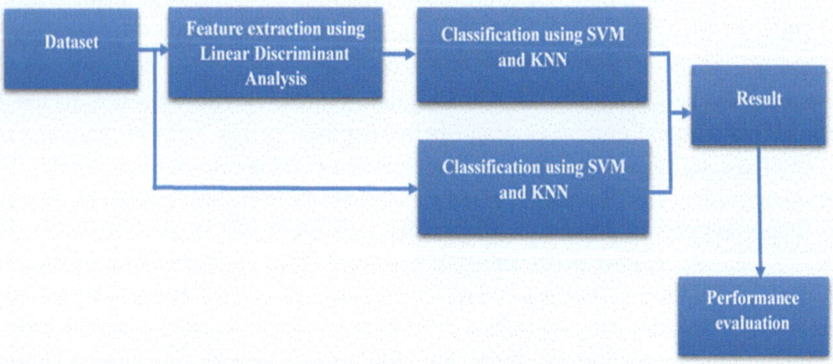

Fig. 1. Flow diagram of the proposed model

From the flow diagram of the proposed model, the process is of two stages. The first process is by passing the data through feature extraction technique (LDA) before being classify. While the second stage is by passing the data directly to the classifiers (SVM and KNN) without passing through (LDA) feature extraction [22, 23].

2.2 Dataset

A dataset from the Kaggle (https://www.kaggle.com/kmader/skin-cancer-mnist-ham 10000) is used in this study. This dataset includes image samples of skin cancer and predicts what type of skin cancer it is. The dataset includes 10,000 occurrences, 7 classes and no values that are missing [24, 25].

The dataset contains 7 different classes to predict each type of the skin cancer. The 7 classes include as shown in Table 1:

Table 1. Classes of dataset used

Classes	Attributes dataset type
0	Actinic keratoses and intraepithelial carcinoma ('akiec')
1	Basal cell carcinoma ('bcc')
2	Benign keratosis-like lesions ('bkl')
3	Dermatofibroma ('df')
4	Melanocytic nevi ('nv')
5	Pyogenic granulomas and hemorrhage ('vasc')
6	Melanoma ('mel')

Table 1 display the data consists of melanoma dataset with 10,000 samples, as well as attributes such as the image name of the skin images and dataset type.

2.3 Model Evaluation of Algorithm

After the model has been fully implemented, the performance of the model will be evaluated using some specific metrics.

To understand these matrices, it is important to know the following terms:

i. True positive: When the model correctly predicts the positive class, this is referred to as a true positive [25, 26].
ii. True negative: This refers to when the model effectively predicts the negative class.
iii. False positive: This refers to when the model wrongly predicts the right class.
iv. False negative: This refers to when the model wrongly predicts the negative class.

The following are the metrics that will be used for the model evaluation:

1) **Precision:** Precision is the level of effectively characterized instances

$$\text{Precision} = \frac{True\ Positive}{True\ Positive + False\ Positive}$$

Precision of positive surveys is depicted as the quantity of reviews effectively evaluated as certain over the complete number of reviews.

2) **Accuracy:** Accuracy is a measure related to the total number of elements correctly identified in relation to a total amount of data in the analysis you are making. In other words, is a percent of the amount of the things that have gone right in your process.

$$Accuracy = \frac{True\ positive\ + True\ negative}{positive + negative}$$

3) **Recall (Sensitivity):** It has to do with the extent of real positives that were recognized effectively.

$$Recall = \frac{True\ Positive}{True\ Positives + False\ Negatives} = \frac{True\ Positive}{Positives}$$

4) **F1 score:** The F1 score can be defined as a weighted average of precision and recall, with the highest incentive at 1 and the lowest at 0.

$$F = \frac{2^{*}precision^{*}recall}{precision + recall}$$

5) **Specificity:** Specificity is referred to as the percentage of genuine negatives that are accurately detected as opposed to positive, known as selectivity or real negative rate (TNR).

$$\text{Specificity} = \frac{\boldsymbol{TN}}{\boldsymbol{FP + TN}}$$

3 Result and Discussion

The melanoma dataset was obtained from Kaggle repository (https://www.kaggle.com/kmader/skin-cancer-mnist-ham10000) and were implemented on Goggle Collaboratory, thereafter classification techniques were performed. Specifically, this section presents the results of the studies for the proposed model. This study implements a melanoma prediction from skin images using Linear Discriminant Analysis with SVM and KNN [26, 27]. The data consists of melanoma dataset with 10,000 samples, as well as attributes such as the image name of the skin images and dataset type. This model is implemented on Google Collaborator IDE using Python Programming Language.

3.1 Loading the Dataset

The melanoma dataset is imported and loaded into the Google Collaborator IDE. The loaded dataset comprises of 10,000 instances of melanoma dataset.

3.2 Data Pre-Processing

The dataset is cleaned by applying a dimensionality reduction technique known as Linear Discriminant Analysis (LDA) and the data is transformed to be used in the models.

3.3 Support Vector Machine Classifier

The Support Vector Machine classifier is used to classify the pre-processed dataset using a technique called kernel trick to transform the data and then finds an optimal boundary between the possible outputs based on the transformations. It does some fairly complicated data transformations before determining how to separate your data using the labels or outputs that have been specified as display in Table 2.

The SVM classifier obtained 71% of classification accuracy when passed into the dataset. As shown in Table 2 above the result for the support vector machine classifier. Figure 2 shows the confusion matrix plot using the Support Vector Machine classifier where TP = 2637 FP = 322 FN = 322 TN = 17,429.

Table 2. Classification using Support vector machine

Classification Report for SVM algorithm				
Classes	Precision	Recall	F1-score	Support
0	0.06	0.38	0.10	16
1	0.34	0.56	0.42	100
2	0.28	0.55	0.37	180
3	0.00	0.00	0.00	0
4	0.98	0.74	0.84	2653
5	0.00	0.00	0.00	0
6	0.10	0.55	0.16	56
Accuracy			0.71	3005
Macro avg.	0.25	0.40	0.27	3005
Weighted avg.	0.89	0.71	0.78	3005

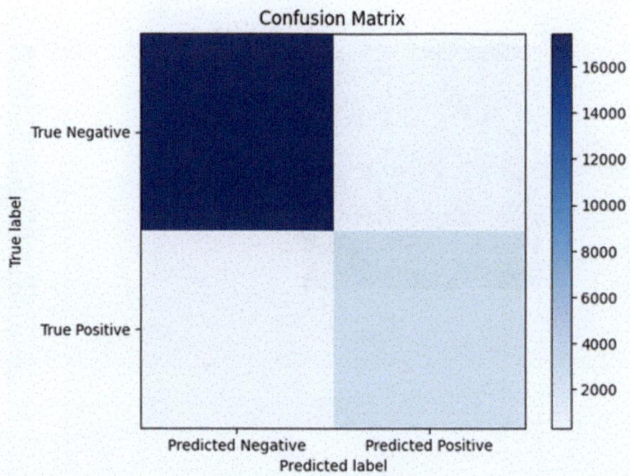

Fig. 2. Skin cancer data confusion matrix classification using support vector machine

3.4 SVM Using Linear Discriminant Analysis

The data is first pre-processed using Linear Discriminant Analysis where there is a combination of features which characterizes or separates the different classes. The SVM classifier is then used on the dataset with a technique called kernel trick to transform the data and then finds an optimal boundary between the possible outputs based on the transformations [28–30]. Table 3 shows the SVM algorithm used with LDA.

Table 3. Classification using support vector machine and Linear Discriminant Analysis

Classification report for SVM algorithm with LDA				
Classes	Precision	Recall	F1-score	Support
0	0.93	0.91	0.92	101
1	0.84	0.95	0.89	146
2	0.70	0.86	0.77	281
3	0.89	0.89	0.89	28
4	0.97	0.90	0.93	2154
5	0.86	0.94	0.90	32
6	0.65	0.81	0.72	263
Accuracy			0.89	3005
Macro avg.	0.83	0.89	0.86	3005
Weighted avg.	0.91	0.89	0.90	3005

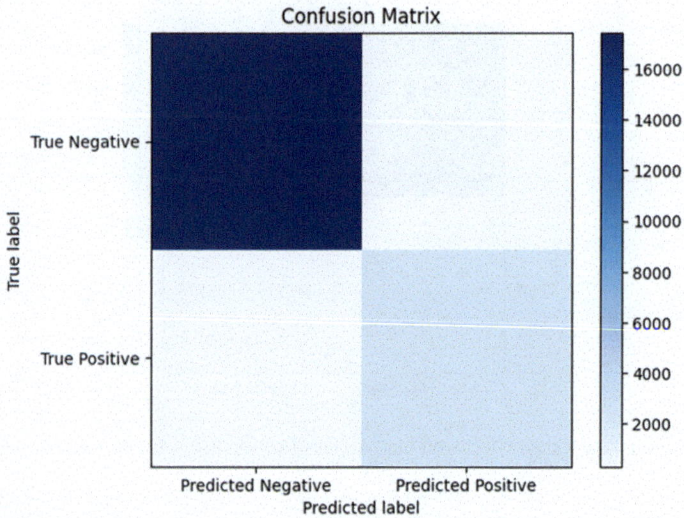

Fig. 3. Skin cancer data confusion matrix classification using support vector machine and linear discriminant analysis

The Linear Discriminant Analysis and SVM classifier obtained 89% of classification accuracy when passed into the dataset as shown in Table 3 above result for the LDA and SVM. Figure 3 shows the confusion matrix plot using the Support Vector Machine classifier and Linear Discriminant Analysis where TP = 2637 FP = 322 FN = 322 TN = 17,429.

3.5 K-Nearest Neighbour Classifier

The k- nearest neighbour classifier passed into the dataset. KNN is based on the idea that every data point that is close to each other belongs to the same class. Basically, it uses similarity to classify a new data point. The KNN algorithm chooses a number, k as the nearest neighbour to the data point to be classified. Figure 4 display classification algorithm of K nearest neighbour. Therefore, k is set to 5, so it will search for the 5 closest Neighbours to the data point.

Table 4. Classification using K-Nearest neighbour algorithm

Classification report for KNN algorithm

Classes	Precision	Recall	F1-score	Support
0	0.20	0.35	0.26	51
1	0.38	0.39	0.38	130
2	0.36	0.42	0.39	289
3	0.08	0.43	0.14	7
4	0.95	0.77	0.85	2499
5	0.00	0.00	0.00	1
6	0.03	0.39	0.06	28
Accuracy			0.71	3005
Macro avg.	0.29	0.39	0.30	3005
Weighted avg.	0.84	0.71	0.77	3005

The KNN classifier obtained 71% of classification accuracy when passed into the dataset. Table 4 shows the result of the KNN classifier. Figure 4 shows the confusion matrix plot using the K-Nearest Neighbour classifier where TP = 2781 FP = 162 FN = 43 TN = 13,561.

3.6 K-NN using Linear Discriminant Analysis

The data is first pre-processed using Linear Discriminant Analysis where there is a combination of features which characterizes or separates the different classes. The k-nearest neighbour classifier is passed into the dataset. The KNN algorithm chooses a number, k as the nearest neighbour to the data point to be classified. KNN classifier with LDA feature extraction techniques result is display in Table 5. Therefore, k is set to 5, so it will search for the 5 closest neighbours to the data point.

The KNN classifier obtained 86% of classification accuracy when passed into the dataset. Table 5 shows the result of the KNN classifier and LDA. Figure 5 shows the confusion matrix plot using the K-Nearest Neighbour classifier and LDA where TP = 2581 FP = 209 FN = 409 TN = 18,236.

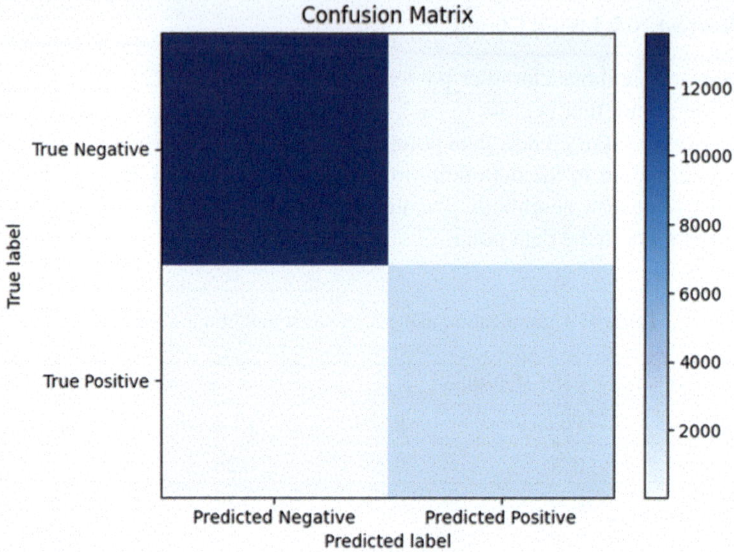

Fig. 4. Skin cancer data confusion matrix classification using K-nearest neighbour

Table 5. Classification using the KNN classifier and linear discriminant analysis

Classification report for KNN algorithm with LDA				
Classes	Precision	Recall	F1-score	Support
0	0.83	0.93	0.88	81
1	0.80	0.81	0.81	134
2	0.63	0.74	0.68	281
3	0.84	0.89	0.86	35
4	0.95	0.88	0.92	2174
5	0.85	1.00	0.92	35
6	0.62	0.81	0.70	265
Accuracy			0.86	3005
Macro avg.	0.79	0.87	0.82	3005
Weighted avg.	0.88	0.86	0.87	3005

3.7 Performance Metrics Evaluation

Table 6 below shows the average performance of SVM, KNN and SVM with LDA and KNN with LDA using accuracy, precision, recall and F1-score performance metrics to evaluate which of the algorithm outperform the other.

From Table 7, it was shown that SVM with LDA outperform other algorithms in terms of accuracy with 0.89 accuracy compare to conventional SVM and KNN. It can

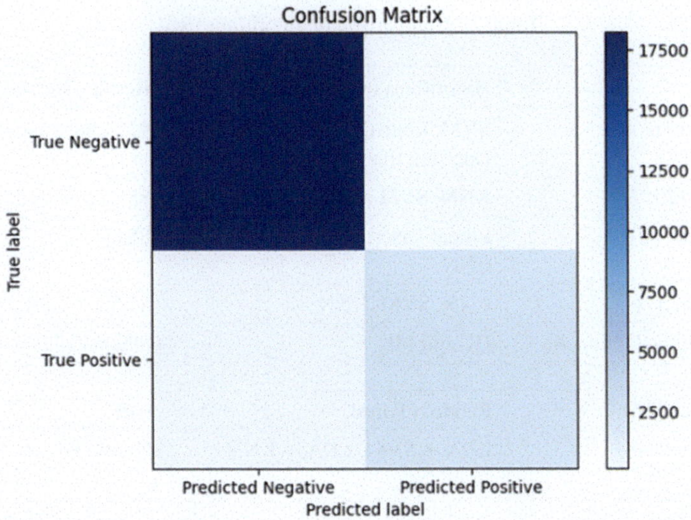

Fig. 5. Skin cancer data confusion matrix classification using K-nearest neighbour and linear discriminant analysis

Table 6. Average performance evaluation

	LDA and SVM	SVM	LDA and KNN	KNN
Accuracy	0.89	0.71	0.86	0.71
Precision	0.83	0.25	0.79	0.29
Recall	0.89	0.40	0.87	0.39
F1-Score	0.86	0.27	0.82	0.30

be deduced that when SVM is used with feature extraction techniques, it will perform better and produce and efficient result. Table 8 shows the accuracy comparison of the results obtained with the state-of-the-art results.

The comparison of each algorithms shows that some algorithms produced a higher accuracy compare to the proposed algorithms such as Narayanan (2017) [6, 16, 31], however, they used small dataset for the classification compare to our study that used large quantity of dataset. With the huge dataset used in this study, we were still able to produce an accuracy of 89.0% and 86.0% respectively. Thus, our study still perform effectively in the prediction of melanoma skin cancer.

Table 7. Comparison of algorithms used

Authors	Algorithms or methods used	Result (Accuracy)
Jaworek-Korjakowska and Kłeczek [6]	SVM, Logistic Regression and Decision tree	92%, 88% and 80%
Narayanan (2017)	KNN, SVM and CNN	85%, 96% and 98%
Majtner et al. [10]	Linear Discriminant Analysis + CNN	85.8%
Daghrir et al. [1]	CNN, SVM, KNN	85.5%, 71.8%, 57.3%
Bechelli and Delhommelle [17]	DL and ML	88% and 72%
Damian et al. [16]	Random Forest	94%
Proposed method	LDA + SVM, LDA + KNN	89.0%, 86.0%

4 Conclusion

The study made use of a skin cancer dataset which was obtained from Kaggle repository. The skin cancer dataset was pre-processed and split into the training and testing set. The training set (70%) was used to determine, or learn, the optimal combinations of variables that will generate a good predictive model and the testing set (30%) was used to provide an unbiased evaluation of a final model fit on the training dataset. The data was passed into the Support Vector Machine and K-Nearest Neighbour classification algorithms and results were obtained with accuracy of 71% and 71% respectively. Then the data was passed through the Linear Discriminant Analysis algorithm where there is a combination of features that separates the classes and then the pre-processed data is then passed into the Support Vector Machine and K-Nearest Neighbour classification algorithms to improve the accuracy of the system and the results were obtained with the accuracy of 89% and 86% respectively. The results were also evaluated with performance metrics such as sensitivity, precision, F1-score and an accuracy comparison of the results obtained with the state-of-the-art was made with related works.

This study is of advantage to the medical sector in melanoma prediction. It can be used to improve on predicting skin cancer and assist them in proper diagnosis with an easy approach. This work contributes to the research area of skin lesion classification for several reasons. Firstly, it focuses on specifying the exact type of a skin lesion and not only on categorization of skin lesions as benign or malignant. Finally, the refinement of current approaches and development of new techniques and methods will help to improve the ability to diagnose skin moles more precisely and to achieve the goal of the significant reduction in melanoma mortality rate.

References

1. Daghrir, J., Tlig, L., Bouchouicha, M., Sayadi, M.: Melanoma skin cancer detection using deep learning and classical machine learning techniques: a hybrid approach. In: 2020 5th International Conference on Advanced Technologies for Signal and Image Processing (ATSIP), September, pp. 1–5. IEEE (2020)
2. Ayo, F.E., Ogundokun, R.O., Awotunde, J.B., Adebiyi, M.O., Adeniyi, A.E.: Severe acne skin disease: a fuzzy-based method for diagnosis. In: Computational Science and Its Applications–ICCSA 2020: 20th International Conference, Cagliari, Italy, July 1–4, 2020, Proceedings, Part VI 20, pp. 320–334. Springer International Publishing (2020)
3. Gautam, D., Ahmed, M.: Melanoma detection and classification using SVM based decision support system. In: 2015 Annual IEEE India Conference (INDICON), pp. 1–5. (2015). https://doi.org/10.1109/indicon.2015.7443447
4. Global Cancer Observatory. (2020, June 20). World Health Organization, International Agency for Research on Cancer. https://gco.iarc.fr/
5. Oladipupo, M.A., Obuzor, P.C., Bamgbade, B.J., Adeniyi, A.E., Olagunju, K.M., Ajagbe, S.A.: An automated python script for data cleaning and labeling using machine learning technique. Informatica **47**(6) (2023)
6. Jaworek-Korjakowska, J., Kłeczek, P.ł.: Automatic classification of specific melanocytic lesions using artificial intelligence. BioMed Res. Int. 1–17 (2016). https://doi.org/10.1155/2016/8934242
7. Awotunde, J.B., Misra, S., Adeniyi, A.E., Abiodun, M.K., Kaushik, M., Lawrence, M.O.: A feature selection-based K-NN model for fast software defect prediction. In: International Conference on Computational Science and Its Applications, July, pp. 49–61. Cham, Springer International Publishing (2022)
8. Linsangan, N.B., Adtoon, J.J.: Skin cancer detection and classification for moles using K-nearest neighbor algorithm. In: Proceedings of the 2018 5th International Conference on Bioinformatics Research and Applications, pp. 1–5. (2018). https://doi.org/10.1145/3309129.3309141
9. Awotunde, J.B., Adeniyi, A.E., Ajagbe, S.A., González-Briones, A.: Natural computing and unsupervised learning methods in smart healthcare data-centric operations. In: Cognitive and Soft Computing Techniques for the Analysis of Healthcare Data, pp. 165–190. Academic Press (2022)
10. Majtner, T., Yildirim-Yayilgan, S., Hardeberg, J.Y.: Optimised deep learning features for improved melanoma detection. Multimedia Tools and Appl. **78**(9), 11883–11903 (2018). https://doi.org/10.1007/s11042-018-6734-6
11. Matthews, N.H., Li, W.-Q., Qureshi, A.A., Weinstock, M.A., Cho, E.: Epidemiology of melanoma. Cutaneous Melanoma: Etiology and Therapy, 3–22 (2017). https://doi.org/10.15586/codon.cutaneousmelanoma.2017.ch1
12. Melanoma—Statistics. (September, 2020). Cancer.Net. https://www.cancer.net/cancer-types/melanoma/statistics
13. Nowroozipour, F., Raahemifar, K.: Skin lesion segmentation techniques for melanoma diagnosis: Comparat. Stud. **30**(2), 129–138 (2016)
14. Perera, E., Gnaneswaran, N., Jennens, R., Sinclair, R.: Malignant melanoma. Healthcare **2**(1), 1–19 (2013). https://doi.org/10.3390/healthcare2010001
15. Rigel, D.S., Carucci, J.A.: Malignant melanoma: prevention, early detection, and treatment in the 21st century. CA A Cancer J. Clinicians **50**(4), 215–236 (2000). https://doi.org/10.3322/canjclin.50.4.215
16. Damian, F.A., Moldovanu, S., Moraru, L.: Melanoma detection using a random forest algorithm. In: 2022 E-Health and Bioengineering Conference (EHB), November, pp. 1–4. IEEE (2022)

17. Bechelli, S., Delhommelle, J.: Machine learning and deep learning algorithms for skin cancer classification from dermoscopic images. Bioengineering **9**(3), 97 (2022)
18. Van der Weyden, L., Patton, E.E., Wood, G.A., Foote, A.K., Brenn, T., Arends, M.J., Adams, D.J.: Cross-species models of human melanoma. J. Pathol. **238**(2), 152–165 (2015). https://doi.org/10.1002/path.4632
19. Watson, M., Holman, D.M., Maguire-Eisen, M.: Ultraviolet radiation exposure and its impact on skin cancer risk. Semin. Oncol. Nurs. **32**(3), 241–254 (2016). https://doi.org/10.1016/j.soncn.2016.05.005
20. Farhaoui, Y.: Design and implementation of an intrusion prevention system. Int. J. Netw. Secur. **19**(5), 675–683 (2017). https://doi.org/10.6633/IJNS.201709.19(5).04
21. Farhaoui, Y. et al.: In: Big Data Mining and Analytics, vol. 6(3), pp. I–II. (2023). https://doi.org/10.26599/BDMA.2022.9020045
22. Farhaoui, Y.: Intrusion prevention system inspired immune systems. Indones. J. Electri. Eng. Comput. Sci. **2**(1), 168–179 (2016)
23. Farhaoui, Y.: Big data analytics applied for control systems. Lecture Notes in Netw. and Syst. **25**, 408–415 (2018). https://doi.org/10.1007/978-3-319-69137-4_36
24. Farhaoui, Y. et al.: In: Big Data Mining and Analytics vol. 5(4), pp. I–II. (2022). https://doi.org/10.26599/BDMA.2022.9020004
25. Alaoui, S.S., Farhaoui, Y.: Hate speech detection using text mining and machine learning. Int. J. Decis. Support Syst. Technol. **14**(1), 80 (2022). https://doi.org/10.4018/IJDSST.286680
26. Alaoui, S.S., Farhaoui, Y.: Data openness for efficient e-governance in the age of big data. Int. J. Cloud Comput. **10**(5–6), 522–532 (2021). https://doi.org/10.1504/IJCC.2021.120391
27. El Mouatasim, A., Farhaoui, Y.: Nesterov step reduced gradient algorithm for convex programming problems. Lecture Notes in Netw. Syst. **81**, 140–148 (2020). https://doi.org/10.1007/978-3-030-23672-4_11
28. Tarik, A., Farhaoui, Y.: Recommender system for orientation student. Lecture Notes in Netw. Syst. **81**, 367–370 (2020). https://doi.org/10.1007/978-3-030-23672-4_27
29. Sossi Alaoui, S., Farhaoui, Y.: A comparative study of the four well-known classification algorithms in data mining. Lecture Notes in Netw. Syst. **25**, 362–373 (2018). https://doi.org/10.1007/978-3-319-69137-4_32
30. Farhaoui, Y.: Teaching computer sciences in Morocco: an overview. IT Professional **19**(4), 12–15, 8012307 (2017). https://doi.org/10.1109/MITP.2017.3051325
31. Farhaoui, Y.: Securing a local area network by IDPS open source. Proc. Comput. Sci. **110**, 416–421 (2017). https://doi.org/10.1016/j.procs.2017.06.106

Representation of a GED Functionality in the Transformation of the BPMN Model to the UML Model Using the MDA Approach

Soufiane Hakkou[1,2](✉), Redouane Esbai[1,2], and Lamlili El Mazoui Nadori Yasser[1,2]

[1] Department of Computer Science, Pluridisciplinary Faculty of Nador, Mohammed the 1st University, Nador, Morocco
soufianehakkou@yahoo.fr

[2] MASI Laboratory, Mohammed the 1St University, Oujda, Morocco

Abstract. One of the main advantages of EDM lies in its many functionalities which contribute to the smooth running of electronic document management. Digitization, which consists of dematerializing physical media via the scanner, is one of the main functionalities of EDM. It is necessary to ensure that the digitization provides exploitable image files which would make it possible to recognize their content and facilitate their indexing. In this article, we propose a new transformation from the BPMN (Business Process Model and Notation) representation to the UML (Unified Modeling Language) representation through the QVT (Query View Transformation) language. This gives users the best of both worlds and keeps things simple. Readability of BPMN, and precision of UML. Moreover, applying the solution on top of the MDA (Model Driven Architecture) framework can be saves time and money, and also improves the quality of the Program.

Keywords: EDM · BPMN · UML · MDA · QVT

1 Introduction

With the broad development of the innovation and its utilize in numerous industries [1] and companies, electronic record administration gets to be an commitment, in arrange to keep up the digitization and diminish the utilize of paper because it turns into an obsolete strategy to oversee or, stock data since it is more troublesome to track information in it comparing to digital records that may be recorded, and after that looked effectively. In this setting we thought of using one of the most functionalities of an EDM [2] framework which is the digitization of records, to assist the client get to data effectively.

Since it gets to be a need to discover a straightforward representation that is simple and easy to understand by the distinctive team of an organization to guarantee the leading collaboration and improve productivity inside the endeavor, we had the thought of employing a modeling dialect that's simple to execute, utilize and get it by the diverse performing artists in a company which is the BPMN [3], it could be a standard modeling dialect embraced by the OMG [4] (Object Management Group) considering its productivity, effortlessness to utilize since it is based on clear graphical representations.

Y. Farhaoui et al. (Eds.): ICAISE 2023, LNNS 837, pp. 339–344, 2024.
https://doi.org/10.1007/978-3-031-48465-0_43

The OMG made the MDA [5] standard to encourage the movement of the Information framework from an ancient one with ancient or obsolete innovations to a unused one with unused advances without losing information, time, or vitality in that relocation.

As an architecture, MDA consists of a set of levels such as the CIM (Computing independent model) and the PIM (Platform independent model), it is suggested to utilize the UML modeling in the PIM level, it could moreover be utilized within the CIM one, but in our case, we are proposing to utilize the BPMN show since it'll rearrange the understanding of the primary stage to clients from different specialties in a company.

2 Related Work

In [6] the creators present an information change of UML lesson graph by utilizing MDA, the creators utilized ANDROMDA generator taking after the MDA designs, on a multi-level showcasing, the technique announced by the creators will spare time and exertion, but the creators center on producing the code without taking in thought the EDM functionalities or counting the information portion.

In [7] the creators present a unused modeling approach utilizing BPMN and ODM as CIM models within the to begin with step at that point change the representation into an Interaction Stream Modeling Dialect graph, the creators connected the proposed model-based advancement handle on a case ponder of e-health. Be that as it may, the creators center on the unique portion for modeling taking in thought the CIM metamodels and the PIM metamodels.

In [8] the creators a change show from BPMN show to adjusted UML over an MDA approach, the commerce prepare show at the CIM level was modeled with the BPMN and the PIM level was modeled through a UML utilize case chart, the change proposed on this approach will offer an undertaking productivity and straightforwardness of creating a modern data framework or relocating to ancient one in arrange to create it.

In [9] The creators propose a change from BPMN into UML, in arrange to change over from the commerce handle (BP) to the movement graph (Advertisement), this transformation will protect the semantics part of the setting connected on, it'll moreover bring numerous benefits to the commerce prepare and the modeling.

2.1 EDM Overview

EDM stands for Electronic Document Management [10] and describes the document management process within an enterprise. This process involves various steps (Fig. 1):

1. Acquisition Part
2. Indexing and classifying Part
3. Storage Part
4. Access and distribution of documents Part
5. Archiving Part.

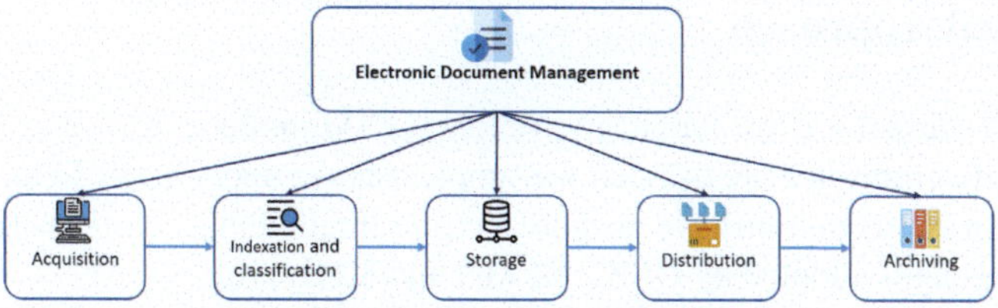

Fig. 1. The 5 primordial steps in EDM

2.2 MDA

Model-Driven Architecture (MDA) [11] is the approach to software development adopted by the OMG that provides procedures and guidelines for organizing software specifications presented as models. MDA consists of three different models (Fig. 2):

CIM: This model describes the system's functionality and requirements from a high-level perspective without specifying implementation details.
PIM: PIM models are the analysis and design models of the application.
PSM: The PSM is a more detailed model that specifies how the PIM is implemented using specific technologies and platforms.

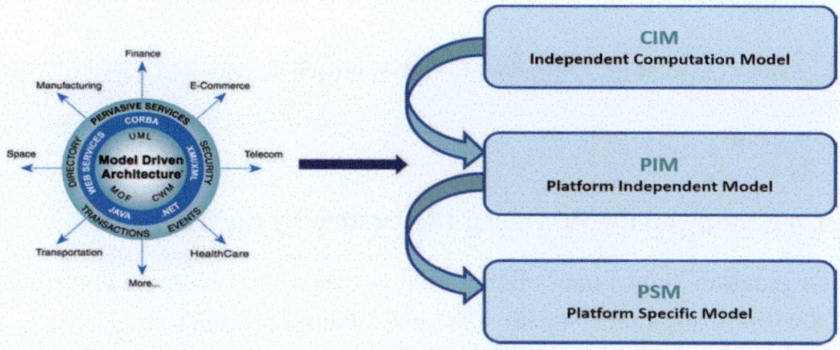

Fig. 2. MDA models

This article will focus on the CIM and PIM model. It describes the method of passing between them, implementing BPMN at the first level and UML at the second.

2.3 BPMN Standard

BPMN [12–15] represents a standard business process model and standard notation. This makes it easier for organizations to understand their internal plans by providing a graphical notation that facilitates understanding of different employees from different specialties within the company.

3 Proposed Approach

In addition to using the UML language to express CIM models,

Use case diagrams that are difficult for simple users to understand. To turn this into a PIM defined in class diagrams, we propose a representation that combines two different modeling languages adopted by OMG: BPMN and UML [16].

Let's consider the case of EDM's main features. Document-acquisition.

First, we used the BPMN language as a source model for inputs to present a clear and understandable process diagram for various collaborators. After the transformation, we reach the UML class, which is the PIM layer.

This transformation is accomplished using the QVT language [17] which has the advantage of being an OMG standard (Fig. 3).

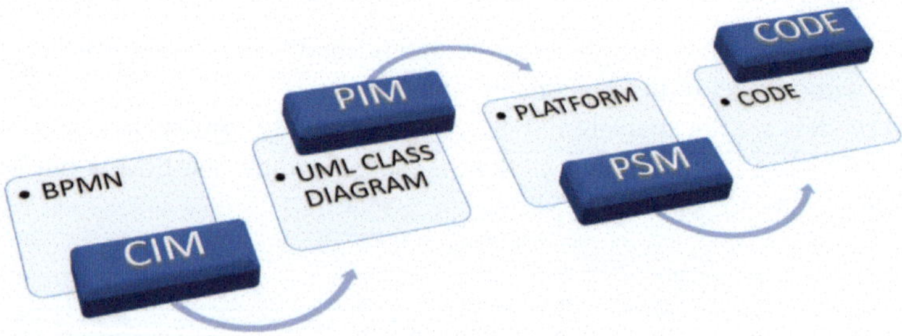

Fig. 3. Proposed MDA approach models

3.1 Use Case Study on the Process of Digitization by the BPMN

Our BPMN diagram shows the collaboration between the two main contributors that influence the digitization mechanism: User and running machine.

Once entered, users who are members of the group have the right to digitize the data. For paper documents, we authenticate the file and send a request to the machine for digitization. The latter executes the request and sends the results to the user to check the quality and integrity of the document.

If the digitized document at the output does not meet the user's needs, the action ends and the process has to start over. Otherwise it is up to the user. Sends a request to the device to save and classify a file as an electronic document in a specified location (Fig. 4).

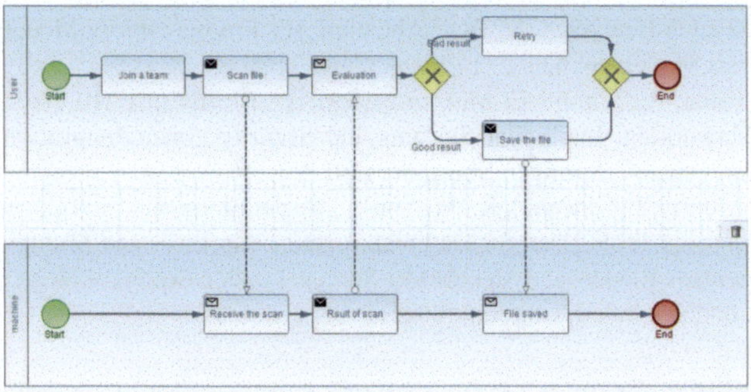

Fig. 4. BPMN diagram of our use case study

4 Conclusion

This article presents a new approach to introducing BPMN into the MDA framework in order to facilitate conversion between the CIM layer using BPMN and the PIM layer using what is recommended in MDA's UML class diagram. I suggested. A framework that considers the first feature of EDM systems: document capture. BPMN makes this possible due to its graphical simplicity, and the UML part is the part for more experienced users, so this proposal helps make the software easier to understand for simple users. As a user in the IT field, the work was done in his CIM - PIM level, no code was generated. In a subsequent study, using QVT he generated code to automatically enable conversion between CIM and PIM. This allows us to work with the next layer of the MDA framework, the PSM. Ultimately, we plan to tackle more complex features of EDM beyond the document capture part.

References

1. Guo, F., Jahren, C.T., Turkan, Y.: Electronic document management systems for the transportation construction industry. Int. J. Constr. Educ. Res. 52–67 (2021)
2. Kurganova, M.V.: Enterprise knowledge base based on EDM system. In: Ashmarina, S.I., Horak, J., Vrbka, J., Suler, P. (eds.) Economic Systems in the New Era: Stable Systems in an Unstable World. IES 2020. Lecture Notes in Networks and Systems, vol 160 (2021)
3. Chinosi, M., Trombetta, A.: BPMN: an introduction to the standard. Comput. Stand. Interfaces (2011)
4. Colomb, R., et al.: The object management group ontology definition metamodel. In: Calero, C., Ruiz, F., Piattini, M. (eds.) Ontologies for Software Engineering and Software Technology (2006)
5. Bucchiarone, A., Cabot, J., Paige, R.F., et al.: Grand challenges in model-driven engineering: an analysis of the state of the research. Softw. Syst. Model. **19**, 13 (2020)
6. Deeba, F., Kun, S., Shaikh, M., Dharejo, F.A., Hayat, S., Suwansrikham, P.: Data transformation of UML diagram by using model driven architecture. In: IEEE 3rd International Conference on Cloud Computing and Big Data Analysis (ICCCBDA), pp. 300–303 (2018)
7. Laaz, N., Kharmoum, N., Mbarki, S.: Combining domain ontologies and BPMN models at the CIM level to generate IFML models. Procedia Comput. Sci. **170**, 851–856 (2020)

8. Khlif, W., Elleuch Ben Ayed, N., Ben-Abdallah, H.: From a BPMN Model to an Aligned UML Analysis Model, pp. 623–631 (2018)
9. Habri, M., Esbai, R., Lamlili El Mazoui Nadori, Y.: BPMN to UML class diagram using QVT. In: Networking, Intelligent Systems and Security. Smart Innovation, Systems and Technologies, vol. 237. Springer, Singapore (2021)
10. Keating, P., Murray, J., Schenkel, K., Merson, L., Seale: Electronic Data Collection, Management and Analysis Tools Used for Outbreak Response in Low- and Middle-Income Countries: A Systematic Review and Stakeholder Survey. London School of Hygiene and Tropical Medicine, London, United Kingdom (2021)
11. https://www.mda.org/
12. https://www.bpmn.org/
13. Farhaoui, Y., et al. Big Data Min. Anal. **6**(3), I–II (2023). https://doi.org/10.26599/BDMA.2022.9020045
14. Farhaoui, Y., et al.: Big Data Min. Anal. **5**(4), I–II (2022). https://doi.org/10.26599/BDMA.2022.9020004
15. Farhaoui, Y.: Big data analytics applied for control systems. Lect. Notes Netw. Syst. **25**, 408–415 (2018). https://doi.org/10.1007/978-3-319-69137-4_36
16. Sbai, R., Arrassen, I., Meziane, A., Erramdani, M.: QVT transformation by modeling. Int. J. Adv. Comput. Sci. Appl. **2**(5), (2011)
17. Esbai, R., Elotmani, F., Belkadi, F.Z.: Model-driven transformations: toward automatic generation of column-oriented NoSQL databases in big data context. Int. J. Online Biomed. Eng. (iJOE) **15**(9), (2019)
18. Petrasch, R., Kongkachandra, R.: Towards enterprise data management: data model mapping. In: IEEE International WIE Conference on Electrical and Computer Engineering (WIECONECE), pp. 72–74 (2018)
19. Rhazali, Y., Hadi, Y., Mouloudi, A.: Third World Conference on Complex Systems (WCCS) (2015)
20. Boujrad, M., Lamlili El Mazoui Nadori, Y.: A New Artificial Intelligence-Based Strategy for Digital Marketing Reinforcement, Innovations in Smart Cities Applications Volume 4. SCA 2020. Lecture Notes in Networks and Systems, pp. 689–699. Karabuk, Turkey (2021)
21. Anh, P.V., Khanh, N.D.T., Dat, T.M., Van Dan, P.: Improved OCR quality for smart scanned document management system. J. Sci. Technique-Le Quy Don Tech. Univ. **210**, 9 (2020)
22. Montalvo, A., Parra, P., Rodriguez Polo, O. et al.: Model-driven system-level validation and verification on the space software domain. Softw. Syst. Model. (2021)

A Comprehensive Performance Analysis of Pretrained Transfer Learning Models for Date Palm Disease Classification

Abdelaaziz Hessane[1(✉)], Ahmed El Youssefi[1], Yousef Farhaoui[1], and Badraddine Aghoutane[2]

[1] STI Laboratory, T-IDMS Faculty of Sciences and Techniques of Errachidia, Moulay Ismail University of Meknès, Meknes, Morocco
{a.hessane,ah.elyoussefi}@edu.umi.ac.ma, y.farhaoui@fste.umi.ac.ma

[2] IA Laboratory, Department of Computer Science, Faculty of Sciences, Moulay Ismail University of Meknès, Meknes, Morocco
b.aghoutane@umi.ac.ma

Abstract. Artificial Intelligence (AI) has emerged as a game-changer in the field of agriculture, revolutionizing various aspects of crop management. Among its many applications, disease detection and classification have gained significant attention due to their direct impact on crop health and productivity. This study focuses on a comprehensive performance analysis of six pre-trained deep-learning models specifically designed for a stage-wise classification of white-scale date palm disease (WSD). By evaluating key metrics such as accuracy, sensitivity towards the amount of data used for training, and inference time, the study aims to identify the most effective model for accurately identifying and categorizing different stages of WSD. The results highlight the MobileNet model as the top performer, demonstrating superior accuracy and inference time compared to the other models. Moreover, the MobileNet model achieves high classification accuracy while using only 60% of the data for training. By leveraging the power of deep learning, this study contributes to enhancing disease management practices in date palm agriculture, leading to improved crop yield, reduced losses, and sustainable food production.

Keywords: Deep learning · Transfer learning · Agriculture · Disease detection · Classification · Date palm · Phoenix dactylifera

1 Introduction

Agriculture is a cornerstone of global sustenance, providing the foundation for food security and ensuring the well-being of a world characterized by a growing population and increased food demand. However, one of the persistent challenges farmers face is managing crop diseases, which can have devastating effects on crop yield and quality. Recent years have witnessed the emergence of artificial intelligence as a transformative

Y. Farhaoui et al. (Eds.): ICAISE 2023, LNNS 837, pp. 345–353, 2024.
https://doi.org/10.1007/978-3-031-48465-0_44

technology, holding tremendous promise for addressing these challenges, particularly in disease detection and classification [1]. By harnessing the power of neural networks and leveraging large-scale datasets, deep learning models have demonstrated exceptional capabilities in accurately identifying and categorizing diseases across various crops [2].

In agriculture, date palm cultivation holds immense economic and cultural significance. The health and vitality of date palm trees are crucial, making detecting and classifying diseases at different stages a paramount concern. In this context, deep learning technology has shown great potential in computer vision tasks, including disease detection and classification [3–5]. However, Deep learning models present two principal challenges. First, they rely on large, diverse datasets, and second, they require substantial computational resources for efficient results. Collecting and curating such datasets can be challenging while training deep learning models using these datasets can be computationally intensive and time-consuming. Transfer learning techniques have emerged as a valuable solution to address these challenges.

This study aims to undertake a comprehensive performance analysis of pre-trained transfer learning models designed explicitly for date palm disease classification. The primary objective is to assess the effectiveness of these models in accurately identifying and categorizing the different stages of the White Scale Disease (WSD) affecting date palm trees [6, 7]. This solution is supposed to address the limits of traditional methods used for detecting the WSD, such as the inability to detect early infestations, the time and effort required for visual inspection, the susceptibility to human error, and the potential for delayed diagnosis. The mean contributions of the present research are:

1. To propose a new solution for an automatic stage-wise classification of WSD based on a transfer learning technique.
2. To conduct a comprehensive performance analysis of six pre-trained convolutional neural networks regarding the task of WSD classification.
3. To identify the most effective model in terms of various metrics.

The remainder of this paper is structured as follows. Related works are discussed in Sect. 2. Section 3, "Methods and Experiments," details the dataset, and the investigated transfer learning-based models. Section 4, "Results and Discussion," presents the findings, including performance metrics and analysis. Finally, in Sect. 5, "Conclusions," the essential findings and their implications are summarized, and future research directions are suggested.

2 Related Work

Transfer learning has received much attention for crop disease identification and classification in recent years. Numerous studies have investigated its application and effectiveness. For instance, authors in [8] propose a deep Gaussian transfer learning model for classifying cassava leaf diseases. The model achieved an accuracy of 98.5% on the cassava leaf disease dataset, outperforming other state-of-the-art models. The pre-trained transfer learning models used in this study were MobileNet V2 and Inception-ResNet V2, and the hybrid kernel was used to improve the model's performance. The Imagenet dataset was used to train pre-trained weights for the modified VGG16 architecture used in [9]. The authors employed state-of-the-art data augmentation, transfer learning, and fine-tuning methods to develop a deep learning model for classifying olive leaf diseases. The model achieved high accuracy, with 100, 99.93, and 100% accuracy in the training, validation, and test sets. The F1-Score on the validation set was reported to be 1, 0.9901, and 0.9899 for the Nutritional Deficiencies, Fumagina, and Virosis olive leaf diseases, respectively. In [10], a novel approach that uses transfer learning and multi-level dimensional reduction is proposed. The objective is to improve the performance of the VGG16 model for classifying tomato plant leaf diseases. The proposed model was trained on the tomato leaf dataset obtained from the Plant Village dataset (https://doi.org/10.48550/arXiv.1511.08060) and achieved an accuracy of 95.68% and a weighted average F1 score of 0.9566, outperforming the three other models tested. In [11], the authors present a plant disease classification and pest recognition method using the EfficientNet-V2 model coupled with Transfer Learning and progressive learning techniques. The results show that EfficientNet-V2 achieves high accuracies of 99.5, 97.5, and 80.1% for Cassava, PlantVillage, and IP102 datasets, respectively, surpassing the performance of Inception-V3 and Vision Transformer models. Authors in [12] propose a framework that combines deep learning features and hand-crafted features based on feature fusion and transfer learning to accurately and efficiently identify plant diseases using leaf images. The study uses three datasets of diseased plant leaves and applies data augmentation techniques to increase the size of the datasets. The proposed method uses an adapted Convolutional Neural Network (CNN) model based on transfer learning and the auxiliary discriminative constraint, achieving high recognition accuracy rates of 99.79, 82.59, and 97.12% on the two apple leaf datasets and one coffee leaf dataset, respectively. The authors of [13] tackle the problem of accurately diagnosing tomato leaf diseases, which is crucial for maintaining tomato crop production. To solve this problem, the authors used transfer learning and feature concatenation to develop deep learning models that can accurately classify tomato leaf diseases. The authors evaluated the performance of their models using a dataset of tomato leaf images and compared their results to other existing techniques. The evaluation metric used was classification accuracy, and the authors achieved an accuracy of 98.5% using their proposed method, which outperformed other existing techniques. The authors also applied Kernel PCA to reduce the dimensionality of the extracted features, which significantly reduced the training and testing time of the classifiers.

3 Methods and Experiments

Within this study, six pre-trained deep learning models belonging to four different families were evaluated. The models included VGG16 and VGG19 from the VGG family [14], Xception as an independent architecture [15], ResNet50 and ResNet50v2 representing the ResNet family [16, 17], and MobileNet from the MobileNet family [18]. Comparing the performance of different pre-trained deep learning model families is valuable for model selection, benchmarking, understanding trade-offs, assessing generalization, and driving research and innovation. Moreover, the tested models were mainly designed to perform a stage-wise classification of the WSD. In this section, we present an overview of the dataset used and the proposed methodology to achieve the sus mentioned objectives.

3.1 Dataset

The dataset [19] comprises 1091 images of date palm leaflets categorized into four classes: healthy, low-degree WSD infestation, medium-degree infestation, and high-degree infestation. It consists of 320 healthy leaflet images, 311 images with a low degree of infestation, 350 with a medium degree of infestation, and 110 with a high degree of infestation. The necessary preprocessing steps were applied using the preprocessing module provided by Keras,[1] a popular deep-learning library. This module offers various preprocessing functions, such as data normalization, resizing, and data augmentation techniques. Data augmentation techniques were employed to address the risk of overfitting and enhance the model's generalization capability. These techniques involve applying random transformations to the training images, such as rotation, translation, scaling, and flipping.

3.2 Transfer Learning

By leveraging pre-trained models on large-scale datasets, transfer learning enables fine-tuning on smaller, task-specific datasets. This approach reduces the need for extensive data and computational resources, saving time and enhancing performance by transferring relevant knowledge from related domains. Transfer learning has become a powerful tool in accelerating the development and deployment of deep learning models, making them accessible to a broader range of applications and practitioners. Figure 1 demonstrates the proposed transfer learning-based solution.

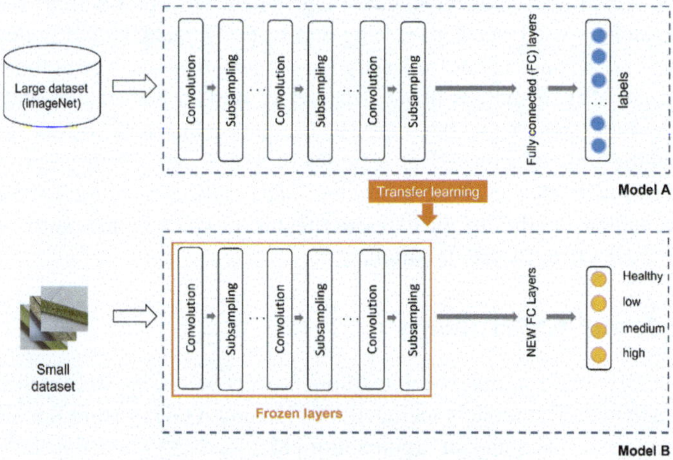

Fig. 1. Transfer learning-based approach for WSD classification

All the models were trained using the ImageNet dataset [20–22]. Moreover, the last fully connected layers of each model were modified to meet the need of the new

[1] https://www.tensorflow.org/api_docs/python/tf/keras/preprocessing.

task of the stage-wise classification of WSD. Essential performance metrics, such as accuracy precision, recall, and f1-score were calculated to assess the various models' effectiveness. Furthermore, inference time and the model's sensibility towards the used training data will be meticulously evaluated to identify the most suitable model for practical implementation. All the experiments were conducted using the gradient tools provided by Paperspace,[2] utilizing the machine specifications outlined in Table 1.

Table 1. Technical specifications of the training environment

GPU	GPU RAM	CPU	RAM	Storage
Nvidia M4000 and P5000	8 Gb	8 × vCPU	30 Gb	5 Gb

4 Results and Discussion

In general, as the amount of training data increases from 60 to 80%, the performance of most models improves. Among the models, MobileNet consistently achieves the highest scores for average precision, average recall, and average F1 score across all three amounts of training data. Moreover, Xception and ResNet50 demonstrate strong performance across all three amounts of training data, consistently achieving high scores for all three metrics. Regarding the VGG16 and VGG19 models, they exhibit slightly lower scores compared to the other models, especially for average precision and average recall. Finally, the performance of ResNet50v2 varies, achieving competitive scores in some cases but slightly lower scores in others. Table 2 displays the average precision, average recall, and average F1-score metrics for different models trained on varying amounts of training data. The test set used represents 10% of the dataset, and it was fixed to be the same while varying the amount of data training.

Regarding the sensibility towards the amount of used training data, VGG16, VGG19, and Xception exhibit sensitivity to the amount of training data used. Among them, VGG19 is particularly sensitive to the training data size, as it shows the most significant improvement (6.24 and 4.16%) when the dataset is reduced by 10% and 20%, respectively. This sensitivity can be attributed to the model's tendency to overfit, which emphasizes the importance of employing regularization techniques and data augmentation when training with VGG19. On the other hand, models such as ResNet and MobileNet demonstrate less sensitivity to changes in the training set size. These models consistently perform at similar levels across different training set sizes, with improvement ranging from −0.92% to 0.94%. This consistency indicates reliability, robustness, and the ability to effectively learn from available data, regardless of the specific amount used (Fig. 2).

The inference time analysis demonstrates that the MobileNet tends to have lower inference time across the different training datasets sizes. The best time recorded is 0.0337 s per image, enhancing the model's computational efficiency. It achieves this

[2] https://www.paperspace.com/gradient.

Table 2. Performance metrics of different models across various amounts of the training set

Training Data Amount (%)	Model	Average Precision	Average Recall	Average F1-score
60	VGG16	0.9411	0.9233	0.9301
	VGG19	0.9292	0.9232	0.9247
	Xception	0.9727	0.9688	0.9692
	ResNet50	0.9720	0.9851	0.9780
	ResNet50v2	0.9853	0.9844	0.9847
	MobileNet	**0.9924**	**0.9922**	**0.9922**
70	VGG16	0.9570	0.9389	0.9468
	VGG19	0.9406	0.9388	0.9394
	Xception	0.9865	0.9844	0.9850
	ResNet50	**0.9865**	**0.9844**	**0.9850**
	ResNet50v2	0.9640	0.9779	0.9705
	MobileNet	0.9853	0.9844	0.9844
80	VGG16	0.9453	0.9233	0.9321
	VGG19	0.8973	0.8933	0.8946
	Xception	0.9702	0.9694	0.9695
	ResNet50	**0.9931**	**0.9922**	**0.9925**
	ResNet50v2	0.9803	0.9766	0.9774
	MobileNet	0.9853	0.9844	0.9844

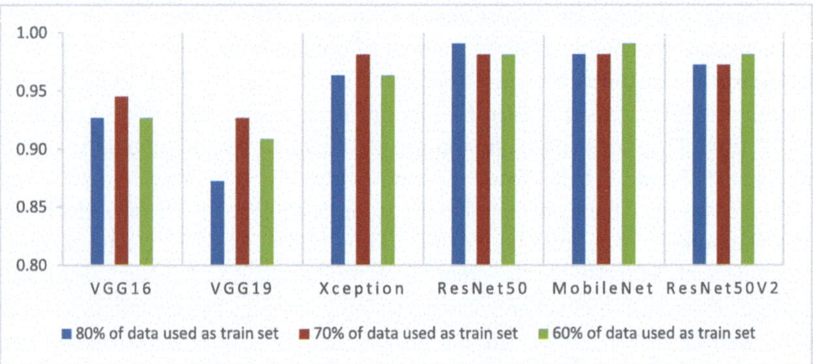

Fig. 2. Accuracy of different models with varying training dataset sizes

through depthwise separable convolutions, a lightweight architecture, parameter efficiency, and model optimization techniques. Using these strategies, MobileNet reduces the computational load and model complexity, resulting in faster predictions. Moreover, MobileNet is designed to provide efficient deep-learning models for mobile and embedded devices (Fig. 3).

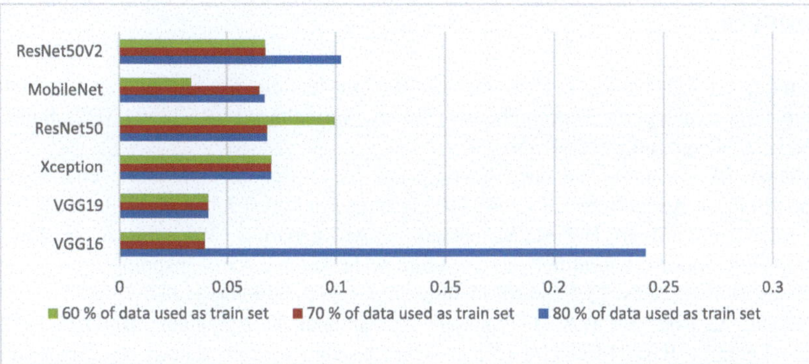

Fig. 3. Inference time of different models with varying training dataset sizes

5 Conclusions

Based on the findings presented in this study, transfer learning demonstrates to be an effective technique for accurately detecting and classifying different stages of WSD affecting date palm trees. Our comprehensive performance analysis of six pre-trained convolutional neural networks revealed that the MobileNet model outperformed the other models regarding various evaluation metrics, sensitivity to the amount of data used for training, and inference time. This study offers significant benefits for the agriculture industry and crop management practices, providing farmers and researchers with a reliable tool for early disease detection and effective management strategies. Future research directions include exploring the potential of transfer learning for other crop diseases and pests and investigating the use of different deep-learning models and techniques for improved accuracy and efficiency. While this study provides valuable insights into the effectiveness of the proposed approach, there are some limitations to consider. One limitation is using a single dataset, which may only be representative of some date palm diseases and their various stages. Additionally, the study only evaluated six pre-trained convolutional neural networks, and other models and techniques may offer improved accuracy and efficiency. In terms of perspectives, future research could explore the potential of transfer learning for other crop diseases and pests and investigate the use of different deep-learning models and techniques for improved accuracy and efficiency. Additionally, efforts could be made to collect and curate larger and more diverse datasets to enhance the performance of deep-learning models for crop disease detection and classification. Finally, developing user-friendly tools and interfaces could facilitate the adoption of deep learning technology in the agriculture industry and improve crop management practices. The insights gained from this study can pave the way for the implementation of advanced disease monitoring systems, aiding in the timely intervention and control of diseases, thus ensuring the overall health and vitality of agricultural ecosystems.

References

1. Domingues, T., Brandão, T., Ferreira, J.C.: Machine learning for detection and prediction of crop diseases and pests: a comprehensive survey. Agriculture **12**(9), 1350 (2022). https://doi.org/10.3390/agriculture12091350
2. Albahar, M.: A survey on deep learning and its impact on agriculture: challenges and opportunities. Agriculture **13**(3), 540 (2023). https://doi.org/10.3390/agriculture13030540
3. Gibril, M.B.A., Shafri, H.Z.M., Shanableh, A., Al-Ruzouq, R., Wayayok, A., Hashim, S.J.: Deep convolutional neural network for large-scale date palm tree mapping from uav-based images. Remote Sens. **13**(14), (2021). https://doi.org/10.3390/rs13142787
4. Hessane, A., Boutahir, M.K., El Youssefi, A., Farhaoui, Y., Aghoutane, B.: Deep-PDSC: A Deep Learning-Based Model for a Stage-Wise Classification of Parlatoria Date Scale Disease, pp. 207–219 (2023)
5. Hessane, A., El Youssefi, A., Farhaoui, Y., Aghoutane, B., Qaraai, Y.: Artificial Intelligence Applications in Date Palm Cultivation and Production: A Scoping Review, pp. 230–239 (2023)
6. Primary Industries and Regions SA (PIRSA)—Government of South Austalia: Parlatoria Date Scale Parlatoria blanchardii [Fact sheet], 2020. [Online]. Available: https://pir.sa.gov.au/__data/assets/pdf_file/0011/299468/Fact_Sheet_-_Parlatoria_Date_Scale_-_June_2020.pdf
7. Abbas, M., et al.: Date palm white scale (Parlatoria blanchardii T): a new threat to date industry in Pakistan. J. Entomol. Zool. Stud. **2**(6), 49–52 (2014)
8. Emmanuel, A., Mwangi, R.W., Murithi, P., Fredrick, K., Danison, T.: Classification of Cassava Leaf Diseases Using Deep Gaussian Transfer Learning Model (2023). https://doi.org/10.1002/eng2.12651
9. Osco-Mamani, E.F., Chaparro-Cruz, I.N.: Highly accurate deep learning model for olive leaf disease classification: a study in Tacna-Perú. Int. J. Adv. Comput. Sci. Appl. **14**(4), 851–860 (2023). https://doi.org/10.14569/IJACSA.2023.0140494
10. Borugadda, P., Lakshmi, R., Sahoo, S.: Transfer learning VGG16 model for classification of tomato plant leaf diseases: a novel approach for multi-level dimensional reduction. Pertanika J. Sci. Technol. **31**(2), 813–841 (2023). https://doi.org/10.47836/pjst.31.2.09
11. Sandhya Devi, R.S., Vijay Kumar, V.R., Sivakumar, P.: EfficientNetV2 model for plant disease classification and pest recognition. Comput. Syst. Sci. Eng. **45**(2), 2249–2263 (2023). https://doi.org/10.32604/csse.2023.032231
12. Fan, X., Luo, P., Mu, Y., Zhou, R., Tjahjadi, T., Ren, Y.: Leaf image based plant disease identification using transfer learning and feature fusion. Comput. Electron. Agric. **196** (2022). https://doi.org/10.1016/j.compag.2022.106892
13. Al-Gaashani, M.S.A.M., Shang, F., Muthanna, M.S.A., Khayyat, M., Abd El-Latif, A.A.: Tomato leaf disease classification by exploiting transfer learning and feature concatenation. IET Image Process. **16**(3), 913–925 (2022). https://doi.org/10.1049/ipr2.12397
14. Simonyan, K., Zisserman, A.: Very Deep Convolutional Networks for Large-Scale Image Recognition, Sep. 2014 [Online]. Available: http://arxiv.org/abs/1409.1556
15. Chollet, F.: Xception: Deep Learning with Depthwise Separable Convolutions, Oct. 2016, [Online]. Available: http://arxiv.org/abs/1610.02357

16. He, K., Zhang, X., Ren, S., Sun, J.: Deep Residual Learning for Image Recognition, Dec. 2015 [Online]. Available: http://arxiv.org/abs/1512.03385
17. He, K., Zhang, X., Ren, S., Sun, J.: Identity Mappings in Deep Residual Networks, Mar. 2016 [Online]. Available: http://arxiv.org/abs/1603.05027
18. Howard, A.G., et al.: MobileNets: Efficient Convolutional Neural Networks for Mobile Vision Applications, Apr. 2017 [Online]. Available: http://arxiv.org/abs/1704.04861
19. Date Palm Data | Kaggle: https://www.kaggle.com/hadjerhamaidi/date-palm-data. Accessed 30 Jan 2022
20. Deng, J., Dong, W., Socher, R., Li, L.-J., Li, K., Fei-Fei, L.: ImageNet: a large-scale hierarchical image database. In: 2009 IEEE Conference on Computer Vision and Pattern Recognition, pp. 248–255 (2009). https://doi.org/10.1109/CVPR.2009.5206848
21. Farhaoui, Y., et al.: Big Data Min. Anal. **5**(4), I–II (2022). https://doi.org/10.26599/BDMA.2022.9020004
22. Farhaoui, Y., et al.: Big Data Min. Anal. **6**(3), I–II (2023). https://doi.org/10.26599/BDMA.2022.9020045

Risks of Energy-Oriented Attacks on Mobile Devices

Ayyoub El Outmani[1,4(✉)], Jaara El Miloud[1], and Mostafa Azizi[2,3]

[1] LARI, FSO, Oujda, Morocco
{ayyoub.eloutmani,e.jaara}@ump.ac.ma
[2] MATSI Research Lab, ESTO, Oujda, Morocco
azizi.mos@ump.ac.ma
[3] University Mohammed First (UMP), Oujda, Morocco
[4] Research Center, High Studies of Engineering School, EHEI, Oujda, Morocco

Abstract. Even with the rapid and exponential growth of innovative apps for mobile devices, the capacity and autonomy of their batteries remain limited and unable to keep pace with the high demands imposed by the increasingly energy-hungry new apps. Over the course of this decade, progress in battery technology has plateaued at a certain level, unlike computing powers that have sensibly reached upper ceilings and gained users' satisfaction. These devices could be vulnerable to a new trend of attacks, that we call Denial of Energy Service (DoES) attacks. This is our vision of an energy-based attack that we present in this paper; it aims to block the apps by cutting the system access to energy, or dissipating it on mobile devices such as smartphones. To our knowledge, no previous studies have been carried out on this subject. Within this context, we will adopt a specific approach aimed at detecting energy efficiency and overheating issues, while mitigating the impact of attacks that seek to excessively drain the device's battery. By combining accurate detection of energy-related problems with appropriate countermeasures, we can prevent and alleviate the detrimental consequences of these attacks on our mobile devices.

Keywords: Denial of Energy Service (DoES) · AppsDrain · Energy attack · Battery drain · Excessive power/energy usage · Power depletion · Energy overconsumption

1 Introduction

By the end of 2023, it is estimated that nearly 6.71 billion smartphones will be in use worldwide [1]. Unfortunately, these devices are equipped with batteries that restrict their autonomy. Recent smartphones are now incorporating more energy-efficient components [2], such as screens, processors, and graphics processing units (GPUs), among others, demonstrating an enhanced emphasis on energy conservation in their design. Battery life and performance are among the key factors to consider when purchasing a new smartphone. When choosing a new mobile device, considering two key factors

Y. Farhaoui et al. (Eds.): ICAISE 2023, LNNS 837, pp. 354–359, 2024.
https://doi.org/10.1007/978-3-031-48465-0_45

proves important: the device's performance and battery life. These parameters significantly influence the experience provided by the device. Firstly, high performance in smartphones ensures a smooth experience by managing demanding apps and multitasking without slowdowns. Moreover, their power enables the completion of complex tasks such as high-quality photo/video editing and seamless use of professional apps. However, it's crucial to note that the simultaneous pursuit of high performance and extended battery life is an extremely intricate issue. Increasing a device's power can decrease its battery life, and vice versa. Finding the optimal balance between these two aspects remains a challenge for leading smartphone manufacturers, aiming to ensure a high-quality experience while addressing mobility requirements. Up to now, no revolutionary technology has provided an absolute solution to this issue, and the sole avenue to enhance battery capacity is its size [2], constrained by the device's form factor. This is why scientists are consistently striving to find ways to address this problem, particularly due to the heightened energy demands of recent apps, which are becoming increasingly energy-intensive [3], necessitating frequent device recharging [4].

In 2022, according to data provided by Statista, a reputable source for compiling and analysing information on apps available on the main platforms, it was observed that during the 3rd quarter of 2022, a wide range of 3.55 million downloadable mobile apps were offered to Android users through the Google Play Store [5], solidifying it as the app store providing the highest diversity of choices. In parallel, Apple's App Store featured around 1.6 million apps available to iOS users, securing the second position in terms of catalogue size. This trend of a steady growth in the number of available apps has persisted over the years. Unfortunately, according to a recent study, it has been found that 18% of Android apps available on the Google Play Store platform experience issues related to energy efficiency [3]. This issue affects both free and commercial apps. With millions of apps in the Play Store, given the large number of apps in these shops, energy security has become as important as the performance and quality of use of the apps themselves, App developers have come to understand that user feedback on the platform might adversely affect their rating. An examination of distinctions between the Android and iOS operating systems reveals that the Android community emphasises energy efficiency more prominently [6].

In this paper, we delve into a set of mechanisms that could be exploited to create disruptions in the energy supply while inducing confusion for the user. Building upon these identified mechanisms, we introduce novel attack strategies, named "denial of energy service" or DOES for short, capable of discreetly consuming energy available on Android smartphones (or even other mobile devices). We qualify these strategies as energy-based attacks. Unfortunately, current methods for monitoring energy consumption do not pay attention to such attacks on the Android platform. Our initial approach enables the estimation of energy consumed by each app, thus we end up by concurrently identifying the responsible components for any excessive consumption. Although no energy attack of this nature has been recorded to date, our objective is to simulate such attacks in order to enhance future protection. This simulation will be accompanied by suitable measures and precautions, thereby providing a protective method against potential threats of this nature.

The remainder of this paper is structured as follows: Sect. 2 provides an exploration of the potential Denial of Energy Service (DoES) attack. Next, Sect. 3 presents a detailed examination of Energy Malware. Afterwards, before concluding, Sect. 4 explains our approach to counter the DoES attack.

2 Exploration of the Potential Attack: Denial of Energy Service (DoES)

Energy Denial of Service (DoES) attacks take various forms, but they share a common objective: to disrupt system access for legitimate users by reducing its availability. This can also extend to mobile devices such as smartphones, tablets, etc., which could be targeted by such attacks to rapidly deplete their batteries using active mechanisms designed to drain the charge. This can be achieved by intentionally cutting off the system's power supply or by altering it. Thus, attackers seek to render the device non-operational by leaving it with a depleted battery [7, 13]. Whether physical or host-based, these attacks demand enhanced security measures and robust authentication mechanisms for effective counteraction. To gain a deeper understanding of the energy consumption behaviour of various apps, researchers have delved into the detection of program code that leads to abnormally high energy consumption [2]. In the context of our study, we will adopt a specific approach aiming to identify energy efficiency and overheating issues while mitigating the impact of attacks that seek to excessively deplete the device's battery. By combining precise detection of energy-related issues with appropriate countermeasures, we can prevent and alleviate the detrimental consequences of such attacks on our devices. It is crucial to note that many apps may encounter energy efficiency issues. Unfortunately, recent statistics reveal that the majority of apps developed thus far have overlooked this vital aspect [3]. These issues manifest in the form of energy bugs, which are programming errors leading to excessive energy consumption. Furthermore, there are also energy hotspots, which are specific areas within the apps that consume unusually high amounts of energy. We consider these issues as energy-based attacks, requiring appropriate detection and mitigation. In our study, we will delve into defense strategies in detail. To achieve this, we propose a technique based on our AppsDrain approach. This approach facilitates the detection of energy bugs and/or energy hotspots while mitigating the effects of attacks aiming to excessively drain the battery. We will also delve deeper into the analysis of defence strategies and measures to be taken against these attacks. However, it is important to emphasise that software patches and updates do not provide complete protection against energy-oriented DoES attacks. Taking preventive measures and closely monitoring the energy behaviours of installed apps on mobile devices are crucial. An energy attack can occur without malicious intent from the developer when an app abnormally consumes system energy. Therefore, it is essential to detect and rectify energy bugs and limit energy hotspots to ensure optimal and secure operation.

3 Energy Malware

In this section, we focus more on energy malware. Its objective is to drain the battery by targeting one or more components of mobile devices, such as the CPU, GPU, screen, cellular networks, Wi-Fi, GPS, and memory card. These components, being the most energy-intensive, are directly targeted due to their high-power consumption. Both the CPU and GPU perform computation and graphics rendering tasks, requiring a significant amount of energy. Additionally, the screen, cellular networks, Wi-Fi, and memory card also demand substantial energy consumption in their operations. Attacks aimed at disrupting their operation and overloading them can result in exaggerated energy usage by the device. When a virus infects a mobile device, it exploits system vulnerabilities or uses social engineering techniques to persuade the user to download suspicious apps or visit malicious websites. Once operational, the virus activates one or more high-energy-consuming components. Malicious code can trigger resource-intensive operations on the CPU and/or GPU, leading to excessive processor usage. Similar to a Fork Bomb (a computer attack technique aimed at overloading a system by repetitively creating a large number of processes or tasks that monopolise system resources, including the CPU, memory, and other general resources), this aggression can have devastating consequences on the system by depleting its resources and rendering normal operation impossible [8]. Furthermore, this breach can also be adapted to target mobile devices, where the goal is to overload the CPU, RAM, and other device resources, resulting in excessive energy use. The smartphone relies on its screen to display visual information and facilitate user interaction through a touch interface. It accounts for approximately 35.5% of a smartphone's energy consumption, and the displayed content can impact this consumption. For instance, an all-black display can consume up to 74.2 mW [9]. To prevent the device from going into sleep mode, a WAKE LOCK is used, which is a feature present in Android devices. It keeps the device active by preventing it from going to sleep [10]. Several smartphone components (like screen, CPU, and sensors) can acquire a WAKE LOCK. For example, a streaming app may use a "FULL_WAKE_LOCK" to prevent the device from going to sleep until the video being played is finished. When the screen lock is not released properly, it can result in an energy bug where the device remains active even when not necessary, leading to high consumption. To avoid this waste, it is essential for apps using a wake lock to release it when it is no longer needed. For example, the authors of E-Greendroid [11] reported that in some cases, the educational app RedBlackTree requires an active screen lock (WAKE LOCK) and fails to release it even after the app has ended [11]. This issue can occur with all smartphone components that can acquire a wake lock without releasing it, including cameras, sensors, GPS, gyroscope, and so on. Horvath and his team [12] made a surprising discovery during their experiments regarding the accuracy and reliability of battery monitoring tools. The experimental results revealed a surprising finding: smartphones operated much longer than what battery monitoring apps predicted. This difference is explained by the fact that these tools assume the device remains constantly active, with the screen always on, without any other background-running apps, including sensors (Wi-Fi, Bluetooth, etc.). However, by disabling other apps, battery life can be extended up to three times longer [12]. Similarly, for sensors or by switching to aeroplane mode, battery life can be multiplied by five, with very reduced baseline consumption [12, 14].

4 Defense Approach

To counter DoES attacks, it is mandatory to implement certain preventive measures. First of all, it is highly recommended to keep the device's OS up to date with the latest security patches and exercise caution when downloading apps. Additionally, implementing energy consumption monitoring mechanisms and limiting the use of high-consumption apps, especially when the battery level is low, is recommended. In this context, our approach relies mainly on using our app called AppsDrain [3].

Developers and users have the ability to closely monitor the energy consumption of each app and component, as presented in Fig. 1. AppsDrain provides an effective solution to address issues related to battery drain attacks. The app conducts a thorough identification and control of each installed app on the respective device. Furthermore, it identifies the component responsible for excessive consumption within that app. AppsDrain offers data analysis, as clearly depicted in the Fig. 1, making diagnosis easier. For example, we observe that the Microsoft app has higher energy consumption at the CPU level compared to other components. In the case of Microsoft, this consumption is normal given the nature of the app. However, if abnormal battery drain is noticed, or if the user does not regularly use this app, it is advisable to take precautionary measures. In such situations, users are recommended to uninstall this suspicious app from their device. This helps prevent any negative impact on battery life and ensures more efficient energy utilisation. It is crucial to remain vigilant and make informed decisions when it comes to app usage to preserve performance and battery life. In the case of developers in the process of creating an app, it is essential to focus on optimising the energy efficiency of the app. A recommended approach, such as using AppsDrain, plays a key role in identifying the suspect app's component which is responsible for consuming the most energy and conducting thorough energy testing before publishing the app on the Play Store.

Fig. 1. Screenshot of AppsDrain

5 Conclusion

In this article, we have presented our approach to counter DoES attacks on mobile devices, emphasising the importance of preventive measures. Our approach, which relies on the use of AppsDrain, provides an effective solution to address issues related to battery depletion attacks. It allows developers and users to closely monitor the energy consumption of each app and component, thereby facilitating the diagnosis of energy-related problems. What we have introduced here is like the first stage of this ongoing project on DoES attacks. As future works, we continue next stages by developing real prototypes of DoES and anti-DoES, and collect data regarding this issue to fine-tune our perception of energy consumption.

References

1. Mobile Network Subscriptions Worldwide 2028: Statista. https://www.statista.com/statistics/330695/number-of-smartphone-users-worldwide/. Accessed 11 Aug 2023
2. El Outmani, A., Jaara, E.M., Azizi, M.: Literature review of energy consumption modeling for mobile devices. In: International Conference on Advanced Intelligent Systems for Sustainable Development, pp. 99–112. Springer, Cham (2023)
3. El Outmani, A., Jaara, E.M., Azizi, M.: Monitoring energy consumption of android apps with AppsDrain. In: Farhaoui, Y., Rocha, A., Brahmia, Z., Bhushab, B. (eds.) Artificial Intelligence and Smart Environment, pp. 635–641. Springer International Publishing, Cham (2023)
4. Hamzaoui KI, Berrajaa M, Azizi M, Lipari G, Boulet P (2017) An experimental methodology for modeling the energy consumption of mobile devices. In: 2017 First International Conference on Embedded & Distributed Systems (EDiS). IEEE, Oran, pp. 1, 6
5. Biggest App Stores in the World 2022: Statista. https://www.statista.com/statistics/276623/number-of-apps-available-in-leading-app-stores/. Accessed 25 Aug 2023
6. Cruz, L., Abreu, R.: Catalog of energy patterns for mobile applications. Empir Software Eng **24**, 2209–2235 (2019). https://doi.org/10.1007/s10664-019-09682-0
7. Carl, G., Kesidis, G., Brooks, R.R., Rai, S.: Denial-of-service attack-detection techniques. IEEE Internet Comput. **10**, 82–89 (2006). https://doi.org/10.1109/MIC.2006.5
8. Berlot, M., Sang, J.: Dealing with process overload attacks in UNIX. Inf. Secur. J. Glob. Perspect. **17**, 33–44 (2008). https://doi.org/10.1080/19393550801929547
9. Hamzaoui, K.I.: Contribution à la modélisation de la consommation d'énergie dans un dispositif mobile. Thèse de doctorat, Université de Lille (2018–2021) (2018)
10. Keep the Device Awake: Android Developers. https://developer.android.com/training/scheduling/wakelock. Accessed 14 Jun 2023
11. Wang, J., Liu, Y., Xu, C., Ma, X., Lu, J.: E-greenDroid: effective energy inefficiency analysis for android applications. In: Proceedings of the 8th Asia-Pacific Symposium on Internetware, pp. 71–80. ACM, Beijing China (2016)
12. Horvath, Z., Jenak, I., Brachmann, F.: Battery consumption of smartphone sensors. J. Reliab. Intell. Environ. **3**, 131–136 (2017). https://doi.org/10.1007/s40860-017-0034-1
13. Idrissi, I., Azizi, M., Moussaoui, O.: A lightweight optimized deep learning-based host-intrusion detection system deployed on the edge for IoT. Int. J. Comput. Digit. Syst. **5**(3), 189–196 (2021)
14. Farhaoui, Y., et al.: Big Data Mining and Analytics, vol. 6, no. 3, pp. I–II (2023). https://doi.org/10.26599/BDMA.2022.9020045

Solar Irradiation Prediction and Artificial Intelligence for Energy Efficiency in Sustainable Buildings, Case of Errachidia, Morocco

Imad Laabab[1(✉)], Said Ziani[2], and Abdellah Benami[1]

[1] Laboratory of Mechanics, Energy Efficiency and Renewable Energies (M3ER), Department of Physics, Faculty of Sciences and Techniques-Errachidia FSTE, Moulay Ismail University of Meknes UMI, Errachidia, Morocco
im.laabab@edu.umi.ac.ma

[2] ENSAM, Mohammed V University, Rabat, Morocco

Abstract. Photovoltaic electricity is a type of electricity based on a remarkable technology that converts sunlight into electricity. The knowledge of the parameters influencing this process will allow us to anticipate the measures to be taken to optimize the energy to be consumed. Artificial intelligence (AI) approaches are used to predict solar irradiation and the results are positive. The purpose of this study is to present an algorithm using artificial neural networks (ANNs) for solar irradiation prediction. The accessibility and direct use of many data sources simplify this process. The importance of this work lies in the fact that it provides us with an effective method for predicting solar irradiation. This is the purpose of this paper which makes a chronological synthesis of the use of ANNs applied to the estimation of solar irradiation for a building supposed to be built in Errachidia. Finally, the results highlight our method. We hope that the work presented here can facilitate the improvement of energy optimization to ensure sustainable development in the Moroccan region of Errachidia.

Keywords: Solar irradiation prediction · Artificial neural networks · Energy optimization · Sustainable development

1 Introduction

One of the top African countries for the utilization of sun-oriented energy is Morocco. Considering this, photovoltaic boards are filling in fame in various provincial areas. Furthermore, innovative, and monetary limits affect various factors, including geographic area, the accessibility of fuel sources, and the framework's energy yield, which raises the general expense of photovoltaic energy. The ongoing work is led by utilizing Errachidia geographical information (see Table 1 and Fig. 1).

Some AI applications for the optimization of solar energy in Errachidia are solar panel placement optimization, predictive maintenance, solar energy storage, smart grid management, and solar forecasting [1]. This last one means that AI can be used to predict solar energy production in a city based on weather forecasts and historical data, allowing for better management of the energy grid and reducing the need for backup fossil fuels.

Y. Farhaoui et al. (Eds.): ICAISE 2023, LNNS 837, pp. 360–366, 2024.
https://doi.org/10.1007/978-3-031-48465-0_46

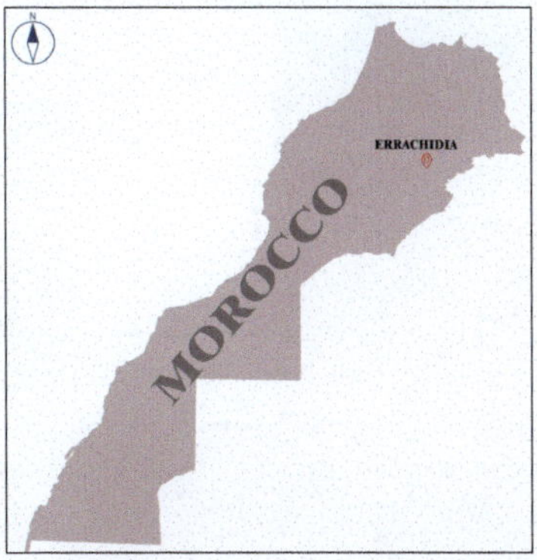

Fig. 1. Geographical location map of the study site in Errachidia, Morocco.

Moreover, machine learning algorithms can be used to evaluate data from temperature sensors on the panels and forecast when overheating is most likely to happen as one method of achieving this.

Table 1. Studied location

Source	National solar radiation database
Latitude	31°55′48″ N
Longitude	– 04°25′12″ W
Elevation	1027 m

For example, the National Solar Radiation Database (NSRDB) is a platform that gives us the average forecasting for daily/annual solar Global Horizontal Irradiance GHI in Moroccan north areas [2] (see Fig. 2).

This may enable the implementation of preventative steps, such as modifying the panels' angles or turning off the power to panels. Additionally, by adjusting the power output to match the energy consumption, AI can be used to improve the efficiency of solar panels [3–5].

In sum, the sustainable building behaves more dynamically toward its occupants by integrating AI techniques.

2 Methods

AI techniques are able to acquire the main models of the systems being examined without having to worry about the mathematical relationships that link the inputs and outcomes of these systems. To improve energy efficiency, we can use ANNs. They are a type of machine learning model that is inspired by the structure and function of the human brain. Also, they are made up of interconnected nodes, or artificial neurons, that are

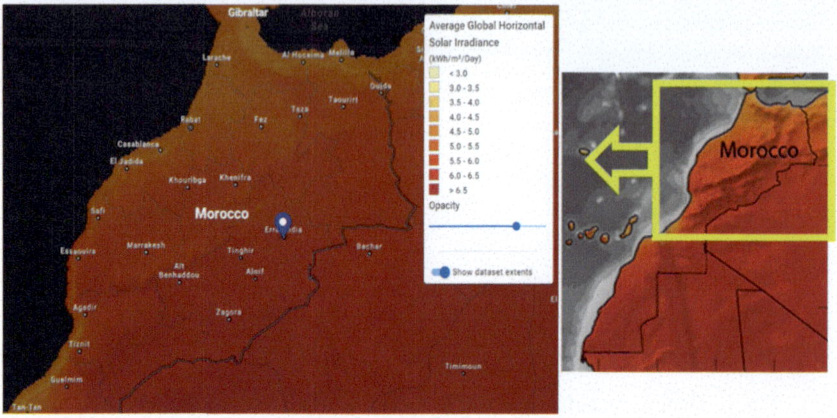

Fig. 2. GHI of solar radiation in Moroccan north areas (NSRDB [2], 2023).

grouped into layers. Feedforward neural networks are the most basic sort of ANNs, with information flowing in only one direction, from input to output layers. This kind is employed in tasks like regression and classification.

ANNs have two important characteristics: they are inspired by the brain and they learn from data. The growing success of ANNs is mainly due to their main properties: parallelism, learning capacity, and generalization capability.

ANNs can be used to predict solar radiation [6], to analyze data from smart meters and other meteorological systems, to monitor energy production in buildings, and to identify patterns and opportunities for energy savings.

We used the following reasoning to develop our ANN prediction model. The steps are described in the bottom figure (see Fig. 3).

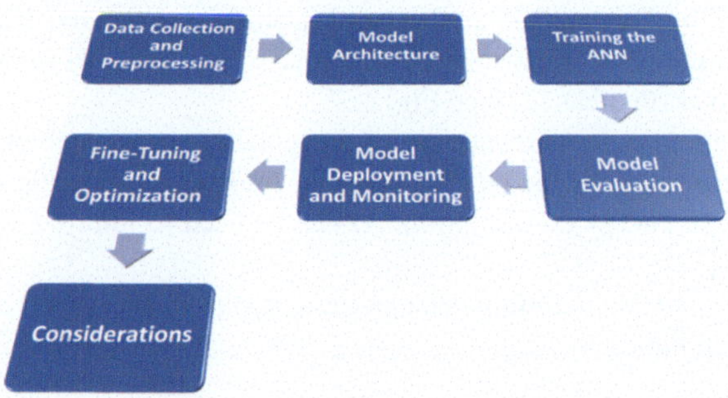

Fig. 3. Prediction model steps.

The detailed description of the data that we used for this investigation was generated by NSRDB from 2017 to 2019. We used Python code to predict solar irradiation for the city of Errachidia. The Keras library was used to build the simple feedforward neural network. Furthermore, it uses data and ANN methods to predict solar irradiation.

The working methodology consisted of reviewing the following elements:

1. Import necessary libraries;
2. Read historical weather data;
3. Create a data frame with the input features and the target variable;
4. Split the data into training and testing sets;
5. Scale the input features to improve the performance of the neural network;
6. Build a simple feedforward neural network model;
7. Compile the model and specify the loss function, optimizer, and metrics to use;
8. Train the model on the training set;
9. Evaluate the model on the testing set;
10. Make a prediction for a new set of input features.

The code creates a data frame using the input features (temperature, humidity, wind speed, and pressure) and the goal variable (solar irradiation) after first reading historical weather data for Errachidia City. The input features are then scaled using StandardScaler to increase the neural network's performance once the data has been divided into training and testing sets.

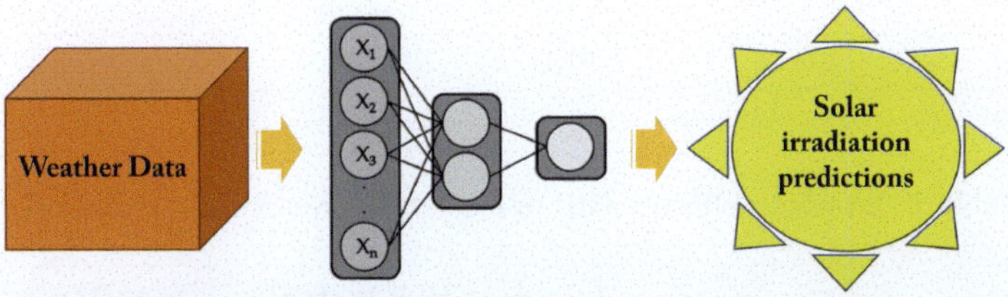

Fig. 4. Methodology adopted for the implementation.

Using Keras, we create a straightforward feedforward neural network model with a single input layer, two hidden layers, and a single output layer. We construct the model, describe its loss function, optimizer, and metrics, and then train it for 50 iterations on the training data (see Fig. 4).

On the test set, we evaluated the model and produced the test loss and metric. The trained neural network model is then used to predict a new set of input features, such as the weather the next day. Predict solar irradiation under certain circumstances (in this case, we assume temperature is 30 degrees Celsius, humidity is 21%, wind speed is 3 mph, and pressure is 1007 hPa).

3 Results

The suggested methodology shows that this model loads and collects data using relatively simple techniques, which simplifies the process of creating our prediction model. The resulting long-term monthly and annual averages are as follows (Table 2).

After a comparison of the approximate and effective GHI results presented in Table 3, we can see in Fig. 5, that our method gives us a promising result for accurate prediction once the model has been properly trained.

Table 2. Errachidia long-term averages (monthly/annual) GHI sum results for 2023.

Month	Predicted GHI [kWh/m^2]
January	117
February	135
March	184
April	212
May	233
June	234
July	228
August	209
September	178
October	149
November	118
December	107
Year	2102

Table 3. Errachidia effective GHI results for the seven months of 2023 [7].

Month	Effective GHI [kWh/m^2]
January	128
February	130
March	193
April	224
May	239
June	240
July	211

4 Discussion

GHI may be more important for building-integrated solar systems such as solar water heaters or passive solar designs because they catch both direct and diffuse solar energy. The use of AI-based methods to forecast solar irradiation has several drawbacks. The models must be retrained because this solution needs a large amount of training data for model construction and the preservation of prediction quality. Forecasting techniques and AI-based methodologies can be coupled if the application is made simpler in terms of both the type and quantity of input parameters. The performance of the prediction model could be greatly enhanced by the incorporation of more data.

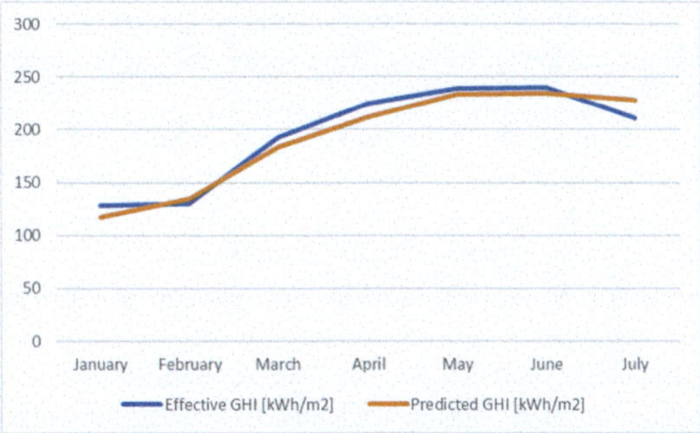

Fig. 5. Comparison between the predicted and effective results.

The use of AI, particularly its capacity for prediction, can help allay these worries. But because these issues are directly related to time, they are made more difficult. As systems become increasingly complicated and scattered, it is projected that the use of AI technologies in the energy sector will expand [8]. Even though AI applications are currently sporadic on networks, the energy transition may someday benefit from them. Additionally, it could be useful to validate the algorithm by comparing its predictions to actual solar irradiation measurements.

5 Conclusion

We conclude that there are numerous ways the energy sector can utilize AI. Two crucial tasks are principally made easier by its predictive capability. First, AI enables us to develop adaption plans and predict solar irradiation at various scales. Second, AI gives us the ability to recognize and correct monitored items, such as smart buildings [9, 10], using baseline behavior. However, only if there is a significant and useful amount of data can each of these predictions be produced. The proposed model would lack training without this data. To improve the accuracy of the forecasts, it could be helpful to use satellite data or to collect additional information from nearby weather stations to supplement the ground-based measurements.

We intend to contrast our efforts with other contributions in the future.

Acknowledgment. We express our thankfulness to all who helped make this work an achievement.

References

1. Liu, Z., et al.: Artificial intelligence powered large-scale renewable integrations in multi-energy systems for carbon neutrality transition: challenges and future perspectives. Ener. AI **10**, 100195 (2022). https://doi.org/10.1016/j.egyai.2022.100195
2. National Solar Radiation Database–Errachidia solar data information. https://nsrdb.nrel.gov/. Website visited on April 27th, 2023, at 10:21 pm

3. Laabab, I., Ziani, S., Benami, A.: Solar panels overheating protection—a review. Indonesian J. Electr. Eng. Comput. Sci. **29**(1), 49–55 (2023). ISSN: 2502-4752. https://doi.org/10.11591/ijeecs.v29.i1.pp49-55
4. Laabab, I., Ziani, S., Benami, A.: A literature review of solar cell overheating control. In: Farhaoui, Y., Rocha, A., Brahmia, Z., Bhushab, B. (eds.) Artificial Intelligence and Smart Environment. ICAISE 2022. Lecture Notes in Networks and Systems, vol 635. Springer, Cham (2023). https://doi.org/10.1007/978-3-031-26254-8_7
5. Boutahir, M.K., Farhaoui, Y., Azrour, M.: Machine learning and deep learning applications for solar radiation predictions review: Morocco as a case of study. In: Yaseen, S.G. (eds.) Digital Economy, Business Analytics, and Big Data Analytics Applications. Studies in Computational Intelligence, vol. 1010. Springer, Cham (2022). https://doi.org/10.1007/978-3-031-05258-3_6
6. Laabab, I., Ziani, S., Benami, A.: A review of the application of artificial intelligence for weather prediction in solar energy: using artificial neural networks. In: Farhaoui, Y., Rocha, A., Brahmia, Z., Bhushab, B. (eds.) Artificial Intelligence and Smart Environment. ICAISE 2022. Lecture Notes in Networks and Systems, vol. 635. Springer, Cham (2023). https://doi.org/10.1007/978-3-031-26254-8_17
7. Meteonorm 8 Software Version 8.0, September 2020. Global Meteorological Database Data for Engineers, Planners and Education
8. De-linchant, B., Nascimento, G.F.M., Laranjeira, T., Vu, T.T.H., Shadid, M.S., Wurtz, F.: Machine learning on buildings data for future energy community services. In: SGE 2021, Symposium de Génie Electrique 2021, Nantes 6–8 Juillet 2021 (2021)
9. Basnayake, B.A.D.J.C.K., Amarasinghe, Y.W.R., Attalage, R.A., Udayanga, T.D.I., Jayasekara, A.G.B.P.: Artificial intelligence based smart building automation controller for energy efficiency improvements in existing buildings. Int. J. Adv. Inform. Sci. Technol. (IJAIST) **40**, 150–156 (2015). ISSN: 2319:2682
10. Martin Nascimento, G. F., Wurtz, F., Kuo-Peng, P., Delinchant, B., Jhoe Batistela, N.: Outlier detection in buildings' power consumption data using forecast error. Energies **14**(24), 8325 (2021). MDPI AG. Retrieved from https://doi.org/10.3390/en14248325

Enhancing Conducted EMI Mitigation in Boost Converters: A Comparative Study of ZVS and ZCS Techniques

Zakaria M'barki(✉), Ali Ait Salih, Youssef Mejdoub, and Kaoutar Senhaji Rhazi

Laboratory of Networks, Computer Science, Telecommunication, Multimedia (RITM), Higher School of Technology ESTC, Hassan II University, Casablanca, Morocco
mbarki.ensem@gmail.com

Abstract. This article highlights the growing demand for efficient power conversion technologies in the context of increasing electric vehicle (EV) adoption. Boost converters are crucial in elevating battery pack voltage to propel EV motors but often generate electromagnetic interference (EMI) due to high-frequency power device switching, potentially causing malfunctions in other electronic systems. To mitigate EMI, zero-voltage-switching (ZVS) and zero-current-switching (ZCS) methods are compared in this study. The findings reveal that both ZVS and ZCS effectively reduce EMI emissions compared to conventional hard-switching techniques. Nonetheless, ZVS soft switching excels in efficiency and EMI reduction at higher loads, making it a more promising solution for EV boost converters, though the choice should consider specific operating conditions.

Keywords: Boost converter · Zero voltage switching · Zero current switching · Conducted EMI

1 Introduction

The rising demand for electric vehicles (EVs) has spurred significant research and development efforts in power electronics converters to meet their high-power requirements. Among these converters, the boost converter plays a pivotal role by elevating battery voltage for motor operation [1]. However, a drawback of boost converters is the electromagnetic interference (EMI) generated during their operation [2].

To tackle this issue, soft-switching techniques like zero-current switching (ZCS) and zero-voltage switching (ZVS) have emerged as promising solutions for EMI reduction [3, 4]. ZCS ensures the main switch current turns off before voltage reaches zero, while ZVS turns off the voltage across the switch.

This study aims to compare the effectiveness of ZCS and ZVS in mitigating conducted EMI in boost converters for EVs. It evaluates EMI levels, efficiency, power loss, and switching frequency, considering varying conditions like load current and switching frequency. While previous research explored soft-switching techniques in power electronics converters [5, 6], this study addresses the gap by providing a comparative analysis of ZCS and ZVS in the context of boost converters for EVs.

Y. Farhaoui et al. (Eds.): ICAISE 2023, LNNS 837, pp. 367–374, 2024.
https://doi.org/10.1007/978-3-031-48465-0_47

In summary, this document provides an overview of the choice between the soft switching techniques ZCS and ZVS, which will be beneficial for researchers and engineers working on improving power electronics for EV applications. It begins by explaining the characteristics of soft switching and then delves into the spectral analysis of conducted electromagnetic disturbances generated by a Boost converter, with the aim of assessing the utility of the proposed techniques in terms of electromagnetic compatibility (EMC).

2 The Boost Converter: Modeling and Practical Implementation

Boost converters are essential in electric vehicles (EVs) for elevating battery pack voltage to meet motor controller requirements. In our study, we investigate boosting the voltage from 24 to 48 V in a standard 48 V EV using a rapid input voltage switch, inductor, and capacitor. A Matlab-based schematic is shown in Fig. 1. The boost converter maintains stable output voltage, even with input and load fluctuations, contributing to improved power electronics efficiency in EVs.

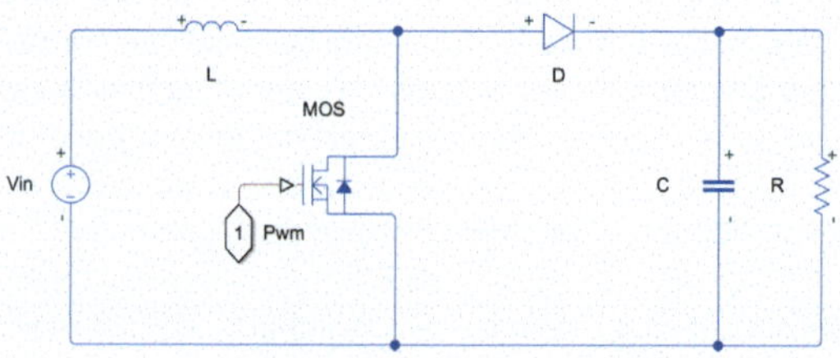

Fig. 1. Fundamental structure of the boost converter.

The construction of the assessed Boost converter is aimed at aligning with the specifications detailed in Table 1.

Table 1. Simulation parameters

Parameter	Value
Input DC voltage V_{in}	24 V
Output voltage V_o	48 V
Inductance L	150 uH
Capacitance C	10 uF
Load resistance R_{load}	12 Ω
Duty cycle D	0.5
Switching frequency F_s	100 kHz

3 Soft Switching Approach for the Boost Converter

Soft switching is a crucial technique in power electronics to mitigate rapid changes in voltage or current within electronic circuits, and to curtail the spread of conducted and radiated disturbances [7]. It entails the incorporation of a damping circuit, such as a capacitor in parallel with the turn-off switch (ZVS) to slow down voltage rise, or an inductance in series with the turn-on switch (ZCS) to moderate current rise.

In fact, there are several soft switching topologies, such as:

Zero Voltage Switching (ZVS): This technique ensures that the voltage across the power switch (such as a MOSFET transistor) is zero during the switching process.

Zero Current Switching (ZCS): Another technique, ZCS ensures that the current through the power switch is zero at the moment of switching. Both techniques effectively reduce energy losses and disturbances associated with switching.

In general, the design of quasi-resonant converters operating at zero voltage or current is characterized by the following properties:

1. Normalized resonance frequency $f_n = \frac{f_s}{f_r}$ with $f_r = \frac{1}{2\pi\sqrt{L_R C_R}}$.
2. Characteristic impedance $Z_n = \sqrt{\frac{L_R}{C_R}}$.
3. Normalized load resistance $Q = \frac{R_L}{Z_n}$.
4. Conversion ratio $M = \frac{V_{out}}{V_{in}}$.

3.1 Soft Switching: ZVS

In this scenario, we assume that the filter's inductance (L) and capacitance (C) dominate over resonance components. Consequently, we replace the series-connected voltage source (V_{in})) with a direct current source (I_{in}) and simulate the output with a direct voltage source (V_o), as depicted in Fig. 2a. The schematic of the previously modeled Boost converter is divided into four operational modes (Fig. 2b), determined by the state of the main switch and the freewheeling diode.

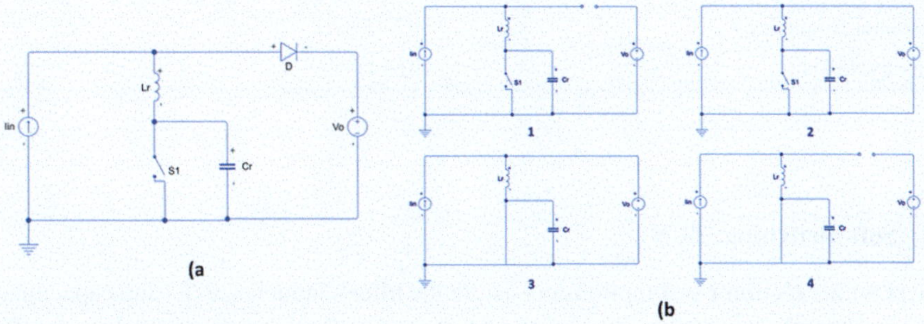

Fig. 2. ZVS Quasi-Resonant Boost Converter with (a) Internal schematic and (b) Equivalent circuits for operating modes.

The theoretical waveforms of the resonant tank are shown in Fig. 3.

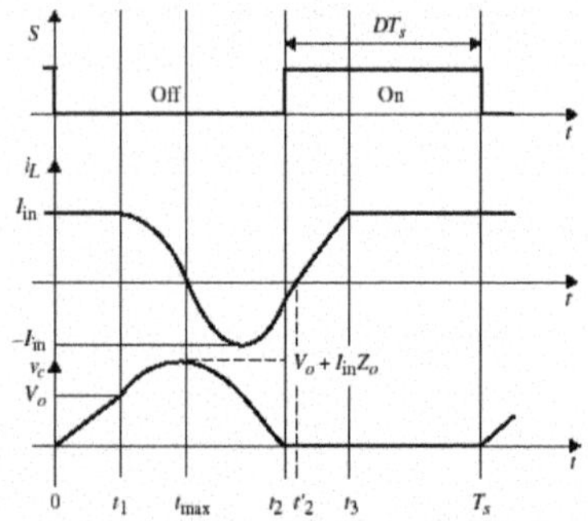

Fig. 3. Voltage and current for the ZVS method.

We can express the voltage conversion ratio M of the quasi-resonant ZVS converter using the concept of energy conservation as follows:

$$M = \left[\frac{f_n}{2\pi} \left(\frac{Q}{2M} + \pi + \sin^{-1}\left(\frac{Q}{M}\right) + \frac{M}{Q} \right) \left(1 + \sqrt{1 - \left(\frac{Q}{M}\right)^2} \right) \right]^{-1} \tag{1}$$

Subsequently, by examining the graph depicting the conversion M as a function of normalized frequency for various values of charge Q, we obtained the following results, which are summarized in Table 2.

Table 2. The boost ZVS-QRC converter parameters.

Component	Normalized value
Inductance L_r	4.7 uH
Capacitance C_r	150 nF
Resonant frequency F_r	197.75 kHz
Duty cycle D	~ 0.54

3.2 Soft Switching: ZCS

The previous section's assumptions persist,ith the filter's inductance (L) and capacitance (C) significantly exceeding those of the resonant elements, as shown in Fig. 4a. The schematic of the previously analyzed Boost converter is segmented into four operational modes (Fig. 4b), contingent on the status of the main switch and the freewheeling diode.

The theoretical waveforms of the resonant tank are shown in Fig. 5.

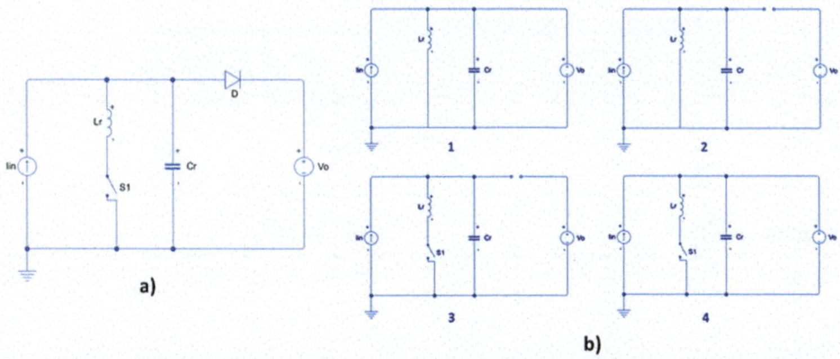

Fig. 4. ZCS Quasi-Resonant Boost Converter with (a) Internal schematic and (b) Equivalent Circuits for Operating Modes.

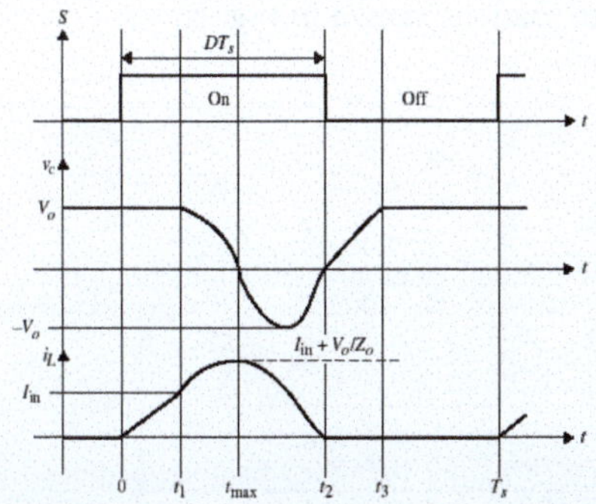

Fig. 5. Voltage and current for the ZCS method.

We can express the voltage conversion ratio M of the quasi-resonant ZCS converter using the concept of energy conservation as follows:

$$M = \left[1 - \frac{f_n}{2\pi}\left(\frac{M}{2Q} + 2\pi - \sin^{-1}\left(\frac{M}{Q}\right) + \frac{Q}{M}\left(1 - \sqrt{1 - \left(\frac{M}{Q}\right)^2}\right)\right)\right]^{-1} \quad (2)$$

Subsequently, by examining the graph depicting the conversion M as a function of normalized frequency for various values of charge Q, we obtained the following results, which are summarized in Table 3.

4 Simulation Results and Discussion

4.1 Conducted Electromagnetic Interference Measurement in a Boost Converter

The boost converter's switching cell generates significant conducted-mode electromagnetic disturbances within the 150 kHz to 30 MHz frequency range [8]. To assess these interferences, the Line Impedance Stabilization Network (LISN) is commonly used. The LISN acts as a filter, isolating the power supply from the device under test and preventing disturbances in common-mode and differential-mode.

Table 3. The Boost ZCS-QRC converter parameters.

Component	Normalized value
Inductance L_r	4.7 uH
Capacitance C_r	150 nF
Resonant frequency F_r	197.75 kHz
Duty cycle D	~ 0.46

This section focuses on quantifying conducted-mode EMI [9] in the conventional Boost converter. It involves a setup with a DC power source, a conducted emissions measurement device (LISN), a control structure, and a switching cell equipped with a filter and a resistive output load, as shown in Fig. 6.

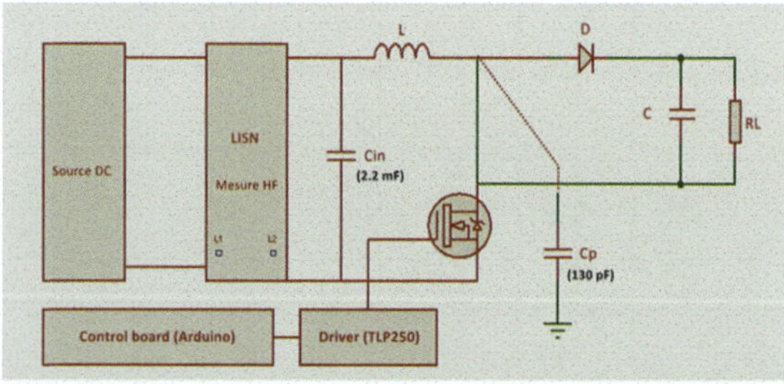

Fig. 6. The conducted emissions testing framework for Boost converter.

The simulation design of the model illustrated in Fig. 1 is conducted in the Matlab environment using the parameters listed in Table 1. We examined the spectral composition of the voltage V_{lisn}, which represents disturbances in differential and common modes [10, 11]. The main objective is to keep the frequency content of this voltage (Power Spectral Density) as low as possible within the range of (150 kHz, 30 MHz).

The interest in these projected techniques (ZVS and ZCS) and their performances in terms of reducing conducted electromagnetic disturbances created by the conventional Boost converter (Hard switching) will now be discussed. The simulation results are as follows (Fig. 7).

The simulation results highlight the distinct performances of the soft switching methods ZVS (Zero Voltage Switching) and ZCS (Zero Current Switching) in reducing conducted electromagnetic disturbances. ZVS stands out for its pronounced reduction of the high-frequency spectrum through its smooth transitions that dampen rapid voltage variations. In practical terms, during switching, the voltage across the switches remains close to zero, curbing rapid voltage transitions and constraining the generation of high-frequency harmonics. Conversely, ZCS proves particularly effective at low frequencies by minimizing current spikes during switching transitions. This approach aims to reduce disturbances stemming from abrupt current changes, crucial for mitigating emissions at lower frequencies. In comparison to the conventional hard switching, where transitions

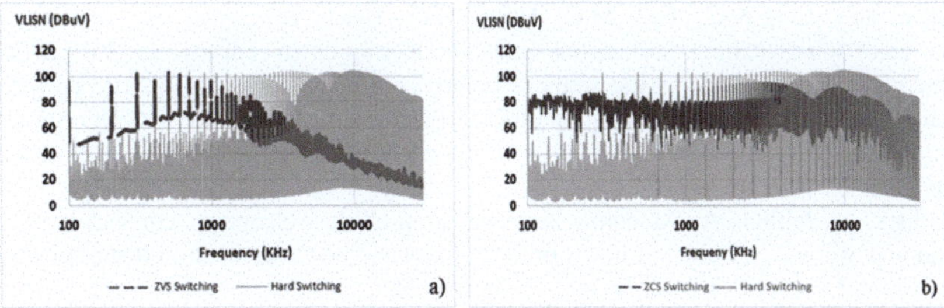

Fig. 7. Spectral content of the voltage V_{lisn} for (a) ZVS method and (b) ZCS method.

are abrupt and lead to undesirable harmonics, both ZVS and ZCS methods offer substantial advantages in terms of electromagnetic compatibility and interference reduction. These characteristics position them as promising choices for effectively managing conducted electromagnetic disturbances, thereby enhancing the quality and integrity of the electrical signal in high-performance demanding applications.

5 Conclusion

In conclusion, the simulations conducted in this study demonstrate the effective mitigation of EMI in EV boost converters using ZVS and ZCS techniques. These techniques contribute significantly to reducing the power spectral density of disturbances across the frequency range of [10 kHz, 30 MHz], leading to substantial gains in EMC. ZVS focuses on high-frequency EMI reduction with smooth voltage transitions, while ZCS targets lower frequencies by minimizing current spikes during transitions. Both outperform traditional hard-switching methods, indicating potential for improved efficiency and broader applications in EV power electronics. This study can also encompass the investigation of radiated emissions, which are a significant concern in the automotive and aerospace sectors. This contributes to enhancing the reliability and efficiency of power electronics tools.

References

1. Al Sakka, M., Van Mierlo, J., Gualous, H.: DC/DC Converters for Electric Vehicles (2011)
2. Muttaqi, K.M., Haque, M.E.: Electromagnetic interference generated from fast switching power electronic devices. Int. J. Innov. Ener. Syst. Power **3**(1) (2008)
3. M'barki, Z., Rhazi, K., Mejdoub, Y.: A proposal of structure and control overcoming conducted electromagnetic interference in a buck converter. Int. J. Power Electron. Drive Syst. **13**, 380–389 (2022). https://doi.org/10.11591/ijpeds.v13.i1.pp380-389
4. M'barki, Z., Rhazi, K., Mejdoub, Y.: A novel fuzzy logic control for a zero current switching-based buck converter to mitigate conducted electromagnetic interference. Int. J. Electr. Comput. Eng. (IJECE) **13**(2), 1423–1436 (2023)
5. M'barki, Z., Rhazi, K.: Optimization of electromagnetic interference conducted in a devolver chopper. In: Hajji, B., Mellit, A., Marco Tina, G., Rabhi, A., Launay, J., Naimi, S. (eds.) Proceedings of the 2nd International Conference on Electronic Engineering and Renewable Energy Systems, ICEERE 2020. LNEE, vol. 681, pp. 523–529. Springer, Singapore (2021)

6. Chen, P.C., Lin, W.C., Tsai, M.C., Hsieh, G.C., Hsieh, H.I.: Analysis, simulation and design of soft-switching mechanisms in DC-to-DC step-down converter. In: 2019 IEEE 4th International Future Energy Electronics Conference (IFEEC), 25–28 Nov 2019, pp 1–5
7. Fang Lin, L., Hong, Y.: Investigation of EMI, EMS and EMC in power DC/DC converters. In: The Fifth International Conference on Power Electronics and Drive Systems, vol. 1, pp. 572–577 (2003)
8. Farhadi, A., Jalilian, A.: Modeling and simulation of electromagnetic conducted emission due to power electronics converters. In: 2006 International Conference on Power Electronic, Drives and Energy Systems, pp. 1–6, 12–15 Dec 2006
9. Nassireddine, B., Abdelber, B., Nawel, C., Abdelkader, D., Soufyane, B.: Conducted EMI prediction in DC/DC converter using frequency domain approach. In: 2018 International Conference on Electrical Sciences and Technologies in Maghreb (CISTEM), pp. 1–6, 28–31 Oct 2018
10. M'barki, Z., Rhazi, K.S., Mejdoub, Y.: Practical implementation of pseudo-random control in step-down choppers and its efficiency in mitigating conducted electromagnetic emissions. In: Cham, Y.F., Rocha, A., Brahmia, Z., Bhushab, B. (eds.) Artificial Intelligence and Smart Environment, 2023. Springer International Publishing, pp. 674–682
11. M'barki, Z., Mejdoub, Y., Rhazi, K.S.: Implementing pseudo-random control in boost converter: an effective approach for mitigating conducted electromagnetic emissions. Indonesian J. Electr. Eng. Inf. (IJEEI) **11**, (2023). https://doi.org/10.52549/ijeei.v11i3.4832

On the Finitude of the Tower of Quadratic Number Fields

Said Essahel(✉)

Sciences and Engineering Laboratory, Sidi Mohammed Ben Abdellah University, Polydisciplinary Faculty of Taza, Taza-Gare PB, 1223 Taza, Morocco
essahel69@yahoo.fr

Abstract. In this paper we will construct some quadratic number fields having an infinite Hilbert class field tower. Then, we will select numerically some examples of these fields that have infinite class field tower and finite r-class field tower for all prime integer r.

Keywords: Class field · Class group · Quadratic fields · Units

MSC: 11R11 · 11R21 · 11R29 · 11R37

1 Introduction

Let k be a number field. The maximal unramified abelian extension of k denoted $k^{(1)}$ is called the Hilbert class field of k (We use HCF for abbreviation). According to the Artin reciprocity law, the Galois group of $k^{(1)}/k$ is isomorphic to $Cl(k)$ the class group of k.

We define inductively the fields $\left(k^{(n)}\right)_{n\in\mathbb{N}}$ by

$$\begin{cases} k^{(0)} = k \\ k^{(n+1)} = \left(k^{(n)}\right)^{(1)} \ \forall n \in \mathbb{N} \end{cases}$$

Then we have

$$k = k^{(0)} \subset k^{(1)} \subset \cdots \subset k^{(n)} \subset k^{(n+1)} \subset \cdots$$

The sequence $\left(k^{(n)}\right)_{n\in\mathbb{N}}$ is called the Hilbert class field tower (We use HCFT for abbreviation). We say that the HCFT is finite if and only if it is a stationary sequence. If we denote $k^{\infty} = \underset{n\in\mathbb{N}}{\cup} k^{(n)}$, then k^{∞}/k is a Galois extension and the HCFT of k is finite iff the extension k^{∞}/k has finite degree.

The finiteness of the HCFT for a number field is, in general, an unsolvable problem. The Mathematicians waited until 1964 when Golod and Shaverevich established an inequality bearing their names that helped them to find for the first time a field with infinite HCFT [2]. To simplify this problem, the mathematicians returned to the p-HCFT that we define next.

Y. Farhaoui et al. (Eds.): ICAISE 2023, LNNS 837, pp. 375–379, 2024.
https://doi.org/10.1007/978-3-031-48465-0_48

Let $\mathbb{P}$ be the set of positive prime integers and $p \in \mathbb{P}$. The maximal unramified abelian p-extension of k denoted $k_p^{(1)}$ is said to be the p-Hilbert class field of k (p-HCF for abbreviation). One can easily see that $k_p^{(1)} \subset k^{(1)}$. We denote $Cl_p(k)$ the p-sylow subgroup of $Gal(k^{(1)}/k)$. Then $k_p^{(1)}$ is the subfield of $k^{(1)}$ such that $Gal(k_p^{(1)}/k) = Cl_p(k)$. We define inductively the p-HCFT of k by

$$\begin{cases} k_p^{(0)} = k \\ k_p^{(n+1)} = \left(k_p^{(n)}\right)_p^{(1)} \ \forall n \in \mathbb{N} \end{cases}$$

Then we have

$$k = k_p^{(0)} \subset k_p^{(1)} \subset \cdots \subset k_p^{(n)} \subset k_p^{(n+1)} \subset \cdots$$

The p-HCFT of k is said to be finite if the sequence $\left(k_p^{(n)}\right)_{n\in\mathbb{N}}$ is stationary. If we denote $k_{(p)}^{\infty} = \underset{n\in\mathbb{N}}{\cup} k_{(p)}^{(n)}$, then the p-HCFT is finite iff $k_{(p)}^{\infty}/k$ has finite degree.

If, for some $p \in \mathbb{P}$, the p-HCFT is infinite, then the HCFT is infinite. Using this property and their inequality, Golod and Shaverevich proved that the field $\mathbb{Q}(\sqrt{-3 \times 5 \times 7 \times 11 \times 13 \times 17 \times 19})$ has an infinite HCFT.

2 Useful Results

Let k be a number field and $p \in \mathbb{P}$.

Proposition 1. *If $Cl_p(k)$ is cyclic, then the p-HCFT of k is finite.*

Proof. See [6]

Proposition 2. *If the HCFT of k is finite, then the p-HCFT of k is finite too.*

Proof. By definitions of $k_p^{(1)}$ and $k^{(1)}$, we can see that $k_p^{(1)} \subset k^{(1)}$. By induction, we can easily prove that $k_p^{(n)} \subset k^{(n)}$ for all $n \in \mathbb{N}$. Then $k_p^{(\infty)} \subset k^{(\infty)}$. From this, we deduce the proposition.

Remark 1. The reciprocal of this proposition is false in general. In the next section, we will give examples of number fields that do not verify the reciprocal.

Proposition 3. *Let K/k be an abelian unramified extension of number fields. The HCFT of k is finite iff the HCFT of K is finite.*

Proof. We will prove that $k^{(\infty)} = K^{(\infty)}$. Since K/k is an abelian unramified extension of number fields, then $K \subset k^{(1)}$. On the other hand, $k^{(1)}/K$ is abelian unramified, then $k^{(1)} \subset K^{(1)}$ and moreover $K^{(1)}/k^{(1)}$ is abelian unramified. We deduce that $K^{(1)} \subset k^{(2)}$. Then we have

$$k^{(1)} \subset K^{(1)} \subset k^{(2)}$$

We can prove inductively that

$$\forall j \in \mathbb{N}, \quad k^{(j)} \subset K^{(j)} \subset k^{(j+1)}$$

We deduce that $k^{(\infty)} = K^{(\infty)}$.

Proposition 4. (Golod-Shafarevich inequality) *Let G be a p-group having a minimal presentation defined by d generators and r relations. If G is finite, then we have*

$$r > \frac{d^2}{2}$$

Proof. See [5].

Remark 2. The above inequality is a refinement of the initial Golod-Shafarevich inequality which was $r > \frac{(d-1)^2}{2}$. This refinement is due to H. Koch [3] and E.B Vinberg [7].

For a group G, let G^{ab} be its abelianization. If $p \in \mathbb{P}$, then the group $G^{ab}/(G^{ab})^p$ can be seen as a vector space over $\mathbb{F}_p = \mathbb{Z}/p\mathbb{Z}$. Its dimension is named the p-rank of G. We denote it $rank_p(G)$.

Let E_k be the group of units of $\mathcal{O}_k$. In [1], page 233, P. Roquette gave a suitable version of the Golod-Shavarevich inequality for number fields.

Proposition 5. *Let p be a prime number. If the p-HCFT of k is finite, then we have:*

$$rank_p(Cl(k) < 2 + 2\sqrt{rank(E_k) + 1}$$

3 Some Number Fields with Infinite HCFT and Finite *r*-HCFT for All Prime *r*

Let $p \in \mathbb{P}$ verifying $p \equiv 3 \mod (4)$ and $F = \mathbb{Q}(\sqrt{-p})$ has a class number $h_F \geq 15$. In [5], using proposition 5, R. Schoof proved that for a prime integer q that splits completely in $F^{(1)}(i)$, the field $k = \mathbb{Q}(\sqrt{-pq})$ has infinite HCFT.

Lemma 1. *Let $q \in P$. With the above assumptions on p, if $q \equiv 1 \mod (4)$ and q is of the form $q = x^2 + xy + \frac{p+1}{4}y^2$ where $(x, y) \in \mathbb{Z}^2$, then q totally splits in $F^{(1)}(i)$.*

Proof. Let $q \equiv 1 \mod (4)$ and q is of the form $q = x^2 + xy + \frac{p+1}{4}y^2$ where $(x, y) \in \mathbb{Z}^2$. Since $q \equiv 1 \mod (4)$, then q splits completely in $\mathbb{Q}(i)$. On the other hand, we have $-p \equiv 1 \mod (4)$, then $\mathcal{O}_F = \mathbb{Z}[\frac{\sqrt{-p}+1}{2}]$. We have $q = x^2 + xy + \frac{p+1}{2}y^2 = N(z)$ where $z = x + \frac{\sqrt{-p}+1}{2}y$ and N the norm map on $F/\mathbb{Q}$. Since $z \in \mathcal{O}_F$, then $\mathfrak{q} = z\mathcal{O}_F$ is a principal ideal of $\mathcal{O}_F$. We have $N(\mathfrak{q}) = q \in \mathbb{P}$, then $\mathfrak{q}$ is a principal prime ideal of $\mathcal{O}_F$. We deduce that $\mathfrak{q}$ totally splits in $F^{(1)}$, then we conclude that q totally splits in $F^{(1)}(i)$.

Theorem 1. *The field $k = \mathbb{Q}(\sqrt{-239 \times 15377})$ has infinite HCFT and for all prime integer r, the r-HCFT of k is finite.*

Proof. The integers 239 and 15377 are prime. Moreover $239 \equiv 3 \mod (4)$, $15377 \equiv 1 \mod (4)$ and they satisfy the conditions in the above lemma ($15377 = 1^2 + 1 \times 16 + \frac{239+1}{4} 16^2$). Since the class number of $F = Q(\sqrt{-239}) = 15$, then the HCFT of k is infinite. Using Pari/GP, we have $Cl(k) \simeq \mathbb{Z}/912\mathbb{Z}$. Then $Cl(k)$ is cyclic. We deduce that $\forall r \in \mathbb{P}$, $Cl_r(k)$ is cyclic. By proposition 1, the r-HCFT is finite.

4 Numerical Examples

Let us take $p = 239$, $q \in \mathbb{P}$ and put $k = \mathbb{Q}(\sqrt{-239q})$. According to the previous section, if $q \equiv 1 \mod (4)$ and q is of the form $q = x^2 + xy + \frac{p+1}{4} y^2$ where $(x, y) \in \mathbb{Z}^2$, then the HCFT tower of k is infinite. And if moreover, $Cl(k)$ is cyclic, then, for all prime integer r, the r-HCFT of k will be finite. Based on these arguments, we made a program in Pari/Gp [4] that determines some values of q such that the field $k = \mathbb{Q}(\sqrt{-239q})$ has an infinite HCFT and finite r-HCFT for all prime integer r. This program gives some values of q and gives the integers x and y verifying $q = x^2 + xy + 60y^2$. The results are given in the following tables. In the last column, $[n]$ designates the cyclic group with n ments.

x	y	q	$Cl(\mathbb{Q}(\sqrt{-239q}))$
1	16	15377	[912]
1	36	77797	[1420]
1	48	138289	[4616]
1	52	162293	[896]
1	60	216061	[3408]
1	96	553057	[7752]
7	40	96329	[2076]
7	76	347141	[1820]
7	96	553681	[12028]
11	8	4049	[788]
11	12	8893	[500]
11	36	78277	[1380]
11	48	138889	[7008]
11	60	216781	[4288]
11	80	385001	[5028]
13	4	1181	[152]
13	16	15737	[628]
13	48	139033	[5988]
13	76	347717	[1288]
13	96	554377	[12048]

x	y	q	$Cl(\mathbb{Q}(\sqrt{-239q}))$
17	12	9133	[596]
17	24	35257	[3056]
17	32	62273	[1256]
17	56	189401	[4128]
17	60	217309	[2308]
19	28	47933	[412]
19	64	247337	[3588]
19	84	425317	[2140]
19	88	466673	[3672]
23	20	24989	[476]
23	44	117701	[1276]
23	56	189977	[2844]
23	60	217909	[2560]
23	80	386369	[5496]
23	96	555697	[7716]
29	12	9829	[404]
29	24	36097	[2252]
29	44	118277	[888]
29	48	140473	[4388]
29	68	280253	[900]

According to these calculations, we can conjecture that there exist infinitely many values of the prime $p \equiv 1 \mod (4)$ and infinitely many values of the prime $q \equiv 3 \mod (4)$ such that de field $k = Q(\sqrt{-pq})$ has infinite HCFT and finite r-HCFT for all prime integer r.

References

1. Cassels, J.W.S, Frohlich, A.: Algebraic Number Fields. London (1967)
2. Golod, E.S., Safarevich, I.R.: On class field towers (Russian). Zzv. Akud. Nauk. SSSR, **28**(1964), 261–272. English translation in Am. math. Sot. Transl. (2) **48**, 91–102
3. Koch, H.: Satz von Golod-schafarewitsch. Math. Nach. **42**, 321–333 (1969)
4. The Pari Group, Univ. Bordeaux, Pari/GP version 2.9.3, 2017. http://pari.math.ubordeaux.fr/
5. Schoof, R.: Infinite class field towers of quadratic fields. J. Reine Angew. Math. **372**, 209–220 (1986)
6. Taussky, O.: A remark on the class field tower. J. London Math. Soc. **12**, 82–85 (1937)
7. Vinberg, E.B.: On the dimension theorem of associative algebras. Izv. Ak. Nauk. SSSR **29**, 209–214 (1965)

Miniaturized 2.45 GHz Metamaterial Antenna

Abdel-Ali Laabadli(✉), Youssef Mejdoub, Abdelkebir El Amri, and Mohamed Tarbouch

RITM Laboratory, CED Engineering Sciences, Higher School of Technology, Hassan II University of Casablanca, Casablanca, Morocco
aali.laabadli@gmail.com

Abstract. This paper presents a reduced patch antenna which resonate at 2.45 GHz. The miniaturization technique used in this work is Metamaterial (MTM); two symmetric metamaterial square unit cells based on split ring resonator (SRR) was printed on the edge of the radiating element of a conventional patch antenna which resonating at 3.54 GHz, this design technique shifts the initial resonance frequency from 3.54 to 2.45 GHz. Therefore, the resulted MTM antenna is 51.48% smaller than a 2.45 GHz standard rectangular patch antenna and has a reflection coefficient of – 32.6 dB, a gain of 2.03 dB and a bandwidth of 186.4 MHz. The materials used in the design are cooper annulled for conducting elements and Epoxy FR-4, with permittivity 4.4, as a substrate. The simulation of the antenna was done with CST solver.

Keywords: Miniaturization · Split ring resonator (SRR) · Conventional patch · Metamaterial

1 Introduction

Connecting small objects to WLAN network will not be possible without miniaturized antennas, because of the close dependency of the antenna size and the resonance frequency. For this reason, antenna designers should apply the appropriate techniques to conceive a reduced size antennas with acceptable performances, like gain, ..., bandwidth.

In this paper we present the methodology to design a reduced size antenna by using the metamaterial technique of miniaturization. Before starting the design steps, you find here after, some works reported in literature in where metamaterial technique is used to design miniaturized antennas [1, 2]. In [1]: the author designs two metamaterial unit cells with double negative permittivity and permeability in the vicinity of a half loop antenna, they shifted the resonance frequency by 20%. In [3]: the author presented a rectangular miniaturized patch antenna loaded by six negative permeability unit cells metamaterial which are made of a spiral and three wires printed on both sides of a dielectric, the reduction achieved was about 40% and 30%. In [4]: The author proposed a miniaturized metamaterial rectangular microstrip patch antenna by etching on the ground plane of conventional patch two CSRR unit cells symmetrical to the microstrip line, the rate reduction achieved was about 45,7%. In [5]: The author miniaturized a square patch

Y. Farhaoui et al. (Eds.): ICAISE 2023, LNNS 837, pp. 380–386, 2024.
https://doi.org/10.1007/978-3-031-48465-0_49

antenna by using concentric complementary split ring resonator (CSRR) structures in between the patch and ground plane, he reduced the normal patch antenna by 25%. In [2]: the author presents in his paper two metamaterial (MTM) unit cells with negative permittivity and permeability placed above the patch antenna, he reduced the antenna with 43%.

2 Antenna Design Methodology

In this section we start with the design of conventional patch antenna which resonate at 3.6 GHz, then we design and study the behavior of the SRR unit cell which resonate at 2.45 GHz and finely we embed the metamaterial unit cell in the top of the 3.6 GHz conventional patch antenna to affect it to resonate at the resonance frequency of the unit cell 2.45 GHz. By this why we shift the resonance frequency of the conventional patch from 3.6 to 2.45 GHz which means that we conceive an antenna that resonating at 2.45 GHz with a dimension of conventional patch that resonate at 3.6 GHz.

2.1 Design of the 3.6 GHz Conventional Patch Antenna

(1) *Calculating the length and width of the patch*

Basing on the TLM (Transmission Line Model) explained in [6], the resonance frequency (fr) of the patch antenna, Fig. 1a, is approximated by the Eq. (1)

$$f_r = \frac{c}{2W\sqrt{\frac{\varepsilon r+1}{2}}} \tag{1}$$

c = 3.10e8 ms^{-1}, ε_r = The relative permittivity, W: The width of the patch.

From Eq. (1), the width is WP = 25.34 mm if fr = 3.6 GHz and $\varepsilon_r = 4.4$.

The length of the patch is LP = 19.4 mm, this is obtained by substituting W, εr, fr, and h (the thickness of the substrate), by their values in the below formulas.

$$\varepsilon reff = \frac{\varepsilon r+1}{2} + \frac{\varepsilon r-1}{2} \times \left[1+12\frac{h}{W}\right]^{-\frac{1}{2}} \tag{2}$$

$$Leff = \frac{\lambda}{2\sqrt{\varepsilon reff}} \tag{3}$$

$$\lambda = \frac{c}{fr} \tag{4}$$

$$\frac{\Delta L}{h} = 0.412 \times \frac{(\varepsilon reff+0.3) \times \left(\frac{W}{h}+0.264\right)}{(\varepsilon reff-0.258) \times \left(\frac{W}{h}+0.8\right)} \tag{5}$$

$$Leff = L + 2 \times \Delta L \tag{6}$$

(2) *Calculating the width of the feeding microstrip line*

The rectangular patch microstrip antennas have several feeding techniques, such as the microstrip line, coaxial probe, proximity coupling, and CPW.

Referring to the comparison between those feeding techniques done in [7], microstrip line is the most suitable for this work even its low bandwidth, it is simple to design, directional and offers high gain. Basing on the on Eq. (7) [8], the width of the microstrip line that has an impedance of 50 Ω is 3.083

$$Z_0 = \frac{120\pi}{\sqrt{\varepsilon_r}\left(\frac{W}{h} + 1.393 + 0.667\ln\left(\frac{W}{h} + 1.44\right)\right)} \tag{7}$$

h: The high of the substrate, Z0: the impedance and W the width of the microstrip line.

To verify the previous theoretical study, the conventional microstip patch antenna with microstrip line feeding was simulated with CST. The antenna is printed on the substrate Epoxy FR-4 with relative permittivity εr = 4.4, thang loss equal to 0.025 and a thickness of 1.6 mm. The other parameters are WP = 25.34 mm, LP = 19.4 mm, WS = WP, LS = 2*LP, WF = 3.083 mm, and Yf = 5 mm, Fig. 1.

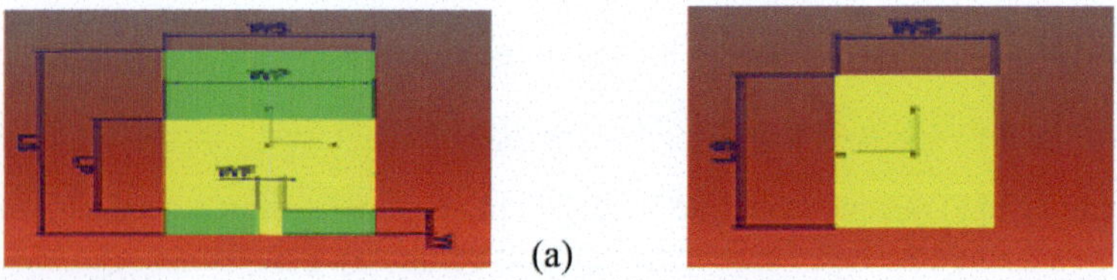

Fig. 1. Conventional microstrip patch antenna at 3.54GHz, (a): top view and (b): back view.

Figure 2 presents the S11 of the conventional patch antenna, the simulated resonance frequency is 3.54 GHz, it is lower than the resonance frequency 3.6 GHz assumed in TLM, CST and other commercial software based on a full wave method like: Finite Element Method (FEM), Finite Integral Technique (FIT), and the Method of the Moment (MoM) are more accurate compared to the TLM method. Hence, we consider that the resonance frequency of the studied conventional patch is 3.54 GHz.

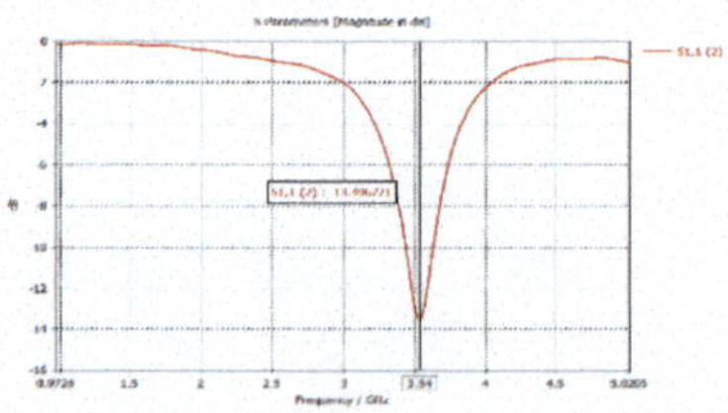

Fig. 2. S11 of the conventional patch simulated with CST basing on TLM.

To demonstrate the effect of the length of the microstrip line feeding on the Impedance matching, a parametric study was done; Fig. 3 shows the variation of S11 versus the length Yf of the feedline. For 3.54 GHz the better impedance matching is obtained when the feedline length is Yf = 1 mm.

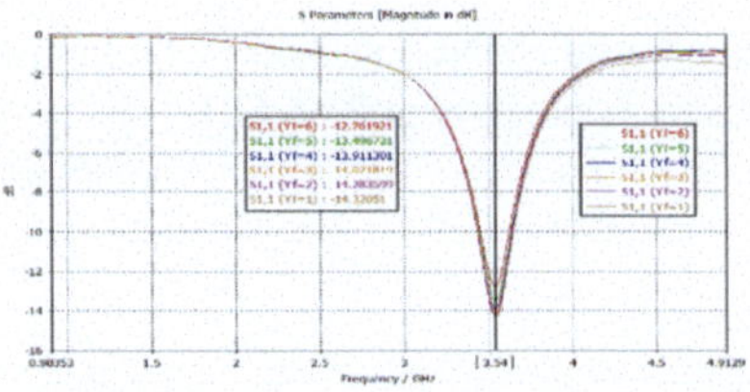

Fig. 3. Parametric study of S11vs feedline Yf

2.2 Design and Study of the SRR Unit Metamaterial Unit Cell

As shown on the Fig. 4a. The unit cell used in this work is an SSR, it consists of two concentric copper annulled square rings with splits on the opposite sides, loaded on Epoxy FR-4 with permittivity = 4.4. To study the behavior of the permeability and permittivity of the metamaterial structure, the unit cell is placed between two waveguide ports (1,2) located at the negative and positive part of the Y axis. In the X and Z axes, perfect electric conductor (PEC) and perfect magnetic conductor (PMC) boundaries are applied respectively, Fig. 4b. CST Microwave Studio electromagnetic simulator has been used to simulate the reflection coefficient (S11) and transmission coefficient (S21) which are used to extract the effective permittivity and permeability of the SRR.

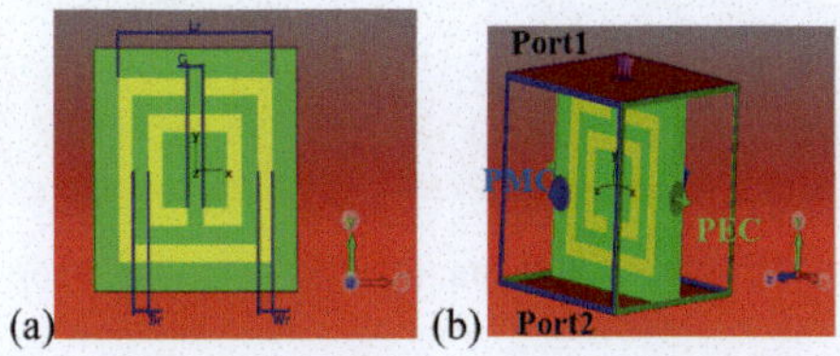

Fig. 4. The Square SRR and its boundry condition.

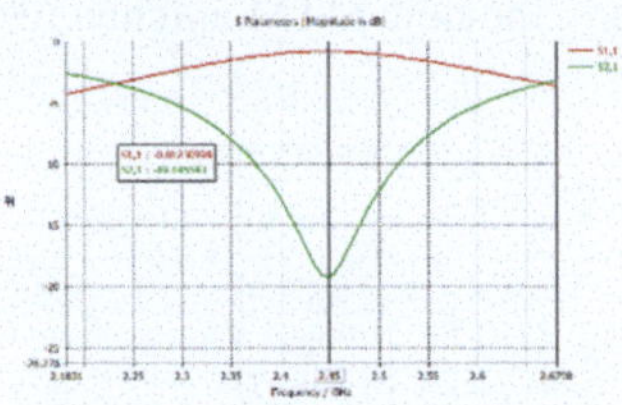

Fig. 5. S Parametric of SRR.

The desired resonant frequency of the unit cell is 2.45 GHz, as shown in the Fig. 5, it is obtained by adapting the parameters of the unit cell which are the length lr = 10 mm, the width of the rings is Wr = 1 mm, the spacing between rings is Sr = 1 mm and the split gap is Gr = 1 mm.

Figure 6 present the extracted effective permeability and permittivity of the SRR unit cell. From those figures (Fig. 6a, b), we deduce that our studied metamaterial unit cell is a Mu negative metamaterial (MNG) at 2.45 GHz; the permeability is negative at 2.45 GHz, but the permittivity is positive at 2.45 GHz.

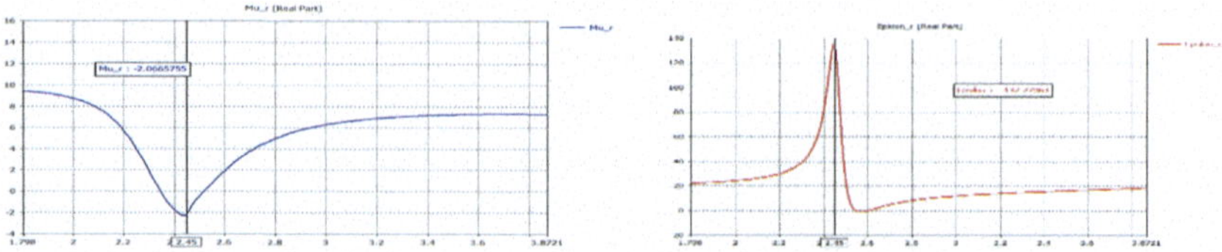

(a) (b)

Fig. 6. Effective permeability and permittivity of the SRR: (a): permeability, (b): permittivity.

2.3 Design of the Metamaterial Antenna

As shown in Fig. 7b, at the edge of the radiating elements of the conventional patch which resonate at 3.54 GHz, previously studied, we print two symmetric metamaterial unit cells, the unit cell studied in the Sect. 2.2, both SRR has a distance D between them.

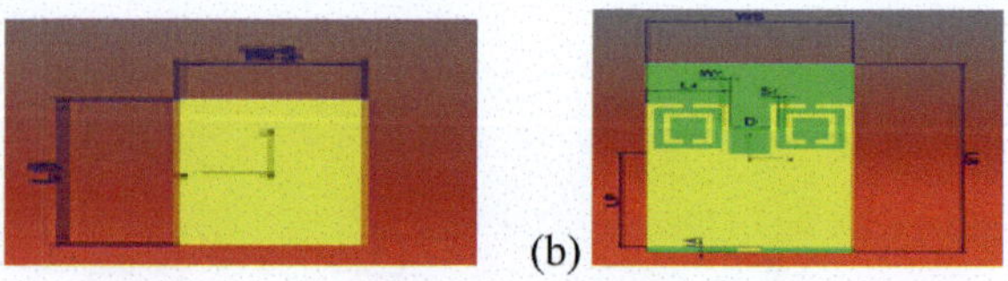

(a) (b)

Fig. 7. MTM patch antenna: (a): Back view, (b): Top view

To define the distance D between the two symmetric SRR that results to the suitable resonance frequency 2.45 GHz, a parametric study about the distance D was done, Fig. 8. We notice that at D = 4.84 mm, the new MTM antenna resonate at 2.45 GHz.

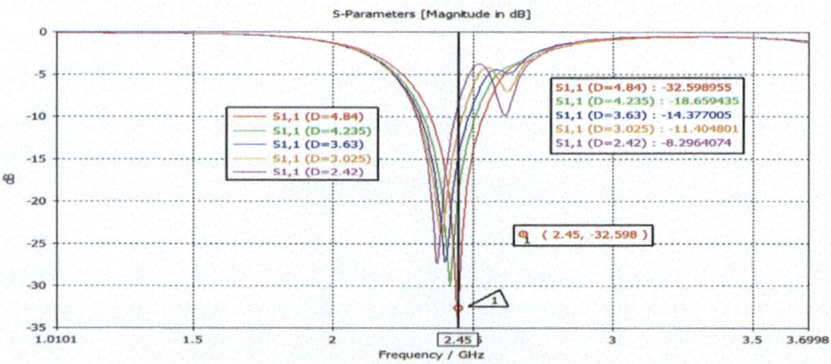

Fig. 8. S11 Parametric study about D of MTM antenna.

3 Simulation Results and Discussion

Figure 9 illustrates the comparison of the reflection coefficients of the 3.54 GHz conventional patch antenna without metamaterial structure and the conventional patch with the two SRR unit cells which printed at its radiating element edge. We notice that after printing the two symmetric SRR metamaterial unit cells on the edge of the radiating element of the conventional patch antenna, its resonance frequency was shifted from 3.54 to 2.45 GHz, which means that we have conceived a reduced size antenna at the resonance frequency 2.45 GHz with the dimension of 3.54 GHz antenna which 38.8 × 25.34 × 1.6 mm^3.

To calculate the reduction rate achieved, first we define the dimensions of the standard 2.45 GHz patch antenna. Basing on the equations of the TLM (Transmission Line Model), Sect. 2.1. If the fr = 2.45 GHz the length and width of the patch antenna are: Lp = 28.81 mm and Wp = 37.23 mm. After simulation in CST, Fig. 10 (red curve), we notice that the resonance frequency is 2.376 GHz different to the one assumed with TLM. Because of the full wave method used in CST is more exact than the TLM, the design was optimized until we found 2.45 GHz, Fig. 7 (green curve), and therefore, the new dimensions of the 2.45 GHz conventional patch are: Lp = 27.99 mm and Wp = 36.2 mm. So the dimension of the 2.45 GHz conventional patch antenna is 55.98 × 36.2 × 1.6 mm^3, and therefore, we notice that the proposed MTM antenna at 2.45 GHz is 51.48% smaller than the 2.45 GHz conventional patch antenna.

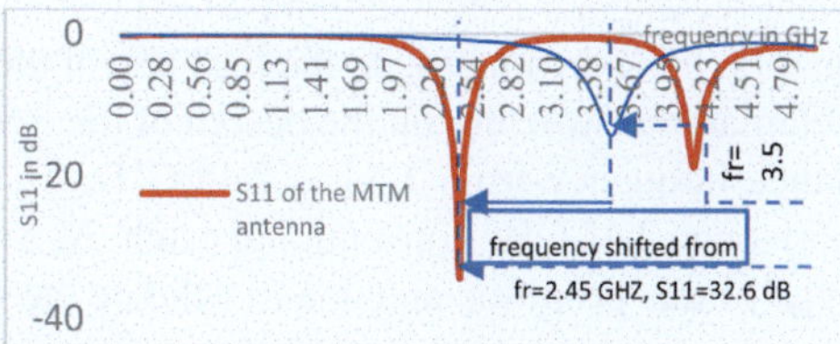

Fig. 9. S11 of the MTM and conventional antenna.

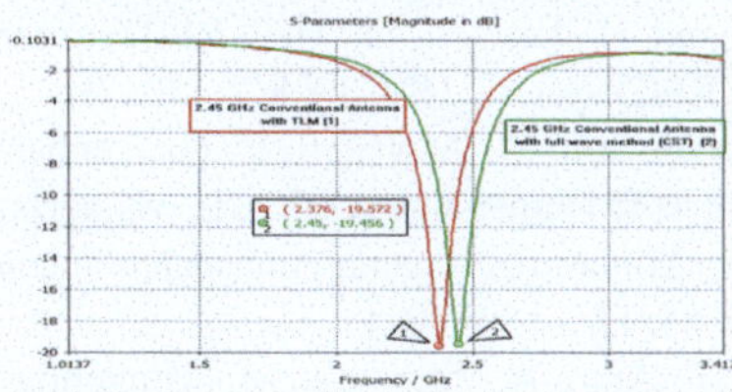

Fig. 10. S11 of the 2.45 GHz conventional patch.

Figure 11 shows the 2D radiation pattern of the new designed metamaterial antenna at its resonance frequency fr = 2.45 GHz. We observe that the radiation pattern is almost omnidirectional in H-Plane (red graph) and bidirectional in E-Plane (bleu graph).

Figure 12 presents the reflection coefficient of the MTM antenna, it is – 32.6 dB at 2.45 GHz, and the bandwidth is 186.4 MHz, it is from 2.5353 to 2.3489 GHz.

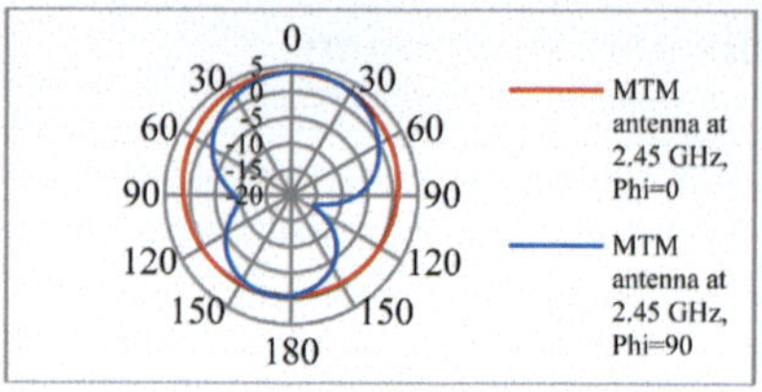

Fig. 11. Radiation pattern of the proposed (MTM) antenna at 2.45 GHz

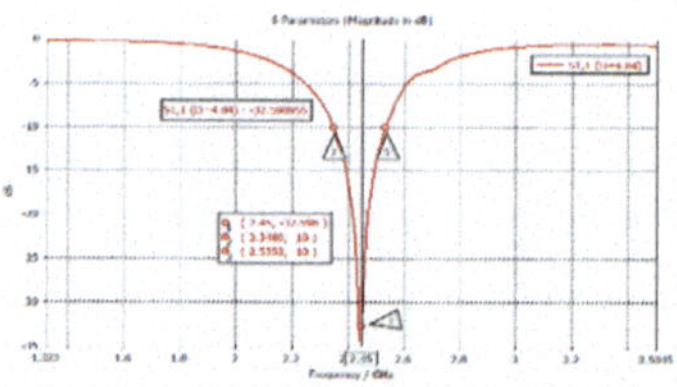

Fig. 12. S11 of the designed MTM antenna

4 Conclusion

In this work, we have designed by simulation a miniaturized antenna which has a bandwidth of 186.4 MHz and a gain of 2.03 dB, this is by printing on the edge of the radiating element of a conventional rectangular patch antenna which resonates at 3.54 GHz two squares split ring resonators (SRR) symmetric to the feed line of the conventional patch. The intrusion of the metamaterial structure on the design of the 3.54 GHz conventional patch, shifts the resonance frequency from 3.54 to 2.45 GHz. The achieved reduction rate is 51.48% in comparison to 2.45 GHz conventional patch. As a next step to this work is to conceive a prototype of this proposed antenna to validate the simulation results.

References

1. Zamali, M., Osman, L., Regad, H., Latrach, M.: UHF RFID reader antenna using novel planer metameterial structure for RFID system. IJACSA **8**(7) (2017)
2. Gudarzi, A., Bazrkar, A., Mahzoon, M., Mohajeri, F.: Gain enhancement and miniaturization of microstrip antenna using MTM superstrates. ICCCE, Malaysia, 3–5 July 2012
3. Lafmajani, I.A., Rezaei, P.: Miniaturized rectangular patch antenna loaded with spiral/wires metamaterial article. EJSR **65**, 121–130 (2011)
4. Laabadli, A., Mejdoub, Y., El Amri, A., Tarbouch, M.: Miniaturized metamaterial antenna for 2.45 GHz services. IJMOT **18**(4), 349–358 (2023)
5. Ramzan, M., Topalli, K. A Miniaturized Patch Antenna by Using a CSRR Loading Plane. Bilkent University, Ankara
6. Balanis, C.A.: Antenna Theory: Analysis and Design, 3rd edn. John Wiley, Hoboken (2005)
7. Barrou, O., El Amri, A., Reha, A.: Comparison of feeding modes for a rectangular microstrip patch antenna for 2.45 GHz applications. ISUNet 457–469 (2016)
8. Huang, Y., Boyle, K.: Antennas: From Theory to Practice. Wiley, Chichester (2008)

Miniaturized Dual Band Antenna for WLAN Services

Abdel-Ali Laabadli(✉), Youssef Mejdoub, Abdelkebir El Amri, and Mohamed Tarbouch

RITM Laboratory, CED Engineering Sciences, Higher School of Technology, Hassan II University of Casablanca, Casablanca, Morocco
aali.laabadli@gmail.com

Abstract. In This paper we present a miniaturized dual band antenna suitable for WLAN applications, 2.45 and 5.8 GHz. The miniaturization and the upper frequency were obtained by two-unit cells metamaterial which were etched on the ground plane of a conventional patch antenna initially resonating at 3.38 GHz. The simulation of the antenna was done with CST solver. Epoxy FR-4, with permittivity 4.4, loss tangent of 0.025 and thickness of 1.6 mm, is chosen as a substrate and the copper annulled is used as material for the ground and the patch of the antenna. This proposed antenna has a gain 1.39 dB at 2.45 GHz and 2.1 dB at 5.8 GHz, the bandwidths are 76.4 MHz for the inferior band and 160 MHz for the superior band.

Keywords: Miniaturization · Dual band · Unit cell · Conventional patch · CSRR · Metamaterial

1 Introduction

The trend in the antenna design is to conceive a single antenna structure which have the capability to operate at different frequency bands. WLAN (Wireless Local Area Network) the most popular network technology which links, within a limited area, different devices (computers or peripherals) to each other. WLAN is defined by the IEEE 802.11 standard to operate in the frequency bands 2.45 and 5.8 GHz.

In this work we aim to design a dual compact antenna for WLAN application with metamaterial technique of miniaturization. Before starting in the design methodology, we present in the following some works extracted from literature in where metamaterial technique is used to design miniaturized antennas and create multiband antennas [1, 2]. In [1]: The author presents two metamaterial unit cells with double negative permittivity and permeability beside of a half loop antenna, the resonance frequency was shifted by 20%. In [3]: The author conceived a miniaturized metamaterial rectangular patch antenna by etching on the ground plane of conventional patch two CSRR unit cells symmetrical to the microstrip feed line, the achieved rate reduction was about 45,7%. In [4]: The author places above the patch antenna two metamaterial (MTM) unit cells with

Y. Farhaoui et al. (Eds.): ICAISE 2023, LNNS 837, pp. 387–392, 2024.
https://doi.org/10.1007/978-3-031-48465-0_50

negative permittivity and permeability and therefore he achieves to reduce the conventional patch antenna by 43%. In [5]: The author presents a new dual band metamaterial printed antenna for RFID. He applied two technics of antenna design, he etched two L-shaped slots in the radiating element for dual band operation and a complementary split ring resonator (CSRR) in the ground plane for size miniaturization. In [6]: The author proposed a metamaterial inspired antenna with tapered patch for WLAN/WiMAX applications by modifying the structure of the patch and etching two rectangular unit cells metamaterial (CSRR) on it. In [2]: The author proposed a dual compact metamaterial microstrip antenna based on two metamaterial unit cells formed by CSRR etched on the ground plane of the antenna and a T-shaped slot etched on the radiating elements.

2 Antenna Design Methodology

2.1 Design of the 3.6 GHz Conventional Patch Antenna

(1) *Calculating the length and width of the patch*

The initial dimensions of the conventional patch antenna Fig. 1 were calculated from the below Eqs. (1)–(6) of the TLM (Transmission Line Model) which explained in [7]:

$$f_r = \frac{c}{2W\sqrt{\frac{\varepsilon r+1}{2}}} \tag{1}$$

$$\varepsilon reff = \frac{\varepsilon r + 1}{2} + \frac{\varepsilon r - 1}{2} \times \left[1 + 12\frac{h}{W}\right]^{-\frac{1}{2}} \tag{2}$$

$$Leff = \frac{\lambda}{2\sqrt{\varepsilon reff}} \tag{3}$$

$$\lambda = \frac{c}{fr} \tag{4}$$

$$\frac{\Delta L}{h} = 0.412 \times \frac{(\varepsilon reff + 0.3) \times (\frac{W}{h} + 0.264)}{(\varepsilon reff - 0.258) \times (\frac{W}{h} + 0.8)} \tag{5}$$

$$Leff = L + 2 \times \Delta L \tag{6}$$

where, c is the speed of light, εr is the relative permittivity of the dielectric, W is the width of the conventional patch antenna, L is the antenna length and h is the thickness of the substrate.

If the resonance frequency is fr = 3.38 GHz, εr = 4.4 and h = 1.6 mm, the width of the patch is WP = 26.99 mm, and the length of the patch is LP = 20.71 mm.

(2) *Calculating the width of the feeding microstrip line*

The feeding technique chosen in our antenna design is microstrip line, because it is simple to design and offers directional pattern and high gain, thanks to the comparison of the feeding techniques (microstrip line, coaxial probe, proximity coupling, and CPW) of the patch antennas already done in the work [8].

The width of the microstrip line that has an impedance of 50 Ω is 3.083, Eq. (7) [9]

$$Z_0 = \frac{120\pi}{\sqrt{\varepsilon_r}\left(\frac{W}{h} + 1.393 + 0.667\ln\left(\frac{W}{h} + 1.44\right)\right)} \tag{7}$$

With: Z0: the characteristic impedance of the microstrip line.

W: The width of the microstrip line and h: The thickness of the substrate.

To verify the antenna dimensions calculated with TLM, the microstrip conventional patch antenna, Fig. 1, was simulated with CST solver. The dielectric Epoxy FR-4 with relative permittivity εr = 4.4, thang loss equal to 0.025 and a thickness of 1.6 mm was chosen as substrate of the antenna. The copper annulled was chosen as a material for conducting elements (patch and ground plane), The width of the patch is WP = 26.99 mm, the length of the patch is LP = 20.71 mm, The width of the substrate is WS = 2*WP, the length of substrate is LS = 2*LP, the width of the feed line is WF = 3.083 mm, and the inset coordinates are isny = 5 mm and insx = 1 mm.

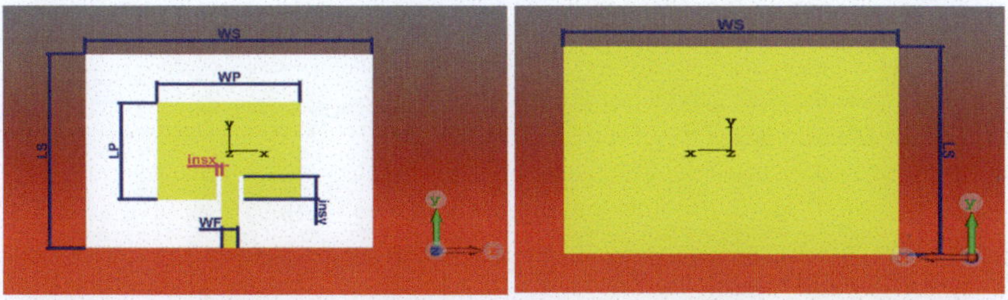

(a) (b)

Fig. 1 Conventional microstrip patch antenna, (a): top view and (b): back view

Figure 2 presents the S11 of the conventional patch antenna, the simulated resonance frequency is 3.33 GHz (The bleu curve), it is lower than the resonance frequency 3.38 GHz assumed in TLM. CST simulator is based on the full wave method, Finite Integration Technique (FIT), it is more accurate than the TLM method. Hence, we adjust the dimensions WP and LP of the patch until we find the resonance frequency equal to 3.38 GHz (The red curve). The new dimensions of the 3.38 GHz conventional patch are WP = 26.082 and LP = 20.431.

2.2 Design of the Metamaterial Antenna

As shown in Fig. 3, In the ground plane of the conventional patch antenna previously studied which resonate at 3.38 GHz, we etch two metamaterial unit cells, both of them have a distance X1 and X2 from the center of the ground plane of the patch, the parameters values of the two metamaterial structures are in (mm):

The first unit cell has G1 = 0.5, W1 = 0.5, S1 = 0.5, L1 = 10 and D = 7.

The second unit cell has G2 = 0.6, W2 = 0.6, S2 = 0.6, L2 = 4.5

In order to define the right values of X1 and X2 for which both unit cells were excited and make the new MTM antenna resonate at 2.45 and 5.8 GHz, a parametric study about

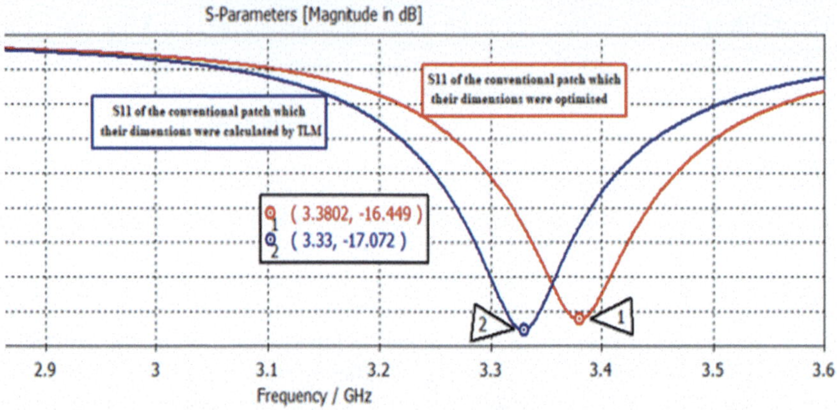

Fig. 2 S11 of the conventional antenna with TLM and full wave method

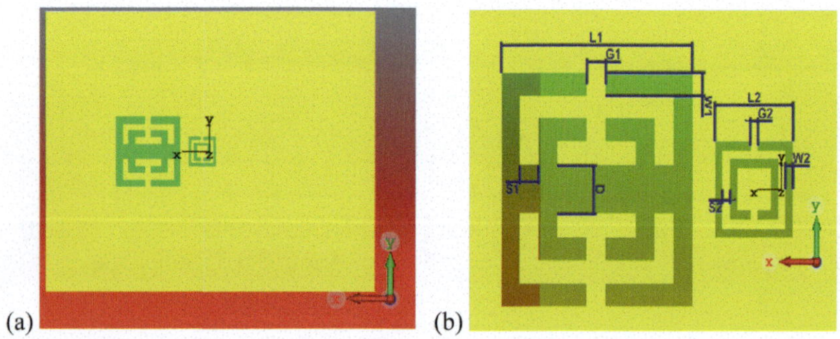

Fig. 3 (a): Back view of the MTM patch antenna, (b): the dimensions of the MTM unit cells

X1 and X2 was done, Fig. 4. We noticed that at X1 = 9.5 mm and X2 = 0.75 mm, the new MTM antenna resonate at the two main frequencies 2.45 and 5.8 GHz.

3 Simulation Results and Discussion

Figure 5 presents a comparison between the reflection coefficients curves of the 3.38 GHz conventional patch antenna without metamaterial structures and the conventional patch with the two metamaterial unit cells which embedded at its ground plane. We notice that after etching the two metamaterial unit cells on the ground plane of the conventional patch antenna, its resonance frequency was shifted from 3.38 to 2.45 GHz, and another band around 5.8 GHz was created. By this way of design, we have conceived a new dual band miniaturized antenna.

Following the same steps done before on the Sect. 2.1, we define, with TLM (Transmission Line Model), the dimensions of the conventional patch at 2.45 GHz.

Table 1 sums up the dimensions of both 2.45 GHz antennas, the conventional patch approximated by TLM and the proposed MTM antenna, as we can observe from the table, the achieved approximated reduction rate is 50%.

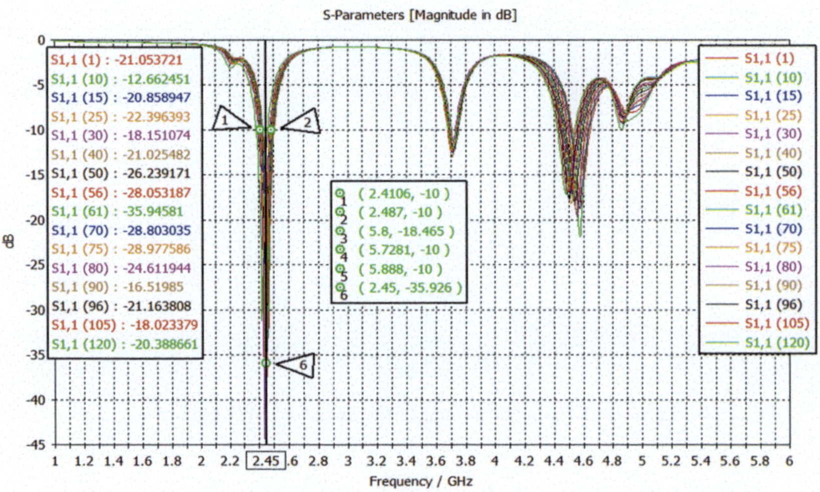

Fig. 4 Parametric study about X1 and X2 vs the S11 of the MTM antenna

Table 1 The dimensions of the 2.45 GHz conventional patch antenna and MTM antenna

Antenna	Volume (LS × WS × h) in mm^3
Conventional patch at 2.45 GHz (approximated by TLM)	57.62 × 74.46 × 1.6
The proposed antenna at 2.45 GHz	40.862 × 52.164 × 1.6

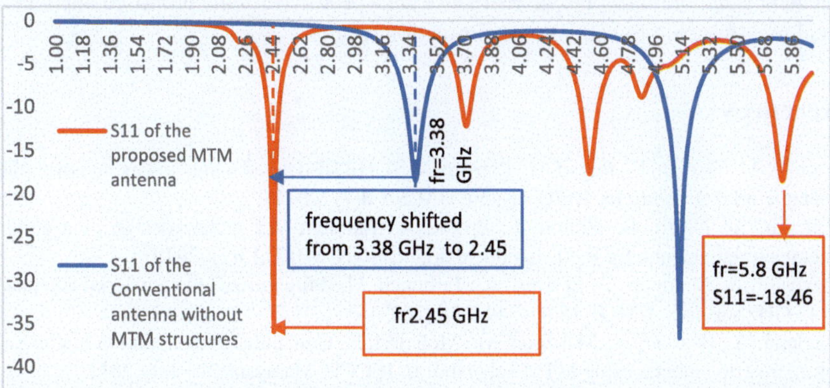

Fig. 5 Reflection coefficient (S11) of the MTM antenna and the conventional antenna

Figure 6 illustrates the reflection coefficient curve of the proposed antenna, in the inferior band, the bandwidth is 76.4 MHz, it is from 2.4106 to 2.487 GHz, and the S11 is − 35.946 dB at 2.45 GHz, in the superior band, the bandwidth is 160 MHz, it is from

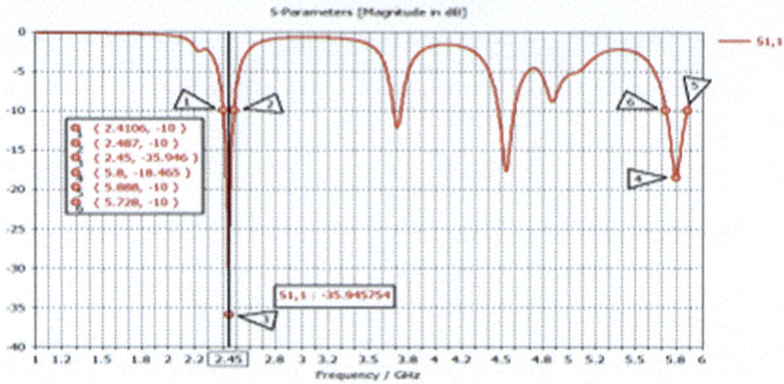

Fig. 6 Reflection coefficient of the designed MTM antenna

5.728 to 5.888 GHz, and the reflection coefficient for its central resonance frequency 5.8 GHz is − 18.465 dB.

4 Conclusion

In this work, we have designed by simulation a compact dual band metamaterial antenna, the lower band has a bandwidth of 76.4 MHz and a gain of 1.39 dB, the upper band has a bandwidth of 160 MHz and a gain of 2.1 dB. Those bands 2.45 and 5.8 GHz, are obtained by etching in the ground plane of a conventional patch antenna resonating at 3.38 GHz, two different metamaterial structures. The first unit cell shifts the resonance frequency from 3.38 to 2.45 GHz and the second unit cell generate the superior band. In the next stage of this work, we will fabricate a prototype of the proposed antenna and do real measurement to validate the simulation results.

References

1. Zamali, M., Osman, L., Regad, H., Latrach, M.: UHF RFID reader antenna using novel planer metameterial structure for RFID system. IJACSA **8**(7) (2017)
2. Ennajih, A., Nasiri, B., Zbitou, J., Errkik, A., Latrach, M.: A new design of a compact metamaterial antenna for RFID handheld applications. MJEE **13**(2) (2019)
3. Laabadli, A., Mejdoub, Y., El Amri, A., Tarbouch, M.: Miniaturized metamaterial antenna for 2.45 GHz services. IJMOT **18**(4), 349–358 (2023)
4. Gudarzi, A., Bazrkar, A., Mahzoon, M., Mohajeri, F.: Gain enhancement and miniaturization of microstrip antenna using MTM superstrates. ICCCE, Malaysia, 3–5 July 2012
5. Ennajih, A., Zbitou, J., Latrach, M., Errkik, A., Mandry, R.: A new dual band printed metamaterial antenna for RFID reader applications. IJECE **7**(6), 3507–3514 (2017)
6. Geetharamani, G., Aathmanesan, T.: A metamaterial inspired tapered patch antenna for WLAN/WiMAX applications. WPC **113**, 1331–1343 (2020)
7. Balanis, C.A.: Antenna Theory: Analysis and Design, 3rd edn. John Wiley, Hoboken (2005)
8. Barrou, O., El Amri, A., Reha, A.: Comparison of Feeding Modes for a Rectangular Microstrip Patch Antenna for 2.45 GHz Applications. ISUNet 457–469 (2016)
9. Huang, Y., Boyle, K.: Antennas: From Theory to Practice. Wiley, Chichester (2008)

Sentiment Analysis by Deep Learning Techniques

Abdelhamid Rachidi[1(✉)], Ali Ouacha[1], and Mohamed El Ghmary[2]

[1] Computer Science Laboratory (LRI), Computer Science Department, FSR, Mohammed V University, Rabat, Morocco
{abdelhamid.rachidi,a.ouacha}@um5r.ac.ma

[2] Department of Computer Science, FSDM, Sidi Mohamed Ben Abdellah University, Fez, Morocco
mohamed.elghmary@usmba.ac.ma

Abstract. This study focuses on sentiment analysis in Arabic texts using the embedding approach of AraBERT and various neural network architectures, including CNN, LSTM, GRU, BI-LSTM, and BI-GRU. The objective was to predict the sentiments associated with each comment. According to the findings, combining the embedding of AraBERT with the GRU and BI-GRU models demonstrated superior performance in precision, recall, F1 score, and accuracy measures. Nonetheless, it's worth noting that the AraBERT + LSTM model distinguishes itself due to its notably shorter learning time. This research opens new perspectives for sentiment analysis in Arabic using deep learning.

Keywords: Sentiment analysis · AraBERT · Word embedding · Deep learning · NLP · CNN · LSTM · GRU · BI-LSTM · BI-GRU

1 Introduction

The increasing importance of sentiment analysis in the field of social media, specifically focusing on sentiment analysis in Arabic texts, is noteworthy. Social media has generated a vast amount of data expressing views, opinions, and sentiments on various subjects. Sentiment analysis enables the extraction and classification of these sentiments as positive or negative, providing valuable insights for businesses and organizations. However, most studies in this field have primarily focused on the English language, leaving a research gap in Arabic sentiment analysis. The significance of sentiment analysis in Arabic as a research field has grown due to the rise of Arabic social media platforms and the language's increased complexity. The specificities of Arabic, such as complex verb forms, dialectal variations, and specific syntactic structures, require tailored approaches and models for accurate sentiment analysis. To address this challenge, the study utilizes the AraBERT embedding, combined with different neural network architectures such as LSTM, BI-LSTM, GRU, BI-GRU and CNN. The rest of this paper is organized as follows: in the second section, we preprocess the data. We provide an explanation of the simulation outcomes in the fourth section, and in the final section, we offer a conclusion.

Y. Farhaoui et al. (Eds.): ICAISE 2023, LNNS 837, pp. 393–398, 2024.
https://doi.org/10.1007/978-3-031-48465-0_51

2 Related Work

AI has gained significant attention in sentiment analysis research in recent years. Scholars have employed diverse AI methods, such as natural language processing (NLP), machine learning, and deep learning, to enhance sentiment analysis accuracy across various languages, including Arabic.I want to mention two approaches:

Machine learning approach: Different studies have used machine learning approaches such as Naïve Bayes, K-Nearest Neighbor, Support Vector Machine, Logistic Regression, and Decision Tree for sentiment analysis of Arabic tweets [1–4]. The results of these studies demonstrated that SVM and logistic regression achieved the highest accuracy rates in their respective domains [1, 2]. Additionally, some studies have specifically addressed the challenges related to the linguistic and cultural characteristics specific to the Arabic language [3].

Deep learning approach: In the realm of sentiment analysis for Arabic text, the authors of [5] introduced an approach that involved combining CNN and LSTM models, achieving an F1 score of 64.46% on the ASTD dataset. Mohammed and Kora in [6] evaluated the CNN, LSTM, and RCNN models, reporting an average accuracy of 81.31% for LSTM. Researchers in [7] evaluated different types of Arabic-specific embeddings, which led to improved sentiment analysis performance. Abu Kwaik et al. in [8] utilized the LSTM-CNN model, surpassing benchmark models with high accuracy rates. Elfaik and Nfaoui in [9] developed a bidirectional LSTM model, surpassing existing approaches with the use of deep contextualized embeddings. Fouad et al. in [10] proposed ArWordVec word embedding models, demonstrating high efficiency in word similarity tasks and sentiment analysis. Despite these significant contributions, none of these research efforts specifically addressed the weaknesses of the approach we present in this paper. In our work, we strive to bridge this gap by incorporating the pre-trained AraBERT model, combined with various neural network architectures such as LSTM, BI-LSTM, GRU, BI-GRU, and CNN. This approach stands out due to its reliance on pre-trained language models designed specifically for Arabic text, coupled with diverse neural network architectures, all aimed at enhancing sentiment analysis accuracy for Arabic data. Through the integration of these advanced techniques, we aim to overcome the challenges posed by the complex structure of the Arabic language, dialect variations, and limited resources, thereby contributing significantly to the field of sentiment analysis for Arabic text.

3 Proposed Methodology

3.1 Datasets

The dataset [11] includes 93,700 hotel reviews in both Modern Standard and Dialect Arabic, collected from Booking.com in June and July 2016. It is noteworthy for its balanced distribution of ratings, with 46,850 reviews in both positive and negative categories, ensuring linguistic diversity and fair sentiment class representation. This dataset stands out for its inclusion of both Modern Standard and dialectal Arabic, reflecting linguistic diversity in real-world Arabic text. Additionally, it focuses on hotel reviews, a domain featuring specialized language and unique sentiment expressions.

3.2 Preprocessing

Reviews were pre-processed to eliminate terms that had no impact on sentiment analysis, reducing the complexity of word encoding and making the model easier to train. Preprocessing steps include removing diacritics, punctuation (including Arabic commas), multiple spaces, emojis, Arabic stop words, repeated characters, HTML tags, HTML entities, characters non-Arabic, and lines containing the latter. Additionally, Arabic normalization was performed, followed by sequence vectorization with AraBERT [12]. The data was split into three sets (60% training, 20% validation, 20% testing) to ensure balanced representation, optimal generalization, and to prevent model overfitting.

3.3 AraBERT

The AraBERT model has been incorporated into our deep learning architectures using its pre-trained embeddings, thus effectively representing Arabic words. AraBERT was chosen because of its specificity to the Arabic language, capable of handling linguistic and dialectal complexities. Its proven performance in Arabic text processing and contextualized embeddings have been assets for accurate contextual sentiment analysis in Arabic text.

3.4 AraBERT CNN Architecture

CNNs are used to extract local features in text sequences, which is useful for pattern detection in comments. They use convolution filters to scan text and extract relevant information. The AraBERT CNN model is an architecture that combines the AraBERT model with 1D Convolutional Neural Network (1D-CNN) layers for text processing. It uses three parallel blocks of convolutional layers with different kernel sizes and filters, followed by Global-max-pooling layers. The outputs of the convolutional layers are then concatenated. Next, the model employs fully connected layers with a Dropout layer for regularization. Finally, an output layer with a sigmoid activation function is added for binary classification. This combination allows for the processing of large-scale structures and unstructured texts, leveraging both the features extracted by AraBERT and the contextual information captured by the convolutional layers (see Fig. 1).

3.5 AraBERT Recurrent Neural Networks Architecture

RNNs enable the model to understand text sequentially and capture essential dependency relationships, thereby contributing to a better understanding of the nuances and complexity of sentiments expressed in the Arabic language. The model is an architecture that combines the AraBERT model with a recurrent neural network layer (LSTM, GRU, BiLSTM, BiGRU). Then, a dense layer and a ReLU activation function are added to introduce non-linearity into the model and capture complex relationships between features. A dropout layer is applied to regularize the model and reduce the risk of overfitting. A second dense layer and ReLU activation are added, followed by another dropout layer. Finally, an output layer with one unit and a sigmoid activation function is used for binary classification (see Fig. 1).

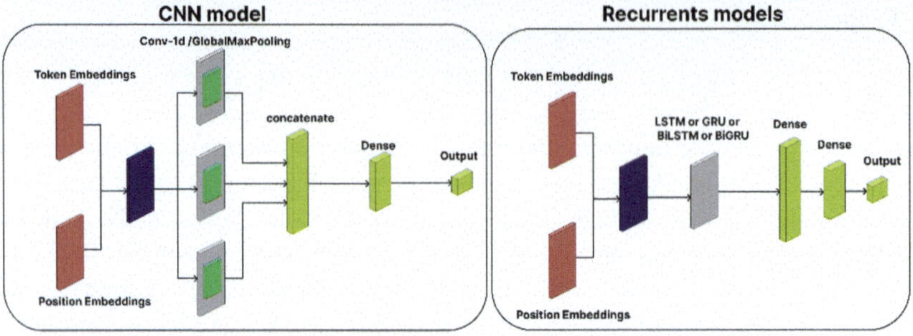

Fig. 1 AraBERT CNN and Recurrent Neural Architecture

4 Experiment and Results

This study focuses on sentiment analysis in Arabic texts using a deep learning approach. The approach adopted is based on the use of the pre-trained model Rabert, combined with different architectures of neural networks. The goal was to predict the sentiments associated with each comment. The simulation environment used was Google Colab. The results showed that the Arabert + GRU and Arabert + BiGRU models obtained the best performance, thus opening new perspectives.

4.1 Simulation and Hyperparameters

The simulation environment used was Google Colab Pro+, a platform that offers advanced features beyond the free version. These improvements include increased computing capabilities, expanded memory capacity, extended execution times, and priority access to valuable resources. Several hyperparameters were carefully tuned to optimize model performance. The maximum sequence length was set to 64 to define the maximum size of text sequences. The learning rate, set to 0.0001, controlled the rate at which the weights adapted during training. The Adam optimizer was used for weight updates, while activation functions included RELU for the hidden layers and Sigmoid for the output layer. Additionally, the Arabert model had 12 encoder layers. The batch size was 64, determining the number of training samples processed in each iteration, and the training was performed over 20 epochs. A dropout rate of 0.5 was applied to reduce the risk of overfitting. The training durations specific to each architecture were documented in the study (see Table 1).

4.2 Simulation Results

Table 1 and Fig. 2 present the performance results of the different models. all models showed good performance in binary sentiment classification for Arabic texts. The AraBERT + GRU and AraBERT + BiGRU models performed best in terms of accuracy, precision, recall, and F1 score. However, it is important to note that the AraBERT + LSTM model required the least amount of time to train. The results demonstrate the effectiveness of deep learning, particularly the AraBERT model, for sentiment analysis in Arabic texts, revealing its ability to capture and predict sentiment patterns in a complex linguistic context.

Table 1 Models results

Models	Accuracy (%)	Precision (%)	Recall (%)	F1-Score (%)	Training time (s)
AraBERT + CNN	93,32	91,338	95,101	93,182	7420,97
AraBERT + LSTM	92,98	92,957	93,001	92,979	6642,81
AraBERT + GRU	93,39	90,884	95,675	93,218	8524,39
AraBERT + BiLSTM	93,08	90,780	95,158	92,917	6734,34
AraBERT + BiGRU	93,42	92,294	94,422	93,346	7515,75

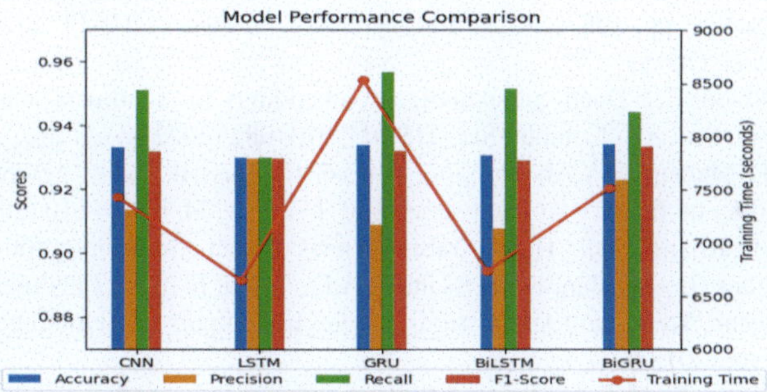

Fig. 2 Models results

5 Conclusion

This research focuses on sentiment analysis in Arabic texts using deep learning techniques. The main objective was to evaluate the effectiveness of the AraBERT model combined with different neural network architectures to predict the sentiments associated with each comment. The results show that the AraBERT + GRU and AraBERT + BiGRU models achieve the best performance. However, the AraBERT + LSTM model stands out for its shorter learning time. These results open the way to several interesting perspectives, including the use of alternative evaluation measures, the comparison with other traditional Arabic sentiment analysis approaches, the exploration of larger datasets or classification tasks. more complex, as well as adapting the models to other languages or similar tasks.

References

1. Boudad, N., Faizi, R., Thami, R.O.H., Chiheb, R.: Sentiment classification of Arabic tweets: a supervised approach. J. Mob. Multimed. **13**, 233–243 (2017)
2. Bolbol, N.K., Maghari, A.Y.: Sentiment analysis of Arabic tweets using supervised machine learning. In: Editor, F., Editor, S. (eds.) ICPET 2020, LNCS, vol. 9999, pp. 89–93. IEEE, Jerusalem, Palestine (2020)
3. Alwakid, G., Osman, T., Hughes-Roberts, T.: Challenges in sentiment analysis for Arabic social networks. Procedia Comput. Sci. **117**, 89–100 (2017)
4. Shoukry, A., Rafea, A.: Sentence-level Arabic sentiment analysis. In: 2012 International Conference on Collaboration Technologies and Systems (CTS), pp. 546–550. IEEE, Denver, Colorado, (2012)
5. Heikal, M., Torki, M.: Sentiment analysis of Arabic tweets using deep learning. Procedia Comput. Sci. **142**, 114–122 (2018)
6. Mohammed, A., Kora: Deep learning approaches for Arabic sentiment analysis. Social Netw. Anal. Min **9**(52), 99–110 (2019)
7. Barhoumi, A., Camelin, N., Aloulou, C., Esteve, Y., Hadrich Belguith, L.: An empirical evaluation of arabic-specific embeddings for sentiment analysis. In: Smaïli, K. (ed.) Arabic Language Processing: From Theory to Practice, pp. 34–48. 7th International Conference, ICALP 2019, Proceedings. Springer, France (2019)
8. Kwaik, K.A., Saad, M., Chatzikyriakidis, S., Dobnik, S.: LSTM-CNN deep learning model for sentiment analysis of dialectal Arabic. In: Smaïli, K. (ed.) Arabic Language Processing: From Theory to Practice, pp. 108–121. 7th International Conference, ICALP 2019, Proceedings. Springer, France (2019)
9. Elfaik, H., Nfaoui, E.: Deep contextualized embeddings for sentiment analysis of Arabic book's reviews. Procedia Comput. Sci. **215**(01), 973–982 (2022)
10. Fouad, M.M., Mahany, A., Aljohani, N., Abbasi, R., Hassan, S.-U.: ArWordVec: efficient word embedding models for Arabic tweets. Soft. Comput. **24**(1), 99–110 (2020)
11. Elnagar, A.M., Khalifa, Y.S.: Hotel Arabic-reviews dataset construction for sentiment analysis applications. In: Shaalan, K., Hassanien, A.E., Tolba, F. (eds.) Intelligent Natural Language Processing: Trends and Applications, Studies in Computational Intelligence, pp. 35–52. Springer, Egypt (2018)
12. Antoun, W., Baly, F., Hajj, H.M.: AraBERT: transformer-based Model for Arabic language understanding. In: Al-Khalifa, H., Magdy, W., Darwish, K., Elsayed, T., Mubarak, H. (eds.), Proceedings of the 4th Workshop on Open-Source Arabic Corpora and Processing Tools, with a Shared Task on Offensive Language Detection, pp. 9–15. American University of Beirut, Marseille, France (2020)

Examples of Diophantine Equations

Said Essahel[1(✉)] and Mohamed M. Zarrouk[2]

[1] Sciences and Engineering Laboratory, Sidi Mohammed Ben Abdellah University, Polydisciplinary Faculty of Taza, Taza-Gare PB, 1223 Taza, Morocco
essahel69@yahoo.fr

[2] Department of Mathematics, FSTE, University of Moulay Ismail, 52000 Meknes, Morocco

Abstract. In this paper we investigate in the Diophantine equations and we mention some contributions of these equations in the development of many area of mathematics. Then, by examples, we present some methods to solve Diophantine equations and for a prime integer p, we give a family (E_n) of these equations that have no solution in $\mathbb{Z}_p$ but have a solution modulo p^n for n as large as desired.

Keywords: Diophantine equations · Units · Group of units

MSC: 11D45 · 11D72 · 11D88

1 Introduction

A Diophantine equation is any polynomial equation with one or many variables in which the unknowns are integers or rational numbers. This definition may be generalized to any ring of integers of a number field. Although simple in expression and easy for the general public to understand, Diophantine equations are generally very difficult to solve. To solve them, we may need a number of mathematical tools, such as algebraic number theory, algebraic geometry, elliptic curves,modular forms.

The Diophantine equations have a great contribution in the development of mathematics. Around 1637, P. Fermat has conjectured that for any integer $n \geq 3$, the Diophantine equation

$$x^n + y^n = z^n$$

has no trivial solution $(x, y, z) \in \mathbb{Z}^3$. This famous equation was at the origin of the development of many branches of mathematics as algebraic number theory, class field theory and algebraic geometry. It also created a link between elliptic curves and modular forms.

Y. Farhaoui et al. (Eds.): ICAISE 2023, LNNS 837, pp. 399–405, 2024.
https://doi.org/10.1007/978-3-031-48465-0_52

2 Local Method for Resolving a Diophantine Equation

2.1 Introductory Examples

Let the Diophantine equation

$$(E_1) \quad x^2 + 7y^2 + 14z^3 = 10009$$

Let $(x, y, z) \in \mathbb{Z}^3$. If (x, y, z) is a solution of (E_1), then (x, y, z) will be a solution of (E_1) modulo 7. Then $\overline{x}^2 = -\overline{1}$ in $\mathbb{Z}/7\mathbb{Z}$. Since $\left(\frac{-1}{7}\right) = -1$ ($\left(\frac{*}{*}\right)$ denotes the Legendre symbol), then -1 can not be a square in $\mathbb{Z}/7\mathbb{Z}$. We conclude that (E_1) has no solution in $\mathbb{Z}^3$. We can generalize this method: Let (E) be a Diophantine equation. If it exists a prime integer p such that (E) has no solution modulo p, then (E) has no solution.

Let consider now the Diophantine equation

$$(E_2) \quad x^2 + 9y^2 + 9z^2 = 3$$

We can see that $(\overline{0}, \overline{1}, \overline{1})$ is a local solution modulo 3. Using Maple, we can see that (E_2) has no solution modulo 9:

```
> L:={}:
    for x from 1 to 4 do
        for y from 1 to 4 do
            for z from 1 to 4 do
                if irem(x^2+9*y^2+9*z^2,9)=3 then
                  L:=L union{(x,y,z)}:
                fi
            od:
        od:
   od;
   L;
```

$$\{\}$$

We conclude that (E_2) has no solution in $\mathbb{Z}^3$. So, for a prime integer p, it's possible that a Diophantine equation has a solution modulo $p, p^2, p^3, ...$ and p^n, but it has no solution modulo p^{n+1}. The local study of the equation at p signify to see if it has a solution modulo p^n for every $n \in \mathbb{N}$. There is a set denoted $\mathbb{Z}_p$ which was invented for this.

2.2 Recalls on $\mathbb{Z}_p$

Let $i, j \in \mathbb{N}^*$ such that $i \leq j$, and consider the canonical projection $f_{j,i}$: $\mathbb{Z}/p^j\mathbb{Z} \longrightarrow \mathbb{Z}/p^i\mathbb{Z}$, $x \mod (p^j) \longmapsto x \mod (p^i)$. The system $(\mathbb{Z}/p^i\mathbb{Z}, f_{j,i})$ is projective. We define $\mathbb{Z}_p$ as the inverse limit of this system. Then $\mathbb{Z}_p$ can be seen as

$$\mathbb{Z}_p = \{(x_i)_{i\in\mathbb{N}^*} \in \prod_{i\in\mathbb{N}^*} \mathbb{Z}/p^i\mathbb{Z} / f_{j,i}(x_j) = x_i, \forall j \geq i\}$$

There is another construction of $\mathbb{Z}_p$. In fact the map $\mathbb{Z} \longrightarrow \mathbb{R}$, $z \longmapsto p^{-v_p(z)}$, where $v_p(z)$ is the p-adic valuation of z, is an absolute value. $\mathbb{Z}_p$ is the completion of $\mathbb{Z}$ for this absolute value.

We recall some properties of $\mathbb{Z}_p$:

◇ $\mathbb{Z}_p$ is a principal ring with only one maximal ideal which is $p\mathbb{Z}_p$.
◇ There is a valuation $v_p : \mathbb{Z}_p \longrightarrow \mathbb{N}$ and for all $z \in \mathbb{Z}_p$, we have

$$z \text{ is a unit of } \mathbb{Z}_p \Longleftrightarrow v_p(z) = 0$$

◇ $\mathbb{Z}_p$ can be defined as the set of the formel sum $\sum_{n\in\mathbb{N}} a_n p^n$ with $a_n \in \{0, 1, ..., p-1\}$ for all $n \in \mathbb{N}$. The valuation v_p is now defined for $z = \sum_{n\in\mathbb{N}} a_n p^n$ by $v_p(z) = \inf(\{n \in \mathbb{N} / a_n \neq 0\})$. This definition of $\mathbb{Z}_p$ allows us to see $\mathbb{Z}$ as a subring of $\mathbb{Z}_p$, and generalize the dveloppement of an integer in the p-adic basis of numeration.
◇ A Diophantine equation (E) admits a solution in $\mathbb{Z}_p$ if and only if it admits a solution in $\mathbb{Z}/p^n\mathbb{Z}$ for all $n \in \mathbb{N}^*$.

For more information about $\mathbb{Z}_p$ you can see [2,3,5,6] or [8].

2.3 New Examples by Local Method

We can easily see that if a Diophantine equation (E) has a solution in $\mathbb{Z}$, then it will have a solution in $\mathbb{Z}_p$ for all prime integer p. We deduce that if it exists some prime integer p and integer $n \in \mathbb{N}^*$ such that (E) has no solution modulo p^n, then (E) has non solution in $\mathbb{Z}$. For this reason, to show that (E) has no solution, we try to find a prime p and an integer n such that (E) has no solution modulo p^n, but this is not easy to do in general. Next, for all prime integer p, we construct a family of Diophantine equations (E_n) such that $\forall n \geq 2$, (E_n) has a solution modulo $p, p^2, ...$ and p^{n-1} but has no solution modulo p^n.

First let $p = 3$. We have already defined (E_2) above. Now, let the Diophantine equation

$$(E_3) \quad x^3 + 27y^2 + 27z^2 = 9$$

It's clear that (E_3) has a solution modulo 3 and 9. Using Maple, we can see that (E_3) has no solution modulo 27.

```
> L:={}:
    for x from 1 to 26 do
      for y from 1 to 14 do
        for z from 1 to 14 do
          if irem(x^3+27*y^2+27*z^2,27)=9 then
            L:=L union{(x,y,z)}:
```

```
          fi
        od:
      od:
    od;
  L;
```

{}

For $n = 4$, let the Diophantine equation

$$(E_4) \quad x^4 + 81y^2 + 81z^2 = 27$$

It's clear that (E_3) has a solution modulo 3 and 9 and 27. Using Maple, we can see that (E_4) has no solution modulo 81.

```
> L:={}:
    for x from 1 to 40 do
      for y from 1 to 40 do
        for z from 1 to 40 do
          if irem(x^4+81*y^2+81*z^2,81)=27 then
             L:=L union{(x,y,z)}:
          fi
        od:
      od:
    od;
  L;
```

{}

Now, we generalize the above examples:

Theorem 1. *Let p be a prime integer and n a positive integer. Consider the Diophantine equation*

$$(E_n) \quad x^n + p^n y^2 + p^n z^2 = p^{n-1}$$

Then (E_n) has locally a solution modulo p^k for $k = 1, 2, ..., n-1$, but has no solution modulo p^n, and then (E_n) has non solution in $\mathbb{Z}^3$.

Proof. Let $n \in \mathbb{N}^*$ and $k \in \{1, 2, \ldots, n-1\}$. It's clear that $(0, 1, 1)$ is a solution of (E_n) modulo p^k. If (x, y, z) is a solution modulo p^n, then $x^n \equiv p^{n-1} \mod (p^n)$. We deduce that p divides x. Let $x = pa$ with $a \in \mathbb{Z}$. Then it exists $b \in \mathbb{Z}$ / $a^n p^n = p^{n-1} + bp^n$, and thus p divides 1 and this is impossible. We conclude that (E_n) has no solution modulo p^n.

3 Examples of Diophantine Equations

3.1 Method Using Geometrical Arguments

Let consider the Diophantine equation

$$(E): \quad x^2 + y^2 + 4z^4 = 10000$$

If (x, y, z) is a solution of (E) then, we have:

- $(-x, y, z)$, $(x, -y, z)$, $(x, y, -z)$, $(-x, -y, z)$, ... are also solutions of (E). We deduce that we can study (E) only for x, y and z positive.
- (y, x, z) is a solution of (E).
- $|x| \leq 100, |y| \leq 100$ and $|z| \leq 7$.

Then, even if it's difficult to solve (E) algebraically, we can easily solve it numerically. In what follows, we give a program in Maple solving (E)

```
> for x from 0 to 100 do
        for y from 0 to 100 do
             for z from 0 to 7 do
                 if x^2+y^2+4*z^4=10000 then
                    print(x,y,z):
                 fi:
             od:
        od:
  od;

                         0, 100, 0
                         28, 96, 0
                         60, 80, 0
                         80, 60, 0
                         96, 28, 0
                         100, 0, 0
```

In conclusion, the solutions (x, y, z) in $\mathbb{Z}^3$ of (E) with x, y and z positive are $(0, 100, 0), (28, 96, 0), (60, 80, 0), (100, 0, 0), (96, 28, 0)$ and $(80, 60, 0)$. The other solutions can be deduced by symmetries.

3.2 Methods Derived from Algebraic Number Theory Arguments

Let consider now the Diophantine equation

$$(E) \quad x^3 + 2y^3 + 4z^3 - 6xyz = \pm 1$$

Let $K = \mathbb{Q}(\sqrt[3]{2})$. The ring of integers of K is $O_K = \mathbb{Z}[\sqrt[3]{2}]$, and an integral basis of O_k is $B = (1, \sqrt[3]{2}, \sqrt[3]{4})$. Let $u \in O_k$. Then, there are 3 integers $x, y, z \in \mathbb{Z}$ such that

$$u = x + y\sqrt[3]{2} + z\sqrt[3]{4}$$

Let $f_u : O_K \longrightarrow O_k$ defined by $f_u(v) = uv$. f_u is a linear map and its matrix relatively to B is

$$M_u = \begin{pmatrix} x & 2z & 2y \\ y & x & 2z \\ z & y & x \end{pmatrix}$$

We deduce that the norm of u is $N_{K/\mathbb{Q}}(u) = det(M_u) = x^3 + 2y^3 + 4z^3 - 6xyz$, and (x, y, z) is a solution of (E) if and only if $N_{K/Q}(u) = \pm 1$. It's well known that for any number field L, a number $v \in L$ is a unit of L if and only if $v \in O_L$ and $N_{L/Q}(v) = \pm 1$ (see [7]). It comes that

$$(x, y, z) \text{ is a solution of } (E) \Longleftrightarrow u = x + y\sqrt[3]{2} + z\sqrt[3]{4} \text{ is a unit of } K = Q(\sqrt[3]{2})$$

Then to finalize the resolution of (E) we have to determine U_K, the group of units of K. The next theorem is well known in number theory and we can find it in ([1, 7]).

Theorem 1. *If K is a number field and (r, s) is the signature of K, then the group U_K of units of K is isomorphic to $W \times \mathbb{Z}^{r+s-1}$ where W is the group of elements of K^* with finit order.*

Applying this theorem to $K = \mathbb{Q}(\sqrt[3]{2})$, we have $W = \{-1, 1\}$ and $(r, s) = (1, 1)$. We deduce that $U_K \simeq \{-1, 1\} \times \mathbb{Z}$. Using Pari/Gp [4], we can determine the fundamental unit of K:

```
 gp > P=x^3-2
%1 = x^3 - 2
 gp > K=nfinit(P);              \\  K is the number field defined by P.
 gp > K.zk                      \\ To give an integral basis of K.
%3 = [1, x, x^2]
 gp > K.sign
%4 = [1, 1]
 gp > L=bnfinit(K);
 gp > L.fu
%5 = [Mod(x - 1, x^3 - 2)]
```

Then the fundamental unit of K is $w = \sqrt[3]{2} - 1$, and then $U_k =< -1, w >= \{\pm w^n,\ n \in \mathbb{Z}\}$. We conclude that for all $n \in \mathbb{Z}$, there are two solutions s_n and $-s_n$ of (E) corresponding to the units w^n and $-w^n$. as examples:

- For $n = 1$, we have $s_1 = (-1, 1, 0)$ and $-s_1$.
- For $n = 2$, we have $s_2 = (1, -2, 1)$ and $-s_2$.

References

1. Janusz, G.J.: Class field theory (1995)
2. Katok, S.: p-adic analysis compared with real. Student Mathematical Library, 37. American Mathematical Society, Providence, RI; Mathematics Advanced Study Semesters, University Park, PA (2007)
3. Neukirch, J.: Algebraic Number Theory. Springer (1999)
4. The Pari Group, Univ. Bordeaux, Pari/GP version 2.9.3 (2017). http://pari.math.ubordeaux.fr/
5. Robert, A.M.: A Course in p-adic Analysis. Springer-Verlag (2000)
6. Schikhof, W.: Ultrametric calculus: an introduction to p-Adic analysis (Cambridge Studies in Advanced Mathematics). Cambridge University Press, Cambridge (1985). https://doi.org/10.1017/CBO9780511623844
7. Samuel, P.: Thorie algbrique des nombres. Amer. Math. Monthly **78**, 806 (1971)
8. Serre, J.P.: Cours d'arithmtique. Math. Gaz. **55**, 342 (1971)

Query Optimization Using Indexation Techniques in Datawarehouse: Survey and Use Cases

Mohamed Ridani[1](✉) and Mohamed Amnai[2]

[1] Mohamed Ridani Laboratory of Research in Informatics (LaRI), Faculty of Science, Ibn Tofail University, Kenitra, Morocco
mohamed.ridani@uit.ac.ma

[2] Mohamed Amnai Laboratory of Research in Informatics (LaRI), Faculty of Science, Ibn Tofail University, Kenitra, Morocco
mohamed.amnai@uit.ac.ma

Abstract. The query optimization process is the big challenge of most of decision support systems around the world, such systems based on data warehouse structure to store all data to be analyzed. Moreover, the volume and variety of data on the data warehouse are very important which involves more effort and investment to minimize the cost of refreshment and data update. Our aim in this paper is to describe the state of the art of research works dealing with the optimization approaches in big data environments especially, the techniques based on indexes selection problems. We will explain the principle of each technique and its particularities approved by our experimental data. In addition, we'll conclude by the comparison of all these approaches developed in the literature.

Keywords: Query optimization techniques · OLAP · Data warehouse · Index problem selection

1 Introduction

In contrast to traditional systems, which are frequently referred to as transactional systems, which were initially designed to process and record data in real time without taking into account the needs of analysis, decision support systems (DSS) were initially developed to simplify the process of online analysis OLAP data (On-Line Analytical Processing). As a result, these systems are focused on decision-supporting applications and are organized around a "Data warehouse," which is a set of Datamart data stores, each of which represents an extract from the warehouse dealing with a specific theme. A data warehouse is a collection of integrated data that is subject-oriented, non-volatile, historical, summarized, and available for interrogation.

However, as big data, the Internet of Things (IOT), and the rapid growth of data proliferate, queries are growing more complicated and response time is crucial. To address this, a number of optimization techniques, including the usage of materialized views,

Y. Farhaoui et al. (Eds.): ICAISE 2023, LNNS 837, pp. 406–412, 2024.
https://doi.org/10.1007/978-3-031-48465-0_53

indexing strategies, and others, have been developed in the literature. We will provide an overview of query optimization using index approaches in this paper using our experimental use case.

The remainder of this paper is structured as follows: The global context of the query optimization problem is defined in the second section. The third section provides a brief overview of some approaches to solving the query optimization problem based on indexation techniques. The fourth section then presents our practice result which present the summary comparison values with graphics presentation. Finally, we provide some perspectives and a conclusion in the final part.

2 Query Optimization Problem in Datawarehouse

Indeed, as business needs continue to grow, data warehouse tasks become more challenging in terms of performance, particularly when looking for data, and this has an impact on user experience [1]. In this fact, it is essential to have methods for fast gathering crucial data, allowing for multiple analyses, while also converting the collected data into information the business can use. Several query optimization strategies have been described in the literature. These strategies are mainly materialized view [2, 3], vertical fragmentation, horizontal fragmentation, indexes, and cache mechanism, however the fragmentation is a technique [4, 5].

Furthermore, the index approaches such as B-Tree Indexes, Hash indexes, Bitmap index and Join index consists of additional structures in order to speed up query evaluation by the fast location of data. Regardless of the optimization technique used, the addition of these objects in a setting with a lot of data necessitates a greater investment in terms of storage space and the performance impact, which can be one of the limitations that have made the object subject to the optimization method itself. As a result, the next part implements new strategies that allow for the discovery of subsets that have a favorable impact on performance while taking into consideration all the limitations associated with each proposed strategy. Index-based optimization approaches in the big data environment surrounding the data warehouse are the only ones we have focused on in this research.

In theory, it is possible to index on the most used columns, however, this slows down data insertion and deletion and requires a much larger amount of memory for storing indexing data, but in practice it is very important to balance such a trade-off, moreover, the size of the database remains a challenge.

In the literature there is many types of indexes techniques like: (1) B-Tree Indexes is a directed tree structure, (2) Hash index is a technique, (3) Bitmap index which are most appropriate on low distinct cardinality data as opposed to B-Tree indexes. However, the Bitmap index will require less space than the B-Tree index. Bitmap indexes are very useful when created on columns with low cardinality, used with the AND & OR operator in the query condition.

3 Related Works

With the advent of big data, the data warehouse field has undergone a major evolution in recent years, both in terms of data and associated optimization techniques. In our previous works Mohamed and Amnai [2, 3]. We have been proposed a survey of optimization techniques based on materialized views in data warehouse, in this study we have been described many approaches using different algorithms to solve a problem of selection of materialized views MVP to be selected in order to reduce the query execution process and query response time. Indeed, due to space limitations and the variety of approaches developed based on indexing techniques, we will focus our work in this paper on the two techniques B-tree index and Bitmap index with both simple and compressed versions. other approaches will be the subject of our future work.

Several approaches have been proposed in the research field, including Bayer et al. in Bayer and McCreight [6] are among the researchers of the 70 of the last twentieth century who invented the B-Tree indexing technique for the organization and maintenance of indexes in random access files. O'Neil and al in O'Neil and Quass [7] proposed two approaches based on two different indexing structures respectively: Projection indexing and Bit-Sliced indexing, Wu et al. in this paper proposed an encoded bitmap indexing [8]. Paradias et al. in their paper Papadias [9], describe the framework for supporting OLAP operation over spatiotemporal Datawarehouse by modeling the spatial and temporal dimensions and purpose several novel approaches to deal with spatio-temporal query called aRB-tree (aggregate R-B-tree), aHRB-tree (aggregate Historical R-B-tree) which combines the concepts of aRB-trees and HR-trees and compared it's with Data Cube from olap (column scanning). Dehne et al. have been presented a distributed multi-dimensional ROLAP indexing scheme which is practical to implement, requires only a small communication volume, and is fully adapted to distributed disks [10].

Wu et al. in [11] gives an analytical comparison of bitmap indexes, most of which are in the class of multi-component bitmap indexes and incorporate the effects of compression on their performance. Sharma et al. focuses on explanation of five indexing techniques used in data warehouse and their use (B-Tree index, Pure Bitmap index, Encoded bitmap index, Bitmap Join Index, and Projection index) [12]. Zhu et al. in their article proposes a novel hash-based method named sparse hashing to focus on generating binary codes and encoding unseen data effectively and efficiently to implement approximate similarity search. [13]. Bornaz et al. presents a new n-dimensional cube indexing algorithm derived from B-Tree index (The n-Tree index algorithm which is a generalized B-tree index [14]. Yildiz et al. propose an approach based on bitmap indexes to perform a membership queries (Identifying whether a given set of values is a subset of Attribute values, which is discrete in nature and has classification as its characteristics) [15].

Abdalla et al. In this paper, purpose a new indexing algorithm which the data clustering based on the Sparse Fuzzy-c-means (Sparse FCM) algorithm [16]. Agrawal el al proposed in their approaches [17] an automated tools for selection of materialized views and indexes for SQL workloads based on one architecture with four modules. Naeem et al. presents a memory optimization algorithm for rapidly joining streaming data with persistent master data in order to reduce data latency during the transformation phase of ETL (Extraction, Transformation, and Loading) [18]. Raza et al. in their study propose a cluster-based autonomic performance prediction framework using a case-based reasoning approach that determines the performance metrics (such as: precision, recall, accuracy, and relative error rate) of the data warehouse in advance by incorporating autonomic computing characteristics [19]. The paper, argue for reconsidering prefix trees as in-memory index structures and present the generalized tries, which is a prefix tree with variable prefix length for indexing arbitrary data types of fixed or variable length [20].

4 Experimental Result: Views and Interpretations

In fact, we consider for our experimental area, the star schema composed of fact table and 4 dimensions like base structure of our data warehouse with initial various dataset, to evaluate the performances in our schemes, we consider the workload with 12 queries. Moreover, the queries have been differentiated by the cardinality of attribute used to filter the data to be searched, and to study and evaluate the performance of query using two variation of the index techniques such as: B-tree index and basic Bitmap index. Additionally, the goal is to explain the impact of using or not of the indexes technique in the global storage space and the response time of each query. Also, we have considered four levels cardinality attributes used in the creation index process: Low level, Medium and high to evaluate the query performances using different indexes. Figures 1 and 2 shows respectively the Data Set size for our dimensions Before/After using B-Tree Indexing Techniques and the comparison of B-Tree Index and Bitmap Index according to the cardinality of indexed columns. The bleu diagram refers to the size of dimension before creation of B-Tree index and red one present the size of associate index, The chart indicates that for all cases the size of index is at least greater than a quart of dataset i.e., 23, 25, 33, and 33% for DIM_CUSTOMER, DIM_DATE, DIM_GEOGRAPHY and DIM_PRODUCT.

Figures 3, 4, and 5 illustrates the performance evaluation measured by the response time using two indexing techniques: B-TREE and Basic Bitmap index for three query conditioning operator—equality, bound, interval—respectively beginning from lower cardinality as shown in Fig. 3 to high cardinality as illustrated in Fig. 5.

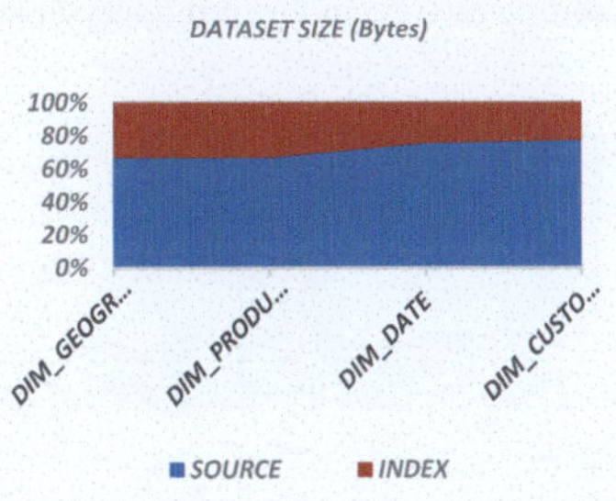

Fig. 1 Data Set size for our dimensions Before/After Indexing Techniques (B-Tree index)

Finally, after having seen the impact of using the indexing technique via the B-TREE index and the basic Bitmap index, the results of our experiment show that these two techniques are very limited in terms of performance, with the exception of some benefits concerning low cardinality for the bitmap index and high cardinality for the BTREE indexes, depending on the type and size of the data in our experiment. In our

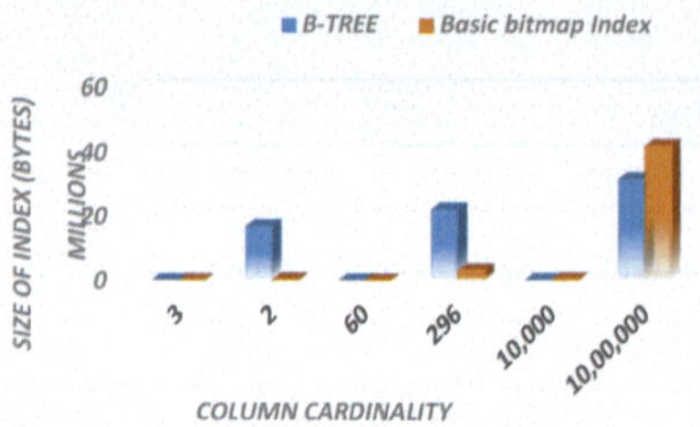

Fig. 2 Index Size for Column indexed according to cardinality Level using Indexing Techniques (B-Tree index and Bitmap Index)

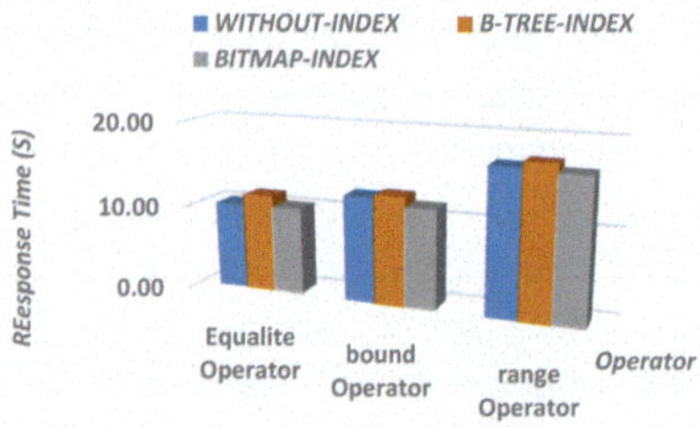

Fig. 3 Response Time according of type index and query operator for Low Cardinality

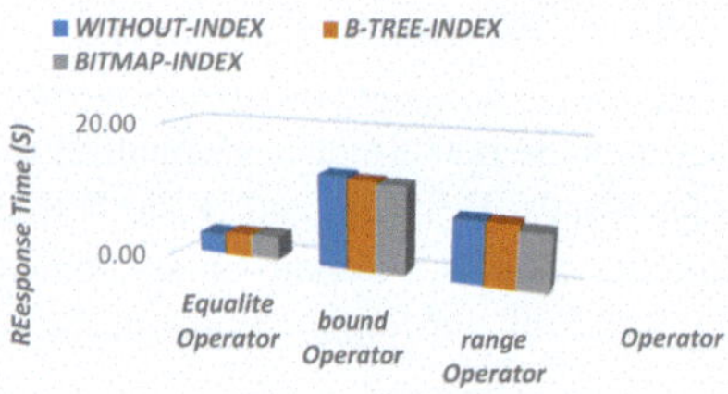

Fig. 4 Response Time according of type index and query operator for High Cardinality

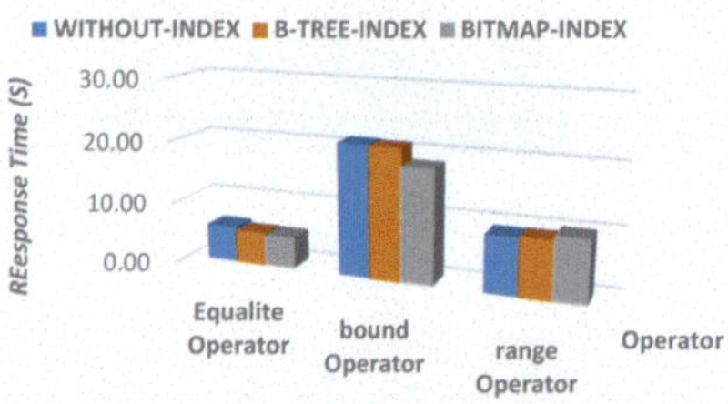

Fig. 5 Response Time according of type index and query operator for Medium Cardinality

future works we will focus the research axis on the frameworks that could out perfume better in the choice of columns to be indexed and the type of index that can improve the performance according to the cardinality, nature and size of data trying to use data collected from the Internet to model the diversity of data.

5 Conclusion

In this paper an overview of approaches dealing with the indexing techniques selection problems has been checked, starting with the determination of the global context of the indexing techniques in the literature followed by study the impact of these indexes' methods on the performance's issues such as the request response time and a strong influence on storage constraint. Our works has been limited on the BTREE and basic Bitmap index in order to initiate the study for other techniques' and to define the behaviors of and the parameters used in the optimization environment. Consequently, new challenges have been embarking to choose hybrid techniques with combination of materialized views and columns to be indexed.

References

1. Hassan, C.A.U., et al.: Optimizing the performance of data warehouse by query cache mechanism. IEEE Access **10**, 13472–13480 (2022)
2. Mohamed, R., Amnai, M.: Optimization challenge in decision supporting systems: an overview. In: 2022 IEEE 3rd International Conference on Electronics, Control, Optimization and Computer Science (ICECOCS). IEEE
3. Ridani, M., Amnai, M.: Materialized views selection problem in decision supporting systems: issues and challenges. J. Comput. Commun. **10**(9), 96–112 (2022)
4. Bellatreche, L., Boukhalfa, K.: An evolutionary approach to schema partitioning selection in a data warehouse. In: Data Warehousing and Knowledge Discovery: 7th International Conference, DaWaK 2005, Copenhagen, Denmark, 22–26 Aug 2005. Proceedings 7. Springer, Berlin (2005)
5. Noaman, A.Y., Barker, K.: A horizontal fragmentation algorithm for the fact relation in a distributed data warehouse. In: Proceedings of the Eighth International Conference on Information and Knowledge Management (1999)
6. Bayer, R., McCreight, E.: Organization and Maintenance of Large Ordered Indices. In: Proceedings of the 1970 ACM SIGFIDET (Now SIGMOD) Workshop on Data Description, Access and Control (1970)
7. O'Neil, P., Quass, D.: Improved query performance with variant indexes. In: Proceedings of the 1997 ACM SIGMOD International Conference on Management of Data (1997)
8. Wu, M.-C., Buchmann, A.P.: Encoded bitmap indexing for data warehouses. In: Proceedings 14th International Conference on Data Engineering. IEEE (1998)
9. Papadias, D., et al.: Indexing spatio-temporal data warehouses. In: Proceedings 18th International Conference on Data Engineering. IEEE (2002)
10. Dehne, F., Eavis, T., Rau-Chaplin, A.: Parallel multi-dimensional ROLAP indexing. in CCGrid 2003. In: 3rd IEEE/ACM International Symposium on Cluster Computing and the Grid, 2003. Proceedings. IEEE (2003)
11. Wu, K., Shoshani, A., Stockinger, K.: Analyses of multi-level and multi-component compressed bitmap indexes. ACM Trans. Database Syst. (TODS) **35**(1), 1–52 (2008)
12. Sharma, N., Panwar, A.: A comparitive study of indexing techniques in data warehouse. Int. J. Multidiscipl. Sci. Eng. **3**(7) (2012)
13. Zhu, X., et al.: Sparse hashing for fast multimedia search. ACM Trans. Inform. Syst. (TOIS) **31**(2), 1–24 (2013)
14. Bornaz, L.: Optimized data indexing algorithms for OLAP systems. Database Syst. J. **1**(2), 17–26 (2010)

15. Yildiz, B., et al.: Parallel membership queries on very large scientific data sets using bitmap indexes. Concurr. Comput. Pract. Exp. **31**(15), e5157 (2019)
16. Abdalla, H.B., Ahmed, A.M., Al Sibahee, M.A.: Optimization driven MapReduce framework for indexing and retrieval of big data. KSII Trans. Internet Inform. Syst. (TIIS) **14**(5), 1886–1908 (2020)
17. Agarawal, S., Chaudhuri, S., Narasayya, V. (2000). Automated selection of materialized views and indexes for SQL databses. In: Proceedings of 26th International Conference on Very Large Databases, Cairo, Egypt
18. Naeem, M.A., et al.: Big data velocity management-from stream to warehouse via high performance memory optimized index join. IEEE Access **8**, 195370–195384 (2020)
19. Raza, B., et al.: Autonomic performance prediction framework for data warehouse queries using lazy learning approach. Appl. Soft Comput.Comput. **91**, 106216 (2020)
20. Boehm, M., et al.: Efficient in-memory indexing with generalized prefix trees. Datenbanksysteme für Business, Technologie und Web (BTW) (2011)

Virtual Machine Selection in Mobile Edge Computing: Computing Resources Efficiency

Sara Maftah[1(✉)], Mohamed El Ghmary[2], and Mohamed Amnai[1]

[1] Department of Computer Science,Faculty of Sciences, Ibn Tofaïl University, Kenitra, Morocco
sara.maftah@uit.ac.ma, mohamed.amnai@uit.ac.ma

[2] Department of Computer Science, FSDM,Sidi Mohamed Ben Abdellah University, Fez, Morocco
mohamed.elghmary@usmba.ac.ma

Abstract. IoT devices receive an enormous amount of data every second, necessitating storage space for secure data retention and scalable computational resources. Mobile Edge Computing has emerged as an efficient approach to meet real-time computation needs without excessive energy consumption. MEC utilizes computation offloading to transfer a portion of work to the Edge server, enabling the selection of virtual machines that optimize energy and time in distributed and heterogeneous environments. This study focuses on virtual machine selection within Mobile Edge Computing, aiming to enhance the efficiency of computational resources. Through the evaluation of virtual machine performance using computation offloading, we aim to identify strategies and algorithms that optimize energy and time for task execution. The findings provide valuable insights for improving the efficiency of computational resources in MEC-based IoT systems.

Keywords: Mobile edge computing · Computation offloading · Internet of things · Energy efficiency · Virtual machine selection

1 Introduction and Literature

Mobile Edge Computing (MEC) has emerged as a transformative paradigm, bringing computational capabilities closer to mobile devices and end users, enabling low-latency and high-bandwidth applications. MEC harnesses edge network resources like servers and base stations to offload compute-intensive tasks, fostering reduced latency, scalability, and resource optimization. A critical aspect in MEC's optimization is selecting suitable virtual machines (VMs). VM selection, influenced by factors like processing power, memory, and connectivity, dynamically aligns computational needs with available resources. However, VM selection faces unique challenges due to resource heterogeneity, dynamic

Y. Farhaoui et al. (Eds.): ICAISE 2023, LNNS 837, pp. 413–419, 2024.
https://doi.org/10.1007/978-3-031-48465-0_54

application requirements, and network conditions. Addressing these, we aim to pioneer algorithms that optimize VM allocation, considering MEC's dynamic environment and diverse app needs. Improved VM selection enhances MEC performance, scalability, and resource use, propelling advanced edge applications. Through rigorous experiments, we seek to validate our algorithms' efficacy in real-world MEC settings, contributing to MEC's evolution and intelligent resource management design for future deployments.

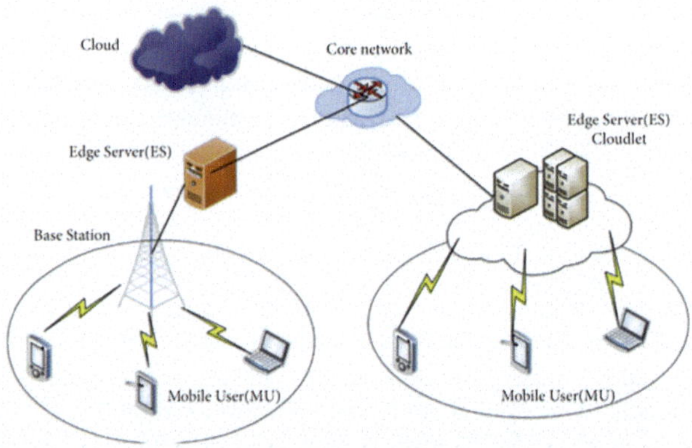

Fig. 1. MEC architecture

An effective edge computing approach for intelligent IoT applications was studied by [1] to integrate a delay-aware task graph partition algorithm and an optimal virtual machine selection method to minimize edge resource occupancy and meet QoS requirements, with performance evaluations confirming its effectiveness. Furthermore, a cost-effective mobile edge computing solution was introduced [2], it employs an intelligent decision model to reduce processing, memory, and energy while enhancing virtual mobile instance performance on physical devices, along with a redesigned genetic-based method to expedite offloading evaluations. Latency consumption minimization is emphasized in mobile edge computing networks [3]. Joint proposals for computation offloading, content caching, and resource allocation reduce latency and are demonstrated through enhanced allocation and decision strategies [3]. In a multi-user mobile edge computing system, the joint optimization of computation offloading, data compression, and resource allocation surpasses benchmark schemes in terms of energy consumption and finite MEC computation capacity [4]. Mobile edge computing also involves service caching, computation offloading, and resource allocation, with simulation results showcasing effective system cost minimization [5]. Edge computing balances server loads, improves system robustness, and reduces user delays, especially in emerging applications like mobile payments and facial recognition [6,7]. Enhancing user experience is highlighted through mobile edge computing and network function virtualization, innovative optimization problems are formulated, and promising algorithms are developed for experimental simulations [8]. Energy cost and packet congestion are further minimized through mobile edge computing optimization [9]. Authors in [10,11] adopted a Simulated Annealing algorithm to decide the tasks' offloading and it resource allocation. A joint approach to computation offloading and transmission is optimized to reduce energy consumption and meet latency requirements [12]. Moreover, mobile edge computing holds the promise of enhancing QoS and user experience, the authors in [13] proposed an algorithm that optimizes computation offloading and scheduling. By combining edge computing with cloud computing, predictive mechanisms and resource allocation are used to achieve efficient resource utilization [14]. Real time services are also enabled through server partitioning based on clustering [15]. An allocation strategy is proven to optimize the utility of edge computing servers [16]. Innovative resource allocation schemes demonstrate significant latency cost reductions [17]. In this paper, We introduced the theme in Sect. 1, followed by a state of art in Sect. 2. In Sect. 3 we elaborated energy-based VM selection approaches in Mobile Edge Computing. Simulation and results are presented and discussed in Sect. 3. Finally, Sect. 4 represents the paper's conclusion.

2 Energy-Based VM Selection Approaches in MEC

Energy consumption modeling is a pivotal aspect in optimizing the efficiency and performance of distributed computing systems. In this context, MEC has emerged as a promising paradigm that brings computing capabilities closer to mobile devices, enabling low-latency and high-bandwidth applications. The efficient allocation of computational tasks to appropriate virtual machines (VMs) is a critical challenge in MEC environments. To address this challenge, a comprehensive approach involves considering both the energy consumption and resource utilization of VMs to minimize overall energy usage while ensuring effective task execution. One approach to energy consumption modeling involves the calculation of the energy demand for each task. The formulated Eq. 1 considers the total resource requirements of the task, including RAM and CPU, as well as the task's latency constraint. By evaluating the inverse proportionality between the latency constraint and the energy demand, a calculated energy demand value is derived. This value provides insights into the energy requirements of the task and con-tributes to informed decision-making during VM selection. Moreover, the selection of the optimal VM for each task is a crucial step in energy-efficient task execution. The process integrates multiple considerations, including the VM's power consumption and its resource capabilities. A refined approach can utilize a formulated Eq. 2 that minimizes the combination of power consumption and resource shortage. This enables the selection of VMs that not only have lower energy consumption but also better match the resource requirements of the tasks, ultimately leading to enhanced energy efficiency and task performance.

$$Energy \quad demand = \frac{(RAM Requirement + CPU Requirement)}{Latency Constraint} \tag{1}$$

$$BestVM = argminVMs[PowerConsumption(VM) * (1 - ResourceShortageScore(VM, Task))] \tag{2}$$

By combining these energy consumption modeling and VM selection strategies, distributed computing systems can achieve substantial improvements in energy efficiency. This integrated approach empowers the system to dynamically allocate tasks to VMs, optimizing both energy consumption and resource utilization. In Mobile Edge Computing environments, where resource availability and energy constraints are critical, this modeling framework offers valuable insights for designing intelligent resource management mechanisms. As distributed computing systems continue to play a pivotal role in modern computing landscapes, the development and application of energy consumption models are instrumental in realizing sustainable and high-performance computing solutions.

3 Simulation and Results

In this section, we delve into the results obtained from our simulation experiments designed to address the optimization challenge of virtual machine (VM) selection within the MEC paradigm. Leveraging the power of the CloudSim 6 simulator, we conducted a series of comprehensive simulations to evaluate the performance of different VM selection algorithms in a dynamic MEC environment. Continuing our analysis, Fig. 2 illustrates the distribution of energy consumption across the spectrum of VMs. Each distinct VM, denoted by its VM ID, is rep-resented by an individual segment in the stacked area graph. The vertical axis quantifies the energy consumption, and the horizontal axis corresponds to the "Finish Time" associated with each VM. The stacking of areas provides a clear depiction of the dynamic evolution of energy consumption for each VM over time, allowing us to discern trends and patterns in their usage. Furthermore, Fig. 3 offers a visual summary of the proportional energy consumption attributed to each VM through a pie chart. The angles of the segments within the circle correspond to the share of energy consumption for each VM. This concise representation swiftly conveys the relative significance of each VM in the overall energy consumption landscape. Figure 4 introduces a scatter plot that plots the correlation between energy consumption and latency for individual tasks. Each data point, labeled with its Task ID, is positioned according to its energy demand on the vertical axis and its latency constraint on the horizontal axis. This visualization enables us to explore potential connections between energy consumption and latency, thereby aiding in informed decision-making regarding task optimization strategies. Figure 5 presents a new perspective with a clustered column chart, depicting the count of tasks assigned to each VM. The VM IDs are plotted along the horizontal axis, while the vertical axis represents the task count. This visualization allows us to quickly identify the distribution of tasks among different VMs, offering insights into task allocation patterns. While these figures provide a preliminary insight into the simulation results, it is essential to note that the presented data is based on a limited sample. With a more extensive dataset, these visualizations can potentially uncover more intricate trends and relationships, contributing to a more comprehensive understanding of the dynamic interplay between energy consumption, latency, and VM selection strategies within the Mobile Edge Computing environment.

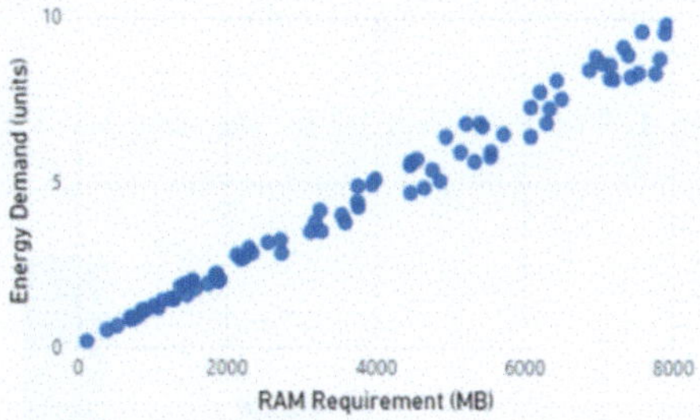

Fig. 2. RAM requirement and energy demand by task ID

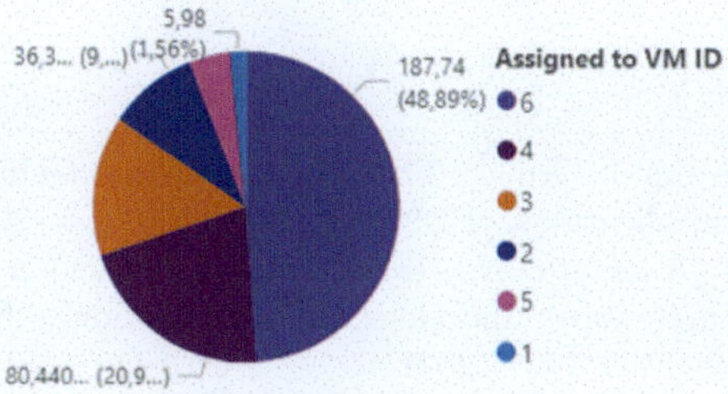

Fig. 3. Energy demand by the assigned VM

4 Conclusion

In conclusion, this study has delved into the critical issue of optimal virtual machine (VM) selection within MEC systems. The experimental findings unveiled in this endeavor the performance of diverse VM selection algorithms operating within the dynamic MEC environment. By harnessing the capabilities of the CloudSim 6 simulator, we meticulously executed simulations to scrutinize the success rates and utilization trends of VMs. These discernments have empowered us to pinpoint VMs that exhibit exceptional task success rates and those that are prominently engaged in the MEC system. Consequently, these insights significantly contribute to enhancing the overall efficiency of task processing.

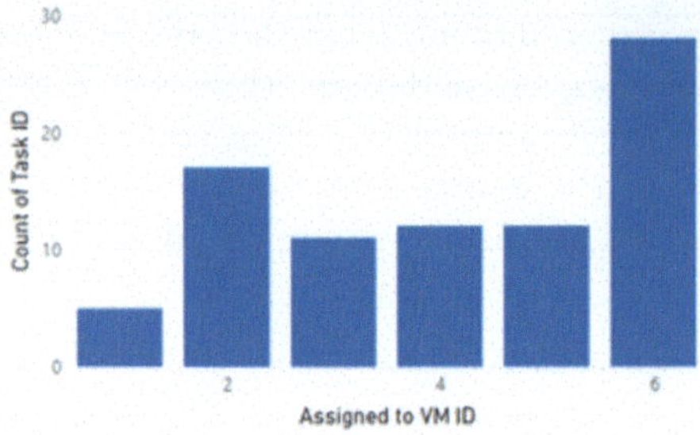

Fig. 4. Count of task ID by the assigned VM

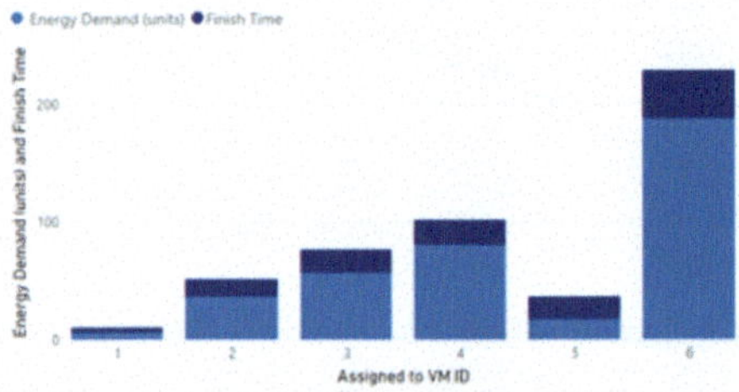

Fig. 5. Energy demand and finish time by the assigned VM

References

1. Chen, X., Shi, Q., Yang, L., Xu, J.: ThriftyEdge: Resource-efficient edge computing for intelligent IoT applications. IEEE Network **32**, 61–65 (2018)
2. Tout, H., Mourad, A., Kara, N., Talhi, C.: Multi-persona mobility: joint cost-effective and resource-aware mobile-edge computation offloading. IEEE/ACM Trans. Network. **29**, 1408–1421 (2021)
3. Zhang, J., Hu, X., Ning, Z., Ngai, E., Zhou, L., Wei, J., Cheng, J., Hu, B., Leung, V.: Joint resource allocation for latency-sensitive services over mobile edge computing networks with caching. IEEE Internet Things J. **6**, 4283–4294 (2018)
4. Xu, D., Li, Q., Zhu, H.: Energy-saving computation offloading by joint data compression and resource allocation for mobile-edge computing. IEEE Commun. Lett. **23**, 704–707 (2019)
5. Zhang, G., Zhang, S., Zhang, W., Shen, Z., Wang, L.: Joint service caching, computation offloading and resource allocation in mobile edge computing systems. IEEE Trans. Wireless Commun. **20**, 5288–5300 (2021)
6. Zhang, Z., Li, C., Peng, S., Pei, X.: A new task offloading algorithm in edge computing. EURASIP J. Wireless Commun. Network. **2021**, 1–21 (2021)
7. Chanyour, T., El Ghmary, M., Hmimz, Y., Cherkaoui Malki, M.: Energy-efficient and delay-aware multitask offloading for mobile edge computing networks. Trans. Emerg. Telecommun. Technol. **33**, e3673 (2022)
8. Ma, Y., Liang, W., Li, J., Jia, X., Guo, S.: Mobility-aware and delay-sensitive service provisioning in mobile edge-cloud networks. IEEE Trans. Mob. Comput. **21**, 196–210 (2020)

9. Yang, Y., Ma, Y., Xiang, W., Gu, X., Zhao, H.: Joint optimization of energy consumption and packet scheduling for mobile edge computing in cyber-physical networks. IEEE Access. **6**, 15576–15586 (2018)
10. El Ghmary, M., Chanyour, T., Hmimz, Y., Malki, M.: Efficient multi-task offloading with energy and computational resources optimization in a mobile edge computing node. Int. J. Electr. Comput. Eng. **9**, 4908–4919 (2019)
11. El Ghmary, M., Malki, M., Hmimz, Y. & Chanyour, T. Energy and computational resources optimization in a mobile edge computing node. *2018 9th International Symposium On Signal, Image, Video And Communications (ISIVC)*. pp. 323-328 (2018)
12. Han, D., Chen, W., Fang, Y.: Joint channel and queue aware scheduling for latency sensitive mobile edge computing with power constraints. IEEE Trans. Wireless Commun. **19**, 3938–3951 (2020)
13. Zheng, X., Li, M., Tahir, M., Chen, Y., Alam, M.: Stochastic computation offloading and scheduling based on mobile edge computing. IEEE Access. **7**, 72247–72256 (2019)
14. Chien, W., Lai, C., Chao, H.: Dynamic resource prediction and allocation in C-RAN with edge artificial intelligence. IEEE Trans. Industr. Inf. **15**, 4306–4314 (2019)
15. Li, G., Lin, Q., Wu, J., Zhang, Y., Yan, J.: Dynamic computation offloading based on graph partitioning in mobile edge computing. IEEE Access. **7**, 185131–185139 (2019)
16. Guo, S., Hu, X., Dong, G., Li, W., Qiu, X.: Mobile edge computing resource allocation: A joint Stackelberg game and matching strategy. Int. J. Distrib. Sens. Netw. **15**, 1550147719861556 (2019)
17. Cui, Y., Liang, Y., Wang, R.: Resource allocation algorithm with multi-platform intelligent offloading in D2D-enabled vehicular networks. IEEE Access. **7**, 21246–21253 (2019)

Efficient Virtual Machine Selection for Improved Performance in Mobile Edge Computing Environments

Nouhaila Moussammi[1(✉)], Mohamed El Ghmary[2], and Abdellah Idrissi[1]

[1] Department of Computer Science, Faculty of Sciences, Mohammed V University in Rabat, Rabat, Morocco
nouhaila_moussammi@um5.ac.ma, a.idrissi@um5r.ac.ma

[2] Department of Computer Science, FSDM, Sidi Mohamed Ben Abdellah University, Fez, Morocco
mohamed.elghmary@usmba.ac.ma

Abstract. Mobile applications have experienced rapid growth thanks to the Internet of Things (IoT). However, the constraints of limited resources on mobile devices (MDs) pose challenges such as processing delays, high energy consumption, and security issues. Mobile Edge Computing (MEC) is an effective solution to address these requirements. MEC systems offload tasks from lightweight MDs to Edge servers, where the necessary computations and processing are performed locally. This approach reduces latency, improves the user experience, and saves mobile device resources. MEC systems utilize virtual machines (VMs) sliced into smaller pieces to deliver their services. However, selecting the appropriate VM is a crucial challenge as it impacts overall performance, energy consumption, and user satisfaction. The suitable VM choice must consider the specific requirements of each task, in terms of com- puting capabilities, memory, and other resources. In this work, various approaches and techniques are explored to solve the problem of optimal VM selection in MEC systems. Decision criteria, algorithms, and strate- gies are studied to ensure optimal performance and efficient resource utilization

Keywords: Virtual Machine · Mobile Edge Computing · Computation Offloading

1 Introduction

The number of MDs, such as smartphones, tablets, smartwatches, and Internet of Things (IoT) devices, has significantly increased over the past decades. To meet the growing demand of mobile users, this rapid expansion has been accompanied by an exponential increase in resource-intensive applications offered by various cloud service providers. These mobile applications can now be used in cloud data centers through mobile cloud computing , which utilizes the abundant resources provided by server farms to host multiple applications. However, this method

Y. Farhaoui et al. (Eds.): ICAISE 2023, LNNS 837, pp. 420–426, 2024.
https://doi.org/10.1007/978-3-031-48465-0_55

introduces delays due to the back-and-forth communication between mobile users and the remote cloud via the Wide Area Network (WAN), which poses a significant challenge for applications requiring instantaneous responses. The European Telecommunications Standards Institute (ETSI) recently proposed MEC [1,2] as a solution to the increasing mobile data traffic and the need for fast and low latency services in the edge network. MEC improves the proximity of computation and data storage by pushing resources to the network edge, closer to end users. This extension of the cloud computing paradigm addresses the challenges posed by real-time interactive services like virtual reality [3] and augmented reality [4] that heavily rely on network resources. With the rising traffic, accessing services in a timely manner becomes crucial for enhancing the mobile user experience and meeting strict latency requirements. As shown in Figure 1, by placing servers at the network edge in the MEC environment, the distance between the server and mobile terminals is reduced, resulting in reduced latency [5]. Terminal requests can be processed locally by the MEC server, eliminating the need to transmit data to a remote server. As a result, mobile terminals receive faster responses, significantly improving the user experience.

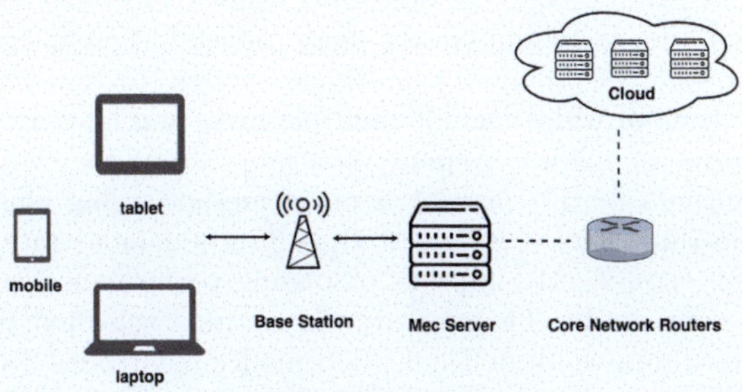

Fig. 1. MEC architecture

By improving the proximity of computation and data storage to end users and leveraging resources at mobile network access points, MEC emerges as a promising approach to address these issues. In this architecture, VMs play a crucial role as they allow for flexible and efficient deployment of services. However, choosing appropriate VMs in an MEC context is quite challenging. To ensure optimal performance, it is essential to consider the unique characteristics of each application, such as processing needs, storage requirements, and latency restrictions. Additionally, constraints imposed by accessible resources at mobile access points, including processing power, storage space, and bandwidth, must be taken into account. Previous research has primarily focused on VM management in edge environments, with particular attention to addressing the issue of energy inefficiency in VM selection. To address this challenge, researchers have developed various algorithms [6], such as the Comprehensive Resource Balance Ranking Scheme, Processing Element Cost Function, Energy Consumption Model, Resource Utilization

Square Model, Variable Item Size Bin Packing Algorithm, RF Aware Algorithm, Sercon Algorithm, and Genetic Algorithm. These algorithms aim to optimize resource utilization and reduce energy consumption. Additionally, a cloud system architecture model has been proposed for efficient task classification and appropriate VM allocation, aiming to enhance energy efficiency and optimize VM placement in cloud data centers [7]. Researchers have also investigated specific challenges related to MEC, including limited latency and bandwidth. To enhance service performance and reduce energy consumption, they have proposed dynamic migration strategies based on multi-agent deep reinforcement learning [8]. Hybrid approaches have been suggested for distributing VMs between the Cloud and Fog nodes, while meta-heuristic algorithms have been utilized to maximize physical server utilization and ensure adherence to service-level agreements [9]. VM selection has been studied using various approaches, including deep learning, multi-objective optimization, and rule-based methods, considering factors such as network connectivity, energy consumption, processing capacity, and resource availability [10,11]. Furthermore, there is a detailed offloading strategy presented for an area featuring numerous IoT devices and edge servers. This strategy successfully attains a 20% cost reduction by employing their SAA-based approach [12]. Additionally, the offloading frameworks introduced in the research [13,14] focus on minimizing energy consumption and latency. In our previous work, we proposed a multi-user system [15] and a single-user system [16], both of which include multiple tasks and high-density computing in order to minimize energy consumption and execution time. Also, the authors of [17] present a novel method using clustering to enhance offloading efficiency. The study's results demonstrate that this approach can significantly improve offloading performance in terms of time and resources. In this context, the optimal selection of VMs is vital for achieving overall optimal performance, controlled energy consumption, and user satisfaction. MEC facilitates the efficient execution of resource-intensive mobile applications, enhancing user experiences while effectively managing resources and energy. This study is structured as follows: Sect. 2 discusses VM selection approaches in the MEC environment, and Sect. 3 presents an overview of simulations and their outcomes. Finally, Sect. 4 concludes the paper by summarizing the findings and contributions.

2 VM Selection Approaches in Mobile Edge Computing

Energy-Based VM Selection: This approach optimizes the selection of VMs and tasks in MEC based on energy consumption. VMs are generated with specific characteristics like processing power, RAM, bandwidth and energy usage, while tasks have their own requirements. The algorithm chooses the VM for each task that minimizes the difference between its energy consumption and the task's energy requirements. If there's a tie, the VM with the lowest energy consumption is preferred. This approach saves energy and extends the lifespan of MDs in MEC.

However, hardware constraints can affect overall system performance. In the following sections, we'll explore other approaches and analyze results to identify tasks and VMs with the execution time and energy consumption.

Resource-Based VM Selection: In MEC, we propose this approach that considers constraints related to available resources in MDs and Edge servers. We analyze task characteristics like computing power, memory, bandwidth and file size. Using appropriate metrics, our algorithm filters eligible VMs and selects the optimal VM based on their resource score. Performance is evaluated through experiments and compared with other VM selection strategies. We also discuss decision criteria, strategies for resource allocation, and security considerations in resource-based VM selection for MEC environments.

Combined VM Selection: This approach aims to minimize the environmental impact and operational costs associated with executing cloudlet tasks by considering the energy consumption of VMs. It identifies VMs that strike a balance between processing capabilities and energy efficiency, thus optimizing the overall energy consumption of the system. Additionally, this approach takes into account the diverse resource requirements of cloudlet tasks, such as computing power, memory, bandwidth and file size. By evaluating available VMs based on these resource metrics, the Combined VM Selection ensures that each cloudlet is assigned to a VM that can effectively meet its specific resource needs. This resource-aware allocation improves overall performance and responsiveness of cloudlet execution, leading to enhanced user experience and reduced task completion times. The effectiveness of the Combined VM Selection approach is evaluated through experiments and simulations in various MEC scenarios. Comparisons with other VM selection strategies allow quantification of energy savings, resource utilization, and task execution efficiency achieved by this approach.

3 Simulation and Results

In our CloudSimPy based simulator, we have developed three distinct approaches for VM selection: one based on energy consumption, another on available resources, and finally a combination of both criteria. Each approach was evaluated through simulations with a large number of tasks (cloudlets), and relevant results were recorded. The outcomes of each approach were visualized using three separate graphs. Figure 2 comprises a bar chart illustrating the comparison of VM selections for individual tasks using three distinct approaches: one based on energy consumption, another on available resources, and a combination of both criteria. Each bar in the chart represents a specific VM. The height of each bar indicates the number of tasks for which that particular VM was chosen. This chart facilitates insights into the frequency of VM assignments across different task selection approaches, allowing for an understanding of task-to-VM distribution. Figure 3 provides a valuable visual representation of the energy-based VM selection approach. It offers a clear insight into the trade-off between energy consumption and execution time for individual tasks. Each point on the plot corresponds to an individual task. This visual depiction allows for the identification of tasks that exhibit efficient energy consumption while maintaining reasonable execution times, aiding in the optimization of VM selection strategies within energy-constrained environments. The scatter plot's value lies in its ability to inform decisions regarding task placement and resource allocation, ultimately contributing to enhanced system performance and resource utilization in MEC and similar environments. Figure 4 consists of a line graph presenting data related to the execution times of tasks for three distinct VM selection approaches: energy, resources, and a combination of criteria. Each curve on the graph corresponds to a specific approach. This graph facilitates the observation of variations in task execution times across the different VM selection strategies, enabling comparisons and pattern recognition. These visualizations provide valuable insights into the performance of each approach and their implications in terms of energy consumption and task execution times. The results offer crucial guidance for making decisions regarding VM selection in MEC environments and optimizing the tradeoff between performance and energy consumption.

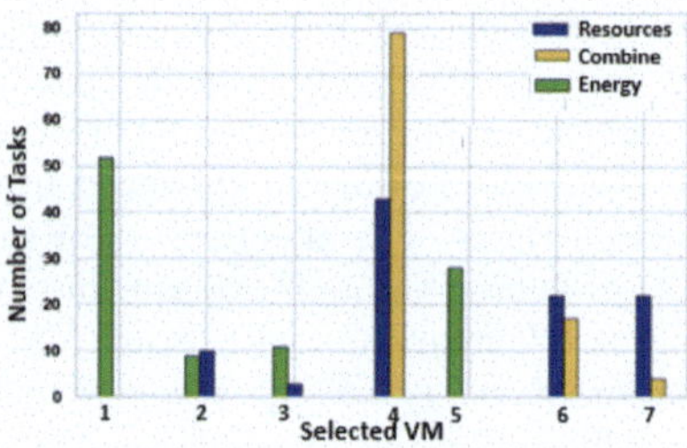

Fig. 2. Comparison of VMs selected for Each task

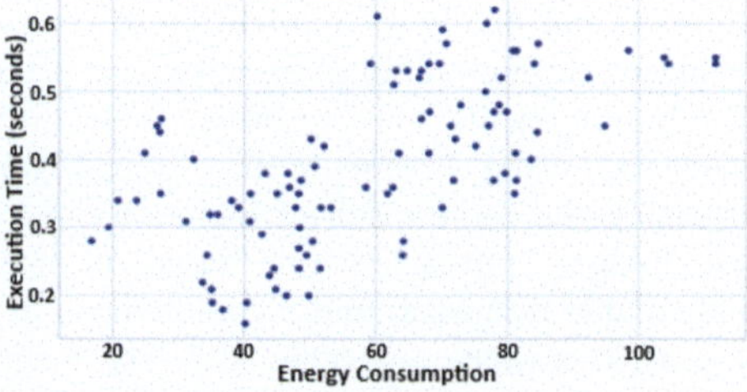

Fig. 3. Energy consumption vs. execution time

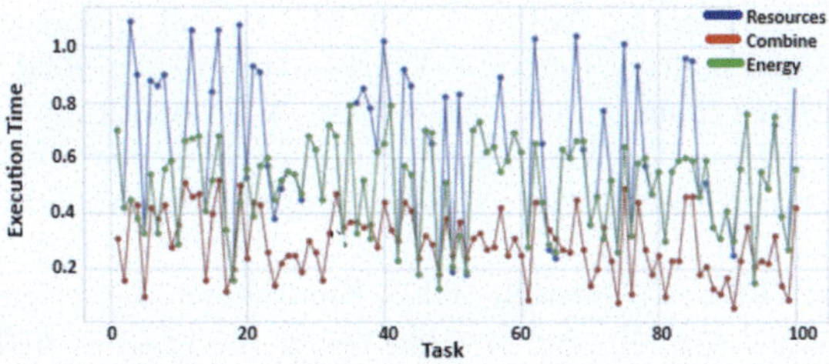

Fig. 4. Execution time for each task

4 Conclusion

In conclusion, this study delved into the pressing issue of optimal VM selection in MEC systems. The growth of mobile applications and IoT devices has emphasized the need for efficient resource utilization and reduced latency. MEC emerged as an effective solution to address these challenges by offloading tasks from lightweight MDs to Edge servers. The experimental results presented in this work shed light on the performance of various VM selection algorithms within the MEC environment. By leveraging the simulator, we conducted detailed simulations to evaluate the success rate and utilization patterns of VMs. These insights enabled us to identify VMs that demonstrated superior task success rates and those that were frequently utilized in the MEC system, contributing significantly to overall task processing efficiency.

References

1. Cao, X., Wang, F., Xu, J., Zhang, R., Cui, S.: Joint computation and communication cooperation for mobile edge computing. In: 16th International Symposium on Modeling and Optimization in Mobile, Ad Hoc, and Wireless Networks pp. 1–6. IEEE (2018)
2. Wang, L., Jiao, L., He, T., Li, J., Mühlhäuser, M.: Service entity placement for social virtual reality applications in edge computing. In: IEEE INFOCOM Conference on Computer Communications, pp. 468–476. IEEE (2018)
3. Ren, J., He, Y., Huang, G., Yu, G., Cai, Y., Zhang, Z.: An edge-computing based architecture for mobile augmented reality. IEEE Netw. **33**(4), 162–169 (2019)
4. Tan, T., Cao, G.: FastVA: deep learning video analytics through edge processing and NPU in mobile. In: IEEE INFOCOM Conference on Computer Communications, pp. 1947–1956 (2020)
5. Liu, H., Cao, G.: Deep reinforcement learning-based server selection for mobile edge computing. IEEE Trans. Veh. Technol. **70**(12), 13351–13363 (2021)
6. Mekala, M.S., Viswanathan, P.: Energy-efficient virtual machine selection based on resource ranking and utilization factor approach in cloud computing for IoT. Comput. Electr. Eng. **73**, 227–244 (2019)

7. Nabavi, S., Wen, L., Gill, S.S., Xu, M.: Seagull optimization algorithm based multi-objective VM placement in edge-cloud data centers. Internet Things Cyber-Phys. Syst. **3**, 28–36 (2023)
8. Dai, Y., Zhang, Q., Yang, L.: Virtual machine migration strategy based on multi-agent deep reinforcement learning. Appl. Sci. **11**(17), 7993 (2021)
9. Alharbi, H.A., El-Gorashi, T.E., Elmirghani, J.M.: Energy efficient virtual machine services placement in cloud-fog architecture. In: 21st International Conference on Transparent Optical Networks (ICTON), pp. 1–6. IEEE (2019)
10. Tao, Z., Xia, Q., Hao, Z., Li, C., Ma, L., Yi, S., Li, Q.: A survey of virtual machine management in edge computing. Proc. IEEE **107**(8), 1482–1499 (2019)
11. Wang, W., et al.: Infrastructure-efficient virtual-machine placement and workload assignment in cooperative edge-cloud computing over backhaul networks. IEEE Trans, Cloud Comput (2021)
12. Zhang, C., Zhao, H., Deng, S.: A density-based offloading strategy for IoT devices in edge computing systems. IEEE Access **6**, 73520–73530 (2018)
13. Hmimz, Y., et al.: Computation offloading to a mobile edge computing server with delay and energy constraints. In: 2019 International Conference on Wireless Technologies, Embedded and Intelligent Systems (WITS), pp. 1–6 (2019)
14. El Ghmary, M., et al.: Energy and processing time efficiency for an optimal offloading in a mobile edge computing node. Int. J. Commun. Netw. Information Secur. **12**(3), 389–393 (2020)
15. Moussammi, N., El Ghmary, M., Idrissi, A.: Multi-task multi-user offloading in mobile edge computing. Int. J. Adv. Comput. Sci. Appl. **13**(12), 2022
16. Moussammi, N., et al.: Multi-task offloading to a MEC server with energy and delay constraint. In: The International Conference on Artificial Intelligence and Smart Environment. Springer International Publishing, Cham, pp. 642–648 (2022)
17. El Ghmary, M., et al.: Time and resource constrained offloading with multi-task in a mobile edge computing node. Int. J. Electr. Comput. Eng. (2088–8708), 10(4) (2020)

A First Order Sliding Mode Control of the Permanent Magnet Synchronous Machine

Hafid Ben Achour[1(✉)], Said Ziani[2], and Youssef El Hassouani[1]

[1] Department of Physics, Faculty of Science and Technology Errachidia, Moulay Ismail University, BP-509 Boutalamine Errachidia, 52000 Meknes, Morocco
h.benachour@edu.umi.ac.ma, y.elhassouani@umi.ac.ma
[2] ENSAM, Mohammed V University, Rabat, Morocco
SAID.ZIANI@univh2.ma

Abstract. One of the most widely used machines in industry is the permanent magnet synchronous machine (PMSM), thanks to its characteristics. The control of this machine has been the focus of much scientific research, with the aim of improving both its performance and its robustness. In this paper, we'll study the control of PMSM using the classic PI controller, followed by the advanced sliding-mode control technique. The results show the preference for sliding-mode control, especially in terms of response time, and robustness against variations in machine parameters.

Keywords: PMSM · Nonlinear control · Sliding mode · PI controller

1 Introduction

The advancement of the industrial sector and its evolving demands have led to a substantial demand for permanent magnet synchronous machines (PMSMs). In the pursuit of enhancing operational efficiency, cost-effectiveness, and seamless real-time implementation of controls, the researchers have dedicated their efforts to elaborate numerous control strategies approaches like PI control or nonlinear techniques such as sliding mode, backstepping, and artificial intelligence-based control for PMSM [1, 2].

The design of sliding mode controllers systematically considers the problems of stability and good performance in its approach, which is divided into the following three main stages: choice of surfaces, establishment of existence and convergence conditions, and determination of control law. Slotine [3] proposes a general form of equation to determine the surface that ensures convergence of a variable to its set point value.

This control technique has several advantages and performances, such as: fast response time, insensitivity to internal variations in machine parameters and to external disturbances variations. But also has the chattering phenomenon, which is a major drawback of sliding-mode control. To minimize the chattering phenomenon and to improve the performance of sliding mode control the researchers have developed several methods: in [4] novel designs for sliding surfaces have been put forth with the aim of elevating

Y. Farhaoui et al. (Eds.): ICAISE 2023, LNNS 837, pp. 427–433, 2024.
https://doi.org/10.1007/978-3-031-48465-0_56

the efficacy of sliding mode control (SMC), departing from the traditional linear sliding surface approach. Utkin and Jingxin [5] introduced Integral Sliding Mode Control (ISMC) to guarantee the robustness at the reaching phase, effectively circumventing the reaching phase and ensuring the enforcement of the sliding phase throughout the complete system response. An alternative approach to nonlinear sliding surface design is the utilization of fractional calculus within the construction of the sliding surface, resulting in the fractional order sliding surface design method [6, 7]. In [8–10] offered terminal sliding mode (TSM) control employs terminal sliding surfaces that incorporate fractional powers, ensuring rapid and finite-time convergence of states within the sliding mode phase.

2 Application of Sliding Mode Controller to PMSM

The state model of the PMSM in the rotate frame (d-q) is presented through the following system:

Figure 1 illustrates the first order sliding mode control (SMC) scheme, employing the principle of cascaded control methodology (three-surface structure). The inner loop facilitates current control, while the outer loop facilitates speed control.

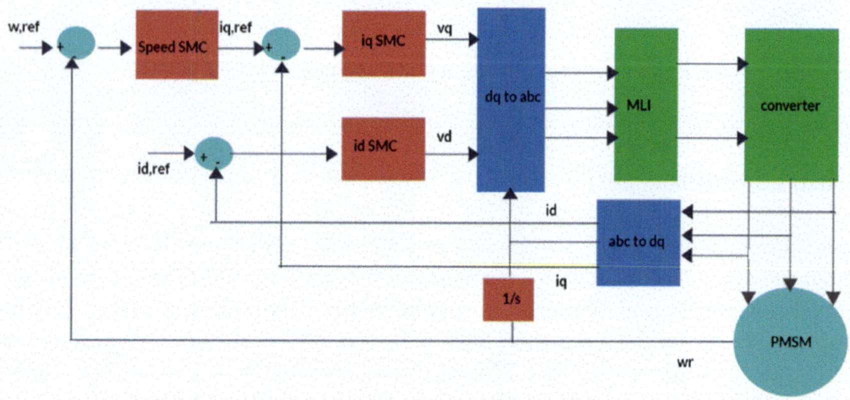

Fig. 1. Sliding mode control scheme

- **Speed Control Surface**

We choose the speed error as the surface (r = 1):

$$s(w_r) = w_{r,ref} - w_r \tag{1}$$

The surface derivative is:

$$\dot{s}(w_r) = \dot{w_{w_{r,ref}}} - \frac{p(L_d - L_q)i_d + p\varphi_f}{J} - \frac{C_r}{J} - \frac{f}{j}w_r \tag{2}$$

The control law is:

$$i_q = i_{q,eq} + i_{q,n} \tag{3}$$

During sliding phase, we have.

$$s(w_r) = 0 \quad \dot{s}(w_r) = 0 \quad i_{qn} = 0 \tag{4}$$

So:

$$i_{q,eq} = \frac{w_{\dot{w}_{r,ref}} + \frac{f}{J}w_r + \frac{C_r}{J}}{p(L_d - Lq)\frac{i_d}{J} + p\frac{\varphi_f}{J}} \tag{5}$$

During the convergence phase, the derivative of Lyapunov function must be negative.

$$\dot{v}(w_r) = \dot{s}(w_r)s(w_r) < 0 \tag{6}$$

$$\dot{s}(w_r) = -\left[\frac{p}{J}(L_d - Lq)i_d + p\varphi_f\right]i_{qn} \tag{7}$$

So: $i_{qn} = k_{wr}sign(s(w_r))$; k_{wr} is a positive constant.
Finally, the control law is given by:

$$i_q = \frac{w_{\dot{w}_{r,ref}} + \frac{f}{J}w_r + \frac{C_r}{J}}{p(L_d - Lq)\frac{i_d}{J} + p\frac{\varphi_f}{J}} + k_{wr}sign(s(w_r)) \tag{8}$$

- **Iq Current Control Surface**

The surface is selected as follows:

$$s(i_q) = i_{q,ref} - i_q \tag{9}$$

The control law is:

$$v_q(t) = L_{q\dot{i}_{q,ref}} + R_s i_q - L_d p\varphi_f i_d + p\varphi_f w_r + k_q sign(s(iq)) \tag{10}$$

With k_q is a positive constant.

- **Id Current Control Surface**

The surface is selected as follows:

$$s(i_d) = i_{d,ref} - i_d \tag{11}$$

The control law is:

$$v_q(t) = L_{d\dot{i}_{d,ref}} + R_s i_d - L_q p i_q + k_d sign(s(id)) \tag{12}$$

With k_d is a positive constant.

3 Simulation Results of Sliding Mode Controller and Classical PI Controller

To validate this controller through simulation in conjunction with speed control of a PMSM, experimentation was conducted within the Matlab/Simulink environment.

Figure 2 illustrates the results of simulation of PI controller applied to PMSM. Since, Fig. 3 shows sliding mode controller results.

In both cases, we apply a load at 0.5 s, then a change in stator resistance (Rs' = 2Rs) at 0.8 s, and finally a speed inversion from 100 rad/s to −100 rad/s at 1 s.

According to the results, a preference for the SMC controller is evident. Figure 2 demonstrates that the PI controller is highly sensitive to variations in both internal and external parameters of the machine, as well as to speed reversal. This is evident from the rotor speed overshot with each speed change, as well as the speed decrease when a load is applied or when the stator resistance is varied (Fig. 2a). The quadrature current and torque exhibit significant changes when a load is applied and/or the stator resistance is varied (Fig. 2b and c).

In contrast, for the sliding mode controller, there is minimal variation in speed when a load is applied or the stator resistance is modified, as shown in Fig. 3a. Similarly, the same holds true for the quadrature current and torque (Fig. 3b and c). Additionally, it can be observed that the SMC has a significantly faster response time compared to the PI controller.

4 Conclusion

In this study, we have highlighted the robustness and performance of sliding mode control. This technique is insensitive to variations in machine parameters and external disturbances. However, a major drawback of this approach is the phenomenon of chattering. Currently, chattering is a challenge for researchers, and several techniques have been developed to mitigate this phenomenon.

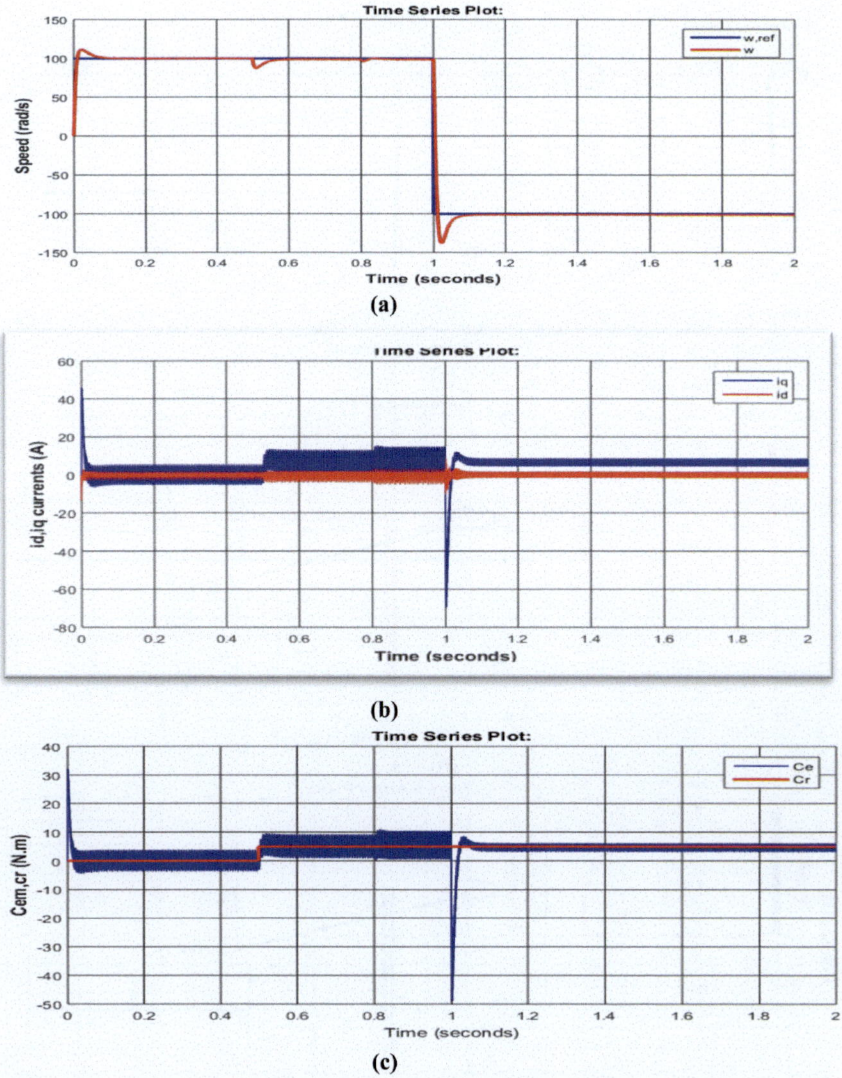

Fig. 2. PI controller results (a) speed, (b) id, iq currents and (c) electromagnetic torque.

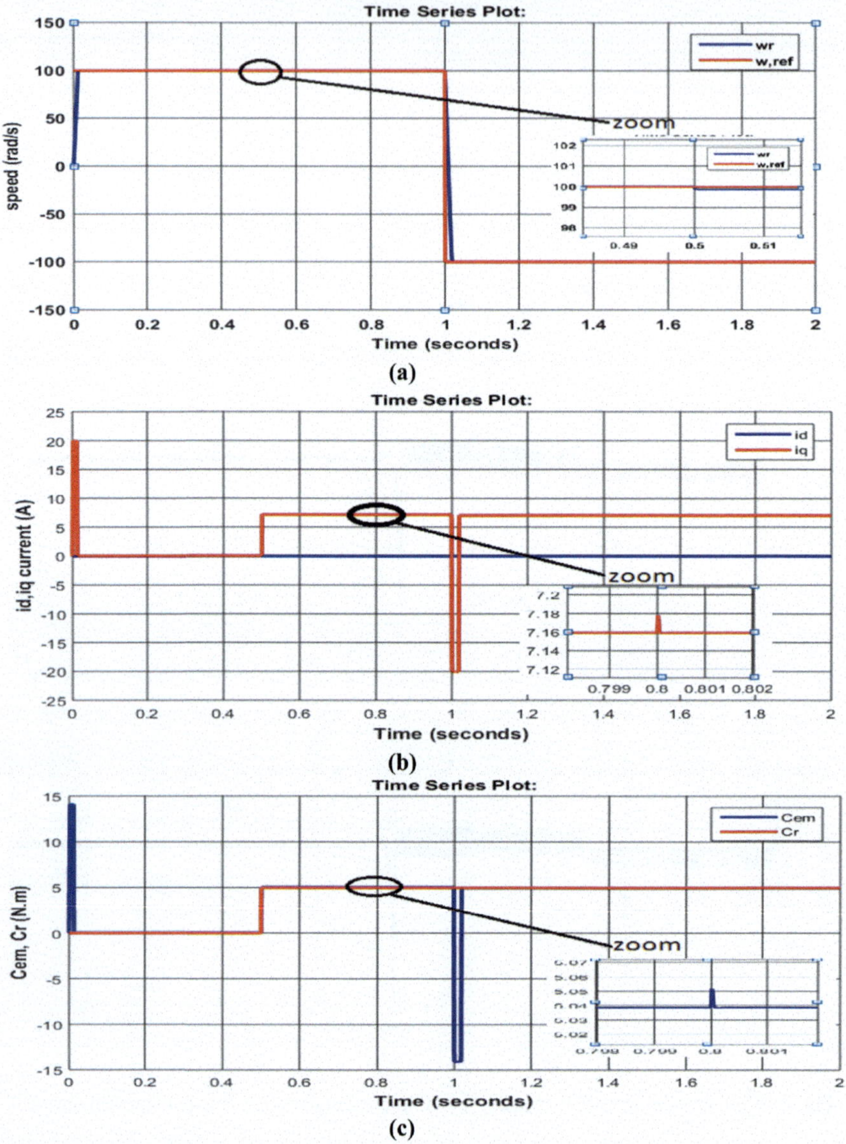

Fig. 3. SMC controller results (a) speed, (b) id, iq currents and (c) electromagnetic torque.

References

1. Ben Achour, H., Ziani, S., Chaou, Y., El Hassouani, Y., Daoudia, A.: Permanent magnet synchronous motor PMSM control by combining vector and PI controller. WSEAS Trans. Syst. Control **17**, 244–249

2. Chaou, Y., Ziani, S., Ben Achour, H., Daoudia, A.: Nonlinear control of the permanent magnet synchronous motor PMSM using backstepping method. WSEAS Trans. Syst. Control **17**, 56–61 (2022)
3. Slotine, J.J., Sastry, S.S.: Tracking control of non-linear systems using sliding surfaces, with application to robot manipulators. Int. J. Control. **38**(2), 465–492 (1983). https://doi.org/10.1080/00207178308933088
4. Utkin, V.I.: Sliding mode control design principles and applications to electric drives. IEEE Trans. Ind. Electron. **40**, 23–36 (1993)
5. Utkin, V., Jingxin, S.: Integral sliding mode in systems operating under uncertainty conditions. In: Proceedings of the 35th IEEE Conference on Decision and Control, Kobe, Japan, 11–13 December 1996, pp. 4591–4596
6. Sun, G., Wu, L., Kuang, Z., Ma, Z., Liu, J.: Practical tracking control of linear motor via fractional-order sliding mode. Automatica **94**, 221–235 (2018)
7. Mujumdar, A., Tamhane, B., Kurode, S.: Observer-based sliding mode control for a class of noncommensurate fractional-order systems. IEEE/ASME Trans. Mechatron. 1–9 (2015)
8. Liu, X., Yu, H., Yu, J., Zhao, L.: Combined speed and current terminal sliding mode control with nonlinear disturbance observer for PMSM drive. IEEE Access **6**, 29594–29601 (2018)
9. Lu, E., Li, W., Yang, X., Xu, S.: Composite sliding mode control of a permanent magnet direct-driven system for a mining scraper conveyor. IEEE Access **5**, 22399–22408 (2017)
10. Mu, C., Xu, W., Sun, C.: On switching manifold design for terminal sliding mode control. J. Frankl. Inst. 353 (2016)

Optimized Schwarz and Finite Volume Cell-Centered Method for Heterogeneous Problems

M. Mustapha Zarrouk(✉)

Department of Mathematics FSTE, University of Moulay Ismaïl, Meknès 52000, Morocco
ms.zarrouk@gmail.com

Abstract. This paper focuses on the optimized Schwarz Waveform Relaxation (OSWR) method for a model problem with discontinuous coefficients. The numerical scheme relies on the finite volume cell-centered (FVCC) method. The domain is divided into two nonoverlapping subdomains, and optimized Robin-type transmission conditions are applied at the interface between these subdomains. Numerical experiments with both equal and different Robin parameters are presented to compare the performance of the methods.

Keywords: Schwarz waveform relaxation · Optimized transmission conditions · Nonoverlapping domains · Finite Volume Method

1 Introduction and Problem Definition

Domain decomposition methods (DDM) offer the opportunity to divide large problems into smaller subproblems, enabling parallel processing, which results in efficient algorithms that can adapt locally in both space and time. In their works [1,2], the authors introduce an approach based on classical transmission conditions. To enhance the convergence of this method, [3–5] have suggested the utilization of optimized transmission conditions at each space-time interface, leading to the development of the Optimized Schwarz Waveform Relaxation (OSWR) method. Furthermore, in their studies [6,7], the authors have conducted an analysis concerning the optimization of Robin or Ventcell parameters, and they have extended this optimization to cases involving discontinuous coefficients. In [8] Optimized Schwarz DDM for nonlinear coupled two species reactive transport system was studied.

In this work, we consider the linear coupled two species reactive transport system

$$\begin{cases} \partial_t c + \rho_\omega \partial_t \bar{c} - \partial_x (D\partial_x c - \beta c) = 0 \text{ in } \Omega \times (0,T] \\ \bar{c} = K_d c \text{ in } \Omega \times (0,T], \\ c(0,.) = c_0 \text{ in } \Omega \end{cases} \tag{1}$$

Y. Farhaoui et al. (Eds.): ICAISE 2023, LNNS 837, pp. 434–439, 2024.
https://doi.org/10.1007/978-3-031-48465-0_57

with $c(x,t)$ denoting the aqueous concentration (in mol/L) of the species. Whereas Ω is a bounded domain and $(0,T]$ is the time interval. The parameters D and ω are respectively the diffusion coefficient and the porosity. Let $\rho_\omega := \frac{\rho(1-\omega)}{\omega}$, and β is the average velocity. The expression $(1)_1$ (see [15, Sect. 2.6.2]) is known as the linear equilibrium isotherm relationship. Inserting equation $(1)_2$ into equation $(1)_1$ and assuming

$$R = 1 + \rho_\omega K_d, \tag{2}$$

in which case problem (1) becomes

$$\begin{cases} R\partial_t c - \partial_x(D\partial_x c - \beta c) = 0 \text{ in } \Omega \times (0,T] \\ c(0,.) = c_0 \text{ in } \Omega \end{cases} \tag{3}$$

The coefficient R refers to the retardation factor, which characterizes the retardation effect caused by the adsorption.

The remainder of this article is organized as follows: In Sect. 2, we present the nonoverlapping Schwarz domain decomposition algorithm for a linear coupled two species reactive transport system. Section 3, determines the optimal transmission conditions of the algorithm. Finally, the simulation results are presented and analyzed in last Section.

2 Schwarz Waveform Relaxation Algorithm

In this section, we introduce a nonoverlapping domain decomposition method for solving problem (3). To keep things simple, we assume a partition Ω into two nonoverlapping subdomains Ω_1 and Ω_2 separated by an interface denoted as Γ. Let k be the iteration index (number of subdomains solve). Utilizing the Jacobi algorithm, gives rise to the following algorithm: for $k \geq 0$, at iteration k, we seek the solutions c_i^k within subdomain Ω_i such that:

$$\begin{cases} R_i\partial_t c_i^k - \partial_x(D_i\partial_x c_i^k - \beta c_i^k) = 0 & \text{in } \Omega_i \times (0,T], \\ B_i(c_i^k) = B_j(c_i^{k-1}) & \text{over } \Gamma \times (0,T], \\ c_i^k(0,\cdot) = c_{0/\Omega_i} & \text{in } \Omega_i, \end{cases} \tag{4}$$

where $i \in \{1,2\}$, $j = 3-i$ and the operators B_j are linear operators to be defined. We start with an initial guess (g_1^0, g_2^0). In the initial step of the algorithm, we solve both problems (4) with transmission conditions replaced respectively by the conditions:

$$B_1(c_2^0) = g_1^0 \quad \textit{and} \quad B_2(c_1^0) = g_2^0. \tag{5}$$

The transmission operators are Robin operators (see for instance [9,10] and the references therein) defined by :

$$B_i = (D\nabla - \frac{\beta}{2}) \cdot n_i + \frac{S_i}{2}, \quad S_i > 0, \quad i = 1, 2. \tag{6}$$

3 Optimal Transmission Conditions

To identify the optimal transmission operators S_1 and S_2, we calculate the convergence factor of the algorithm. This involves using the error equations and applying a Fourier transform in time with a parameter θ. Through this process, we obtain identical ordinary differential equations within each subdomain:

$$R_j i\theta \hat{e}_j - D_j \partial_x^2 \hat{e}_j + \beta \partial_x \hat{e}_j = 0, \quad i = \sqrt{-1},\ j = \{1, 2\} \text{ and } e_j = c_j - c_{/\Omega_j}, \quad (7)$$

with the characteristic roots

$$r_j^{\pm} = \frac{\beta \pm z_j(\theta)}{2D_j}, \quad z_j(\theta) = \sqrt{\beta^2 + 4D_j i\theta R_j}. \quad (8)$$

The complex square root is chosen in such a way that its real part is positive. We refer to σ_j as the corresponding symbol in the Fourier variables of S_j, for $j = 1, 2$. Then, we can define the convergence factor exactly as in [8]:

$$\xi(\theta, \sigma_1, \sigma_2, D_1, D_2, R_1, R_2, \beta) = \left(\frac{\sigma_1 - z_2(\theta)}{\sigma_1 + z_2(\theta)}\right) \left(\frac{\sigma_2 - z_1(\theta)}{\sigma_2 + z_1(\theta)}\right). \quad (9)$$

We approximate the square roots z_j in (9) by parameters s_j which leads to

$$\sigma_1^{app}(\theta) = s_2 \quad and \quad \sigma_2^{app}(\theta) = s_1, \quad \forall \theta \in \mathbf{R}$$

and hence leads to so called two-sided Robin transmission conditions. A better approach is to minimize the convergence factor (see e.g. [11,12] and the references therein), i.e. to solve the following problem

$$\min_{s_1, s_2} \left(\max_{0 \le \theta \le \theta_{max}} | \xi(\theta, s_1, s_2, D_1, D_2, R_1, R_2, \beta) | \right). \quad (10)$$

Since we are using a numerical scheme, it's important to note that the frequencies cannot be excessively high but can be restricted to $\theta_{max} = \frac{\pi}{\Delta t}$. It is clear by inspection of (9) that there is a simpler optimal choice of transmission conditions making the convergence factor vanish for all θ, namely

$$z_1(\theta) = z_2(\theta) = z(\theta) \; and \; \sigma_1^{opt}(\theta) = \sigma_2^{opt}(\theta) = s, \quad s > 0,$$

which leads to so called one-sided Robin transmission conditions (one free parameter to optimize).

4 Numerical Experiments

For numerical experiments, the initial condition is $c_0(x) = 0$ for $0 < x < L$, and the boundary conditions are $c(0, t) = 1$ and a zero diffusive flux at $x = 5$. For the time discretization, we restrict to implicit method, while for space discretization, cell-centered finite volume schemes are applied [13]. Regarding the diffusive flux, we employ two-point diffusion discretization and two different strategies for the advective flux: upwind and hybrid discretization. we refer to [10,14] where a unified handling of advection terms in hybrid finite volumes is described.

4.1 Test Case 1: Transport of an Inert Solute on the Global Domain

As a first study, we consider the transport of inert solute, the concentration profile is compared with the corresponding analytical concentration given by [15] as follows:

$$c(x,t) = \frac{c(0,t)}{2}\left[erfc\left(\frac{Rx-\beta t}{2\sqrt{RDt}}\right) + exp\left(\frac{\beta x}{D}\right) erfc\left(\frac{Rx+\beta t}{2\sqrt{RDt}}\right)\right], \qquad (11)$$

where erfc is the complementary error function.

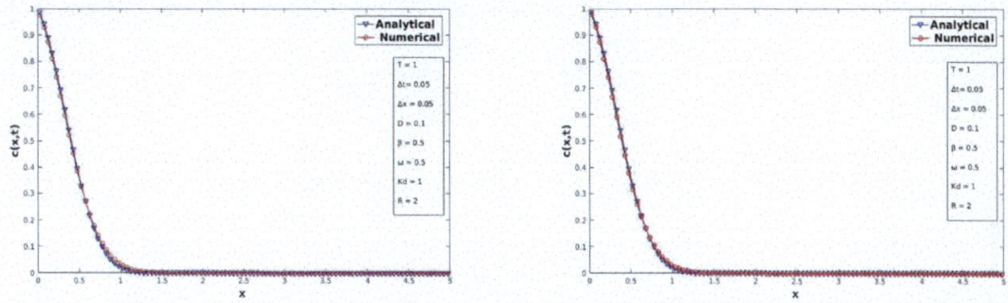

(a) (b)

Fig. 1. Comparison between analytical and numerical solutions. Upwind (on the left) and Hybrid (on the right) finite volume schemes.

Figure 1(a)-(b) shows that the numerical and analytical solutions for both schemes are nearly confounded. Therefore, the upwind and the hybrid finite volume approach are well suited for the numerical solution of the global domain.

4.2 Test Case 2: The Nonoverlapping OSWR

In our second study, we show an example of how our algorithm (4), behaves with the case of discontinuous coefficients. The global domain $\Omega =]0,5[$ is decomposed into nonoverlapping subdomains $\Omega_1 =]0,2.5[$ and $\Omega_2 =]2.5,5[$. The parameters ρ and K_d are constants equal to 1 in each subdomain, the mesh size is $h_1 = h_2 = 0.05$, the velocity $\beta = 0.5$. The diffusion-dispersion coefficient $D_1 = 0.05$ and $D_2 = 0.1$. The porosity $\omega_1 = 0.06$ and $\omega_2 = 0.8$. $T = 0.5$ and the time step $\Delta t = 5.10^{-2}$. We assign a random initial guess on the interface Γ to guarantee a diverse range of potential frequencies in the error.

To verify the performance of the optimized Robin's parameters choices, we fixe an error tolerance of 10^{-6} for the interface values calculated in the discrete L^2 norm in time.

We compare Robin transmission conditions with zero order Taylor approximation of the square root z_j around $\theta = 0$, where in our case $s = \beta = 0.5$ (for more in-depth exploration of various approximation techniques in this context, refer to [16]) and the one-sided Robin and finally the two-sided Robin tarnsmission conditions. The value founded by minimizing the two half-spaces convergence factor (10) is around $s = 1.3179$ for the one-sided Robin and $s_1 = 1.0064$, $s_2 = 3.64$ for the two-sided Robin parameters.

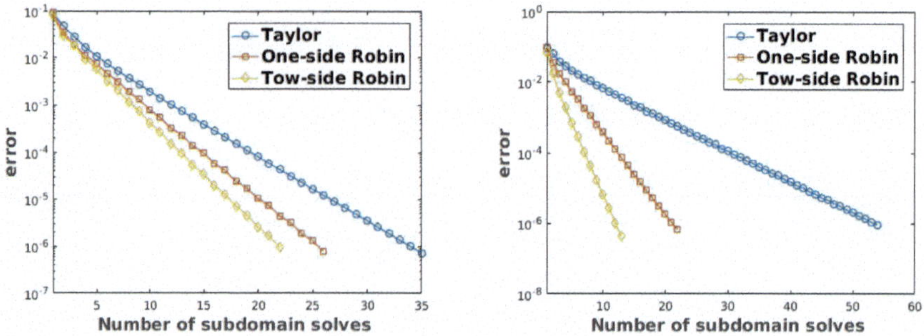

(a) (b)

Fig. 2. Convergence curves for the different algorithms using upwind (on the left) and hybrid (on the right) multidomain schemes.

Figure 2 shows the error (in logarithmic scale) for the different algorithms versus the number of subdomain solves. We see that for discontinuous coefficients, the two-sided Robin conditions provide the best performance together with the one-sided Robin conditions, while that the convergence of the algorithm employing the hybrid scheme is at least as good as that of the upwind scheme.

5 Conclusion

In the frameworks of cell-centered finite volume (FV), we have studied the Optimized Schwarz Wave Relaxation (OSWR) algorithm for linear coupled two species system within heterogeneous medias. Through numerical analysis, we have shown that even though the discrete upwind scheme worked well in a monodomain context, it's not good enough to work satisfactorily in a domain decomposition context. We also have shown how well the two-sided Robin transmission conditions has enhanced the convergence speed in comparison to one-sided Robin and Taylor transmission conditions.

References

1. Gander, M.J., Stuart, A.M.: Space-time continuous analysis of waveform relaxation for the heat equation. SIAM J. Sci. Comput. **19**(6), 2014–2031 (1998)
2. Giladi, E., Keller, H.B.: Space-time domain decomposition for parabolic problems. Numer. Math. **93**(2), 279–313 (2002)

3. Gander, M.J., Halpern, L., Nataf, F.: Optimal convergence for overlapping and non-overlapping Schwarz waveform relaxation. In: 11th International Conference on Domain Decomposition Methods, pp. 27–36 (1999)
4. Gander, M.J., Halpern, L., Nataf, F.: Optimal schwarz waveform relaxation for the one dimensional wave equation. SIAM J. Numer. Anal. **41**(5), 1643–1681 (2003)
5. Martin, V.: An optimized schwarz waveform relaxation method for the unsteady convection diffusion equation in two dimensions. Appl. Numer. Math. **52**(4), 401–428 (2005)
6. Lemarié, F., Debreu, L., Blayo, E.: Toward an optimized global-in-time Schwarz algorithm for diffusion equations with discontinuous and spatially variable coefficients, part 1: the constant coefficients case. Electron. Trans. Numer. Anal. **40**, 148–169 (2013)
7. Japhet, C., Omnes, P.: Optimized Schwarz waveform relaxation for porous media applications. In: Domain Decomposition Methods in Science and Engineering XX (pp. 585–592). Springer (2013)
8. Taakili, A., Zarrouk, M.M.: Optimized Schwarz waveform relaxation for nonlinear reactive transport with adsorption. Submitted to: The Moroccan Journal of Pure and Applied Analysis "MJPAA" (Print ISSN: 2605–6364) (2023)
9. Bennequin, D., Gander, M.J., Gouarin, L., Halpern, L.: Optimized Schwarz waveform relaxation for advection reaction diffusion equations in two dimensions. Numer. Math. **134**, 513–567 (2016)
10. Haeberlein, F.: Time Space Domain Decomposition Methods for Reactive Transport— Application to CO2 Geological Storage. Université Paris-Nord - Paris XIII, Theses (2011)
11. Blayo, E., Halpern, L., Japhet, C.: Optimized Schwarz waveform relaxation algorithms with nonconforming time discretization for coupling convection-diffusion problems with discontinuous coefficients. In: Domain Decomposition Methods in Science and Engineering XVI (pp. 267–274). Springer (2007)
12. Gander, M.J., Halpern, L., Kern, M.: A Schwarz waveform relaxation method for advection-diffusion-reaction problems with discontinuous coefficients and non-matching grids. In: Domain Decomposition Methods in Science and Engineering XVI (pp. 283–290). Springer (2007)
13. Eymard, R., Gallouët, T., Herbin, R.: Finite volume methods. Handb. Numer. Anal. **7**, 713–1018 (2000)
14. Berthe, P.-M.: Méthodes de décomposition de domaine de type relaxation d'ondes optimisées pour l'équation de convection-diffusion instationnaire discrétisée par volumes finis. Ph.D. thesis, Paris 13 (2013)
15. Sun, N.-Z.: Mathematical Modeling of Groundwater Pollution. Springer-Verlag and Geological Publishing House, New York (1996)
16. Gander, M.J., Halpern, L.: Optimized Schwarz waveform relaxation methods for advection reaction diffusion problems. SIAM J. Numer. Anal. **45**(2), 666–697 (2007)

A Maximum Power Point Tracking Using P&O Method for System Photovoltaic

Hafid Ben Achour[1(✉)], Said Ziani[2], and Youssef El Hassouani[1]

[1] Department of Physics, Faculty of Science and Technology Errachidia, Moulay Ismail University, BP-509 Boutalamine Errachidia, 52000 Meknes, Morocco
h.benachour@edu.umi.ac.ma, y.elhassouani@umi.ac.ma
[2] ENSAM, Mohmmed V University, Rabat, Morocco
SAID.ZIANI@univh2.ma

Abstract. This work presents the complete procedure for simulating MPPT (Maximum Power Point Tracking) control, based on the perturbation and observation (P&O) method. A control chain was created between the photovoltaic panel and the load. To do this, we modeled and simulated the system in Matlab/Simulink. The results show that this technique, which is based on the "perturb and observe" algorithm, enables the photovoltaic system to operate optimally.

Keywords: MPPT · MPP · PV · Converter · Power electronic

1 Introduction

With the growing demand for electrical energy, the rising cost of fuels used in conventional power plants, and growing concerns about their impact on the environment, there is a great deal of interest in renewable energy sources, particularly solar power. The latter is perceived as an inexhaustible, widely accessible and low-maintenance green energy resource. Furthermore, thanks to the increasing availability of photovoltaic panels and their acceptable yields, solar electricity is gaining ground.

The operating point of a photovoltaic (PV) panel does not always coincide with the maximum power point (MPP). Therefore, a control mechanism is needed to identify and track the maximum power point, known as Maximum Power Point Tracking (MPPT), so that maximum power is always generated.

To achieve the optimal operation of a photovoltaic system, several MPPT control techniques are developed by researchers to control the maximum power point of the system (PPM). Among these techniques we can cite: constant parameter tracking techniques, the latter is based on several methods, namely: in [1] the constant voltage method is used for MPP, Koutroulis et al. [2] used the voltage method in open circuit To achieve the control of to follow the MPP, another method called short circuit current is studied by Kasa et al. [3], there are other methods concerning this technique [4–6]. Another tracking technique used is the tracking technique with trial error, as we have already seen, this technique is also based on several methods, namely: the disturb and observe (P&O) [7] method, the only current photovoltaic method [8], and other methods [7, 9, 10]. Another MPPT tracking technique is presented in the literature [11–13].

Y. Farhaoui et al. (Eds.): ICAISE 2023, LNNS 837, pp. 440–445, 2024.
https://doi.org/10.1007/978-3-031-48465-0_58

2 Modeling of the Photovoltaic System

- **Modeling of a Photovoltaic Cell**

The mathematical model for the current-voltage characteristic of a PV cell is used to express the intensity of a PV module as a function of the voltage at its terminals and the climatic conditions. This expression is written:

$$I = I_{ph} - I_d - I_{RP} \tag{1}$$

The current I_0 corresponds to the current of the parallel connections is given by:

$$I_{RP} = \frac{V + I_{Rs}}{R_P} \tag{2}$$

Or I_d current:

$$I_d = I_0\left[e^{\frac{V+I_{Rs}}{AV_t}} - 1\right] \tag{3}$$

where diode saturation current which given by:

$$I_0 = K_1 T^3 e^{\frac{E_g}{KT}} \tag{4}$$

So, the expression for characteristic I-V is

$$I = I_{ph} - I_0\left[e^{\frac{V+I_{Rs}}{AV_t}} - 1\right] - \frac{V + I_{Rs}}{R_P} \tag{5}$$

$$I_{ph} = (I_{ph,n} + K_1 \Delta_T)\frac{G}{G_n} \tag{6}$$

where:

A: Ideality factor of the junction ($1 < A < 3$). K: Boltzmann constant (1.38×10^{-23} J/K). k1: Constant (1.2 A/cm^2K^3). Iph: Photo-generated current created within the structure by a portion of radiation. I0: Diode current, representing the internal leakage current of a cell caused by the junction. PN T: Absolute temperature. Eg: Energy gap.

- **Perturb and Observation Algorithm**

The P&O method is generally the most used because of its simplicity and ease of implementation. As the name suggests, this method works by perturbing the system and observing the impact on the power output of the PV generator. This disturbance occurs at the level of the voltage Vpv.

If an increase in voltage causes an increase in power, the operating point is to the left of the PPM, if on the contrary the power decreases, it is to the right. In the same way, one can make a reasoning for a decrease in the voltage. In summary, for a voltage disturbance, if the power increases, the direction of the disturbance is maintained. If not, it is reversed so that the operating point converges towards the PPM.

The process is repeated periodically until the MPP is reached. The system then oscillates around the MPP, which causes power losses. Oscillation can be minimized by decreasing the size of the disturbance. However, too small a disturbance size considerably slows down the pursuit of the MPP. There is then a compromise between precision and response time. Figure 1 presents perturb and observe method algorithm.

3 Simulation Results

For the PV to provide its maximum power, there must be a permanent adaptation of the load with the photovoltaic generator. As shown in the Fig. 1.

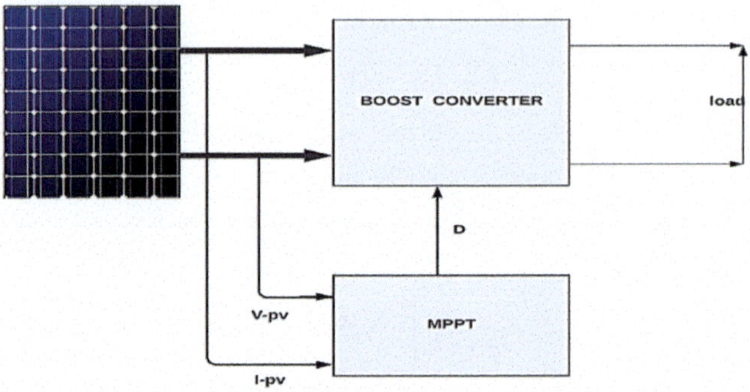

Fig. 1. Photovoltaic conversion chain controlled by an MPPT command on DC load.

The simulation is done for a photovoltaic system with the technical data sheet mentioned in Table 1. The P&O method is used to make the system deliver the maximum power. The simulation is done on the one hand for a constant temperature of 25 °C and an irradiation of 700 W/m then 1000 W/m, on the other hand for a constant irradiation of 1000 W/m and a temperature of 30 °C then 25 °C. Figure 2 shows the P-V and I-V characteristics for the two previous cases.

Table 1. The technical data sheet of photovoltaic system

The maximum operating power PMAX	200 W
The maximum operating voltage VMPP	26.3 V
The maximum operating current IMPP	7.61 A
The circuit course current ISC	8.21 A
The open circuit voltage VOC	32.9 V
Temperature coefficient at ISC	3.18×10^{-3} A/°C
Temperature coefficient at VOC	-1.23×10^{-1} A/°C
Module	54

Figure 3 illustrates the simulation results for the first case (T = 25°C, and irradiation at 7000 W/m^2 then 1000 W/m^2 at 0.5 s). Since Figure 4 shows those of the second case (irradiation at 1000 W/m^2 and T = 30 °C then T = 25 °C at 0.5 s).

From Figure 4 we can know the maximum power, voltage and maximum current of the PV system for each case. The results found in Figure 3 confirm that the PV system delivers the maximum power to the load in case 1, we see that for a temperature of 25 °C

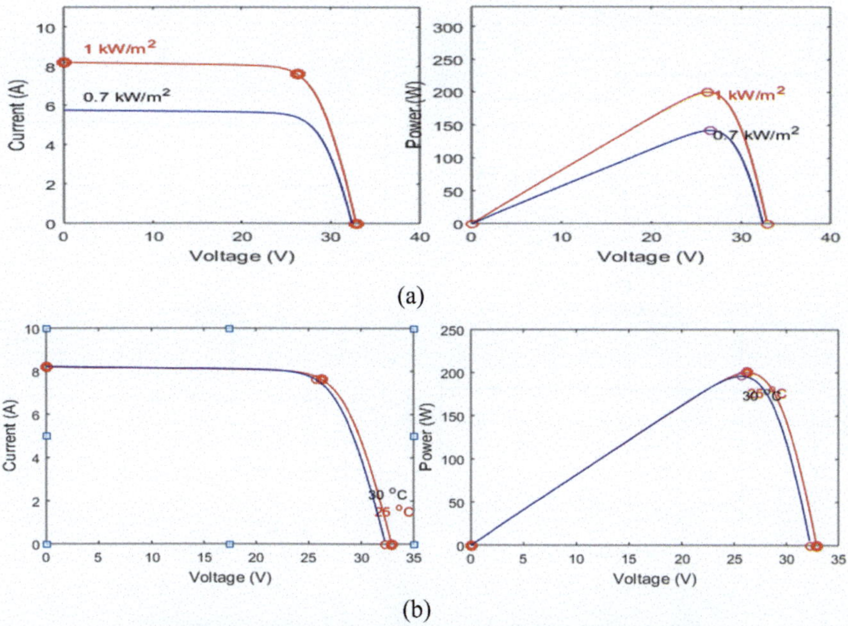

Fig. 2. The P-V and I-V characteristics for case 1 (a), case 2 (b)

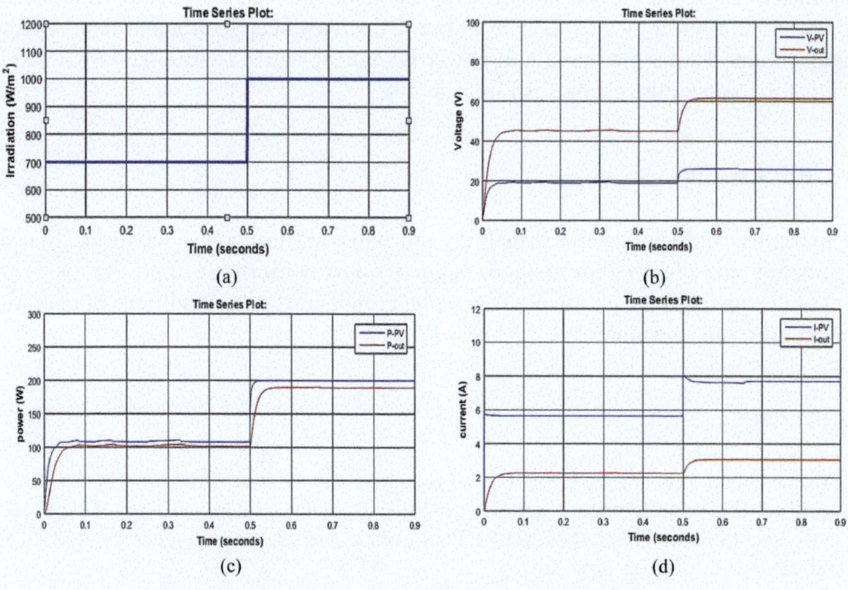

Fig. 3. Irradiation, power, voltage and current for case 1

and an irradiation of 1000 W/m^2, the maximum power is 200 W while that at 700 W/m^2 the maximum power is around 120 W. we see that the power delivered by the PV system

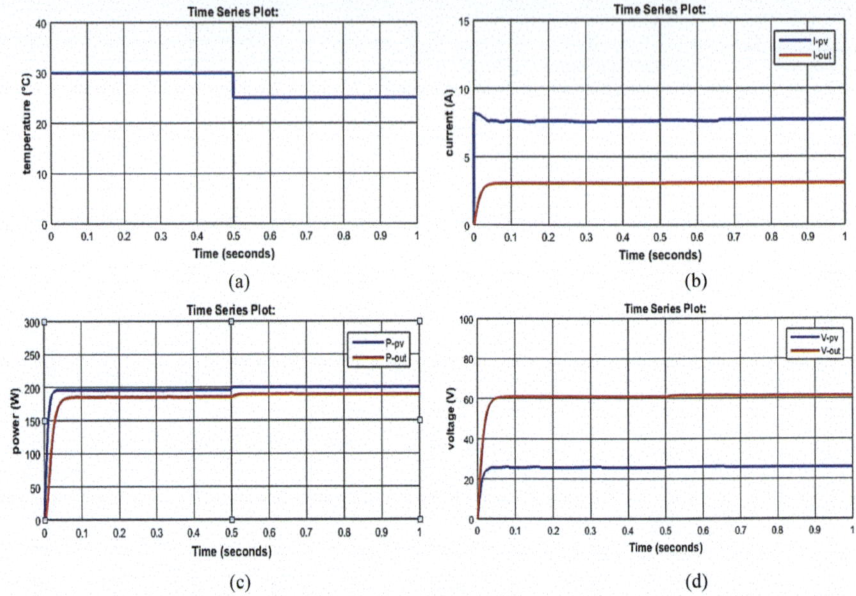

Fig. 4. Temperature, power, voltage and current for case 2

continues its maximum value, in Fig. 3, for each climatic condition itself. Thing for the voltage and the current we observe from Fig. 2 that they keep their maximum. Another remark to note is that the maximum power is more sensitive to irradiation but can be sensitive to temperature change as Fig. 4 shows.

4 Conclusion

We presented and studied the MPPT control based on the power feedback, like the perturbation and observation method, using a Boost converter, to find the maximum point of the power of the photovoltaic generator, under different conditions of operation (temperature and sunshine).

References

1. Yu, G., Jung, Y., Choi, J., Choy, I., Song, J., Kim, G.: A novel two-mode MPPT control algorithm based on comparative study of existing algorithms. In: Proceedings of the Conference Record of the Twenty-Ninth IEEE Photovoltaic Specialists Conference, pp. 1531–1534 (2002)
2. Koutroulis, E., Kalaitzakis, K., Voulgaris, N.: Development of a microcontroller-based, photovoltaic maximum power point tracking control system. IEEE Trans. Power Electron. **16**(1), 46–54 (2001)
3. Kasa, N., Lida, T., Iwamoto, H.: Maximum power point tracking with capacitor identifier for photovoltaic power system. IEE Proc. Electr. Power Appl. **147**(6), 497–502 (2000)

4. Chomsuwan, K., Prisuwanna, P., Monyakul, V.: Photovoltaic grid-connected inverter using two-switch buck-boost converter. In: Proceedings of the Photovoltaic Specialists Conference, pp. 1527–1530 (2002)
5. Park, M., Yu, I.-K.: A study on the optimal voltage for MPPT obtained by surface temperature of solar cell. In: Proceedings of the 30th Annual Conference of IEEE Industrial Electronics Society, IECON, vol. 3, pp. 2040–2045 (2004)
6. Kobayashi, K., Matsuo, H., Sekine, Y.: An excellent operating point tracker of the solarcell power supply system. IEEE Trans. Ind. Electron. **53**(2), 495–499 (2006)
7. Sefa, I., Ozdemir, O.: Experimental study of interleaved MPPT converter for PV systems. In: Proceedings of the Industrial Electronics, IECON, 35th Annual Conference of IEEE, pp. 456–461 (2009)
8. Salas, V., Olias, E., Lazaro, A., Barrado, A.: New algorithm using only one variable measurement applied to a maximum power point tracker. Sol. Energy Mater. Sol. Cells **87**(1–4), 675–684 (2005)
9. Femia, N., Petrone, G., Spagnuolo, G., Vitelli, M.: Optimization of perturb and observe maximum power point tracking method. IEEE Trans. Power Electron. **20**(4), 963–973 (2005)
10. Park, S.-H., Cha, G.-R., Jung, Y.-C., Won, C.-Y.: Design and application for pv generation system using a soft-switching boost converter with sarc. IEEE Trans. Ind. Electron. **57**(2), 515–522 (2010)
11. Messalti, S., Harrag, A., Loukriz, A.: A new variable step size neural networks MPPT controller: Review, simulation and hardware implementation. Renew. Sustain. Energy Rev. **68**(Part 1), 221–233 (2017)
12. Motahhir, S., El Hammoumi, A., El Ghzizal, A.: The most used MPPT algorithms: review and the suitable low-cost embedded board for each algorithm. J. Clean. Prod. **246**, 118983 (2020)
13. Faranda, R., Leva, S., Maugeri, V.: MPPT techniques for PV systems: energetic and cost comparison. In: Proceedings of the Power and Energy Society, pp. 1–6 (2008)

A Review of Non-linear Control Methods for Permanent Magnet Synchronous Machines (PMSMs)

Chaou Youssef[1(✉)], Ziani Said[2], and Daoudia Abdelkarim[1]

[1] Department of Physics, Faculty of Science and Technology Errachidia, Moulay Ismail University, BP-509 Boutalamine Errachidia, 52000 Meknres, Morocco
yo.chaou@edu.umi.ac.ma

[2] ENSAM, Mohammed V University, Rabat, Morocco

Abstract. This scientific review provides a comprehensive overview of the methods of non-linear control applied to Permanent Magnet Synchronous Machines (PMSMs). PMSMs are widely used in various industrial applications due to their high efficiency and precise control capabilities. Traditional linear control techniques have been successful in many cases; however, they may face limitations in handling the complex dynamics and non-linearity inherent to PMSMs. To overcome these challenges, researchers have developed non-linear control methods that exploit the advantages of advanced control theories. This review highlights the principles, advantages, and limitations of various non-linear control techniques for PMSMs, shedding light on the most promising avenues for future research and practical implementation.

Keywords: Nonlinear control · Permanent magnet synchronous machines (PMSMs) · Dynamics systems

1 Introduction

Permanent magnet synchronous machines (PMSMs) have become essential devices in various industrial applications, especially in the fields of electro-mobility, renewable energies, and industrial automation. To fully exploit the potential of these machines, it is imperative to develop advanced control techniques that ensure optimal performance under diverse operating conditions. Controlling MSAPs can be challenging due to their non-linear nature and the complex interactions between different parameters. Traditional linear control approaches, such as proportional-integral-derivative (PID) controllers [1] and field-oriented control (FOC) [2], cannot fully exploit the machine's capabilities and adaptability to varying operating conditions to manage the inherent non-linearity of MSAPs, including load variations, electromagnetic disturbances and magnetic flux saturation phenomena. Non-linear control methods have therefore been developed to optimize the performance and stability of MSAPs. Several projects have been carried out in the fields of non-linear control and vector control. The vector control, including:

Y. Farhaoui et al. (Eds.): ICAISE 2023, LNNS 837, pp. 446–452, 2024.
https://doi.org/10.1007/978-3-031-48465-0_59

robust control of MSAP speed using linearization input control input-output linearization control [3], sliding mode control [4], backstepping control [5–8]and DTC [9]... etc. This study aims to provide a comprehensive review of various nonlinear control methods applied to PMSMs. We will discuss various advanced control approaches that have been proposed in scientific and industrial literature. These approaches include current loop control, speed loop control, flux loop control, vector control, predictive control, sliding mode control, and other advanced control strategies. We will analyze the main advantages and limitations of each nonlinear control method, with a focus on their ability to enhance the dynamic and static performance of PMSMs. Energy efficiency, reference tracking accuracy, and ease of implementation will be critical criteria evaluated in this comparative analysis. Finally, we will examine recent progress in the field of nonlinear control of PMSMs, highlighting technological advancements and future prospects. The ultimate goal of this study is to provide engineers, researchers, and industrialists with a comprehensive overview of the most promising nonlinear control methods for PMSMs.

2 Non-linear Control Techniques

2.1 Sliding Mode Control (SMC)

Sliding mode control is a control strategy widely used in engineering and automation. The aim of this control approach is to maintain a dynamic system within a specific operating zone, called sliding mode, despite any disturbances or parameter variations that may occur. The fundamental principle of Sliding Mode Control is based on the creation of a sliding surface, which is a mathematical function defined in the state space of the system. This sliding surface is designed so that the system rapidly converges on it and then remains confined within it, keeping its states under control. When the system reaches this surface, it enters the sliding mode where the dynamic behaviors are predictable and stable [10].

2.2 Adaptive Control

Adaptive Control is a branch of automation and dynamic systems control that aims to design algorithms capable of automatically maintaining a system's performance despite the disturbances, parameter variations and uncertainties to which it may be subjected. Unlike traditional control approaches, where the control laws are predefined and fixed, Adaptive Control allows a system to adjust in real time to adapt to its environment. One of the main features of Adaptive Control is the ability to recognize and estimate the uncertainties present in the system. This is usually done using mathematical models and system identification techniques. Once the uncertainties are estimated, the Adaptive Control algorithm adjusts the control parameters to compensate for these uncertainties, thus ensuring stable and optimal performance [11].

2.3 Neural Network Control

Neural network control, also known as neural control, is an approach to machine learning and artificial intelligence that uses neural networks to solve control and decision-making problems. These neural networks, inspired by the workings of the human brain, are capable of modeling complex relationships between input data and the actions to be taken in response to that input. In the context of neural network control, inputs can come from a variety of sources, such as sensors, databases or real-time information. The neural network then processes this data to generate actions or decisions adapted to the specific task. An essential aspect of neural network control is learning. Neural networks can be trained on real or simulated data to adjust their weights and internal parameters to improve their performance [12].

2.4 Feedback Linearization Control

Linearized feedback control is an advanced approach in control engineering that aims to stabilize and control non-linear dynamic systems by transforming them into linear systems via an appropriate transformation. This method simplifies the control problem by using classical linear feedback techniques for systems that would otherwise be difficult to control due to their non-linearity. The fundamental principle behind linearized feedback control is to use a mathematical transformation, usually based on differential calculations or non-linear functions, to make the non-linear system equivalent to a linear system [13].

2.5 Backstepping Control

Backstepping Control is a control design method for non-linear dynamic systems. This approach is used to stabilize and follow specific trajectories in complex systems that may be difficult to control using traditional linear control methods. The fundamental principle of Backstepping Control is based on breaking down the dynamics of the system into several sub-systems, each designed to achieve a specific objective. The controller is designed recursively, starting with the outermost 'output' subsystem and progressing through the internal subsystems until it reaches the innermost subsystem [14].

3 Case Studies

3.1 Electric Vehicle Propulsion System

A case study on the implementation of a non-linear control method which is the backstepping technique applied for a PMSM-based electric vehicle propulsion system is presented. The results of simulations of different speed control tests for a PMSM-based electric vehicle are presented in Figs. 1, 2 3 and 4 and are evaluated under different driving conditions in order to determine the performance and effectiveness of the control strategy [7].

The first test concerns the problem of trajectory following. Assuming that the electric vehicle is travelling on a straight road with a reference speed of 80 km/h (see Fig. 1), with a wind speed of 10 m/s and a gradient of 30% at (t = 2.5 s), the following are applied to the road profile.

The second test involves examining the challenge of following a specified trajectory. In this scenario, the vehicle operates on a straight road with varying reference speeds,

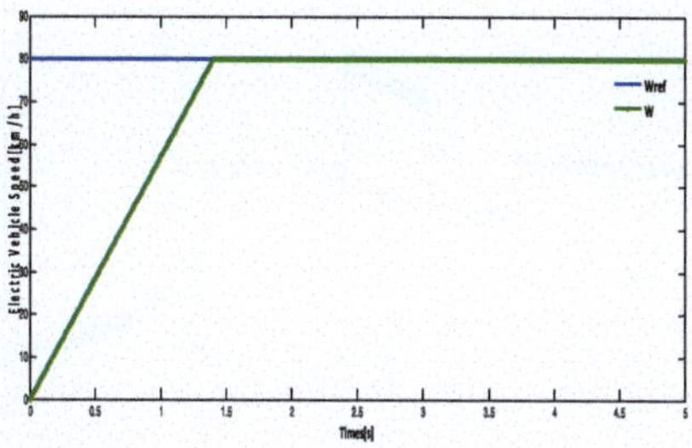

Fig. 1. The speed tracking response for a constant speed reference with a 30% gradient at time (t = 2.5 s)

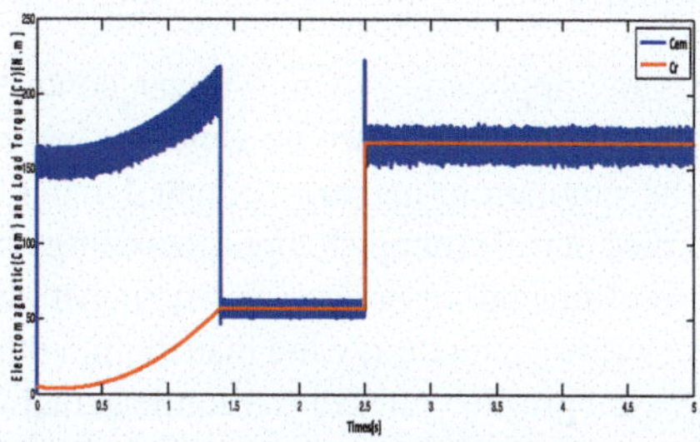

Fig. 2. Response of the driving torque C_r and the electromagnetic torque C_{em} when subjected to a 30% gradient at (t = 2.5 s).

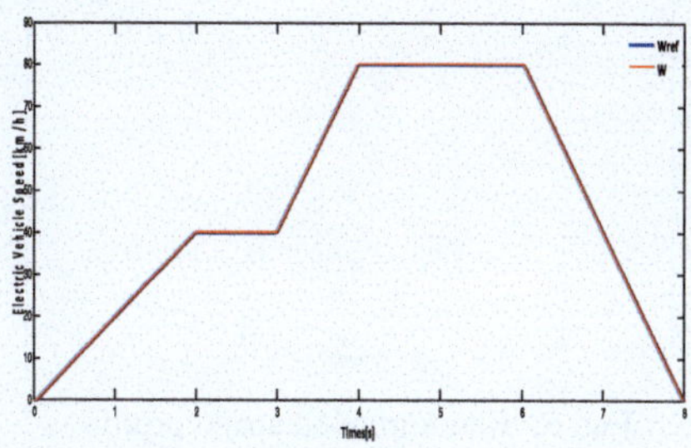

Fig. 3. Response in tracking speed for an alternative reference variable speed

acceleration, and deceleration rates (VE), as depicted in Fig. 3. The wind speed is held constant at 10 m/s, and the road profile remains flat with a 0° slope. As shown in Fig. 3, the vehicle's speed closely tracks the reference speed, while Fig. 4 illustrates the performance of the driving torque (C_r) and electromagnetic torque (C_{em}).

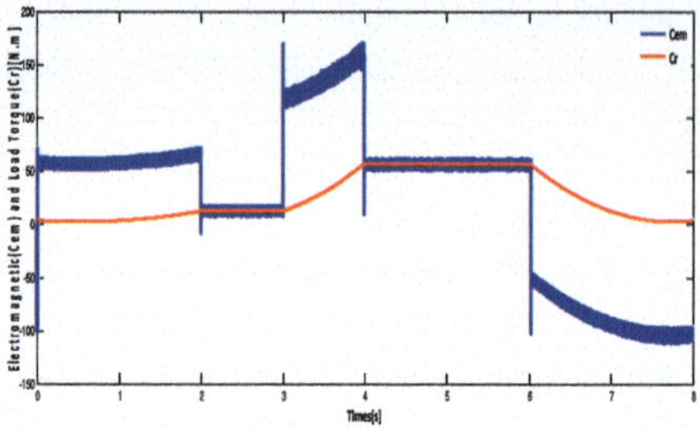

Fig. 4. Performance of the driving torque C_r and the electromagnetic torque C_{em}

3.2 Wind Turbine Generator

This case study examines the application of non-linear control methods that control backstepping to a PMSM used in a wind turbine generator system. Figures 5, 6 and 7 display the outcomes of the simulations and are evaluated under different variable wind conditions. The ability of the control strategy to optimize energy production and manage variable wind conditions is evaluated. In order to verify the validity of the backstepping command, it is necessary to apply a random wind distribution that is closer to the actual evolution, see Fig. 5. The wind profile that is closer to the actual wind was implemented to fit the slow dynamic system tested [6].

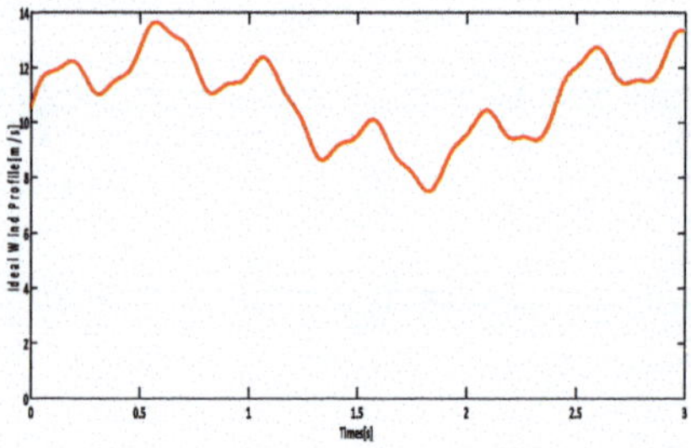

Fig. 5. Random wind speed profile

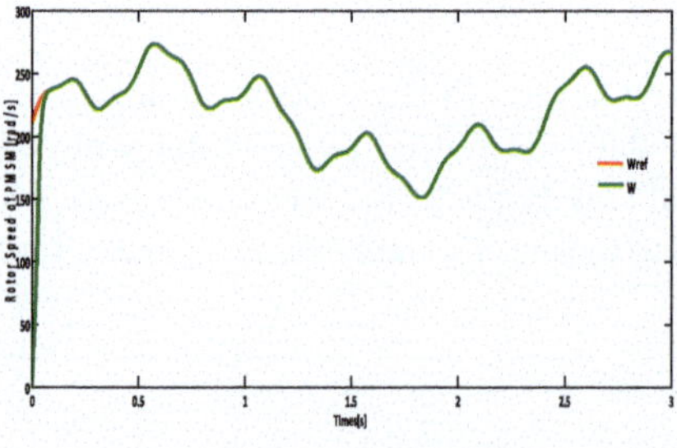

Fig. 6. PMSG's reaction to a fluctuating wind speed profile in terms of tracking the reference speed profile

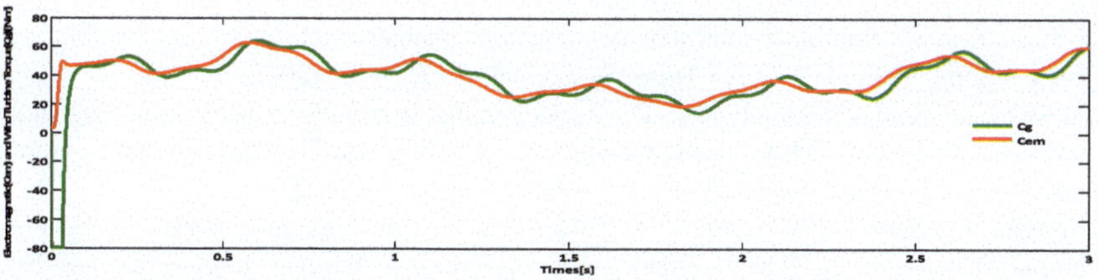

Fig. 7. The response of the driving torque C_g and electromagnetic torque C_{em} under varying wind speed conditions

4 Advantages and Limitations

Each non-linear control method has its own advantages and limitations when applied to PMSMs. Backstepping control excels at handling non-linearity, but its design is not always so straightforward. Neural network control offers excellent approximation capabilities, but can be prone to over-fitting and may require a lot of training data. Model predictive control offers predictive capabilities but can be computationally demanding for real-time applications. Sliding mode and adaptive control offer robustness to uncertainties and disturbances, but can require complex tuning and are more complex to implement.

5 Conclusion

Non-linear control techniques present promising alternatives to traditional linear control methods for Permanent Magnet Synchronous Machines. Each method offers unique benefits in handling the complex dynamics and non-linearity inherent to PMSMs. Researchers and engineers must carefully consider the application requirements and specific characteristics of the PMSM system to select the most appropriate non-linear control strategy. The continuous advancement in non-linear control methodologies promises to unlock the full potential of PMSMs in various industries, contributing to the pursuit of energy-efficient and high-performance systems.

References

1. Willis, M.: Proportional-Integral-Derivative Control, vol. 6. Department of Chemical and Process Engineering University of Newcastle (1999)
2. Merzoug, M., Naceri, F.: Comparison of field-oriented control and direct torque control for permanent magnet synchronous motor (PMSM). Int. J. Electr. Comput. Eng. **2**(9), 1797–1802 (2008)
3. Rebouh, S., et al.: Nonlinear control by input-output linearization scheme for EV permanent magnet synchronous motor. In: 2007 IEEE Vehicle Power and Propulsion Conference. IEEE (2007)
4. Hamida, M.A., et al.: Robust adaptive high order sliding-mode optimum controller for sensorless interior permanent magnet synchronous motors. Math. Comput. Simul **105**, 79–104 (2014)

5. Chaou, Y., et al.: Nonlinear control of the permanent magnet synchronous motor PMSM using backstepping method. WSEAS Trans. Syst. Control **17**, 56–61 (2022)
6. Youssef, C., Said, Z., Abdelkarim, D.: Backstepping control of the permanent magnet synchronous generator (PMSG) used in a wind power system. In: The International Conference on Artificial Intelligence and Smart Environment. Springer (2022)
7. Youssef, C., Said, Z., Abdelkarim, D.: Electric vehicle backstepping controller using synchronous machine. In: The International Conference on Artificial Intelligence and Smart Environment. Springer (2022)
8. Ziani, S., et al.: Developed permanent magnet synchronous motor control using numerical algorithm and backstepping. J. Eng. Sci. Technol. Rev. **16**(1), (2023)
9. Tiitinen, P., Surandra, M.: The next generation motor control method, DTC direct torque control. In: Proceedings of International Conference on Power Electronics, Drives and Energy Systems for Industrial Growth. IEEE (1996)
10. Vaidyanathan, S., Lien, C.-H.: Applications of Sliding Mode Control in Science and Engineering, vol. 709. Springer (2017)
11. Åström, K.J., Wittenmark, B.: Adaptive Control. Courier Corporation (2013)
12. Ge, S.S., Hang, C.C., Zhang, T.: Adaptive neural network control of nonlinear systems by state and output feedback. IEEE Trans. Syst. Man Cybern. Part B (Cybernetics) **29**(6), 818–828 (1999)
13. Krener, A.: Feedback linearization. Mathematical Control Theory, pp. 66–98 (1999)
14. Zhou, J., et al.: Adaptive Backstepping Control. Springer (2008)

Wavelet-Based Denoising of 1-D ECG Signals: Performance Evaluation

Said Ziani[1(✉)], Achmad Rizal[2], Suchetha M.[3], and Youssef Agrebi Zorgani[4]

[1] ENSAM, Mohammed V University in Rabat, Rabat, Morocco
s.ziani@um5r.ac.ma

[2] School of Electrical Engineering Telekomunikasi no.1, Telkom University, Bandung, Indonesia
achmadrizal@telkomuniversity.ac.id

[3] Centre for Healthcare Advancement, Innovation & Research, Vellore Institute of Technology, Chennai 10587, India
suchetha.m@vit.ac.in

[4] Laboratory of Sciences and Techniques of Automatic control & computer engineering (Lab-STA), National School of Engineering, Sfax 10587, Tunisia
econov@uvisfax.com

.

Abstract. ECG signals are essential for diagnosing heart conditions but are often corrupted by noise during acquisition and transmission. This paper focuses on using wavelet-based denoising techniques to improve 1-D ECG signal quality. It primarily evaluates their performance and compares various methods. These techniques aim to extract clear, reliable cardiac data from noisy signals, enhancing diagnostic accuracy and advancing cardiac research.

Keywords: Wavelet · Denoising · ECG Signals

1 Introduction

The electrocardiogram (ECG) stands as a cornerstone in the field of cardiology, serving as a vital tool for the diagnosis and monitoring of diverse cardiac conditions. It provides a real-time electrical snapshot of the heart's activity, offering clinicians invaluable insights into the organ's health, rhythm, and functionality. The ECG's ability to detect irregularities, arrhythmias, ischemic events, and other cardiac anomalies has made it an indispensable asset in healthcare. However, the journey from the patient's body to the medical records is fraught with challenges, as ECG signals are inherently susceptible to a cacophony of noise sources that can distort and obscure the critical information they carry.

Noise in ECG signals can stem from various origins, including muscular contractions, power line interference, baseline wander, electrode contact issues, and patient motion. These sources of noise can introduce artifacts that compromise

Y. Farhaoui et al. (Eds.): ICAISE 2023, LNNS 837, pp. 453–458, 2024.
https://doi.org/10.1007/978-3-031-48465-0_60

the fidelity of the ECG data, potentially leading to misdiagnosis or missed clinical insights. Therefore, mitigating the impact of noise on ECG signals is of paramount importance in ensuring accurate cardiac assessments and facilitating timely interventions.

In response to this challenge, this paper embarks on a comprehensive exploration of wavelet-based denoising techniques as a powerful means to enhance the quality of 1-Dimensional (1-D) ECG signals. The focus here is not only on acknowledging the problem but on providing practical solutions. Wavelet-based denoising methods have gained prominence in recent years due to their capacity to effectively attenuate noise while preserving the essential features of ECG waveforms. This paper's central objectives revolve around evaluating the performance of these techniques and conducting a rigorous comparative analysis to identify their strengths and limitations.

By delving into the intricate world of wavelet transformations and thresholding techniques, this study aims to decipher the intricate dance between mathematics and signal processing. The ultimate goal is to unleash the potential of ECG data, buried beneath layers of noise, for both clinical practitioners and researchers. With an emphasis on empirical results and practical insights, this paper presents a denoising method that provides a significant added value for addressing source separation problems discussed in several works, namely [1–7]. In the following sections, we will delve into the theory behind wavelet-based denoising, describe our experimental methodology, present our findings, and offer a glimpse into future directions for research in this critical domain.

2 Discrete Wavelet Transforms (DWT)

The Discrete Wavelet Transform (DWT) is a mathematical technique extensively employed in signal processing and data analysis. Its primary function is to dissect a signal into distinct frequency components, simplifying data analysis and denoising procedures. The DWT accomplishes this by guiding the signal through a sequence of high-pass and low-pass filters, yielding detailed and approximated representations at various scales. The Discrete Wavelet Transform (DWT) serves a wide array of applications, with a primary focus on signal Denoising, DWT excels in the task of noise reduction by effectively distinguishing between noise and signal components across different scales. This capability results in remarkably efficient noise reduction processes.Furthermore, the flexibility offered by the DWT, including the selection of wavelet families like Haar, Daubechies, or Symlet, along with the adjustment of decomposition levels, allows for tailored solutions. This customization permits a delicate balance between temporal and spectral localization, making DWT an adaptable and versatile tool for signal analysis spanning various domains.

3 Methodology

3.1 Global Algorithm

Denoising an ECG (Electrocardiogram) signal using the Discrete Wavelet Transform (DWT) involves several steps. Here's a methodology for denoising ECG signals using DWT:

1. Preprocessing
Data Acquisition: Acquire the ECG signal data from a reliable source or sensor. Signal Digitization: Ensure that the analog ECG signal is digitized into a discrete time series.

2. Noise Characterization
Identify Noise Sources: Understand the nature of noise in your ECG signal. Common sources include powerline interference (50/60 Hz), baseline wander, and muscle artifacts.

3. Decomposition
Apply DWT: Use the DWT to decompose the ECG signal into detail coefficients (high-frequency components) and approximation coefficients (low-frequency components). Select Wavelet: Choose an appropriate wavelet (e.g., Daubechies, Symlet) based on your signal characteristics and denoising requirements. Determine Decomposition Levels: Decide how many decomposition levels are needed. This depends on the noise level and the desired level of denoising.

4. Thresholding
Threshold Detail Coefficients: Apply a thresholding technique to the detail coefficients. Common thresholding methods include: Hard Thresholding: Zero out coefficients below a certain threshold. Soft Thresholding: Shrink coefficients below a threshold toward zero. Other thresholding techniques like Bayesian, VisuShrink, or SureShrink.

5. Reconstruction
Inverse DWT: Reconstruct the signal using the denoised detail coefficients and the original approximation coefficients. It's essential to note that the effectiveness of ECG signal denoising depends on factors like the choice of wavelet, thresholding method, and the level of decomposition. Fine-tuning these parameters may be necessary to achieve the desired denoising results while preserving important ECG features. Additionally, consulting with experts in ECG analysis can be valuable in clinical applications.

3.2 Mathematical Model

The mathematical representation of an ECG signal can be described using the following mathematical equation:

$$X(t) = P(t) + Q(t) + R(t) + S(t) + T(t) + N(t) \tag{1}$$

where:

X(t) represents the ECG signal as a function of time t.

P(t) + Q(t) + R(t) + S(t) and T(t) are individual components of the ECG signal corresponding to different phases of the cardiac cycle, namely the P-wave, QRS complex, and T-wave. These components are typically represented as mathematical functions.

N(t) represents the noise or interference present in the ECG signal.

Each of these components can be further described mathematically. For instance, the QRS complex, which represents ventricular depolarization, is often modeled as a combination of Gaussian functions:

$$QRS(t) = A_1^{-\frac{(t-t_1)^2}{2\sigma_1^2}} + A_2^{-\frac{(t-t_2)^2}{2\sigma_2^2}} \tag{2}$$

Here, A_1, A_2, are the amplitudes of individual peaks within the QRS complex, t_1, t_2 are the time positions of the peaks, and σ_1, σ_2 are the standard deviations determining the width of the peaks. It's important to note that real ECG signals are often a combination of these components and noise. The goal of ECG signal processing and denoising is to extract and analyze these components accurately.

4 Results and Discussion

A single channel from the Daisy worldwide database is used to derive the fetal electrocardiogram (FECG) from cutaneous recordings as referenced in [8]. In order to construct clear time-frequency and time-scale images, we opted for preprocessing the input abdominal signals using the discrete wavelet transform. This approach was chosen to enhance the quality of the signals prior to further analysis. Figure 1 illustrates the signal before and after the preprocessing process.

In this paper, we have explored the application of wavelet-based denoising techniques for enhancing the quality of 1-D ECG signals. The results obtained from our study reveal several significant insights and observations regarding the effectiveness of wavelet-based denoising methods in the context of ECG signal processing. Metrics for Evaluation: We used metrics like SNR, RMSE, and MAE to quantitatively assess denoising quality.

Wavelet Family Matters: The choice of wavelet family (e.g., Daubechies, Symlet, Haar) significantly impacted denoising performance, with some wavelets excelling in specific noise scenarios.

Decomposition Level: Selecting the right number of decomposition levels is critical, balancing noise reduction with signal preservation.

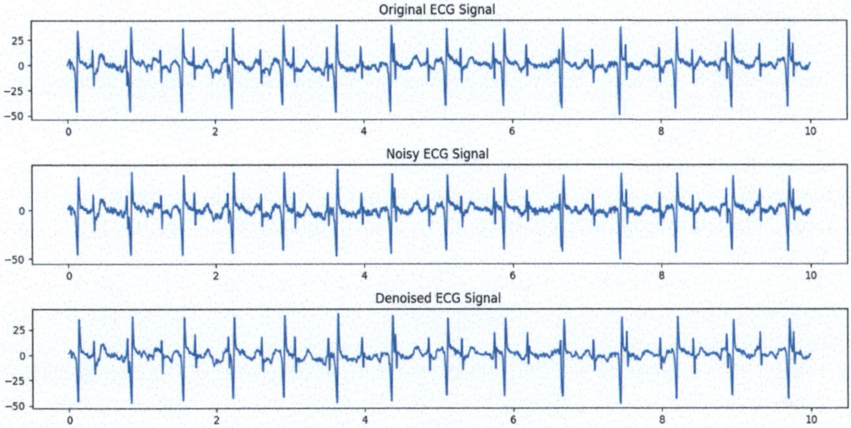

Fig. 1. Denoised ECG signals.

Noise Types: Different noise types require tailored denoising approaches, highlighting the importance of considering noise diversity.

Computational Efficiency: Balancing denoising effectiveness with computational efficiency is essential, especially for real-time applications.

The discrete wavelet transform presented in this paper can yield significant results when combined with optimization methods introduced by artificial intelligence, as indicated in references [9–11]. The importance of this method in denoising high-frequency signals used in electrical machines is also emphasized [12–14]. Future research may focus on adaptive denoising and integrating machine learning for noise classification and removal. In summary, our study enhances the understanding of wavelet-based ECG signal denoising, offering insights into optimization and future directions for improved clinical applications.

5 Conclusion

In this study, we evaluated the effectiveness of wavelet-based denoising techniques for enhancing 1-D ECG signal quality. Our findings highlight the importance of careful parameter selection, including wavelet family and decomposition levels, to achieve optimal denoising results. The choice of wavelet family significantly impacts denoising performance, and tailoring approaches to specific noise types is essential. Collaboration with clinical experts underscores the clinical relevance of denoised signals. This research contributes to the ongoing improvement of ECG signal processing, offering insights for future adaptive denoising and machine learning integration. Ultimately, wavelet-based denoising holds promise for enhancing the clinical utility of ECG signals in diagnostic contexts.

References

1. Ziani, S., El Hassouani, Y.: A new approach for extracting and characterizing fetal electrocardiogram. Traitement du Signal **37**(3), 379–386 (2020). https://doi.org/10.18280/ts.370304
2. Jamshidian-Tehrani, F., Sameni, R., Jutten, C.: Temporally nonstationary component analysis; application to noninvasive fetal electrocardiogram extraction. IEEE Trans. Biomed. Eng. **67**(5), 1377–1386 (2020). https://doi.org/10.1109/TBME.2019.2936943
3. Ziani, S., El Hassouani, Y., Farhaoui, Y.: An NMF based method for detecting RR interval. In International Conference on Big Data and Smart Digital Environment 2019. Springer
4. Tang, S.-N.: Design of a STFT hardware kernel for spectral Doppler processing in portable ulrasound systems. In: 2016 IEEE 5th Global Conference on Consumer Electronics, Kyoto, Japan, 2016, pp. 1–2. https://doi.org/10.1109/GCCE.2016.7800466
5. Aakaaram, V., Bachu, S.: MRI and CT image fusion using synchronized anisotropic diffusion equation with DT-CWT decomposition. In: Smart Technologies, Communication and Robotics (STCR). Sathyamangalam, India, 2022, 1–5 (2022). https://doi.org/10.1109/STCR55312.2022.10009173
6. Qassim, Y.T., Cutmore, T., Rowlands, D.: Multiplier truncation in FPGA based CWT. In: International Symposium on Communications and Information Technologies (ISCIT). Gold Coast, QLD, Australia, 2012, pp. 947–951 (2012). https://doi.org/10.1109/ISCIT.2012.6381041
7. Wang, M., et al.: Low-Latency in situ image analytics With FPGA-based quantized convolutional neural network. IEEE Trans. Neural Netw. Learn. Syst. **33**(7), 2853–2866 (2022). https://doi.org/10.1109/TNNLS.2020.3046452
8. De Moor, B, De Gersem, P., De Schutter, B., Favoreel, W.: DAISY: a database for fiction of systems. J. A, Special Issue on CACSD (Comput. Aided Control Syst. Des.) **38**(3), 4–5 (1997)
9. Ziani, S.: Contribution to single-channel fetal electrocardiogram identification. Traitement du Signal **39**(6), 2055–2060 (2022). https://doi.org/10.18280/ts.390617
10. Ziani, S., Farhaoui, Y., Moutaib, M.: Extraction of fetal electrocardiogram by combining deep learning and SVD-ICA-NMF methods. Big Data Min. Anal. **6**(3), 301–310 (2023). https://doi.org/10.26599/BDMA.2022.9020035
11. Ziani, S.: Fetal electrocardiogram identification using statistical analysis. In: Farhaoui, Y., Rocha, A., Brahmia, Z., Bhushab, B. (eds.) Artificial Intelligence and Smart Environment. ICAISE 2022. Lecture Notes in Networks and Systems, vol. 635. Springer, Cham (2023). https://doi.org/10.1007/978-3-031-26254-8-64
12. Chaou, Y., et al.: Nonlinear control of the permanent magnet synchronous motor PMSM using backstepping method. WSEAS Trans. Syst. Control **17**, 56–61 (2022). 10.37394/23203.2022.17.7
13. Ouhadou, M., et al.: Experimental modeling of the thermal resistance of the heat sink dedicated to SMD LEDs passive cooling. In: Proceedings of the 3rd International Conference on Smart City Applications, 2018
14. Ben, H., et al.: Permanent magnet synchronous motor PMSM control by combining vector and PI controller. WSEAS Trans. Syst. Control **17**, 244–249 (2022)

The Effect of Employment Mobility and Spatial Proximity on the Residential Attractiveness of Moroccan Small's Cities: Evidence from Quantile Regression Analysis

Sohaib Khalid[1](✉), Driss Effina[1], and Khaoula Rihab Khalid[2]

[1] MASAFEQ Laboratory, National Institute of Statistics and Applied Economics, Rabat, Morocco
sohaibxkhalid@gmail.com

[2] LERSRG Laboratory, University of Sultan Moulay Slimane, Béni-Mellal, Morocco

Abstract. The aim of this paper is to investigate the impact of both the mobility of the working population and the spatial proximity to agglomerations on the residential attractiveness of small cities in Morocco. First, we estimated the net migration rate, which is according to the literature, one of the most reliable indicators commonly used to estimate the attractiveness of a territory. Then, using it as a target variable, we studied for small's cities, the impact of both workers mobility and territorial proximity to agglomerations on their ability to be attractive to the population. Estimation of the net migration rate showed that it has a rather dispersed distribution, which is one of the main reasons why we opted for quantile regression modeling. As a result, we have shown that the geographical location of the small cities influences its attractiveness, in the sense that, the closer the city is to the area of an agglomeration, the more its residential attractiveness increases. We've also been able to demonstrate that the high mobility of the workforce and the high migratory attraction are two twin phenomena in small Moroccan cities. Those findings call into question the fact that the residential attractiveness of small cities in Morocco is not only determined by their socio-economic character but is also strongly influenced by their spatial location and the mobility of the workforce.

Keywords: Residential attractiveness · Migration · Labour mobility · Small city · Quantile regression analysis

1 Introduction

In the context of globalization and delocalization, the notion of attractiveness of territories has become a key theme for all those who are interested in territories, economy, social issues, and the evolution of societies [1].

The high mobility of production's factors has very concrete consequences for any territory, whether it is a small or large city, a rural area, a local authority, or a State. Nothing can be taken for granted and there is widespread competition to attract and anchor investors, producers, or skilled labor [2].

Y. Farhaoui et al. (Eds.): ICAISE 2023, LNNS 837, pp. 459–465, 2024.
https://doi.org/10.1007/978-3-031-48465-0_61

While the presence of local resources remains a key factor of competitiveness, territorial dynamism now depends on the capacity of the employment area and its actors to attract external resources [3]. In a context of openness, territories and their actors are called upon to adopt a strategy of reinforcing the attractiveness and competitiveness of territories [4].

Residential attractiveness is considered an integral dimension in the notion of territorial attractiveness. Indeed, the residential attractiveness of a territory is assimilated to its capacity to attract population and revenue; this attraction allows the territory to be the container of positive set externalities, such as the income and skills of the inhabitants [5].

Several empirical works have demonstrated that the residential attractiveness of cities is a very complex phenomenon and is influenced by several economic, social, and spatial determinants [6]. However, it has been shown that in the Moroccan context, the supply of employment remains the most influential factor on the capacity of an urban space to attract the population, especially for a small city [7]. In this precise framework, this scientific contribution is situated as an attempt to investigate and develop both the effect of labour mobility and the spatial proximity to agglomerations, as the main employment basins in any given economy, on the residential attractiveness of small's cities in Morocco.

To do so, the first necessary manipulation is to estimate a relevant variable that reflects the best city's ability to attract population. In this context, according to the literature, the net migration rate is considered, for each territorial unit, as the most reliable indicator to quantify the migration dynamics. Since the High Commission for Planning in Morocco does not publish data on migration, the first statistical exercise was to provide a reliable estimate of the net migration rate for every small city in Morocco and then conduct an appropriate econometrics modeling and make conclusions.

2 Data and Methods

2.1 Target Population, Definition of Small City

Whatever the context studied, the concept of a small city has no single, universal definition. From the same point of view, the diversity of the territories and the variety of urban facts do not allow for the construction of a single territorial typology of the urban space.

Small's city heterogeneous category in terms of number of inhabitants has never yet found a consensus [8]. Moreover, the role of small and medium-sized cities varies according to different criteria, such as accessibility or service provision which does not allow them to be represented as a homogeneous group [9]. In view of their diversity, public policies for these cities are not very well developed, especially in a context that values the metropolitan fact.

In most cases, the approach used to describe this type of urban formation is based on several criteria, such as the administrative, functional, and morpho-statistical criteria.

For this article, the morpho-statistical criterion will be the basis for the definition of a small city. The High Commission for Planning of Morocco specifies that an urban area is a small city if it falls within the category with a threshold of between 1000 and

50,000 residents. According to this criterion, Morocco's urban framework includes 292 small cities, according to the most recent Census of Population and Habitat, conducted in 2014.

2.2 List of Variables

The phenomenon of attractiveness depends on several factors: social, economic, and geographical [7]. As mentioned before, this paper aims to test the causality between the mobility of working population and spatial proximity on the capacity of small cities to attract new migratory flows. On the other hand, we seek to discover whether the attractiveness of a small city is linked to the socio-economic framework it provides, or whether its capacity to attract is also conditioned by other latent phenomena such as labor mobility.

The attractiveness of a territory is theoretically linked to its geographical environment and proximity to employment basins, and for small cities, this fact can have a huge influence on their capacity to polarize and attract people, even though it remains latent and veiled by the other determinants. In fact, a small city can appear attractive, not because of the economic and social conditions that it guarantees, but simply because it is considered to be the place of residence of a demographic group that has a significant impact on the migration phenomenon, namely the active occupied population. Of course, a worker, even if he lives in a given territory, can choose to work outside his home area. The area of residence is considered attractive for this worker, but his choice is probably only based on factors of proximity and the job offer provided by his second work territory.

In order to investigate this matter several indicators related to migration, employment, worker mobility, and territorial proximity were used. Table 1 contains the variables used in this article and their respective descriptions.

Table 1. List of variables

Label	Type	Explanation
Net migration rate	Continu	City's net migration rate, author's estimate
Employment rate	Continu	City's employment rate, official census data
Working_ouside_city	Continu	The share of employed persons living in the city and working outside it, official census data
Agglomeration distance	Categorical < 30 min 30 min $\leq <$ 60 min ≥ 60 min REF	The distance between the city and the nearest agglomeration. The distance is expressed as the mobility time needed between the city and the nearest agglomeration using the shortest route and the maximum speed allowed, official data with author's own manipulation

3 Results

3.1 Estimation

The net migration rate is the most relevant indicator for measuring the territorial dynamics of migration. Given that the High Commission for Planning in Morocco does not publish its own data on migration, estimating this aggregate seems necessary.

Therefore, we first estimate the net migration rate as the difference between the total population growth and the natural balance rate. The natural balance rate is measured as the excess of births to deaths in a city and is derived from fertility and child mortality data between the last two censuses of 2004 and 2014.

The estimation equation can be defined as:

$$\text{Net Migration Rate}_{\text{i}} = \text{Overall Growth Rate}_{\text{i}} - \text{Natural Growth Rate}_{\text{i}} \tag{1}$$

For each city i such as i = {1, 2, … 292}.

$$\text{Natural Growth Rate}_{\text{i}} = \text{Rate Births}_{\text{i}} - \text{Rate Deaths}_{\text{i}} \tag{2}$$

where i refers to city such as i = {1, 2, … 292}.

3.2 Quantile Regression Results

A quantile regression model was designed to capture the interaction of the chosen explanatory variables with the residential attraction phenomenon of small cities, which is characterized by high variability and dispersion. The quantile regression coefficients can change according to the quantile, this characteristic makes it possible to capture the change in the behavior of the explanatory variables with regards to the different levels of the net migration rate.

Our model could be expressed as following:

$$Y = X'\beta_\tau + \varepsilon_\tau \tag{3}$$

$$\begin{aligned} \textit{Net Migration Rate}_\tau = {} & \beta_{0,\tau} + \beta_{1,\tau}\textit{Employment Rate}_\tau \\ & + \beta_{2,\tau}\,\textit{Distance agglomeration}'' < 30\text{ min}''_\tau \\ & + \beta_{3,\tau}\textit{Distance agglomeration}''30 << 60\text{ min}''_\tau \\ & + \beta_{4,\tau}\textit{Working\_outside\_City}_\tau + \varepsilon_\tau \end{aligned} \tag{4}$$

Hence, we exposed below the quantile regression coefficients (9 quantiles) (Table 2).

The analysis of the quantile regression coefficients of the employment rate shows that the supply of employment has a positive influence on the net migration rate. This influence is characterized by an upward trend, the more attractive the city, the greater the contribution of the employment rate becomes.

It should be noted that a territorial entity with a large proportion of employed people does not necessarily mean that it is abundant in terms of job opportunities, it may be simply a place of residence for workers who work outside its spatial boundary.

Table 2. Quantile regression results

	Intercept	Employment rate	Distance agglomeration < 30 min	Distance agglomeration 30 min << 60 min	Working_outside_city
tau = 0.1	−7.446	0.149	0.470	0.455	0.033
tau = 0.2	−6.579	0.139	0.430	0.485	0.037
tau = 0.3	−5.831	0.128	0.630	0.284	0.038
tau = 0.4	−5.822	0.142	0.470	0.152	0.035
tau = 0.5	−5.769	0.150	0.845	0.115	0.029
tau = 0.6	−5.345	0.148	0.900	0.211	0.032
tau = 0.7	−5.252	0.159	0.730	−0.218	0.048
tau = 0.8	−5.707	0.187	0.991	−0.276	0.054
tau = 0.9	−6.820	0.200	0.689	−0.322	0.222

Regarding spatial dimension, the coefficients of the distance variable from a city to agglomeration showed that cities in the second ring (between 30 and 60 min) of agglomerations still have a slightly positive influence on their attractiveness. This effect is quite interesting in the sense that, even at a small scale, agglomerations still exert their influence on the small cities of the second ring.

Worker mobility is the most interactive variable with the city's residential attractiveness. Indeed, the contribution of mobility shows a positive sharp upward trend with the evolution of the ability of the city to attract people. This phenomenon proves that the attractiveness of small Moroccan cities is conditioned by the mobility of workers, and this factor weighs even more heavily when the city is more attractive. This evidence proves that the most attractive small Moroccan cities are marked by the high external mobility of workers. In other words, these cities are undoubtedly under the influence of the proximity of the employment basin, making a large proportion of their workforce related to the jobs generated by other neighboring territorial entities.

4 Conclusions

The results of this article show that the attractiveness of small cities in Morocco is conditioned by their geographical position and the mobility of the labor force. Indeed, a small city can be attractive to population simply because it is a place of residence for the demographic layer of workers. This layer selects to live in a small city while working elsewhere. This finding is confirmed by the fact that the most attractive small cities are those close to the agglomerations. These results lead us to confirm that the attractiveness of small cities in Morocco is more related to other factors beyond the economic and social environment that they guarantee.

As a recommendation, small cities in general should be a subject of common interest among researchers, practitioners, policymakers, and civil society. This interest is aimed at the concretization of relevant policies to promote and develop the small cities' role and integration. The evidence presented in this work shows that the residential attractiveness of small cities in Morocco is deeply influenced by its geographical proximity to the employment basin, i.e. the metropolitan areas. This observation runs the risk of marginalizing these small urban areas and penalizing their balanced and integrated economic development by gradually transforming them into almost dormitory areas. Given this, and to preserve the socio-economic balance,it is recommended that local authorities take responsibility for developing Morocco's small cities and promoting their role in organising the national urban structure. A number of measures can help to improve the situation of small Moroccan cities. These include the creation of local jobs, the provision of quality infrastructure and the enhancement of the living environment.

References

1. European Commission. Joint Research Centre: The future of cities: opportunities, challanges and the way forward. EUR (Luxembourg. Online). Publications Office, LU (2019). Consulté le: 13 avril 2023. [En ligne]. Disponible sur: https://data.europa.eu/doi/10.2760/375209
2. Viatkin, K., Shyshkin, E., Kamieniev, O., Pankeieva, A., Viatkin, R., Moroz, N.: Economic aspects for determining attractiveness of territories. E3S Web Conf. **157**, 03011 (2020). https://doi.org/10.1051/e3sconf/202015703011
3. Servillo, L., Atkinson, R., Russo, A.P.: Territorial attractiveness in EU urban and spatial policy: a critical review and future research agenda. Eur. Urban Reg. Stud. **19**(4), 349–365 (2012). https://doi.org/10.1177/0969776411430289
4. Gavrilova, M.A., Shepelev, V.M., Kosyakova, I.V., Belikova, L.F., Chistik, O.F.: Assessment of entrepreneurial territorial attractiveness by the ranking method. Sci Educ. 10
5. Kavoura, A., Kefallonitis, E., Theodoridis, P. (eds.): Strategic innovative marketing and tourism: 8th ICSIMAT, Northern Aegean, Greece, 2019. In: Springer Proceedings in Business and Economics. Springer International Publishing, Cham (2020). https://doi.org/10.1007/978-3-030-36126-6

6. Sohaib, K., Driss, E., Chaabane, M.S.: The artificial intelligence for modeling and predicting SMALL's cities attractiveness. In: Farhaoui, Y., Rocha, A., Brahmia, Z., Bhushab, B. (eds.) Artificial Intelligence and Smart Environment. Lecture Notes in Networks and Systems, vol. 635. Springer International Publishing, Cham (2023), pp. 763–769. https://doi.org/10.1007/978-3-031-26254-8_110
7. Sohaib, K., Driss, E., Youness, J.: An econometric analysis of the determinants of small's cities' attractiveness: evidence from Moroccan case. In: 2022 8th International Conference on Optimization and Applications (ICOA), Genoa, Italy. IEEE, (2022), pp. 1–6. https://doi.org/10.1109/ICOA55659.2022.9934146
8. Desmond, M., Shollenberger, T.: Forced displacement from rental housing: prevalence and neighborhood consequences. Demography 52(5):1751–1772 (2015). https://doi.org/10.1007/s13524-015-0419-9
9. Khalid, S., Effina, D., Khalid, K.R., Chaabane, M.S.: The artificial intelligence as a decision-making instrument for modeling and predicting small cities' attractiveness: evidence from Morocco. Appl. Math. Inf. Sci. (2023). https://doi.org/10.18576/amis/170517

Detection and Classification of Logos and Trademarks from Images

Assia Ennouni[1], My Abdelouahed Sabri[1(✉)], and Abdellah Aarab[2]

[1] Computer Science Department, Faculty of Sciences, University Sidi Mohamed Ben Abdallah, Fez, Morocco
abdelouahed.sabri@usmba.ac.ma

[2] Physics Department, Faculty of Sciences, University Sidi Mohamed Ben Abdallah, Fez, Morocco

Abstract. In the field of image analysis, various techniques are employed to extract and interpret relevant information from images, such as logos or recurring patterns. Object detection and classification play crucial roles in this process. While the human brain can locate, detect, and recognize an object in images within minutes, machines require substantial time and extensive data to perform the same task. This paper proposes the use of transfer learning as a technique to reduce the time and resources needed. A comparative study of different deep learning-based logo detection algorithms, including Faster RCNN, Mask RCNN, RetinaNet, and Cascade Mask RCNN is presented. The focus is on their application in detecting and localizing brand logos in images from a TV show. The simulations are conducted using the Open Logo challenge dataset, which is widely recognized in this area and comprises 27,000 images from 352 distinct logo classes. The proposed algorithms' performance in detecting and localizing brand logos is evaluated, and the results are reported in terms of precision, recall, and execution time.

Keywords: Image processing · Deep learning · Object detection · Logo detection · Logo recognition

1 Introduction

Object detection in images is an essential requirement for many applications, such as video surveillance, autonomous robotics, and facial recognition. There are several methods for detecting objects, a complete survey can be found in [1]. Convolutional neural networks (CNNs) are becoming more popular in recent years for applications involving object detection [2].

A logo is a visual symbol employed for the purpose of recognizing or symbolizing a corporation, institution, merchandise, or brand [3]. Logos have become an essential element of a company's identity, and a highly familiar logo has the potential to enhance a company's reputation [4]. Logo recognition has been a topic of interest in computer vision for many years, with research dating back to 1993. While the objective of detecting and identifying brand logos in images is well-established, it remains a challenging task in

Y. Farhaoui et al. (Eds.): ICAISE 2023, LNNS 837, pp. 466–471, 2024.
https://doi.org/10.1007/978-3-031-48465-0_62

object recognition and classification due to the absence of a definitive definition for what qualifies as a logo. Logos can appear on various products, backgrounds, or advertising surfaces, and for a specific brand, there may exist different logo types. Additionally, there are instances of inter-class variations, where logos from different brands may share a similar appearance, as depicted in Fig. 1.

Fig. 1. Example of logos of different brands

The process of logo recognition typically involves steps such as image preprocessing, feature extraction, logo matching, and classification. Machine learning algorithms, such as convolutional neural networks (CNNs), are commonly employed to learn the distinguishing features of logos and classify them accurately [2].

Logos recognition approaches can be divided into two main categories: one-stage and two-stage detectors. Two stage detector refers to two independently steps. The first stage involved generating a set of region proposals that were likely to contain objects, and the second stage classified these proposals as either logos or background and refined their bounding boxes. Unlike two-stage detectors, which involve region proposal generation followed by classification and localization, one-stage detectors directly predict the presence and location of logos in a single step. One-stage logo detectors typically leverage convolutional neural networks (CNNs) as their backbone architecture to extract features from the input image. Extracted features are then processed by additional convolutional layers and finally fed into a set of output layers for logo detection.

The remainder of this document is organized as follows. In Sect. 2, we provide an overview of relevant studies and research conducted in the field. Section 3 presents the enhanced architectures of CNN (Convolutional Neural Networks) and lightweight neural networks. In Sect. 4, we assess the performance of existing detectors using a dataset specifically designed for vehicle logo detection. Finally, Sect. 5 offers a succinct conclusion that summarizes the key findings and outlines our future research endeavors in the area of logo detection systems.

2 Related Work

Logo recognition refers to the process of identifying and classifying logos within images or videos. It involves using computer vision techniques and machine learning algorithms to analyze visual features and patterns of a logo and match them against a database of known logos.

Hassanzadeh and Pourghassem proposed to use morphological operators to detect logos from document images [5]. Logos are identified using two specifications including centroid coordinate and intersection logo's bounding box. Canny and Sobel filters have been used to detect lo the corners of logos in document images [6]. To remove noises from images, median filtering has been applied. In [3], Hoi et al. proposes propose a deep learning-based approach for logo detection and brand recognition. The proposed architectures, calls LOGO-Net, was trained on two large datasets which represents 110 brands. Van Trang et al. in [7] studies the impact of using computer vision and machine learning in brand and visibility analysis. Authors in [8] present an approach for vehicle logo detection. Features are extracted based on local patterns and they proceed to features selection based on principal component analysis (PCA). The logo classification was done using a set of machine learning classifier such as K-Nearest Neighbors (KNN), Support Vector Machines (SVM), Random Forest (RF), AdaBoost, and collaborative-representation-based classification (CRC). Authors in [9] proposed a cascaded deep convolutional network to detect and classify vehicle logos without the need of license plates. Hang Chen et al. demonstrated that data augmentation proves to be an effective technic that can improve the logo detection and classification accuracy [10]. They were able to improve the classification accuracy by 8.1% and they were ranked 5th in the ACM MM2021 Robust Logo Detection Grand Challenge. An advanced deep learning approach was proposed by Rahul Kumar Jain et al. where attention mechanism was used [11]. The proposed approach was able to improve the inference time that helps in a real time implementation with an acceptable logo recognition accuracy.

3 Materials and Methods

3.1 Logos Detection Approaches

In this paper, we will conduct a comparative study between four deep learning architectures for logos detection and classification. We will use RetinaNet as one-stage detector and for two-stage detectors we will implement the R-CNN alternatives such as Faster R-CNN, Mask R-CNN, Cascade Mask R-CNN with as backbones respectively ResNet 50, 101, and 152.

3.2 Dataset

It should be noted that the number of publicly accessible datasets of advertising logo images is not very large. Table 1 lists the main publicly accessible logo datasets.

WebLogo-2M is the largest dataset, composed of images from various social networks and constructed automatically [12]. WebLogo-2M is a weakly labeled logo detection dataset, where labels are assigned at the image level rather than at the level of individual object bounding boxes. The dataset consists of over 2 million logo images, encompassing 194 distinct logo classes. However, it should be noted that the dataset contains large-scale but noisy labels for logos due to the nature of web data, which inherently introduces some inaccuracies.

In this paper, we have opted to use the database of the challenge called "Open Logo Detection Challenge; QMUL-OpenLogo" which is an improved version of the WebLogo-2M [13]. Figure 2 presents some logos from the Logo Detection Challenge dataset QMUL-OpenLogo.

Table 1. Main open logo datasets

Dataset	No. of classes	No. of images	Bounding boxes
WebLogo-2M	194	2190,757	0
FlickrLogos-32	32	2240	3404
Logo-In-The-Wild	871	11,054	32,850
BelgaLogos	37	10,000	9842
QMUL-OpenLogo	309	27,083	51,207

Fig. 2. Example of logos from the logo detection challenge dataset QMUL-OpenLogo [10]

3.3 Evaluation Metrics

Object detection task consists of locating the bounding box associated of the object with its corresponding confidence score. In our study, we will use two evaluation metrics the Average Precision (AP) and the mean Average Precision (mAP).

Average Precision (AP) measures the accuracy of the predicted bounding boxes or segmentation masks by comparing them to the ground truth annotations. The equation for Average Precision can be written as:

$$AP = \frac{1}{n} \sum_{r\ in\ \{0,0.1,0.2,\ldots 0;9,1\}} max_{r' \geq r} p(r') \tag{1}$$

where n is the number of recall levels (usually 11, including 0 and 1), r represents each recall value, and p(r′) is the maximum precision value for any recall value r′ greater than or equal to r.

Mean Average Precision (mAP) is a variation of Average Precision (AP) that is commonly used in object detection and other tasks involving multiple object categories. While AP measures the accuracy of object detection for each individual class, mAP provides an overall performance metric by averaging the AP values across all object categories. The equation for calculating mAP can be written as:

$$mAP = \frac{1}{N} \sum_{i=1}^{N} AP_i \tag{2}$$

where N is the total number of object categories, and AP_i represents the AP value for each category i.

4 Experimental Results

The experiments were carried out on a virtual machine in the Google Collaboratory environment, which offers good performance and has the following characteristics: CPU Intel Xeon CPU 2.30GHz, RAM Up to 25.51 GB, GPU NVIDIA Tesla P100-PCIE-16GB, CUDA 10.1, CuDNN 7.6.3, and Disk 68.40 GB.

To evaluate the detection models, we divided the dataset into 80% for learning and 20% for validation. Table 2 presents the four logo detection architectures implementation results using the R 50 FPN and R 101 FPN backbones. The model's Average Precision (AP), mean Precision (mAP), and total loss are the evaluation metrics.

Table 2. Logo detection performance of the implemented deep learning architectures: AP, mAP, and Loss

Name	Backbone	AP (%)	mAP (%)	Loss (%)
Faster R-CNN	R 50 FPN	35.20988	36.7479	**0.10260**
	R 101 FPN	35.50731	37.7648	35.5073
RetinaNet	R 50 FPN	04.23798	05.6390	0.23970
Mask R-CNN	R 50 FPN	32.46427	33.4696	0.25663
	R 101 FPN	33.98278	35.2666	0.19766
Cascade Mask R-CNN	R 50 FPN	**38.33956**	**40.0695**	0.69029

Bold values represent the best obtained values of the metrics AP, mAP, and loss

From the results presented in Table 2, with an AP equal to 38% and mAP equal to 40%, we can conclude that the Cascade Mask R-CNN using Resnet 50 gives the best results. Sure, the validation loss is not better than the Fast R-CNN, but in terms of logo detection precision, Cascade Mask R-CNN is the best.

5 Conclusion

In this study, we considered the application of various deep learning-based object detection algorithms for the detection and localization of brand logos from images. This article highlights the challenges faced by machines in performing object recognition tasks compared to the human brain, which can accomplish such tasks in under a minute. Through a comparative analysis of popular object detection algorithms, including Faster RCNN, Mask RCNN, RetinaNet, and Cascade Mask RCNN, we evaluated their performance in detecting and localizing brand logos. The evaluation was conducted on the widely recognized Open Logo challenge dataset. The results of our study demonstrate the efficacy of deep learning algorithms and especially Cascade Mask R-CNN with the Resnet 50 as backbone for logo detection and localization.

As a future perspective, we intend to expand the dataset used in our study by incorporating additional logo images. Specifically, we aim to include logos from the Moroccan industry.

References

1. Zaidi, S.S.A., Ansari, M.S., Aslam, A., Kanwal, N., Asghar, M., Lee, B.: A survey of modern deep learning based object detection models. Digital Signal Process. **126**, (2022). ISSN 1051-2004. https://doi.org/10.1016/j.dsp.2022.103514
2. Zhiqiang, W., Jun, L.: A review of object detection based on convolutional neural network. In: 36th Chinese Control Conference (CCC), pp. 11104–11109 (2017). https://doi.org/10.23919/ChiCC.2017.8029130
3. Hoi, S., Wu, X., Liu, H., Wu, Y., Wang, H., Xue, H., Wu, Q.: LOGO-Net: Large-Scale Deep Logo Detection and Brand Recognition with Deep Region-Based Convolutional Networks (2015). https://doi.org/10.48550/arXiv.1511.02462
4. Montserrat, D.M., Lin, Q., Allebach, J., Delp, E.J.: Logo detection and recognition with synthetic images. Electron. Imaging **30**(10), 337–341 (2018). https://doi.org/10.2352/issn.2470-1173.2018.10.imawm-337
5. Hassanzadeh, S., Pourghassem, H.: Fast logo detection based on morphological features in document images. In: 2011 IEEE 7th International Colloquium on Signal Processing and its Applications, Penang, Malaysia, pp. 283–286 (2011). https://doi.org/10.1109/CSPA.2011.5759888
6. Beeran Kutty, S., Saaidin, S., Megat Yunus, P.N.A., Abu Hassan, S.: Evaluation of canny and Sobel operator for logo edge detection. In: 2014 International Symposium on Technology Management and Emerging Technologies, Bandung, Indonesia. pp. 153–156 (2014). https://doi.org/10.1109/ISTMET.2014.6936497
7. Van Trang, N.T., Nghiem, T.L., Do, T.M.: Improving the competitiveness for enterprises in brand recognition based on machine learning approach. In: Nguyen, A.T., Hens, L. (eds.) Global Changes and Sustainable Development in Asian Emerging Market Economies Vol. 1. Springer, Cham (2022). https://doi.org/10.1007/978-3-030-81435-9_26
8. Yu, Y., Wang, J., Lu, J., Xie, Y., Nie, Z.: Vehicle logo recognition based on overlapping enhanced patterns of oriented edge magnitudes. Comput. Electr. Eng. **71**, 273–283 (2018). https://doi.org/10.1016/j.compeleceng.2018.07.045
9. Yu, Y., Guan, H., Li, D., Yu, C.: A cascaded deep convolutional network for vehicle logo recognition from frontal and rear images of vehicles. IEEE Trans. Intell. Transp. Syst. **22**, 758–771 (2021). https://doi.org/10.1109/TITS.2019.2956082
10. Chen, H., Li, X., Wang, Z., Hu, X.: Robust logo detection in ecommerce images by data augmentation. In: 29th ACM International Conference on Multimedia, pp. 4789–4793 (2021). https://doi.org/10.1145/3474085.3479227
11. Jain, R.K., Watasue, T., Nakagawa, T., Sato, T., Iwamoto, Y., Ruan, X., Chen, Y.-W.: LogoNet: a robust layer-aggregated dual-attention anchorfree logo detection framework with an adversarial domain adaptation approach. Appl. Sci. **11**(20), 9622 (2021). https://doi.org/10.3390/app11209622
12. Su, H., Gong, S., Zhu, X.: WebLogo-2M: scalable logo detection by deep learning from the web. In: 2017 IEEE International Conference on Computer Vision Workshops (ICCVW), Venice, Italy, pp. 270–279 (2017). https://doi.org/10.1109/ICCVW.2017.41
13. Su, H., Zhu, X., Gong, S.: Open logo detection challenge. In: Proceedings of British Machine Vision Conference (BMVC), Newcastle, UK (2018). https://doi.org/10.48550/arXiv.1807.01964

Skin Lesion Classification Based on Vision Transformer (ViT)

Abla Rahmouni[1], My Abdelouahed Sabri[1(✉)], Asmae Ennaji[2], and Abdellah Aarab[2]

[1] Computer Science Department, Faculty of Sciences, University Sidi Mohamed Ben Abdallah, Fez, Morocco
abdelouahed.sabri@usmba.ac.ma

[2] Physics Department, Faculty of Sciences, University Sidi Mohamed Ben Abdallah, Fez, Morocco

Abstract. Accurate diagnosis of skin lesions plays a critical role in early detection and effective treatment of various dermatological conditions. However, it remains a challenging task due to the complexity and visual variability of skin lesion images. In this study, we address this challenge by exploring the application of Vision Transformer (ViT) for skin lesion classification, aiming to assist dermatologists in improving diagnostic accuracy. Skin lesion classification traditionally relies on methods like Convolutional Neural Networks (CNNs). While CNNs have shown promise, they may struggle to effectively capture complex spatial relationships present in skin lesion images. To overcome this limitation, we investigate the potential of ViT, an architecture based on the self-attention mechanism. The implications of this research are significant, as accurate and efficient skin lesion classification can help dermatologists in making informed decisions and improve patient outcomes. Our findings highlight the potential of ViT to surpass the limitations of traditional approaches and provide a robust framework for skin lesion analysis.

Keywords: Vision Transformer (ViT) · Dermatological conditions · Skin lesion analysis · Skin lesion classification · PH2 · ISIC 2017

1 Introduction

Skin lesion classification and segmentation are of paramount importance. From benign moles to potentially life-threatening conditions like skin cancer, the accurate identification and segmentation of these lesions can make a world of difference [1]. However, the current manual process, reliant solely on dermatologists' expertise, is time-consuming, prone to human error, and, in some cases, inconclusive [2].

By leveraging the unparalleled capabilities of the Vision Transformer ViT [3], originally designed for language tasks, we aim to unlock its potential in image recognition, classification, and segmentation [3]. The implementation of the ViT can enhance the accuracy of both lesion classification and segmentation [4]. Moreover, early detection of

Y. Farhaoui et al. (Eds.): ICAISE 2023, LNNS 837, pp. 472–477, 2024.
https://doi.org/10.1007/978-3-031-48465-0_63

skin cancer made possible by such a system could save lives and significantly improve patient outcomes [5] and can be a captivating intersection of technology and healthcare [6].

2 Related Work

In the ever-evolving realm of dermatology, the challenge of skin lesion classification has persisted as an elusive puzzle, demanding ingenious solutions. The current reliance on manual analysis by dermatologists is riddled with subjectivity, time constraints, and the ever-present specter of human error. According to studies, misdiagnosis rates in skin cancer cases can range from 2% to 38%, highlighting the need for more accurate and reliable methods of analysis [7, 8].

The Vision Transformer (ViT) architecture was first introduced in 2020 by Alexey Dosovitskiy et al. in a work titled "An Image is Worth 16 × 16 Words: Transformers for Image Recognition at Scale" [3]. The research was conducted at Google Research [3]. This seminal paper proposed the ViT architecture, which improved the field of computer vision by applying the Transformer model. Transformer was first proposed for Natural Language Processing (NLP) and then implemented in image recognition tasks. Dosovitskiy et al. demonstrated that ViT could achieve competitive performance on image classification benchmarks, surpassing traditional Convolutional Neural Networks (CNNs) in certain scenarios.

Over time, researchers have made several modifications and improvements to the ViT architecture. Some variations include:

Hybrid architectures: To handle larger images, hybrid architectures combine CNNs with transformers to handle larger images. These architectures reduce computational costs while capturing global dependencies. The paper by Touvron et al. [9] introduces a hybrid model called"T2T-ViT" that converts CNN's feature maps into tokens, allowing efficient processing with transformers. This approach maintains the ability to capture global dependencies while reducing computational requirements.

Patch embeddings: Patch embeddings were introduced in the original ViT architecture by Dosovitskiy et al. [3], where the input image is divided into fixed-size patches transformed into token embeddings. The Swin Transformer model, proposed by Liu et al. [10], adopts a hierarchical approach with overlapping patches to catch effectively global and local contextual information.

Distillation techniques: Distillation techniques in Vision Transformers (ViTs) transfer knowledge from larger to smaller models, reducing computational requirements while maintaining accuracy. Originally proposed by Hinton et al. in 2015 [11], distillation has been applied to ViTs for model compression and efficiency. Touvron et al. [9] introduced distillation techniques specifically for ViTs. For the training process, authors proposed to use a smaller student model to simulate a larger teacher model's outputs and internal representations. This process grants the student model to gather efficiently knowledge's from the teachers.

The exploration of Vision Transformers (ViTs) opens up exciting possibilities for revolutionizing the field of dermatology and advancing skin lesion diagnosis. With the introduction of this technology, a remarkable fusion of transformers and attention mechanisms, a new paradigm emerges in image analysis. ViTs demonstrate their potential by processing images as sequences of patches, harnessing the power of self-attention and multi-head attention mechanisms to extract comprehensive and meaningful representations [12].

3 The Proposed Approach

The Vision Transformer (ViT) is a groundbreaking model that applies the transformer architecture to the field of computer vision. It marks a significant shift from traditional convolutional neural networks (CNNs) and has gained attention for its exceptional performance in image detection and classification tasks.

Figure 1 illustrates how ViT processes a skin lesion image by dividing it into patch units and using them as input. These patches undergo a two-dimensional transformation to understand the connections between them across multi-head self-attention. Afterward, these patches are processed by a learnable linear projection, resulting in two-dimensional patch embeddings as the final output. The resulting embeddings go through successive layers comprising multi-head self-attention, an MLP block, and Layer Normalization (LN). The number of such blocks can vary depending on the application.

Within the embeddings patch that was generated by the encoder, we will use as input for the MLP head only the token. The MLP head is responsible for performing the image classification task, as described in Ref. [13].

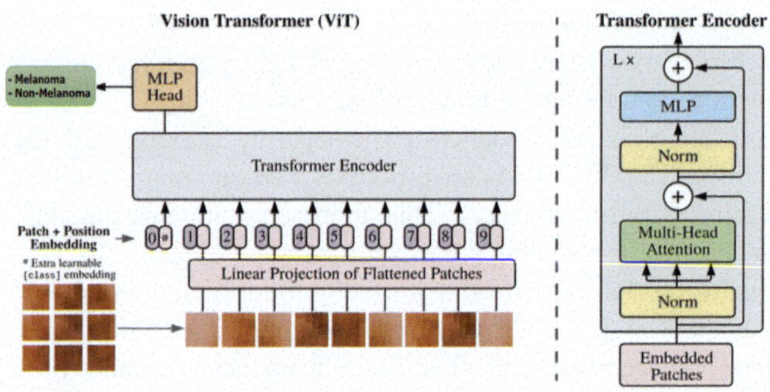

Fig. 1. Architecture of the original ViT applied for skin lesion detection [13].

Model Training: The ViT architecture consists of the following key components:

- Patch Extraction: Input images were divided into 6 × 6 pixel patches to capture local details and spatial information.
- Patch Encoding: Each patch underwent encoding using a projection layer and a position embedding layer. The projection layer elevated the patches to a higherdimensional space, while the position embedding layer added positional information. The PatchEncoder layer executed this encoding process.
- Transformer Blocks: The encoded patches were processed through multiple Transformer blocks. Each block included layer normalization, multi-head attention, a skip connection, and a feed-forward neural network.

- Classification Head: The encoded patches were processed by a multi-layer perceptron (MLP) consisting of dense layers. The MLP extracted high-level features.

Dataset: We train our model on two distinct datasets (PH2 and ISIC2017). Figure 2 presents some examples of melanoma and non-melanoma skin cancer.

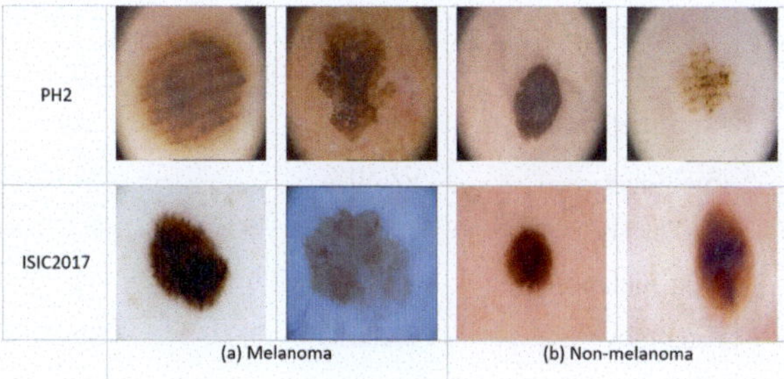

Fig. 2. Examples of cancer from PH2 and ISIC 2017 datasets

The PH2 dataset comprises a total of 200 images, consisting of 160 non-melanomas and 40 cases of melanoma skin cancer. The dataset images are in the RGB (Red, Green, and Blue) color space with a resolution of 764 × 575 pixels. The first row of figure Fig. 2 provides visual examples from this database. With 2000 skin cancer images (1626 non-melanomas and 374 melanomas), the ISIC2017 dataset is used in our implementation. These images are also represented the in the RGB (Red, Green, and Blue) color space with a resolution of 767 × 1022 pixels. The second row of Figure Fig. 2 illustrates examples from this dataset.

4 Experimental Results

In this section, we evaluate our custom-built ViT model for skin lesion classification using two datasets PH2 and ISICI 2017. Table 1 presents the confusion matrices and accuracies for both datasets.

The results presented in Table 1 show that all three models performed well on the PH2 dataset, with accuracies ranging from 92.50% to 95.00%. However, the ISIC 2017 dataset proved to be more challenging for all models, with lower accuracies across the board. Melanoma accuracies ranged from 70.73% to 90.91%, and Non-Melanoma accuracies ranged from 93.64% to 99.87%. The choice of the best model depends on the specific dataset and application requirements, with the VIT model generally performing best but the Transformer model excelling on the ISIC 2017 dataset.

A comparative study will be conducted between ViT and well-known deep learning architectures in skin lesion analysis. Table 2 presents the classification accuracies using PH2 and ISIC 2017 datasets.

From the results presented in Table 2, the ViT-based architecture is a strong contender in skin lesion classification, performing particularly well on the PH2 dataset with an accuracy of 95% and reasonably well on the larger ISIC2017 dataset with an accuracy of 90.5%. In both datasets, the ViT-based architecture outperformed traditional DL

Table 1. The evaluation results of the implemented models using PH2 and ISIC 2017 datasets

Model	Classes	PH2		ISIC 2017	
Self-attention	Melanoma	29	2	29	12
	Non-melanoma	1	8	275	1474
	Accuracy	92.50%		83.97%	
Transformer	Melanoma	28	1	20	2
	Non-melanoma	2	9	284	1484
	Accuracy	92.50%		84.02%	
ViT	Melanoma	29	1	207	73
	Non-melanoma	1	9	97	1413
	Accuracy	**95.00%**		**90.50%**	

Table 2. Comparative study results between ViT based architectures and well-known DL architectures

	Dataset information			Accuracy			
	Size	Melanoma	Non-Melanoma	AlexNet	VggNet	ResNet	ViT
PH2	200	40	160	68	69	70	95
ISIC2017	2000	374	1626	81	80	82	90.5

architectures like AlexNet, VggNet, and ResNet, showcasing the effectiveness of the ViT model for this task.

5 Conclusion

In this research work, we explored the application of the Vision Transformer (ViT) model in the domain of image recognition and classification, specifically focusing on skin lesion analysis. We demonstrated the effectiveness of the ViT model in accurately classifying skin lesions, leveraging self-attention mechanisms to capture both local and global features. Our findings provide valuable insights into enhancing diagnostic capabilities in dermatology. Accurate skin lesion classification plays a crucial role in early detection and diagnosis of skin cancer, leading to improved patient outcomes. Additionally, our research highlights the potential of the ViT model in advancing computer vision and image analysis techniques. Further exploration of the ViT model holds promise for various domains and applications.

References

1. "Skin Cancer Foundation": https://www.skincancer.org/skin-cancerinformation/skin-cancer-facts/

2. Kittler, H., et al.: Identification of clinically featureless incipient melanoma using sequential dermoscopy imaging. Arch. Dermatol. **142**(9), 1113–1119 (2006)
3. Dosovitskiy, A., Beyer, L., Kolesnikov, A., Weissenborn, D., Zhai, X., Unterthiner, T., Houlsby, N.: An image is worth 16 × 16 words: transformers for image recognition at scale (2020). arXiv preprint arXiv:2010.11929
4. Esteva, A., Kuprel, B., Novoa, R.A., Ko, J., Swetter, S.M., Blau, H.M., Thrun, S.: Dermatologist-level classification of skin cancer with deep neural networks. Nature **542**(7639), 115–118 (2017)
5. Wang, D., Khosla, A., Gargeya, R., Irshad, H., Beck, A.H.: Deep learning for identifying metastatic breast cancer (2016). arXiv preprint arXiv:1606.05718
6. Litjens, G., et al.: A survey on deep learning in medical image analysis. Med. Image Anal. **42**, 60–88 (2017)
7. Kittler, H., Pehamberger, H., Wolff, K., Binder, M.J.T.I.O.: Diagnostic accuracy of dermoscopy. Lancet Oncol. **3**(3), 159–165 (2002)
8. Davis, D.M., et al.: American academy of dermatology guidelines: awareness of comorbidities associated with atopic dermatitis in adults. J. Am. Acad. Dermatol. **86**(6), 1335–1336 (2022)
9. Touvron, H., Cord, M., Douze, M., Massa, F., Sablayrolles, A., Jégou, H.: Training data-efficient image transformers distillation through attention. In: International Conference on Machine Learning, pp. 10347–10357. PMLR (2021)
10. Liu, Z., Lin, Y., Cao, Y., Hu, H., Wei, Y., Zhang, Z., Guo, B.: Swin transformer: hierarchical vision transformer using shifted windows. In: Proceedings of the IEEE/CVF International Conference on Computer Vision, pp. 10012–10022 (2021)
11. Hinton, G., Vinyals, O., Dean, J.: Distilling the knowledge in a neural network (2015). arXiv preprint arXiv:1503.02531
12. Vaswani, A., Shazeer, N., Parmar, N., Uszkoreit, J., Jones, L., Gomez, A. N., Polosukhin, I.: Attention is all you need. Adv. Neural Inf. Process. Syst. **30** (2017)
13. Lee, Y., Kang, P.: AnoViT: unsupervised anomaly detection and localization with vision transformer-based encoder-decoder. IEEE Access **10**, 46717–46724 (2022)

Early Detection of Breast Cancer Based on Patient Symptom Data Using Naive Bayes Algorithm on Genomic Data

Agus Perdana Windarto[1](✉), Tutut Herawan[2,3], and Putrama Alkhairi[1]

[1] STIKOM Tunas Bangsa, Pematangsiantar, North Sumatra, Indonesia
{agus.perdana,putrama}@amiktunasbangsa.ac.id
[2] Sekolah Tinggi Pariwisata Ambarrukmo, Yogyakarta, Indonesia
tutut@stipram.ac.id
[3] AMCS Research Center, Yogyakarta, Indonesia

Abstract. The Naive Bayes algorithm is employed to facilitate early detection of breast cancer, a prevalent condition affecting women. Timely detection plays a crucial role in ensuring effective treatment and improved survival rates. Considering the limitations of conventional methods like mammography, we explored the potential of genomic analysis, made possible by recent technological advancements. Genomic data from breast cancer patients were collected and analyzed to identify genetic patterns associated with the disease. Patient symptom data was also incorporated. Utilizing the Naive Bayes algorithm, we processed this dataset and classified patients into either the "breast cancer" or "non-breast cancer" categories based on the probability of symptoms in patients with similar genomic profiles. Our findings indicate that this approach holds promise in accurately detecting breast cancer at an early stage, with the potential to enhance prognosis and treatment outcomes. However, further validation through clinical studies involving larger and diverse samples is necessary. The results demonstrated that the Naive Bayes algorithm achieved a high accuracy rate of 90.4% in classifying early symptoms of breast cancer into two distinct categories.

Keywords: Early detection · Breast cancer · Genomic data · Patient symptoms · Naive Bayes algorithm

1 Introduction

Breast cancer is a prevalent malignancy affecting women globally. Timely identification plays a critical role in improving prognosis and survival rates [1, 2]. Unfortunately, many cases are detected at an advanced stage, making treatment more complex and less effective [3, 4]. Education on early symptoms and routine screenings are important, but accurate diagnosis often requires additional evaluations like histopathological examinations and mammographic imagery [5, 6]. Previous research has highlighted the diverse genomic profiles associated with breast cancer, closely linked with disease attributes and progression [7]. In recent decades, data mining and analysis have become important

Y. Farhaoui et al. (Eds.): ICAISE 2023, LNNS 837, pp. 478–484, 2024.
https://doi.org/10.1007/978-3-031-48465-0_64

tools in medicine for predicting, diagnosing, and treating diseases. The Naive Bayes algorithm, a data mining technique, has shown great potential in classifying and predicting various fields, including medicine [8]. For example, a study by [9] revealed that breast disease caused by consumption is the most dangerous hazard experienced by women worldwide. Nearly 2 million new cases were analyzed in 2018. The study found that an expert doctor can identify cancer with 79% accuracy, while an AI method provides 91% accuracy. The study also employed K-Nearest Neighbors (KNN), Support Vector Machine (SVM), and Decision Tree Classifier (DT) algorithms. The SVM classifier was found to provide more precise results, especially with larger datasets. Another study [10] applied the convolutional neural network (AlexNet) to extract deep features from the BreaKHis dataset for diagnosing breast cancer as benign or malignant. The study conducted four trials with different magnification factors and achieved high accuracy, sensitivity, specificity, and Area Under the Curve (AUC) values.

In this study, we aim to develop an early detection method for breast cancer based on patient symptom data and genomic data using the Naive Bayes algorithm [11]. Patient symptom data will help us understand early symptoms before further diagnostic processes, while genomic data will provide insights into the genetic profile associated with breast cancer [12, 13]. This predictive model is expected to be a useful tool for healthcare professionals, accelerating the diagnostic process and enabling timely treatment [14, 15]. By aligning symptom and genomic data through data mining and analysis, we hope to increase the early detection rate of breast cancer, improving prognosis and treatment outcomes [16].

The paper is organized as follows: Sect. 2 presents the proposed method. Section 3 presents the results and discussion. Finally, Sect. 4 concludes this work.

2 Proposed Method

This research was conducted to implement the Naive Bayes algorithm in the classification of early detection of breast cancer based on patient symptom data on genomic data. From the results of data classification, information can be obtained so that patients can be treated with indications of breast cancer.

2.1 Data Collection

In conducting research to obtain data and information, the methods used in the data collection process are as follows:

(a) Patient symptom data: Data on breast cancer patient symptoms will be collected from various sources, such as medical records, interviews, or questionnaire forms. This data will include initial symptoms reported by the patient
(b) Genomic Data: Genomic data will be collected from patients who have been diagnosed with breast cancer. These data can be gene expression, genetic mutations, or Single Nucleotide Polymorphisms (SNPs) associated with breast cancer.

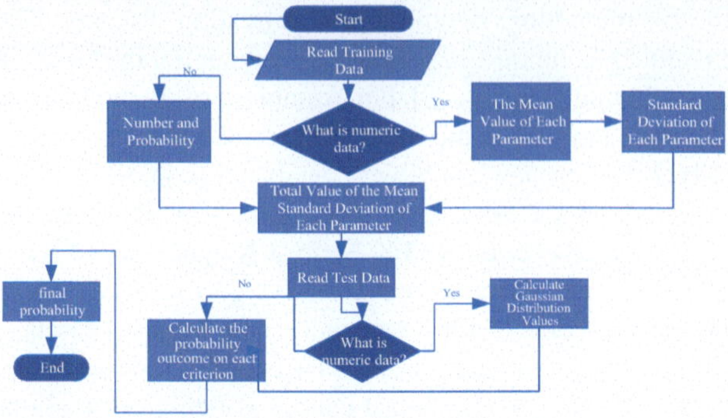

Fig. 1. Naive Bayes algorithm

2.2 Naive Bayes Algorithm

We analyze the factors that cause breast cancer using the Naive Bayes algorithm. The Naive Bayes algorithm framework can be seen in Fig. 1.

The image description entails the following steps: Firstly, the training data is read, and the amount and probability are calculated. If the data contains numeric values, the mean and standard deviation of each parameter are determined, while for non-numeric data, the probability value is calculated for each category. Next, the mean, standard deviation, and probability values are obtained and recorded in a table. Subsequently, if numeric data is present, the Gaussian distribution value is computed for each feature during the data testing process. Following that, the final probability of each class is calculated by entering and evaluating all probability data within the same class. Finally, the probabilities are multiplied together to obtain the final probability. This concise paragraph accurately captures the essence of the provided image description.

3 Results and Discussion

3.1 Research Datasets

The data is grouped using the Naive Bayes method, and the RapidMiner 8.1 application is used for accurate results in early breast cancer detection based on patient symptom data. The dataset consists of 70 samples from Kaggle as training data, with testing data taken from the initial 21 samples. Table 1 lists the variables used in early breast cancer detection based on patient symptom data.

Meanwhile, the dataset can be seen in Table 2:

The data in Table 2 above is taken from Kaggle which has been converted to make it easier to calculate the use of the Naive Bayes algorithm.

3.2 Results

Counting the Number of Same Cases with the Same Class. After looking for the probability value of each criterion, it is obtained from the training data The next step is to

Table 1. Description of features

No.	Variable	Description
1	Age	Age is one of the high-risk factors and is a major factor because, as you get older, your immune system decreases
2	Breast size or shape	Pay attention if there are changes in the size or shape of the breasts for no apparent reason
3	Breast skin	Changes in the skin of the breast: for example, the skin of the breast becomes reddish, feels wrinkled like an orange peel, or has a change in texture
4	Nipples	Nipple that goes inward (retracts), discharges abnormally, or changes shape
5	Skin conditions around the nipples	Peeling or peeling skin around the nipples, like eczema
6	Lumps on the breast	An unusual lump in the breast can be one of the first signs of breast cancer. However, not all lumps are cancer, many conditions are benign

Table 2. Research datasets

No.	Patient	Age	Breast size or shape	Breast skin	Nipples	Skin conditions around the nipples	Lumps on the breast
1	A1	30–39	Normal	Normal	Normal	Moist	No
2	A2	40–49	Right	Redness	Retraction	Peeled off	Yes
3	A3	50–59	Left	Wrinkled	Painful	Scaly	Yes
4	A4	60–79	Left	Redness	Retraction	Peeled off	Yes
5	A5	60–79	Right	Wrinkled	Painful	Scaly	Yes
…	…	…	…	…	…	…	…
68	A68	30–39	Normal	Redness	Normal	Moist	No
69	A69	40–49	Left	Redness	Normal	Peeled off	Yes
70	A70	50–59	Left	Wrinkled	Painful	Scaly	Yes

write down all the results of the Yes and No variables using test data. The following test data used is in Table 3. The classification of the results is not yet known because these variables are the results of predictions from data calculated from each variable.

The calculation process is given as follow:

$$\text{Patient Name(cancer indication)} = \text{Age} * \text{Breast size or shape} \\ * \text{Breast Skin} * \text{Nipples} * \text{Skin Conditions Around the Nipples} \\ * \text{Lumps on the breast}$$

$$\begin{aligned} \text{A1(Yes)} &= 30 - 39 * \text{Normal} * \text{Normal} * \text{Normal} * \text{Moist} * \text{No} \\ &= 0.24 * 0.18 * 0.28 * 0.28 * 0.12 * 0.29 \\ &= 0.00029 \text{ and so on until the last data A21} \end{aligned}$$

Table 3. Calculation results for each variable

No.	Age	Breast size or shape	Breast skin	Nipples	Skin conditions around the nipples	Lumps on the breast	Prediction	Class prediction	Yes	No
1	30–39	Normal	Normal	Normal	Moist	No		No	0.00029	0.04940
2	40–49	right	Redness	Retraction	Peeled off	Yes		Yes	0.00411	0.00000
3	50–59	left	Wrinkled	Painful	Scaly	Yes		Yes	0.00517	0.00000
…	…	…	…	…	…	…		…	…	….
15	30–39	Normal	Normal	Normal	Moist	Yes		No	0.00029	0.04940
16	40–49	right	Wrinkled	Retraction	Scaly	Yes		Yes	0.00255	0.00000
17	50–59	right	Wrinkled	Retraction	Scaly	Yes		Yes	0.00313	0.00000
18	30–39	Normal	Wrinkled	Normal	Peeled off	No		No	0.00153	0.00371
19	40–49	Normal	Normal	Normal	Scaly	No		Yes	0.00094	0.00071
20	50–59	right	Redness	Painful	Scaly	Yes		Yes	0.00550	0.00000
21	40–49	right	Redness	Normal	Scaly	Yes		Yes	0.00285	0.00000

Table 3 above presents a comparison of the results of test data with the results of probability predictions for in classifying breast cancer. Predictive accuracy can be seen in Fig. 2 of the Confusion Matrix below.

accuracy: 90.48%

	true No	true Yes	class precision
pred. No	8	1	88.89%
pred. Yes	1	11	91.67%
class recall	88.89%	91.67%	

Fig. 2. Accuracy performance value

In Fig. 2 above, we can see the results obtained from calculation as following.

$$\begin{aligned}\text{Accuracy} &= \text{Prediction(Yes, Class Yes)} \\ &\quad + \text{Prediction(No, Class No)/Total Number of Predictions} \\ &\quad \text{(Yes, Class Yes)and Predictions(No, Class No)} \\ &= (11 + 8)/21 \\ &= 0.9048 \text{ or } 90.4\%\end{aligned}$$

From the results of these calculations, comparing the probability value between class (P|Yes) and class (P|No) the highest probability value can be concluded that the Customer Ascension Status has an accuracy of 90.4%. The Naive Bayes algorithm was applied for early detection of breast cancer using patient symptom data obtained from the kaggle.com website. The study included 70 patient data for testing, divided into two classes. The Naive Bayes algorithm correctly classified 11 data as "Yes" (indicating breast cancer) and 8 data as "No" (no indication of breast cancer). Only 2 patient data were misclassified. The RapidMiner tool was also used for testing, producing similar results with 11 data classified as "Yes" and 8 data as "No," with 2 misclassified patient data. The Naive Bayes algorithm achieved an accuracy rate of 90.4% in classifying the two classes during testing in RapidMiner 8.1.

4 Conclusion

The Naive Bayes Algorithm has successfully achieved an impressive of 90.4% accuracy in classifying early breast cancer symptoms. By incorporating patient symptom and genomic data, this approach enhances early detection and improves accuracy, benefiting healthcare professionals and contributing to global efforts to improve breast cancer diagnosis and patients' quality of life.

Acknowledgement. This works is supported by Department of Research and Technology, STIPRAM Institute, Indonesia.

References

1. Comander, A., Frates, B., Tollefson, M.: PAVING the path to wellness for breast cancer survivors: lifestyle medicine education and group interventions. Am. J. Lifestyle Med. **15**(3), 242–248 (2021). https://doi.org/10.1177/1559827620986066
2. Burnside, E.S., et al.: Comparing mammography abnormality features to genetic variants in the prediction of breast cancer in women recommended for breast biopsy. Acad. Radiol. **23**(1), 62–69 (2016). https://doi.org/10.1016/j.acra.2015.09.007
3. Kaushal, C., Bhat, S., Koundal, D., Singla, A.: Recent trends in computer assisted diagnosis (CAD) system for breast cancer diagnosis using histopathological images. Irbm **40**(4), 211–227 (2019). https://doi.org/10.1016/j.irbm.2019.06.001
4. Kharya, S., Soni, S.: Weighted Naive Bayes classifier: a predictive model for breast cancer detection. Int. J. Comput. Appl. **133**(9), 32–37 (2016). https://doi.org/10.5120/ijca2016908023

5. Cain, E.H., Saha, A., Harowicz, M.R., Marks, J.R., Marcom, P.K., Mazurowski, M.A.: Multivariate machine learning models for prediction of pathologic response to neoadjuvant therapy in breast cancer using MRI features: a study using an independent validation set. Breast Cancer Res. Treat. **173**(2), 455–463 (2019). https://doi.org/10.1007/s10549-018-4990-9
6. Narod, S.A.: Personalised medicine and population health: breast and ovarian cancer. Hum. Genet. **137**(10), 769–778 (2018). https://doi.org/10.1007/s00439-018-1944-6
7. Xie, J., Liu, R., Luttrell, J., Zhang, C.: Deep learning based analysis of histopathological images of breast cancer. Front. Genet. **10**, 1–19 (2019). https://doi.org/10.3389/fgene.2019.00080
8. Vaka, A.R., Soni, B., S.R.K.: Breast cancer detection by leveraging machine learning. ICT Express **6**(4), 320–324 (2020). https://doi.org/10.1016/j.icte.2020.04.009
9. Hazra, A., Kumar, S., Gupta, A.: Study and analysis of breast cancer cell detection using Naïve Bayes, SVM and ensemble algorithms. Int. J. Comput. Appl. **145**(2), 39–45 (2016). https://doi.org/10.5120/ijca2016910595
10. Senan, E.M., Alsaade, F.W., Al-Mashhadani, M.I.A., Aldhyani, T.H.H., Al-Adhaileh, M.H.: Classification of histopathological images for early detection of breast cancer using deep learning. J. Appl. Sci. Eng. **24**(3), 323–329 (2021). https://doi.org/10.6180/jase.202106_24(3).0007
11. Wibowo, A.H., Oesman, T.I.: The comparative analysis on the accuracy of k-NN, Naive Bayes, and decision tree algorithms in predicting crimes and criminal actions in Sleman Regency. J. Phys. Conf. Ser. **1450**(1) (2020). https://doi.org/10.1088/1742-6596/1450/1/012076
12. Mahvi, D.A., Liu, R., Grinstaff, M.W., Colson, Y.L., Raut, C.P.: Local cancer recurrence: the realities, challenges, and opportunities for new therapies. CA Cancer J. Clin. **68**(6), 488–505 (2018). https://doi.org/10.3322/caac.21498
13. Parada, H., Sun, X., Tse, C.K., Olshan, A.F., Troester, M.A.: Lifestyle patterns and survival following breast cancer in the Carolina breast cancer study. Epidemiology **30**(1), 83–92 (2019). https://doi.org/10.1097/EDE.0000000000000933
14. Chaurasia, V., Pal, S., Tiwari, B.B.: Prediction of benign and malignant breast cancer using data mining techniques. J. Algorithms Comput. Technol. **12**(2), 119–126 (2018). https://doi.org/10.1177/1748301818756225
15. Kaushik, D., Kaur, K.: Application of Data Mining for high accuracy prediction of breast tissue biopsy results. In: 2016 3rd International Conference on Digital Information Processing Data Mining, and Wireless Communications DIPDMWC 2016, pp. 40–45 (2016). https://doi.org/10.1109/DIPDMWC.2016.7529361
16. Chen, S.I., Tseng, H.T., Hsieh, C.C.: Evaluating the impact of soy compounds on breast cancer using the data mining approach. Food Funct. **11**(5), 4561–4570 (2020). https://doi.org/10.1039/c9fo00976k

Information Visualization of Research Evolution on Innovation in Local Wisdom: A Decade Bibliometric Analysis Using the Scopus Database

Tutut Herawan[1,2](✉), Kris Cahyani Ermawati[1,3], John J. O. I. Ihalauw[1], Damiasih Damiasih[1], and Suhendroyono Suhendroyono[1]

[1] Sekolah Tinggi Pariwisata Ambarrukmo Yogyakarta, Jalan Ringroad Timur No. 52, Bantul, Daerah Istimewa Yogyakarta, Indonesia 55198
tutut@stipram.ac.id

[2] AMCS Research Center, Jalan Griya Taman Asri, Yogyakarta, Indonesia 55512

[3] Lembaga Kajian Pariwisata Indonesia, Babadan, Sewon, Bantul, Daerah Istimewa Yogyakarta, Indonesia 55185

Abstract. Studies on Innovation in local wisdom have been widely applied in tourism. This paper presents a bibliometric analysis of innovation in local wisdom in tourism. Bibliometric analysis was used to analyze 116 articles published from 2013 to 2023. The tools used in this bibliometric analysis are Harzing's Publish or Perish for extracting data from Scopus databases and Vos Viewer for data visualization and further used for citation and metrics analysis. The results indicate that most articles related to innovation in local wisdom are published in scientific journals, compared to other publication outlets. Related articles published in scientific journals have received the highest citations compared to others. Based on network visualization, the most dominant terms (often appearing) are rural areas, sustainable development, and villager development. When viewed from the overlay and density visualization, the topic of innovation in local wisdom shows the latest trends in research related to rural areas, sustainable development, village development and urban growth.

Keywords: Tourism · Bibliometric analysis · Scientometrics · Innovation in local wisdom · Vos Viewer

1 Introduction

Local wisdom refers to the traditional wisdom and knowledge of a particular community or culture. It covers various aspects of life, including social norms, beliefs, customs and practices. This knowledge is passed on from generation to generation and is critical in preserving a community's cultural identity [1]. It also promotes sustainable living by utilizing local resources and respecting the environment. It is important to recognize and value this knowledge as it contributes to the diversity and richness of the world's cultures [2, 3]. Innovation in local wisdom refers to the process of applying modern technology and knowledge to traditional practices and beliefs [4]. This approach can help preserve

Y. Farhaoui et al. (Eds.): ICAISE 2023, LNNS 837, pp. 485–491, 2024.
https://doi.org/10.1007/978-3-031-48465-0_65

cultural heritage while improving people's quality of life. Innovation in local wisdom can also involve revitalizing culture through modern marketing techniques. For example, using social media platforms to showcase their culture to tourists. This local wisdom innovation can improve the welfare of the local community [5]. By maintaining traditions and knowledge, the community can create new products without leaving or reducing the existing meaning so that they have a higher selling value. By continuing to develop their traditions, communities can maintain cultural values that are unique and different from other regions [6].

This paper presents a bibliometric analysis of innovation in local wisdom. In summary, this work makes the following contributions:

a. We identify the evolution of innovation in local wisdom from 2013 to 2023, such as: document type, document subjects, and topic trends.
b. We employ bibliometric analysis are Harzing's Publish or Perish for extracting data from Scopus databases and Vos Viewer for data visualization and further used for citation and metrics analysis.

The rest of this paper is organized as follow. Section 2 describes the proposed research method. Section 3 presents the results obtained and is followed by discussions. The Sect. 4 concludes this work and highlight future research works.

2 Proposed Method

In this paper, we adopt method bibliometrics analysis approach [7–10] which consists of six stages i.e.: searching articles, saving data in RIS, entering data into VoS Viewer, bibliometric analysis, presenting results, and visualization mapping as depicted in Fig. 1.

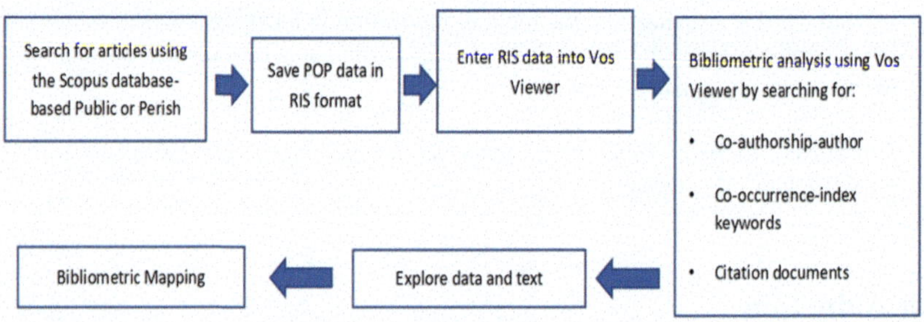

Fig. 1. Flow chart of the proposed method

3 Results and Discussion

This section explains the obtained results of this work, which includes publications, citations, and visualizations.

3.1 Publications and Citations

For knowing comparison of the citation matrix on the data taken via Scopus, then we create a table in which containing number article, number citation, number citations per year, number authors per year, H index, G index, normal hI, and hI yearly at the beginning search and on results repair. Comparison data matrix of initial and improved results could be seen in Table 1.

Table 1. Comparison matrix

Data	Initial search results	Search results repair
Database	Scopus	Scopus
Year publishing	(2013–2023)	(2013–2023)
Year citation	10	10
Number of articles	130	116
Number citation	682	620
Number citation per year	68.20	62.00
Number authors per year	3.34	3.28
H index	12	12
G index	23	23
hI normal	10	10
hI annual	1.00	1.00

Table 2 as follow describe the descriptive statistics of related publications each year.

An overview of the document based on its subject area can be seen more clearly in Fig. 2.

Furthermore, to find out the type of document, it can be seen from the presentation of scientific publications related to the innovation of local wisdom in tourism in Fig. 3.

3.2 Visualization Topics Use VoS Viewer

Figure 4 this show visualization of topic areas using network visualization from VoS Viewer.

As illustrated in Fig. 4, the results of the network visualization show that the most dominant terms (often appearing) are rural areas, sustainable development, and villager development. Figure 5 as follow describes the visualization topic area using overlay visualization from VoS Viewer.

From Fig. 5, viewed from the overlay visualization, the topic of innovation in local wisdom shows the latest trends in research related to rural areas, sustainable development, village development and urban growth. Figure 6 as follow describes the visualization topic area using density visualization from VoS Viewer.

Table 2. The statistics descriptive of publication

Year	TP	% (N = 116)	NCP	TC	C/P	C/CP
2013	8	0.07	6	169	21.13	28.17
2014	6	0.05	5	46	7.67	9.20
2015	0	0.00	0	0	0.00	0.00
2016	3	0.03	3	37	12.33	12.33
2017	8	0.07	4	14	1.75	3.50
2018	10	0.09	7	154	15.40	22.00
2019	10	0.09	9	23	2.30	2.56
2020	30	0.26	18	27	0.90	1.50
2021	23	0.20	13	38	1.65	2.92
2022	14	0.12	8	20	1.43	2.50
2023	4	0.03	1	4	1.00	4.00
	116	100%				

Note TP = total number of publications; NCP = number cited publications; TC = total quotes; C/P = average citations per publication; C/CP = average citation per cited publication

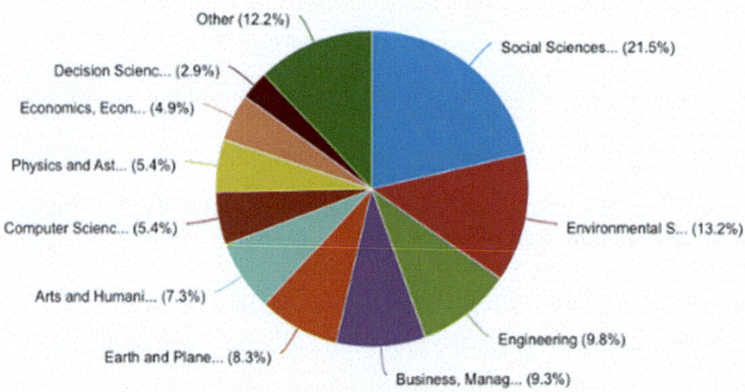

Fig. 2. Documents by subject of selected articles

It can be seen from Fig. 6, that the brighter the colour, the more articles published. Hence, it can be concluded that the topic/keyword about innovation, rural area, sustainable development, and village development becomes the keyword to be further researched.

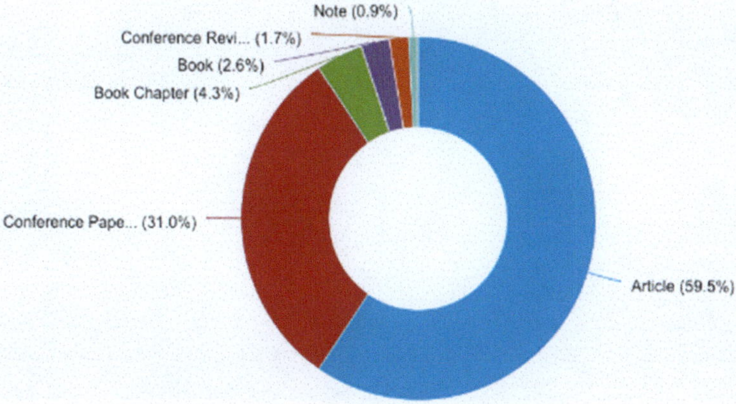

Fig. 3. Documents type of selected articles

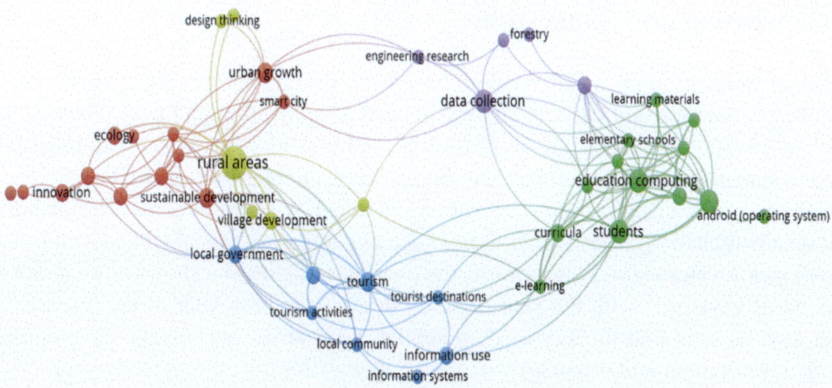

Fig. 4. Visualization topic area using network visualization

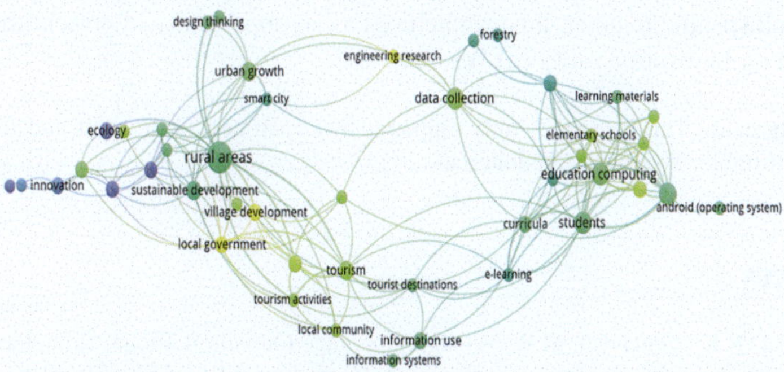

Fig. 5. Visualization topic area using overlay visualization

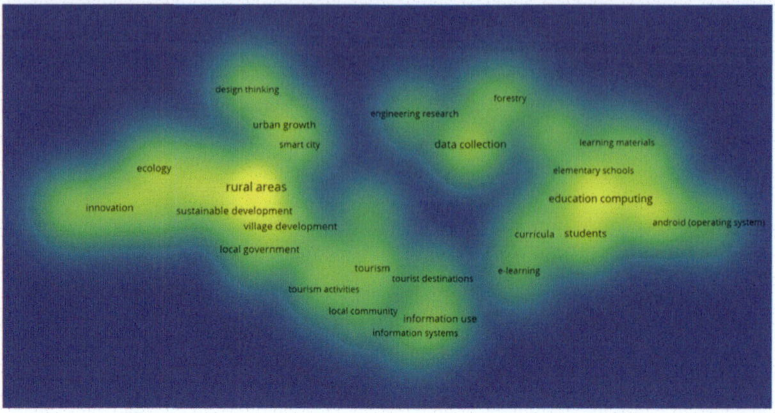

Fig. 6. Visualization topic area using density visualization

4 Conclusion and Suggestions

Studies of Innovation on Local Wisdom have been studied before. This paper has successfully presented a bibliometric analysis related to Innovation on Local Wisdom. The search results show that most articles on the topic of Innovation on Local Wisdom are published in scientific journals compared to books, book chapters, or others. When viewed from the network visualization, Innovation on Local Wisdom becomes the disbursement results related to several other issues such as rural areas, sustainable development, and villager development. Furthermore, the overlay visualization shows more about the updating of keywords with the year of study used as the base year of the research, so it can also be seen that the keywords potential, rural areas, sustainable development, village development and urban growth are the themes for the current study year.

In future research, it is necessary to study more broadly and deeply the marketing of local wisdom in the tourism sector, including in its bibliometric analysis. Since there are not many studies that examine Innovation on Local Wisdom in a more in-depth, focused, and specific manner, this can be used as an opportunity to present the novelty of research on Innovation on Local Wisdom.

Acknowledgment. This research is fully supported by Department of Research and Technology, STIPRAM Institute, Yogyakarta, Indonesia.

References

1. Kausar, D.R.K., Gunawan, M.P.: Managing heritage tourism in Toraja: strengthening local values and improving tourists' experiences. J. Herit. Tour. **13**(6), 550–561 (2018)
2. Wijijayanti, T., Agustina, Y., Winarno, A., Istanti, L.N., Dharma, B.A.: Rural tourism: a local economic development. Australas. Account., Bus. Finance J. **14**(1), 5–13 (2020)
3. Aswita, D., Suryadarma, I.G.P., Suyanto, S.: Local Wisdom of Sabang Island Society (Aceh, Indonesia) building ecological intelligence to support sustainable tourism. Geo J. Tourism Geosites **22**(2), 393–402 (2018)

4. Gupta, H., Kusi-Sarpong, S., Rezaei, J.: Barriers and overcoming strategies to supply chain sustainability innovation. Resour. Conserv. Recycl. **161**, 104819 (2020)
5. Kurniawati, E., Aliman, M.: Participation of green environmental group and Ulur-Ulur local wisdom on Buret Lake ecotourism management in karst area of Tulungagung, Indonesia. Geo J. Tourism Geosites **30**, 889–895 (2020)
6. Ermawati, K.C., Resmi, P.C., Anggraini, F.D., Rahmawati, E., Ariyanto, A., Jhon, J.O.I., Damiasih, D.: Local wisdom in tourism: review and future researches directions. Int. J. Adv. Digit. Lib. Inf. Sci. **3**(2) (2023)
7. Rahmawati, E., Hendratono, T., Sugiarto, S., Pradini, G., Herawan, T.: Railways transport infrastructures for supporting tourism: a bibliometric analysis. In: International Conference on Computational Science and Its Applications (pp. 473–494). Springer Nature Switzerland, Cham (2023)
8. Herawan, T., Sunyoto, S., Widodo, W.I., Disma, F.Z., Budidharmanto, L.P., Damiasih, D., Sari, E.N.: The emergence of robotics in tourism and hospitality: a bibliometric analysis from 2017 to 2023. In: International Conference on Computational Science and Its Applications (pp. 470–488). Springer Nature Switzerland, Cham (2023)
9. Pradini, G., Hendratono, T., Azahari, A., Rahmawati, E., Herawan, T.: Research trends in tourism participation: a bibliometric analysis using the scopus database. In: International Conference on Computational Science and Its Applications (pp. 437–454). Springer Nature Switzerland, Cham (2023)
10. Herawan, T., Arsyad, S., Widodo, W.I., Adiyanti, A.S., Damiasih, D., Ashartono, R., Sari, E.N.: A decade bibliometric analysis of decision making in tourism and hospitality. In: International Conference on Computational Science and Its Applications (pp. 17–36). Springer Nature Switzerland, Cham (2023)

Prediction of Kidney Disease Progression Using *K*-Means Algorithm Approach on Histopathology Data

Agus Perdana Windarto[1(✉)], Tutut Herawan[2,3], and Putrama Alkhairi[1]

[1] STIKOM Tunas Bangsa, Pematangsiantar, North Sumatra, Indonesia
{agus.perdana,putrama}@amiktunasbangsa.ac.id
[2] Sekolah Tinggi Pariwisata Ambarrukmo, Yogyakarta, Indonesia
tutut@stipram.ac.id
[3] AMCS Research Center, Yogyakarta, Indonesia

Abstract. Early diagnosis and prediction of disease progression are crucial for effective medical intervention. This paper presents the employability the *K*-Means algorithm for predicting kidney disease progression as a global health concern, using histopathological data. We conducted an analysis of histopathological data, which includes images of kidney tissue obtained through histological examination, from patients diagnosed with kidney disease. The *K*-Means algorithm was employed to cluster this data into homogeneous groups based on structural and cellular characteristics, subsequently predicting disease progression for each group. These predictions can aid physicians in recognizing disease progression patterns, thereby facilitating informed decisions regarding patient management and care. Our findings suggest that the *K*-Means algorithm can accurately predict kidney disease progression using histopathological data.

Keywords: Kidney disease · Prediction · *K*-Means algorithm · Histopathological data · Early diagnosis

1 Introduction

Kidney disease has a significant impact on patients' quality of life, emphasizing the importance of early detection and prediction of disease progression [1, 2]. Artificial intelligence, particularly the *K*-Means algorithm applied to histopathological data, shows promise in achieving this goal [3, 4]. Histopathological data provides valuable insights into cellular and tissue changes, but manual interpretation is time-consuming and complex, affecting diagnosis accuracy. The *K*-Means algorithm offers an efficient and accurate alternative for analyzing histopathological data [5, 6]. By grouping data into clusters based on feature similarities, the *K*-Means algorithm can identify patterns that are difficult for humans to detect, enabling precise classification of kidney disease progression and improved treatment planning [7, 8]. In a study evaluating the effects of hemodialysis on brain function in end-stage renal failure patients, an improved *K*-means clustering

Y. Farhaoui et al. (Eds.): ICAISE 2023, LNNS 837, pp. 492–497, 2024.
https://doi.org/10.1007/978-3-031-48465-0_66

algorithm was used to scan the brains of routine dialysis patients. The improved algorithm demonstrated better time efficiency and lower error rates in processing rs-fMRI images compared to traditional methods [9]. Another study applied machine learning to digital pathology for assessing kidney function and diagnosing chronic kidney disease. The Clustering-Based Spatial Analysis framework leveraged unsupervised learning to study spatial relationships in kidney tissue, achieving high accuracy in predicting eGFR levels and changes [10]. While previous studies have shown the potential of the *K*-Means algorithm in analyzing histopathological data for medical applications, its use in predicting kidney disease progression is still limited.

This article aims to investigate and apply the *K*-Means algorithm to histopathological data to develop a predictive model for disease progression, with the potential to improve patient quality of life and treatment effectiveness [11]. The methodology includes data processing, implementing the *K*-Means algorithm, and evaluating the model's performance [12, 13]. In summary, the utilization of the *K*-Means algorithm on histopathological data holds promises for predicting kidney disease progression. The results of this research can contribute to improving patient outcomes and the management of kidney disease.

The paper is organized as follows: Sect. 2 presents the proposed method. Section 3 presents the results and discussion. Finally, Sect. 4 concludes this work.

2 Proposed Method

The method used in this research is the data mining method with the *K*-means algorithm [14–17]. This research was conducted to implement the *K*-means algorithm in predicting the development of kidney disease using the *K*-means algorithm approach to histopathological data. From the results of data classification, information can be obtained so that patients can be treated for indications of kidney disease.

2.1 *K*-Means Algorithm

From the problems that have been analyzed, we analyze the factors that cause kidney disease using the *K*-Means Data Mining algorithm. The research design based on *K*-Means algorithm framework can be seen in Fig. 1.

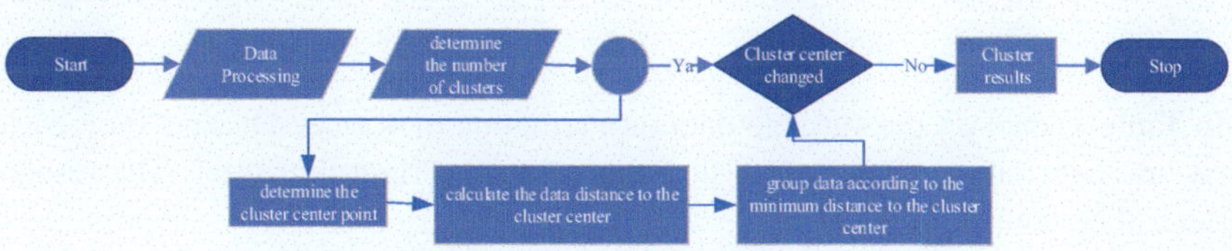

Fig. 1. Research design

The steps depicted in Fig. 1, the process of the *K*-Means method can be explained as follows: Firstly, the number of clusters to be used for data segmentation is determined.

Next, an initial set of centroids is generated randomly, with the number of centroids equal to the desired number of clusters. Each centroid represents the center point or starting point of a cluster. Subsequently, the distance between each data point and the centroid is calculated using the Euclidean Distance equation, aiming to find the closest distance for each data point. The data points are then grouped based on their proximity to the nearest centroid. The centroid values are updated by calculating the mean of the data points within each cluster. If the members of each cluster remain unchanged, the iteration process is considered complete, and the average value of the cluster center (μj) is used as a parameter for determining the data distribution. However, if changes occur, the iteration continues from step (2) to step (5). This summary provides a concise description of the steps involved in the *K*-Means method according to Fig. 1.

3 Results and Discussion

Histopathological data was utilized to study the prediction of kidney disease progression. The data for the study was obtained from the kaggle.com website. To facilitate the research process, RapidMiner 8.1 software was employed to implement the data and obtain accurate results, replacing manual calculations. The kidney disease progression prediction data was converted into an Excel data format and then fed into the *K*-means algorithm formula to generate clusters indicating kidney and non-kidney conditions.

3.1 Datasets

For the process of grouping data, we use the formula in the *K*-means method. Then the Prediction of Kidney Disease Progression Based on Histopathological Data will be processed using the RapidMiner 8.1 application to get accurate results. The data to be processed is 50 data samples on Kaggle.com. The list of variables for detecting kidney disease based on histopathological data is listed in Table 1.

Table 1. Description of features

Variable	age	bp	sg	al	su	rbc	pc	pcc	ba	bgr	bu	sc	sod
Description	–	Blood pressure	Specific gravity	Albumin	Sugar	Red blood cells	Pus cell	Pus cell clumps	Bacteria	Blood glucose random	Blood urea	Serum creatinine	Sodium

In Table 1, data on the variable data on Prediction of Kidney Disease Progression Based on Histopathological Data Using the *K*-means Algorithm used. The research dataset can be seen in Table 2.

Table 2 describes the research dataset taken from Kaggle which has been converted to facilitate calculations using the *K*-means algorithm.

Table 2. Research datasets

id	age	bp	sg	al	su	rbc	pc	pcc	ba	bgr	bu	sc	sod
1	48	70	1.005	4	0	2	1	1	0	117	56	3.8	111
2	53	90	1.02	2	0	1	1	1	0	70	107	7.2	114
3	63	70	1.01	3	0	1	1	1	0	380	60	2.7	131
…	…	…	…	…	…	…	…	…	…	…	…	…	…
49	34	80	1.025	0	0	2	2	0	0	121	27	1.2	144
50	60	80	1.025	0	0	2	2	0	0	131	10	0.5	146

3.2 Results

The results of data clustering in RapidMiner 8.1 show that the results of using tools in conducting research on Predicting the Progression of Kidney Disease using the *K*-Means Algorithm Approach to Histopathology Data. The results obtained from implementation calculations in the RapidMiner application match or are the same (balanced). Clusters can be seen in Fig. 2.

Cluster Model

```
Cluster 0: 30 items
Cluster 1: 20 items
Total number of items: 50
```

Fig. 2. Clustering result data

From Fig. 2, the results obtained are:

a. Cluster 0 (C2) has 30 data which are interpreted as predictions of low risk patients who will be indicated for the development of kidney disease in histopathological data taken and processed from the website kaggle.com
b. Cluster 1 (C1) has 20 data which are interpreted as predictions of high-risk patients who will be indicated for the development of kidney disease in histopathological data taken and processed from the website kaggle.com

For the process of grouping data, we use the formula in the *K*-means method. Then the Kidney Disease Development Prediction data uses the *K*-Means Algorithm Approach. Histopathological data will be processed using the RapidMiner 8.1 application so that accurate results are obtained. Histopathological data were processed from the kaggle.com website.

3.3 Discussion

This sub-section discusses the relationship between the results obtained from manual algorithm calculations vs. the results displayed by the RapidMiner tool. Cluster values

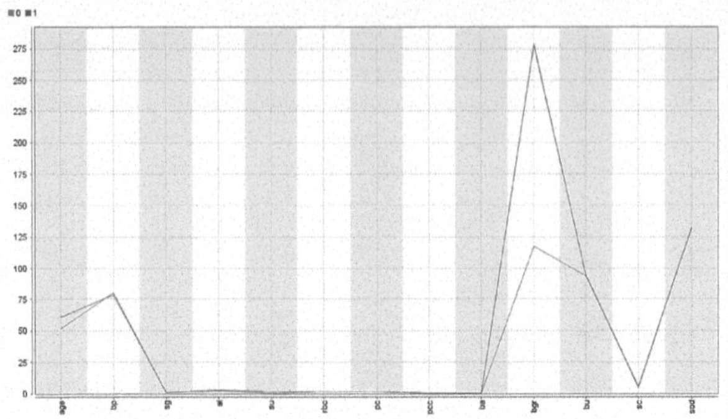

Fig. 3. Kidney disease index graph

are processed using RapidMiner tools as a tool that helps to cluster data more quickly and accurately using Kidney Disease Histopathology Data.

Figure 3 is a Graph of Kidney Disease Index with high and low risk of kidney disease. In Fig. 3, we can see that for cluster 0 the red line has a low risk, and cluster 1 the blue line has a fairly high risk. And the average patient who is in cluster 0 is a population with an average age above 55 years.

4 Results and Discussion

This study uses the *K*-Means algorithm approach on histopathological data to predict kidney disease progression. The results showed that the *K*-Means algorithm was effective in grouping histopathological data and identifying relevant patterns. The developed disease progression prediction model has the potential to improve the management of kidney disease by providing more accurate predictions. The use of artificial intelligence technology in the medical field, especially histopathological data analysis, has a positive impact on improving patient care and providing valuable insights for medical professionals. Nonetheless, challenges in dataset variation and model validation need to be overcome to ensure the reliability and generalizability of the model in the future. It is hoped that the results of this study will encourage further development and implementation of innovative kidney disease progression prediction models for better patient benefit and medical care.

Acknowledgement. This works is supported by Department of Research and Technology, STIPRAM Institute, Indonesia.

References

1. Lee, H.C., et al.: Derivation and validation of machine learning approaches to predict acute kidney injury after cardiac surgery. J. Clin. Med. **7**(10) (2018). https://doi.org/10.3390/jcm7100322

2. Kumar, S., Kumar-Solanki, V., Choudhary, S.K., Selamat, A., Gonzalez-Crespo, R.: Comparative study on ant colony optimization (ACO) and *K*-means clustering approaches for jobs scheduling and energy optimization model in Internet of Things (IoT). Int. J. Interact. Multimed. Artif. Intell. **6**(1), 107 (2020). https://doi.org/10.9781/ijimai.2020.01.003
3. Lee, J., et al.: Unsupervised machine learning for identifying important visual features through bag-of-words using histopathology data from chronic kidney disease. Sci. Rep. **12**(1), 1–13 (2022). https://doi.org/10.1038/s41598-022-08974-8
4. Windarto, A.P., Herawan, T.: *K*-means algorithm with Rapidminer in clustering school participation rate in Indonesia BT - recent trends in mechatronics towards Industry 4.0, pp. 779–794 (2022)
5. Li, L., Wang, J., Li, X.: Efficiency analysis of machine learning intelligent investment based on *K*-means algorithm. IEEE Access **8**, 147463–147470 (2020). https://doi.org/10.1109/ACCESS.2020.3011366
6. Moradi Fard, M., Thonet, T., Gaussier, E.: Deep *K*-means: jointly clustering with *K*-means and learning representations. Pattern Recognit. Lett. **138**(2016), 185–192 (2020). https://doi.org/10.1016/j.patrec.2020.07.028
7. Wayan Suartika Eka Putra, I., S. R., Yudhi, W.A.: Klasifikasi Citra Menggunakan convolutional neural network (Cnn) Pada Caltech 101. J. Tek. ITS **5**(1), 76 (2016) [Online]. Available: http://repository.its.ac.id/48842/
8. Alkhairi, P., Purba, L.P., Eryzha, A., Windarto, A.P., Wanto, A.: The analysis of the ELECTREE II algorithm in determining the doubts of the community doing business online. J. Phys. Conf. Ser. **1255**(1) (2019). https://doi.org/10.1088/1742-6596/1255/1/012010
9. Cheng, Y., Yu, Y.: *K*-means clustering algorithm-based functional magnetic resonance for evaluation of regular hemodialysis on brain function of patients with end-stage renal disease. Comput. Math. Methods Med. **2022** (2022). https://doi.org/10.1155/2022/1181030
10. Wang, Y., Sibaii, F., Lee, K., Gill, M.J., Hatch, J.L.: CluSA: clustering-based spatial analysis framework through Graph Neural Network for Chronic kidney disease prediction using histopathology images. medRxiv **1**(165), 1–13 (2021)
11. Obaid, A.J.: An efficient systematized approach for the detection of cancer in kidney. Int J Sci Eng Res **9728** [Online], ISSN: 2347-971X (Print). Available: www.ijiser.com
12. Pustokhina, I.V., et al.: Automatic vehicle license plate recognition using optimal *K*-means with convolutional neural network for intelligent transportation systems. IEEE Access **8**, 92907–92917 (2020). https://doi.org/10.1109/ACCESS.2020.2993008
13. Waris, S.F., Koteeswaran, S.: Early prediction of heart conditions by *K*-means consensus clustering and convolution neural network 25(3), 6623–6640 (2021) [Online]. Available: http://annalsofrscb.ro
14. Mohd, W.M.B.W., Beg, A.H., Herawan, T., Noraziah, A., Rabbi, K.F.: Improved parameterless k-means: auto-generation centroids and distance data point clusters. In: Information Retrieval Methods for Multidisciplinary Applications (pp. 156–168). IGI Global (2013)
15. Mohd, W.M.W., Beg, A.H., Herawan, T., Rabbi, K.F.: MaxD K-means: a clustering algorithm for auto-generation of centroids and distance of data points in clusters. In: Computational Intelligence and Intelligent Systems: 6th International Symposium, ISICA 2012, Wuhan, China, 27–28 Oct 2012. Proceedings (pp. 192–199). Springer, Berlin Heidelberg (2012)
16. Mohd, W.M.W., Beg, A.H., Herawan, T., Noraziah, A., Chiroma, H.: Multi-dimensional K-means algorithm for student clustering. In: Proceedings of the International Conference on Data Engineering 2015 (DaEng-2015) (pp. 119–128). Springer, Singapore (2019)
17. Prahara, A., Yanto, I.T.R., Herawan, T.: Histogram thresholding for automatic color segmentation based on k-means clustering. In: Recent Advances on Soft Computing and Data Mining: The Second International Conference on Soft Computing and Data Mining (SCDM-2016), Bandung, Indonesia, 18–20 Aug 2016 Proceedings Second (pp. 344–354). Springer International Publishing (2017)

Visualizing Trends in Tourism Entertainment Researches: Bibliometric Analysis Using the Scopus Database

Tutut Herawan[1,2(✉)], Fera Dhian Anggraini[1,3], John J. O. I. Ihalauw[1], Tonny Hendratono[1], Damiasih Damiasih[1], and Suhendroyono Suhendroyono[1]

[1] Sekolah Tinggi Pariwisata Ambarrukmo Yogyakarta, Jalan Ringroad Timur No. 52, Bantul, Daerah Istimewa Yogyakarta, Indonesia 55198
tutut@stipram.ac.id

[2] AMCS Research Center, Jalan Griya Taman Asri, Yogyakarta, Indonesia 55512

[3] Lembaga Kajian Pariwisata Indonesia, Babadan, Sewon, Bantul, Daerah Istimewa Yogyakarta, Indonesia 55185

Abstract. Research related to tourism entertainment has been widely carried out. This paper identifies developments in tourism entertainment research in past decade. Bibliometric analysis was used to analyze 909 articles published from 2013 to 2023. Tourism entertainment is the main keyword used in article titles, abstracts, and keywords to obtain meta data retrieved from the Scopus database. The tool used in the bibliometric analysis is Harzing's Publish or Perish to extract data from the Scopus database and VoS Viewer to visualize data and then use it to analyze citations and metrics. The results show that most related articles are published in scientific journals, compared to other publication outlets. Related articles published in scientific journals received the highest citation compared to others. Based on network visualization, the most dominant terms are development, experience, marketing, and virtual reality. Finally, from the overlay and density visualizations, it is shown the latest trends in researches related to tourist behavior, tourist experience, marketing, cultural tourism, and the tourism industry.

Keywords: Information Visualization · Bibliometric analysis · Scientometrics · Tourism entertainment · VoS viewer · Scopus

1 Introduction

Technological advances in today's digitization systems are growing rapidly. Some of these technological developments occur in several factors, one of which is in the tourism sector. Technological developments have also enabled the emergence of new types of tourism entertainment such as virtual reality and augmented reality which have become increasingly popular by utilizing internet media in recent years. The proliferation of the internet and other technological innovations have changed the structure of the tourism industry and influenced how tourist destinations are perceived and consumed [1]. In addition, internet media is also closely related to cyberspace which is also a form of

Y. Farhaoui et al. (Eds.): ICAISE 2023, LNNS 837, pp. 498–504, 2024.
https://doi.org/10.1007/978-3-031-48465-0_67

development from technological progress which is widely used as a means for the public to obtain tourism entertainment. Cyberspace is Second Life which provides opportunities for companies to communicate marketing messages to the target market and attract the attention of potential customers. Second Life has also been used by the travel and tourism industry as a collaborative and commercial tool to communicate with tourists in marketing tourist destinations [2].

Research on tourism entertainment has been carried out by many previous researchers, as was done by Luo, *et al.*, in 2020, who conducted research on the development of measurement scale for entertainment tourism experience, where they took a case study in Macau [3]. In 2022, Luo, *et al.*, conducted research on supply perspective again with the title analysis of critical success factors for entertainment tourism destinations [4]. Another study was also conducted by Luo, *et al.* in 2021, which took the title exploring the relationship between hedonism, tourist experience, and intention to return to entertainment destinations [5]. In 2021, Abdelmaged conducted research on the implementation of virtual reality in the health, entertainment, tourism, education, and retail sectors [6]. Luo & Fan also researched other matters related to tourism entertainment in 2022, with the research title Exploring entertainment experience in integrated resorts: a process-based perspective [7]. The studies above discussed the relationship between tourism entertainment and entrepreneurship, international housing mobility, hospitality, lifestyle, lifestyle with networks, digital media, product marketing, tourist experience, clusters, and innovation.

This paper identifies developments in tourism entertainment research in past decade. The novelty of this work is that, we explore the VoS Viewer as a tool for bibliometric analysis, in order to obtain popular research topics.

The rest of this paper is organized as follow. Section 2 describes the proposed research method. Section 3 presents the results obtained and is followed by discussions. The last section concludes the work and highlight future research works.

2 Proposed Method

We adopt a bibliometrics analysis method [8–12] which consists of six stages i.e.: searching articles, saving data in RIS, entering data into VoS Viewer, bibliometric analysis, presenting results, visualization mapping as depicted in Fig. 1.

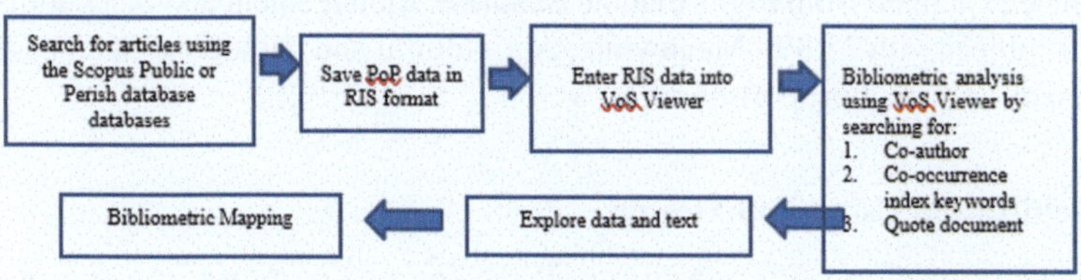

Fig. 1. Flowchart of the proposed method

3 Results and Discussion

This section describes the results obtained from this work, which include publications, citations, and visualizations.

3.1 Publications and Citations

To find out the comparison of the citation matrix on data taken through Scopus, the researcher makes a table which contains the article number, number of citations, number of citations per year, number of authors per year, H index, G index, normal hI, and annual hI. at the start of the search and fix results. Matrix of comparison of initial search results data and search results after improvement can be seen in Table 1.

Table 1. Comparison matrix

Data	Initial search results	Improved search results
Database	Scopus	Scopus
Publishing year	2013–2023	2013–2023
Number of years	10	10
Number of Articles	1163	909
Citations number	9738	8413
Number of citations per year	973.80	841.30
Number of authors per year	2.86	2.76
H index	43	43
G index	71	67
hI Normal	31	29
hi Annual	3.10	2.90

Table 2 describes the descriptive statistics of related publications in each year.

Figure 2 describes the documents by subject from the field.

It can be explained from Fig. 2 that the Business, Management and Accounting is the dominant subject with 25.8%. Meanwhile, Agricultural and Biological Sciences occupy the last position with a proportion of 1.7%.

3.2 Visualization Using VoS Viewer

Figure 3 shows the topic area visualization using the network visualization of the VoS Viewer.

From Fig. 3, the network visualization results show that Tourism is the main term that is at the center of search results related to several other issues such as Tourism Development, Virtual Reality, Heritage Tourism, Experience Economy, and marketing.

Table 2. Publication descriptive statistics

Year	TP	% (N = 909)	NCP	TC	C/P	C/CP
2013	60	0.07	43	1232	20.53	28.65
2014	65	0.07	49	839	12.91	17,12
2015	52	0.06	43	570	10.96	13,26
2016	61	0.07	45	528	8,66	11.73
2017	76	0.08	65	1174	15.45	18.06
2018	86	0.09	74	1139	13,24	15.39
2019	104	0.11	82	950	9,13	11.59
2020	110	0.12	79	950	8,64	12.03
2021	110	0.12	78	434	3.95	5.56
2022	158	0.17	77	316	2.00	4,10
2023	27	0.03	9	18	0.67	2.00
	909	100%				

Note TP = number of publications; NCP = number of cited publications; TC = number of citations; C/P = average citations per publication; C/CP = average citations per publication cited

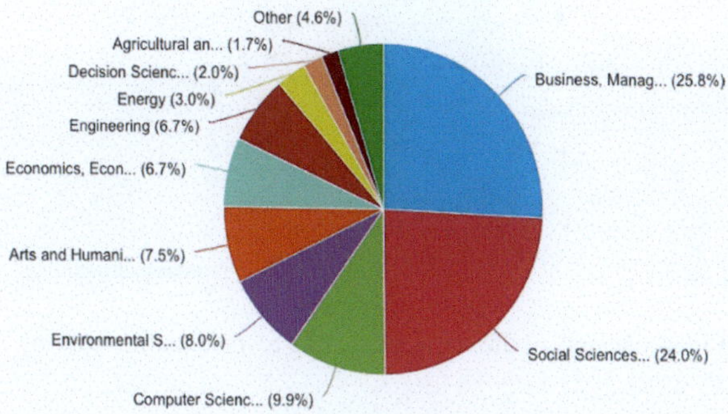

Fig. 2. Documents by subject form fields 2013–2023

Figure 4 describes the topic area visualization using the overlay visualization of the VoS Viewer.

From Fig. 4 it can be seen that the topic of entertainment shows an increasing and thick trend even though it has not yet integrated with tourism. The figure overlay visualization shows more on the updating of keywords with the year of study used as the base year of research, so that it can also be seen that the keywords Covid-19, Tourism Industry, Hotels, and Economic Development are the themes with the most recent year of study. Figure 5 describes the topic area visualization using the density visualization of the VoS Viewer.

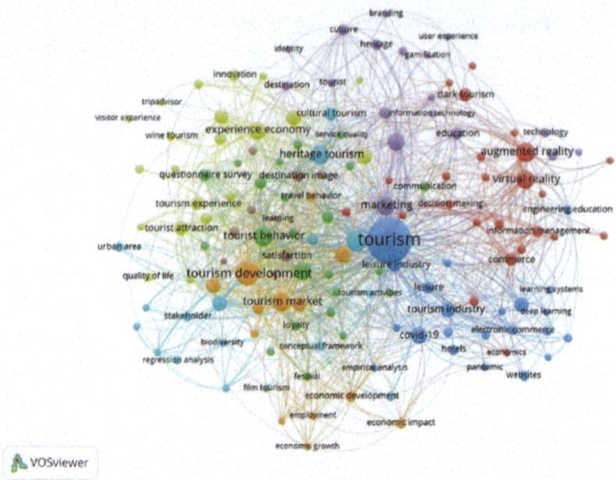

Fig. 3. Topic area visualization using network visualization

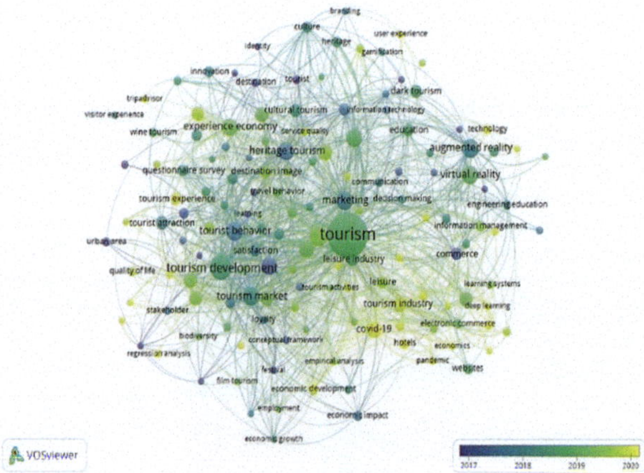

Fig. 4. Visualize the topic area using a visualization overlay

Through Fig. 5, it can be concluded that the topic with the keywords regarding Tourism, Tourism Development, Experience Economy, Augmented Reality, and the Entertainment Industry become most popular keywords to be researched.

4 Conclusions and Recommendations

This paper has successfully in presenting a bibliometric analysis related to the topic of entertainment tourism. In this paper, the results show that articles with related topics are the largest number published in scientific journals when compared to others. In addition, related articles published in scientific journals have obtained the highest citations compared to others. Based on network visualization with the most dominant terms appearing is Tourism is at the center of search results related to several other issues such as Tourism Development, Virtual Reality, Heritage Tourism, Experience Economy, and marketing.

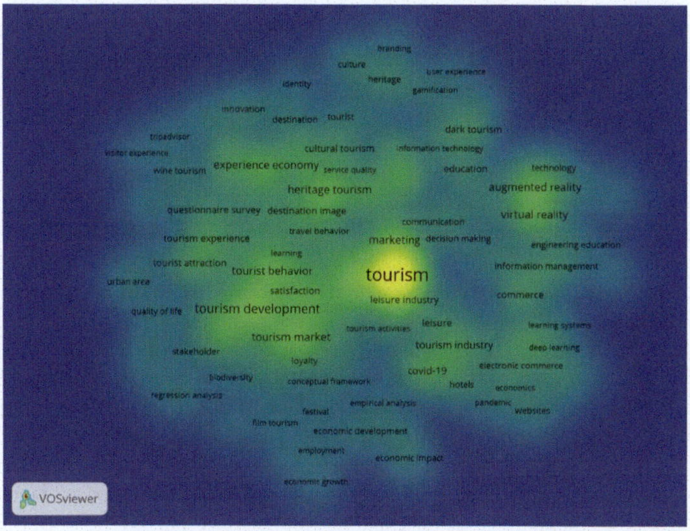

Fig. 5. Visualize topic areas using density visualization

When viewed from the Overlay Visualization it shows more on the updating of keywords with the year of study which is used as the basis for the research base, so that it can also be seen that the keywords Covid-19, Tourism Industry, Hotels, and Economic Development are the themes with the latest year of study. Based on the findings presented in the Network Visualization and Overlay Visualization, it can be concluded that articles on the topic of entertainment tourism have not been studied by many people. This can be used as a motivation for further research.

The suggestion given is that in future research, it should be studied more broadly and in depth on the topic of tourism entertainment, including its bibliometric analysis using ISI WoS database, since it provides high quality of scientific publications results.

Acknowledgement. This research is fully supported by the STIPRAM Institute Yogyakarta, Indonesia.

References

1. Huang, Y.C., Backman, K.F., Backman, S.J., Chang, L.L.: Exploring the implications of virtual reality technology in tourism marketing: an integrated research framework. Int. J. Tour. Res. **18**(2), 116–128 (2016)
2. Huang, Y.C., Backman, S.J., Backman, K.F., Moore, D.: Exploring user acceptance of 3D virtual worlds in travel and tourism marketing. Tour. Manage. **36**, 490–501 (2013)
3. Luo, J.M., Lam, C.F., Fan, D.X.: Development of a measurement scale for the leisure tourism experience: a case study in Macau. Curr. Issue Tour. Issue Tour. **23**(7), 852–866 (2020)
4. Luo, J.M., Fan, Y., Shang, Z.: Analysis of critical success factors for leisure tourism destinations: a supply perspective. J. Qual. Assur. Hosp. Tour. **23**(5), 1159–1182 (2022)

5. Luo, J.M., Lam, C.F., Wang, H.: Exploring the relationship between hedonism, tourist experience, and revisit intentions in entertainment destinations. Open Sage **11**(4), 21582440211050390 (2021)
6. Abdelmaged, M.A.M.: Implementation of virtual reality in healthcare, entertainment, tourism, education, and retail sectors. Munich Personal RePEc Archive (2021)
7. Luo, J.M., Fan, D.X.: Exploring the leisure tourism experience in integrated resorts: a process-based perspective. Anatolia, 1–14 (2022)
8. Rahmawati, E., Hendratono, T., Sugiarto, S., Pradini, G., Herawan, T.: Railways transport infrastructures for supporting tourism: a bibliometric analysis. In: International Conference on Computational Science and Its Applications (pp. 473–494). Springer Nature Switzerland, Cham (2023)
9. Herawan, T., Sunyoto, S., Widodo, W.I., Disma, F.Z., Budidharmanto, L.P., Damiasih, D., Sari, E.N.: The emergence of robotics in tourism and hospitality: a bibliometric analysis from 2017 to 2023. In: International Conference on Computational Science and Its Applications (pp. 470–488). Cham: Springer Nature Switzerland (2023)
10. Pradini, G., Hendratono, T., Azahari, A., Rahmawati, E., Herawan, T.: Research trends in tourism participation: a bibliometric analysis using the scopus database. In International Conference on Computational Science and Its Applications (pp. 437–454). Springer Nature Switzerland, Cham (2023)
11. Herawan, T., Arsyad, S., Widodo, W.I., Adiyanti, A.S., Damiasih, D., Ashartono, R., Sari, E.N.: A decade bibliometric analysis of decision making in tourism and hospitality. In: International Conference on Computational Science and Its Applications (pp. 17–36). Springer Nature Switzerland, Cham (2023)
12. Ajie, A.P., Kurniawati, D., Devylia, D.: A bibliometric analysis on alternative tourism in Indonesia from 2012 to 2022. Int. J. Adv. Tourism Hospitality **3**(2) (2023)

Marketing Based on the Digitalization of Customer Relations: A Priority Orientation for Better Management of Companies' Financial Resources

Khalid Lali[1](✉) and Abdellatif Chakor[2]

[1] Polydisciplinary Faculty, Moulay Ismail University, Meknès/Errachidia, Morocco
khalid.lali1979@yahoo.fr

[2] Faculty of Legal, Economic and Social Sciences, Mohammed V University, Souissi/Rabat, Morocco
abdellatif.chakor@um5s.net.ma

Abstract. Marketing based on the digitalization of the customer relationship reflects the transition from a multichannel strategy to an omnichannel strategy. The latter therefore makes it possible to promote: interactive communication between the company and its customers; the creation, access and operation of a regularly updated customer database; a personalized offer of quality products and brands; and enhancement of the company's brand image. Finally, we note that in order to be able to take full advantage of this marketing approach, the company must always ensure that it is present on social networks or any other intelligent digital communication platform. Thus, thanks for example to CRM, ticketing, live chat, the company manages to better know its customers, their needs, to better analyze their behavior, to anticipate their expectations, to increase its sales and to generate satisfactory financial results which will ensure its resilience.

Keywords: CRM · Digital marketing · Omnichannel strategy · Resilient organization · Digitization of customer relations · Intelligent digital platforms

1 Introduction

In a competitive world characterized by globalization and the voluminous number of commercial exchanges concluded and which is also marked by the acceleration of the pace of use of new information and communication technologies, companies see themselves obliged to follow this trend if they want to demonstrate to their demanding customers that they are always capable of satisfying them and making them happy and this by quickly offering them the high quality products and/or services they desire and this before the other competitors. By doing so, these companies will continue to exist as long as possible on the competitive market which is constantly disrupted by fluctuations emanating mainly from the internal but also external environment of the said companies. In addition and in the same context, the leaders of these companies are aware that the

Y. Farhaoui et al. (Eds.): ICAISE 2023, LNNS 837, pp. 505–515, 2024.
https://doi.org/10.1007/978-3-031-48465-0_68

obsession with customers is a reality that no longer needs to be demonstrated and that they can only achieve by getting involved spontaneously and by investing massively in digital transformation. Indeed, the digital tool is characterized today in the eyes of several specialists in organization and systemic as a key and a determining weapon in the service of the maintenance and the success of the business relations concluded by the companies with a specific range customers in the medium and long term. In the same context, managers are now convinced that thanks to these digital media, they will easily be able to support their strategic positions on the market and open up other niches. In other words, and because of these electronic tools, businesses will be able to stay in touch with their customers around-the-clock. These customers won't just be regular consumers; they'll also serve as customer advisers and brand ambassadors for the businesses they've been doing business with for a while. This means that these new channels will give businesses a way to communicate with their clients, who will be less hesitant to make larger purchases from them and, worse yet, who will refer other clients to these businesses in order to purchase the goods and/or services they are marketing. This will allow businesses to improve their net profit margins and, as a result, produce very satisfactory financial results. Given these realities, therefore, we can say that the main question we are trying to answer in this article is as follows: Can we really affirm that today Moroccan companies installed in the city of Errachidia such Marjane were able to take advantage adequately and effectively of potentialities offered by these digital media, especially in their commercial relations concluded with their customers who are dispersed in several regions of the Kingdom as well as in terms of the favorable effects they have had on these entities' financial performance?

In order to answer this problematic, we established the following two hypotheses:

Hypothesis 1: Working in an environment where processes are increasingly being computerized with high-performance computer applications can only help businesses communicate their ideologies, their ways of thinking, and most importantly, their strategic and operational visions to their employees more easily. It can also help businesses create customer databases that can be consulted and used effectively and frequently, especially when it comes to defining and implementing practical marketing and commercial strategies.

Hypothesis 2: For companies to take full advantage of digitalization and thus achieve satisfactory turnover, it is first necessary to convince employees of the importance of committing to using this digital tool and to having the sufficient material, human and financial resources without forgetting also the importance of always thinking about trying to collect in a first preliminary stage the real needs of the customers and to read their minds and anticipate their expectations.

2 Literature Review on Digital Marketing

Several authors have been interested in defining the concept of digital marketing, in particular Sigala [30], Demuner et al. [10], [1, 3, 9–14, 16–27, 29, 30] and finally [9]. An in-depth examination of the research projects conducted by these authors that are related to the definition of this concept reveals that, for the majority of them, digital marketing is a strategy that can: raise the company turnover (through an interactive and personalized exchange with each client); increase brand awareness; and support customers by ensuring that they are provided with personalized offers that are likely to satisfy them and foster loyalty. The primary distinctive characteristics of this new marketing approach, as determined by these researchers, are explained in the paragraphs that follow. So, for instance, we may conclude that Sigala [30], Teo and Swan Tan [33], and Allen and Fjermestad [1] have highlighted the benefits that the Internet could offer for managers responsible for of creating the salient lines of an effective marketing and/or commercial strategy. These researchers view the Internet as a tool that guarantees a large public's access to high-quality products and/or services in an unlimited manner and at a lower cost, and that successfully promotes good planning and effective management of a high-quality marketing campaign supported by digital media. As we continue to examine the research conducted by other researchers that relates to digital marketing, we can also see that, according to Stokes [31], the distinctive feature of digital marketing that sets it apart from traditional marketing is that it strives to increase the volume of sales of the company in the future by guaranteeing the satisfaction and, more importantly, the loyalty of customers through the individualization of the offer conveyed. These digital tools should make it feasible for the business to maintain permanent, instantaneous touch with clients. Gaining the trust and loyalty of customers, capitalizing on their satisfactory experience, and converting them into customer advocates and ambassadors who can spread positive information about the business, its employees, its products, and its services are, in this author's opinion, the main challenges of digital marketing. This will help businesses to meet the challenge of customer obsession. The primary goal of digital marketing, according to authors like Wymbs [38] and Royle and Laing [27], is to keep customers and thwart their loss. This is done by encouraging direct, customized, measurable, and interactive communication, which is made simple by the use of digital technology. Lemoine and Badot [21], for their part, add that digital marketing refers to the abolition of spatio-temporal barriers through the articulation between traditional physical and electronic channels, with the constant goal being to provide the customer with a product and/or service that is tailored to his needs, preferences, and expectations as well as to always be available to him in order to satisfy him and foster loyalty. On the other hand, it appears from the research of Munshi [24] that digital marketing is seen as a locomotive at the service of real economic growth and that it is therefore able to offer businesses attractive alternative opportunities to operate effectively in the cutthroat market which is constantly in motion. The analysis of the research work of Järvinen et al. [17], also allows us to see that digital marketing is perceived as a new marketing approach which has come to complement that of traditional marketing and which is supported by digital media and which emphasizes direct and personalized exchange and communication with target customers. In addition and according to Chaffey and Ellis Chadwick [4], we can say that digital marketing is a tool that promotes the attraction of new customers and their loyalty, which are considered the main objectives guiding the marketing practices carried out by a marketing manager. These researchers insist on the need to establish a new digital culture based on the use of technologies which allow a greater presence of the company on various online platforms and remarkable visibility on its products and/or services. According to Scheid et al. [29], these writers stressed the significance of marketing managers adopting a strategic orientation that is based on the requirement to use exclusively digital channels and clever technology solutions. According to them, this strategic orientation will allow the company to consolidate and strengthen its brand image, its identity, its reputation, to better understand the preferences and needs of its customers and to anticipate their expectations. These authors also specify that these are the factors that are thus likely to enable these entities to succeed in designing effective digital-based marketing strategies that are especially characterized by serious customer engagement and that are capable of generating growth sales through personalized offers of quality products and/or services. These advantages provided by the digitalization of marketing will therefore help companies to generate new market shares which will allow them not only to establish their strategic positions on the competitive market and to keep them alive as long as possible, but also to prosper and to develop oneself. Subsequently and through the analysis of research established by authors such as Khan and Siddiqui [19], Ryan [28] and Oualidi [25], we noticed according to Khan et al. (2013), as well as Pai and Arnott [26], that the appearance of social networks such as Facebook has been a great catalyst and stimulator of communication between companies and several customers on the products and/or services marketed and which has been considered by marketing managers as an opportunity to seize and a shift to follow and which is likely to allow companies to achieve attractive profit margins. In the same vein, we also observe that Oualidi [25] asserts that digital marketing is a superb interactive communication tool that serves as support for a strong customer targeting strategy based on the internet and other digital media. This author insists on the fact that customers and companies today are led and better than ever to accept to consider digital culture not only as an alternative strategic choice but as a priority whose fallout is significant and positive. For his part, Ryan (2013) argued that digital technology is seen as an intermediary means that will connect the company with its key partners who are also thought of as its targets, in particular customers, and that is likely to encourage both parties to engage in a fruitful business relationship in the medium and long term. This will give a business the opportunity to complete a number of sales transactions for products or services in the future, allowing it to continue to compete and maintain its position in the market for as long as feasible. From another angle, Westerman et al. [37] explain us that digital marketing is a useful tool that supports successful transformation of various modes of operation and/or good change management inside the organization. Another definition that agrees with Taiminen and Karjaluoto [32] asserted, according to which digital marketing is a tool that makes it easier to implement successful marketing strategies and tactics. Additionally, according to Demuner et al. [10], having access to digital technology is not sufficient to successfully utilize digital marketing; rather, everything depends on the path and the plan that are to be followed and practically operationalized. These elements will direct the use and realization of the approved strategic vision and must be predominately focused on the results to be achieved. According to authors like Gutierrez-Leefmans et al. [15] and Autissier and Metais-Wiersch (2016), digital marketing is a type of marketing that aims to promote the design, development, and deployment of applications that are accessible via any electronic medium and are intended to market products and/or quality services, thereby increasing the volume of business sales. The goal is to utilize the resources and possibilities that digital technology offers in order to put into practice the developed marketing strategy, which is much more focused on enhancing corporate performance. Laurent Flores (2016) and Bressolles [3] added that digital marketing is a way to develop direct and personalized relationships with customers based on media, supports and/or channels of communication, exchange and digital interaction, capable of promoting the brands and products marketed by companies. According to them this is just the result of the evolution of the web, which went from web 1.0 to web 4.0 via web 2.0 and web 3.0. According to these two authors, by opting for digital marketing, the customer has the possibility of controlling the flows and expressing his opinion vis-à-vis the products marketed and this through the comments he will leave, for example. On a website or on a social network. These comments are thus considered as a kind of evaluation of his purchasing experience and which should not go unnoticed since marketing managers must take them into consideration to guarantee a good definition of flexible and effective marketing strategies. Finally, we can also deduce from the research work of Dabi-Schwebel [8], that digital marketing is a means at the service of modifying the organization of the company based on electronic tools and this in order to improve its performance. According to all these additional definitions that have been given to this concept, it is clear that there is no universal definition of digital marketing. Given this reality and for what concerns us in the context of this article and as a researcher interested in the systemic approaches adopted and mobilized in the service of flexible and quality marketing strategies and which attach all the importance to the investment in software solutions, in digital platforms and in the digitization of marketing, we thus propose the following definition: digital marketing is a new mode of marketing centered and supported on digital supports such as the Internet, the mobile and the social networks and whose objective is to increase the rate of sales of quality products and/or services likely to create the expected added value, thanks to the implementation of a strategic vision oriented above all towards the sustainability of the fruitful commercial business relationship linking the company with a specific range of customers endowed with specific and distinctive characteristics but which also aims to attract new customers.

2.1 The Opportunities of Digital Marketing

Identifying and listing the opportunities that can be seized by investing in the digitalization of marketing has been the subject of several studies by authors such as Baumeister [2], Glimore et al. (2007), Järvinen et al. [18], Chaffey et al. [5], Tiago and Veríssimo [34], Hirth [16] and Taiminen and Karjaluoto [32]. According to these authors, digital marketing offers the multiplicity and diversity of digital media intended for commercial purposes. Baumeister [2] for his part, emphasized two opportunities that can be seized by digital marketing and which can be summarized as follows:

- Electronic customer relationship management software has facilitated the establishment of personalized relationships with customers, to better understand their misunderstandings, needs, preferences and expectations and therefore to better measure the degree of their satisfaction and loyalty.
- CRM software also makes it possible to instantly report and quickly explore data relating to each customer and each traffic.

According to Hirth [16] and Järvinen et al. [17], digital marketing enables the marketing manager to better exhibit their abilities to analyze, transform, exchange, share, and communicate data related to marketing activities. Finally, according to the research done by Taiminen and Karjaluoto [32] it should also be highlighted that digital marketing and digital technologies support the acquisition of new customers as well as the satisfaction and loyalty of present customers.

2.2 The Threats of Digital Marketing

Many researchers and specialists in digital marketing have been interested in identifying the main threats relating to this new mode of marketing which is based on digital media, in particular Verhoef et al. [35] and Leeflang et al. [20]. According to these authors, the company must constantly worry and be concerned about the users of social networks and their opinions and comments which risk being fatal for the future of the company since in the digital age, they are the ones who now control the reputation of the company and that of these products and brands. In this same context, Verhoef et al. [35] explain to us that sometimes customers, instead of proposing new avenues and/or new ideas for product promotion, sales and distribution turn into destroyers of value and seize every opportunity to disseminate unfavorable information that damages the company's brand image, its history, its reputation and that of its marketed products, services and brands. In addition and in the same direction, [20] add that despite the fact that digital marketing offers the possibility to the company to manage the customer relationship electronically by helping the latter to make purchases, for example online, this new marketing approach – which is based on digital media – requires this entity to review and redefine its traditional business model.

3 Research Methodology

Drawing inspiration from [36] according to which, one cannot delimit all of reality based only on a single methodological approach and also referring to [11], who argued that the limits of each qualitative or quantitative method will be compensated by being used jointly, we based ourselves within the framework of this work on an articulation of mixed methods. These methods have been defined by several authors in particular [7] according to which: « A mixed methods study insolves the collection or analysis of both quantitative and /or qualitative data in a single study in which the data are collected concurrently or sequentially are given a priority and involves the integration of the data at one or more stages in the process of research». It is therefore clear that the use of mixed methods will help to obtain richer data and finally will offer possibilities for better estimating measurement errors and risks, which will only strengthen the confidence that each researcher will have on the results of his research. Therefore, taking into account all these considerations and in order to be able to test the validity of our two pre-established research hypotheses and thus respond to our problematic, the subject of this article, we used an interview guide that we developed and which was supplemented by a questionnaire that we have also carefully drawn up. It is also important to specify in this context that these two privileged data collection techniques were directed towards marketing managers, sales managers, cashiers, storekeepers and some employees working within the Marjane company located in the city of Errachidia. The objective was then to study the degree of contribution of an investment in digital marketing on the management of the customer relationship (based therefore on digital media and software dedicated for this purpose) and consequently on the management and the administration of the financial resources of this organization. This article is also an opportunity to highlight the importance of establishing a digital culture in the mentalities and minds of decision-makers by highlighting the returns on investment and the advantages that digital marketing can provide them, in particular the increase in sales, capitalization of the customer experience, customer loyalty, etc. The data collected was entered, processed and analyzed using SPSS. As for the analysis techniques used, we opted for the factorial analysis of correspondences followed by the analysis in dynamic clusters in two classes and this in order to test the validity of each of our two hypotheses. We specify in this sense, that these two analysis techniques were applied simultaneously to each of the two variables defined for each hypothesis, knowing that each hypothesis corresponds to two variables and that for each variable care was taken to also define sub-variables.

4 Analysis and Discussion of the Main Results

Our study consisted in highlighting the importance of investing in marketing based on the digitalization of the customer relationship, which is considered today as an excellent means for a good allocation of the financial resources of companies. According to our analysis of the data collected from executives, employees and managers working in the company Marjane Market which is located in the city of Errachidia, we were able to achieve the following main results:

– 84% of the people surveyed and working in this company are female, while 16% are male.
– According to the majority of respondents, the average number of customers who shop daily with this company - which opened its doors for the first time in Errachidia in 2002 under the name of ACIMA and which then passed to Marjane Market in 2005 - is 80 people (of which 60% are male and 40% are female).
– According to all the sales managers as well as the marketing managers met, the minimum monthly turnover that the company Marjane Market of Errachidia manages to achieve is 8,000,000 Dhs.
– According to 72.5% of people surveyed, the company's IT department is equipped with software linked to decentralized servers such as G.O.L.D "Ouroumoff Logistic Distribution Group" which makes it possible to process the stock inventory and reference all articles thus facilitating the editing of price tags. In addition, this software offers statistics on the daily turnover collected via the checkouts and the amount of sales made by each department or by product family and it also allows you to make price changes. On the other hand, 27.5% of the people surveyed do not agree at all with this opinion and specify that this software is of no importance and does not bring added value to the company in terms of customer relationship management.
– According to 60% of respondents, the company always ensures that the communication between it and its customers and which is based on digital media is credible. To do this, the company ensures that the messages formulated and transmitted to them are adapted to their needs and expectations, but also that they are in harmony with the

brand image that the company seeks to promote and consolidate. Approximately, 40% of those surveyed have an opinion opposite to that of the first category of respondents.

- According to 56% of people surveyed, the company is now able to better target its customers and easily sell its personalized offer, thanks to the digitalization of customer relations, the direct mail strategy, the implementation of a loyalty card system, the analysis of purchasers' payment vouchers and the many contests organized in the stores.
- For 77% of people surveyed, the digitalization of customer relations has enabled the company Marjane Market of Errachidia to better understand the behavior of its customers, their habits, their needs and to anticipate their expectations. According to these people and thanks to this new marketing approach, the company has come to recognize the different categories of customers and select those whose present value is high. On the other hand, we note that 23% of the people surveyed expressed an opinion opposed to that of the first category.
- 64% of respondents argue that given the fact that competition in supermarkets in the city of Errachidia is very low, with the help of the digitalization of customer relations, Marjane Market is now able to exercise a monopoly on the market with a market share that exceeds 90%.
- According to 89% of those surveyed, the shift from product orientation to customer orientation based on digital media has long been a marketing concern and a major strategic objective and not just a choice. According to these people, the use of means of communication such as personalized mailings, newsletters to inform customers of new product offers and investment in software such as CRM has enabled the Marjane company to gain the customer loyalty challenge.
- According to 97% of people surveyed, thanks to the digitilization of customer relations, Marjane Market can now assess the level of loyalty of its customers and note in this context that:

  - 10% of dissatisfied customers do not hesitate to directly express their dissatisfaction with the product offers offered.
  - On the other hand, 90% of them prefer to share their distress by disseminating unfavorable information on the company, its staff, the service provided and the offer offered, which therefore risks harming its notoriety and its image as a brand.

    Still, according to these people, this new marketing approach based on the digitization of customer relations has favored the creation of a customer database which serves as a support for decision-making, especially with regard to the design and implementation of a flexible and successful marketing strategy.

- According to 81.7% of the people surveyed, the main customer loyalty tools used by Marjane Market of Errachidia in the era of digitalization of customer relations are loyalty cards which are considered as personal cards allowing identify the most loyal customers in order to award them benefits in the form of services, gifts or discounts. On the other hand, we have 18.3% of people surveyed according to whom loyalty cards have become infidelity cards since lured by rewards, some customers prefer to turn into promotion hunters and therefore skip a teach another.

- According to 83% of the people surveyed, marketing based on the digitization of the customer relationship has made it possible to promote: instantaneous and effective interactive communication between the company and its customers; optimizing business performance and building customer loyalty. These people also add that the loyalty sought and achieved by the company Marjane Market was based above all on the trust and credibility of its messages and promises disseminated to these main customers.

5 Conclusion

Companies have now understood that in order to reach the minds and hearts of their customers, who represent their main target, they must boost the visibility of their products, services and brands and make themselves known everywhere, thanks to the transmission and the sharing of various information via the Internet and other digital communication channels. It is clear that by proceeding with a multiplicity of messages sent by advertisers through these digital media, companies will guarantee that their messages are audible and will arouse the curiosity and interest of customers who will no longer hesitate to demonstrate and express themselves concretely vis-à-vis the proposed offers. In general, it can be said that the main objectives of investing in digital marketing can be summarized in the following points:

- Promote the company, its products, its services and its brands on several sites favored by customers and who consult them frequently and regularly.
- Attract customers to provide more information about themselves and their preferences and needs and this by leading them, for example, either to register on a site or to fill out a form, the objective therefore always being to create traffic.
- Mobilize interactive communication on the sites with each Internet user (visitor) to help them express their preferences concretely through well-constructed messages. By doing so, the company seeks to convert a visitor into a buyer.
- Build customer loyalty and ensure their loyalty by planning appropriate actions to encourage them to buy more and repeatedly.

However, and as we have seen throughout this article, in the digital age, the opinions and comments of users who regularly browse via social networks and which relate in particular to the company, its reputation, its business, its main products, services and brands marketed are of paramount importance to marketing managers. The latter must therefore always be concerned about these opinions because they can be fatal for the future of the company and call into question its continuity of existence in a competitive market which is constantly changing. Thus and from this, we can then conclude that the success of a digital marketing strategy is conditioned by the presence of a specific range of customers who are ready to commit to collaborating with companies opting for such an approach. digital-based marketing. In other words, the effectiveness of a digital marketing strategy depends on the availability of customers who try all the time to provide advice and support to the company and its managers and who take care to disseminate information favorable vis à vis the company, its brands and its products, which can only contribute to increasing its sales and consolidating its strategic position. Finally, we emphasize that given that digital marketing offers the possibility of electronically managing the customer relationship in order to capitalize the successful customer experience, the marketing manager is called upon better than ever to plan and prepare in advance the conditions human, financial, material and technical likely to guarantee a good revision and redefinition of the business model of the company.

References

1. Allen, E., Fjermestad, J.: E-commerce marketing strategies: an integrated framework and case analysis. Logist. Inf. Manag. **14**(1/2), 14–23 (2001). https://doi.org/10.1108/09576050110360106
2. Baumeister, H.: Customer relationship management for SME's. Institut fur Informatik, LMU oettingenstr. 67, D-80538 Munchen, Germany, 1–7 (2002)
3. Bressolles, G.: *Le marketing digital* (2e éd). Dunod (2016)
4. Chaffey, D., Ellis-Chadwick, F.: Digital Marketing: Strategy, Implementation and Practice (5. ed). Pearson (2012)
5. Chaffey, D., Ellis-Chadwick, F., Isaac, H., Mercanti-Guérin, M., Volle, P.: Marketing Digital (5e éd). Pearson (2014)
6. Chaffey, D., Smith, P.R.: Emarketing excellence : Planning and optimizing your digital marketing (4th ed). Routledge (2013)
7. Creswell, J.W.: Research Design: Qualitative, Quantitative, and Mixed Methods Approaches (4th ed). SAGE Publications (2014)
8. Dabi-Schwebel, G.: 1 min 30. Récupéré sur Challenge your marketing! (2019). https://www.1min30.com/dictionnaire-du-web/plateforme-collaborative
9. Dabi-Schwebel, G., Couleaud, S.: Marketing automation: Faites plaisir à vos clients, accélérez votre business ! plus de 20 scénarios hyper-détaillés pour faire décoller votre relation client ! 1min30 publishin (2019)
10. Demuner, F. M. R., Nava, R. R. M. y Sandoval, A. R.: TI en la dinámica del conocimiento empresarial. México: Bonobos y Universidad Autónoma de Estado de México (2015)
11. Denzin, N.K.: The Research Act: A Theoretical Introduction to Sociological Methods. AldineTransaction (2009)
12. Florès, L.: Mesurer l'efficacité du marketing digital: Estimer le ROI pour optimiser ses actions (2e éd). Dunod (2016)
13. Gascó Hernández, M., Torres-Coronas, T. (eds.): Information Communication Technologies and City Marketing: Digital Opportunities for Cities Around the World. Information Science Reference (2009)
14. Gilmore, A., Gallagher, D., Henry, S.: E-marketing and SMEs: operational lessons for the future. Eur. Bus. Rev. **19**(3), 234–247 (2007). https://doi.org/10.1108/09555340710746482
15. Gutierrez-Leefmans, C., Nava-Rogel, R.M., Trujillo-Leon, M.A.: Digital Marketing in an Emerging Country: Exploratory Study of the Marketing Mix of SMES with Trust Seal. Revista Brasileira de Marketing **15**(2), 207–219 (2016). https://doi.org/10.5585/remark.v15i2.3117
16. Hirth, J.: Le data marketing : La collecte, l'analyse et l'exploitation des données au coeur u marketing moderne. Eyrolles (2017)
17. Järvinen, J., Tollinen, A., Karjaluoto, H., Jayawardhena, C.: Digital and Social Media Marketing Usage in B2B Industrial Section (2012)
18. Järvinen, et al.: Digital and social media marketing usage in B2B industrial section. Mark. Manage. J. **22**(2), 102–117 (2012). Repéré à https://www.researchgate.net/publication/281117621_Digital_and_social_media_marketing_usage_in_B2B_industrial_section
19. Khan, F., Siddiqui, K.A.: The importance of digital marketing. An exploratory study to find the perception and effectiveness of digital marketing amongst the marketing professionals In Pakistan. J. Inf. Syst. Oper. Manag. **7**, 221–228 (2013)

20. Leeflang, P.S.H., Verhoef, P.C., Dahlström, P., Freundt, T.: Challenges and solutions for marketing in a digital era. Eur. Manag. J. **32**(1), 1–12 (2014). https://doi.org/10.1016/j.emj.2013.12.001
21. Lemoine, J.-F., Badot, O.: La « shazamisation » de l'offre : Modalités, devenir et implications managériales: Management & Avenir, n° **44**(4), 187–196 (2011). https://doi.org/10.3917/mav.044.0187
22. Metais-Wiersch, E., Autissier, D.: La transformation digitale des entreprises: Les bonnes pratiques. Eyrolles (2016)
23. Munshi, A., Munshi, S.: Digital marketing: a new buzz word. Manag. Res. **7** (2012)
24. Munshi, A.: Digital matketing: A new buzz word. Int. J. Bus. Econ. Manage. Res. (2012)
25. Oualidi, H.: Les outils de la communication digitale. Eyrolles (2013)
26. Pai, P., Arnott, D.C.: User adoption of social networking sites: Eliciting uses and gratifications through a means–end approach. Comput. Hum. Behav. **29**(3), 1039–1053 (2013). https://doi.org/10.1016/j.chb.2012.06.025
27. Royle, J., Laing, A.: The digital marketing skills gap: developing a digital marketer model for the communication industries. Int. J. Inf. Manag. **34**(2), 65–73 (2014). https://doi.org/10.1016/j.ijinfomgt.2013.11.008
28. Ryan, D.: The Best Digital Marketing Campaigns in the World II, 2nd edn. Kogan Page (2014)
29. Scheid, F., Vaillant, R., de Montaigu, G.: Le marketing digital: Développer sa stratégie à l'ère numérique. Eyrolles (2012)
30. Sigala, M.: Modelling E-marketing strategies: Internet presence and exploitation of greek hotels. J. Travel Tourism Mark. **11**(2–3), 88–103. https://doi.org/10.1300/J073v11n02_05
31. Stokes, R.: EMarketing: The Essential Guide to Digital Marketing, 4th edn. Quirk eMarketing (2011)
32. Taiminen, H.M., Karjaluoto, H.: The usage of digital marketing channels in SMEs. J. Small Bus. Enterp. Dev. **22**(4), 633–651 (2015). https://doi.org/10.1108/JSBED-05-2013-0073
33. Teo, T.S.H., Swan Tan, J.: Senior executives' perceptions of business-to-consumer (B2C) online marketing strategies: the case of Singapore. Internet Res. **12**(3), 258–275 (2002). https://doi.org/10.1108/10662240210430937
34. Tiago, M.T.P.M.B., Veríssimo, J.M.C.: Digital marketing and social media: why bother? Bus. Horiz. **57**(6), 703–708 (2014). https://doi.org/10.1016/j.bushor.2014.07.002
35. Verhoef, P. C., Beckers, S. F. M., & van Doorn, J.: Understand the Perils of Co-Creation. Harvard Bus. Rev. **91**(9), 28 (2013)
36. Weick, K.E.: The Social Psychology of Organizing, 2nd edn. Random House (1979)
37. Westerman, G., Bonnet, D., McAfee, A.: Leading Digital: Turning Technology into Business Transformation. Harvard Business Review Press (2014)
38. Wymbs, C.: Digital marketing: the time for a new "academic major" has arrived. J. Market. Educ. **33**(1), 93–106 (2011). https://doi.org/10.1177/0273475310392544

Deep Feature-Based Matching of High-Resolution Multitemporal Images Using VGG16 and VGG19 Algorithms

Omaima El Bahi[1](✉), Ali Omari Alaoui[2], Youssef Qaraai[1], and Ahmad El Allaoui[2]

[1] Mathematical and Computer Modelling of Systems Team, Applied Mathematics, Computing and Systems Laboratory, Faculty of Sciences and Techniques Errachidia, Moulay Ismail University of Meknes, Meknes, Morocco
omaima.elbahi@edu.umi.ac.ma, y.qaraai@fste.umi.ac.ma

[2] Decisional Computing and Systems Modelling Team, Engineering Sciences and Techniques Laboratory, Faculty of Sciences and Techniques Errachidia, Moulay Ismail University of Meknes, Meknes, Morocco
al.omarialaoui@edu.umi.ac.ma, a.elallaoui@umi.ac.ma

Abstract. This research focuses on feature-based, high-resolution multi-temporal image matching. The objective is to develop an efficient method for accurately matching features in images captured at different times. The proposed approach relies on the Visual Geometry Group-16 (VGG16) and VGG19 models for feature extraction and calculates Euclidean distances to enable accurate matching. A comparative analysis is carried out with Scale Invariant Feature Transform (SIFT) to evaluate the performance of both models. Experimental results demonstrate the effectiveness of the proposed approach in processing high-resolution multi-temporal imagery, highlighting the advantages of using the VGG19 and VGG16 models in feature-based matching techniques.

Keywords: Feature-Based Matching · VGG16 · VGG19 · High Resolution Images · Feature Extraction · Multi-Temporal Imagery

1 Introduction

High-resolution multi-temporal imagery plays a significant role in several domains, including remote sensing, environmental monitoring, and urban planning. Feature-based matching techniques [1] are algorithms and methods that aim to identify and match distinctive features or keypoints in images, establishing correspondences between them. These techniques, when combined with deep learning models [2], have demonstrated notable potential in achieving precise and robust matching outcomes.

Among convolutional neural network (CNN) models [3], VGG16 [4] and VGG19 [5] have gained attention for their excellent performance in tasks such as image classification and feature extraction. Leveraging the rich hierarchical features learned by these VGG models, our objective is to develop an efficient and robust method for matching features in high-resolution multi-temporal images.

Y. Farhaoui et al. (Eds.): ICAISE 2023, LNNS 837, pp. 516–521, 2024.
https://doi.org/10.1007/978-3-031-48465-0_69

In this study, we propose a feature extraction approach using pre-trained VGG16 and VGG19 models on multi-temporal images. Subsequently, we calculate Euclidean distances between corresponding features for matching. To evaluate the effectiveness of our approach, we conduct a comparative evaluation with the widely used Scale-Invariant Feature Transform (SIFT) [6] method.

The article is structured as follows: Sect. 2 outlines related work in multitemporal image matching. Section 3 presents our deep feature-based matching approach. Section 4 describes our data acquisition and experimental setup. Section 4 discusses our experimental results and compares with existing approaches. Finally, in Sect. 5, we summarize the contributions of our work and highlight future research directions for multi-temporal image matching.

2 Literature Review

The literature on multitemporal image matching includes a variety of approaches for establishing correspondences between images captured at different times. Popular methods include the SIFT algorithm [6, 7], which identifies and describes invariant features, and the Speeded-Up Robust Features (SURF) algorithm [8], known for its efficiency. Techniques based on deep learning, in particular CNNs [3, 9], have also shown promise for feature extraction. Graph-based algorithms, such as the Graph Matching algorithm and Random Sample Consensus (RANSAC) [10, 11], handle geometric transformations and outliers. However, challenges remain in dealing with large-scale feature spaces, viewpoint variations and changes in imaging conditions [12]. Further research is needed to develop more robust and efficient matching techniques for multi-temporal images.

3 Methodology

3.1 Feature Extraction with VGG16 Model

Our process involves using the VGG16 model as shown in Fig. 1, a deep CNN capable of extracting features from pre-processed satellite images. This model has already been trained on a large dataset and has proven its effectiveness in capturing high-level image features. By running pre-processed images through this model, we are able to extract deep discriminative features that are representative of the image content. The VGG16 model has several advantages, such as impressive image classification performance, a simple structure with convolution, pooling and fully connected layers, and deep learning on various images.

3.2 Feature Extraction with VGG19 Model

To extract features using the VGG19 model, shown in Fig. 2, we first import the pre-trained model, which has been trained on large-scale image classification tasks. The weights and architecture of the VGG19 model can be accessed either via a deep learning framework or by loading the pre-trained weights directly. High-resolution multi-temporal images are preprocessed by resizing them to a consistent size and applying the

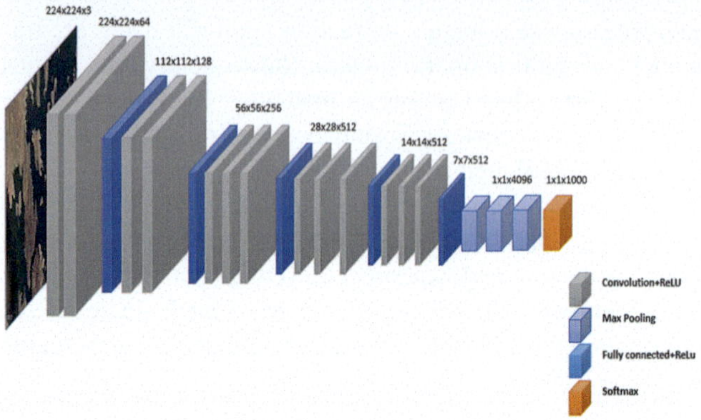

Fig. 1. VGG-16 network architecture for feature extraction.

necessary normalization techniques. These preprocessed images are then run through the VGG19 model to obtain deep features. Features are extracted from several layers of the VGG19 model, capturing different levels of abstraction, from low-level edges to high-level semantic representations. Extracted features are stored as feature vectors for each image, representing different layers of the VGG19 model. This enables a comprehensive representation of images with discriminating and diverse information.

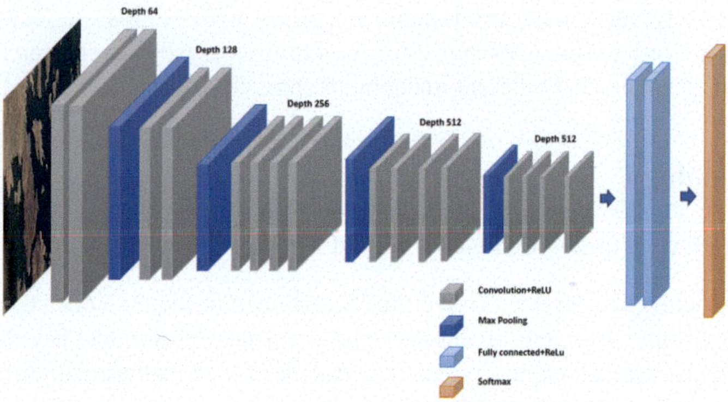

Fig. 2. The architecture of VGG-19 network for feature extraction.

3.3 Feature Matching

In the feature matching step, we calculate the Euclidean distances between the features extracted from the corresponding keypoints in the multi-temporal images. This distance can be computed using the following formula:

$$d = \text{Euclidean} - \text{distance}((\text{Ai}); (\text{Bi})) \tag{1}$$

where d represents the Euclidean distance and Ai and Bi are the corresponding elements of the feature vectors A and B, respectively.

By measuring the distances, we can assess the similarity between features and identify potential matches. Matches are established by selecting key points with the smallest distances, indicating a higher probability of similarity. To guarantee the reliability of matches, we apply a threshold to filter out unreliable matches and eliminate outliers. This threshold serves as a criterion for determining the maximum permissible distance between key points considered to be valid matches. By applying this threshold, we can refine the matches and improve the accuracy and robustness of our feature-based matching approach for multi-temporal images.

4 Experimental Results and Discussion

We used satellite images of agricultural dams. These images were captured from Google Earth, providing high-resolution views of the dams and their surroundings. To achieve image coherence and comparability, we applied a number of pre-processing steps to the acquired images (Fig. 3).

Fig. 3. Example of satellite images of agricultural dams.

The dataset comprised of 9 pairs of images, captured between the years of 2009 and 2023. The selection of images aimed to represent a diverse set of agricultural dams in the region.

For each pair of test images, feature points were extracted and prematched using both methods, VGG16 and VGG19, and the precision was measured to assess the performance and robustness of our feature extraction in obtaining accurate and reliable matching results.

$$\text{Precision} = \text{TP}/(\text{TP} + \text{FP}) \quad (2)$$

where *TP* represents the number of true positive matches and *FP* represents the number of false positive matches. In the context of feature-based matching, precision measures the proportion of correctly matched keypoints out of all the keypoints identified as matches. A higher precision value indicates a higher accuracy and reliability of the matching results.

The numerical results obtained from the test are summarized in Table 1, providing an overview of the precision achieved by our feature matching approach (Fig. 4).

Table 1. Feature matching precision test. Unit: %.

Multi-temporal satellite image	Index	VGG19	VGG16	SIFT
	avg	96.88	94.14	62.74
	max	100.00	100.00	73.73
	min	90.00	87.09	39.32
	std	4.06	5.08	12.09

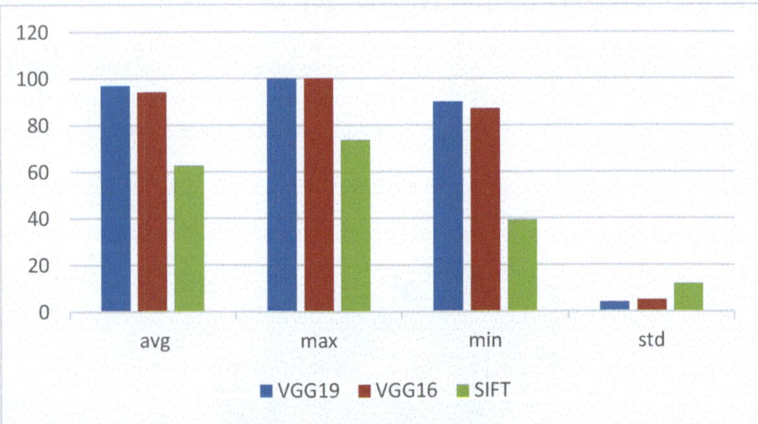

Fig. 4. Comparison of precision statistics.

VGG19 outperformed VGG16 and SIFT in terms of feature matching precision. It generated a higher number of correct correspondences and exhibited a better distribution of feature points across the images. Leveraging the discriminative features from the VGG19 model, our approach achieved superior accuracy and reliability in multi-temporal matching for satellite images.

5 Conclusion

In conclusion, the method we proposed for feature-based matching of high-resolution multitemporal images using the VGG19 model outperformed SIFT and VGG16. It achieved a greater number of exact matches and demonstrated a better distribution of feature points across the images. Based on the discriminating features extracted from the VGG19 model, our approach improved the accuracy and reliability of multi-temporal matching of satellite images. These results validate the effectiveness of our method and its potential for various applications in remote sensing and environmental monitoring.

References

1. Li, J., Hu, Q., Ai, M.: RIFT: multi-modal image matching based on radiation-variation insensitive feature transform. IEEE Trans. Image Process. **29**, 3296–3310 (2020)

2. Shorten, C., Khoshgoftaar, T.M.: A survey on image data augmentation for deep learning. J. Big Data **6** (2019)
3. Kattenborn, T., Leitloff, J., Schiefer, F., Hinz, S.: Review on convolutional neural networks (CNN) in vegetation remote sensing. ISPRS J. Photogramm. Remote Sens. **173**, 24–49 (2021)
4. Mascarenhas, S., Agarwal, M.: A comparison between VGG16, VGG19 and ResNet50 architecture frameworks for image classification (2021)
5. Mateen, M., Wen, J., Nasrullah, Song, S.O., Huang, Z.: Fundus image classification using VGG-19 architecture with PCA and SVD. Symmetry **11**, 1 (2018)
6. Ma, W., et al.: Remote sensing image registration with modified SIFT and enhanced feature matching. IEEE Geosci. Remote Sens. Lett. **14**, 3–7 (2017)
7. Liu, Y.Y., He, M., Wang, Y., Sun, Y., Gao, X.: Farmland aerial images fast-stitching method and application based on improved SIFT algorithm. IEEE Access **10**, 95411–95424 (2022)
8. Wang, R., Shi, Y., Cao, W.: GA-SURF: a new speeded-up robust feature extraction algorithm for multispectral images based on geometric algebra. Pattern Recogn. Lett. **127**, 11–17 (2019)
9. Li, Q., Chen, Y., Zeng, Y.: Transformer with transfer CNN for remote-sensing-image object detection. Remote Sens. **14**, 984 (2022)
10. Jin, Y.-H., Lee, W.-H.: Fast cylinder shape matching using random sample consensus in large scale point cloud. Appl. Sci. **9**, 974 (2019)
11. Li, Y., Gu, C., Dullien, T., Vinyals, O., Kohli, P.: Graph Matching Networks for Learning the Similarity of Graph Structured Objects. Cornell University (2019)
12. Zhu, Z., Qiu, S., Su, Y.: Remote sensing of land change: a multifaceted perspective. Remote Sens. Environ. **282**, 113266 (2022)

Enhancing Solar Power Generation Through Threshold-Based Anomaly Detection in Errachidia, Morocco

Mohamed Khalifa Boutahir[1(✉)], Yousef Farhaoui[1], Benchikh Salma[2], and Mourade Azrour[1]

[1] Engineering Science and Technology Laboratory, Faculty of Sciences and Techniques, IDMS Team, Moulay Ismail University of Meknes, Errachidia, Morocco
moha.boutahir@edu.umi.ac.ma

[2] Advanced Systems Engineering Laboratory, National School of Applied Sciences, Ibn Tofail University, Kenitra, Morocco

Abstract. This research presents an innovative approach to optimize solar power generation in arid regions, focusing on Errachidia, Morocco. The study effectively detects anomalies in solar radiation data by leveraging machine learning techniques such as Isolation Forest (IF), Local Outlier Factor (LOF), and Principal Component Analysis (PCA). These anomalies, driven by dust storms and equipment malfunctions, are classified by severity through threshold-based methods (TBM). The results are promising, with Area Under the Precision-Recall Curve (AUC-PR) scores of 0.8423 for IF, LOF, and PCA. While DBSCAN yields a lower score of 0.0006, the model's overall accuracy is impressive at 99.74%. Robust cross-validation results demonstrate the model's reliability, with a mean accuracy of 99.81%. Notably, feature selection using Random Forest (RF) highlights critical variables, including High Speed (Hi Speed), Outdoor Humidity (Out Hum), Barometric Pressure (Bar), Indoor Humidity (In Hum), Indoor Electromagnetic Compatibility (In EMC), High Solar Radiation (Hi Solar Rad.), Dew Point (Dew Pt), Low Empowerment (Low emp), and Indoor Dew Point (In Dew). The study underscores the potential of machine learning and TBM approaches in solar energy management and offers a replicable framework for regions facing similar climatic challenges.

Keywords: Solar Power Generation · Anomaly Detection · Threshold-Based Classification · Machine Learning

1 Introduction

Integrating solar power into energy systems encounters significant challenges, primarily stemming from the inherent variability and intermittency of solar irradiation [1]. These challenges are particularly pronounced in regions blessed with substantial solar potential but subjected to erratic climatic conditions, necessitating intelligent and adaptive solutions [2].

Y. Farhaoui et al. (Eds.): ICAISE 2023, LNNS 837, pp. 522–530, 2024.
https://doi.org/10.1007/978-3-031-48465-0_70

One such region grappling with these complexities is Errachidia, Morocco, where the juxtaposition of abundant sunlight and unpredictable climate anomalies presents a unique set of obstacles to effective solar energy generation [3]. These anomalies can destabilize renewable energy systems and impair performance [4]. While machine learning techniques have shown promise in solar forecasting, the field of advanced anomaly detection, tailored to the specific climate patterns of Errachidia, is still evolving [5, 6].

The primary goal of this research is to enhance solar power generation in Errachidia by developing climate-adaptive anomaly detection and classification methods. Our approach leverages advanced machine learning techniques, including IF, LOF, and PCA, to analyze historical meteorological data [2, 7]. A key innovation is creating data-driven thresholds finely tuned to Errachidia's unique weather conditions [8, 9]. These thresholds, calibrated to the region's specific solar patterns, outperform static generic models, ensuring that the anomaly detection system is finely attuned to the local context [6, 10].

The subsequent sections of this paper are organized as follows: Sect. 2 provides a comprehensive review of the relevant literature, Sect. 3 elaborates on the methodologies employed, Sect. 4 scrutinizes the obtained results, and Sect. 5 provides the final insights and closing remarks of this study.

2 Related Works

In recent research endeavors, Sarmas et al. [11] devised a meta-learning approach for enhancing one-hour-ahead deterministic forecasts of photovoltaic (PV) systems. They dynamically blended base forecasts from multiple deep learning (DL) models to adapt to different weather conditions and PV system intricacies. This method demonstrated up to 5% accuracy improvements compared to individual base models, showcasing its efficacy in increasing forecast precision. Kaur et al. [12] introduced a novel probabilistic scheme for solar power generation forecasting, addressing data and model parameter uncertainties. They utilized Bayesian bidirectional long short-term memory (BiLSTM) neural networks and variational auto-encoders (VAE) to handle high dimensionality in weight parameters. This approach significantly reduced weight parameters and computational time, leading to remarkable enhancements in forecasting accuracy. Dey et al. [13] tackled the challenge of solar farm voltage anomaly detection using high-resolution μPMU data-driven unsupervised machine learning. They harnessed the Clustering Large Applications (CLARA) algorithm to categorize events, focusing on recognizing solar site-specific behavior patterns and abnormal voltage fluctuations. Their approach demonstrated automatic voltage anomaly detection by processing extensive datasets collected from utility-scale solar power generation sites in England, promising insights for improved solar farm performance and grid behavior understanding. These studies collectively contribute to advancing solar energy forecasting, anomaly detection, and system performance optimization.

In the context of Morocco's scientific contributions, Moulad et al. [14] address security concerns in Wireless Sensor Networks (WSN). These networks play a pivotal role in various applications, including agriculture, military, and hospitality management, yet are susceptible to security threats like Denial Of Service (DOS) attacks. To fortify WSNs, the authors propose a hierarchical hybrid intrusion detection mechanism. This approach, combining anomaly detection with support vector machines (SVM), specifications-based techniques, and clustering algorithms, enhances security without disrupting network performance, particularly in energy management. Boutahir et al. [15] delve into anomaly detection within solar power plants. As solar energy becomes increasingly integral to modern life, the prompt identification of performance issues and equipment anomalies becomes crucial. Leveraging machine learning, the authors introduce the AutoEncoder Long Short-Term Memory (AE-LSTM) method, enhanced with a Genetic Algorithm (GA) for hyperparameter tuning. This AE-LSTM-GA approach outperforms traditional methods like the Isolation Forest (IF) and the Local Outlier Factor (LOF) in anomaly detection in power plants. Belcaid et al. [16] tackle the challenge of change point detection in piecewise constant signals, a problem applicable to various domains, including human activity analysis, speech, image analysis, and genetics anomaly detection. They present an innovative online change point detection algorithm, utilizing a local blanket of a global Markov Random Field (MRF) to represent signals and noisy observations. The algorithm combines local energy minimization and gradient processing to identify change points effectively, demonstrating notable improvements in running time and detection metrics compared to existing methods.

3 Data Source

In this study, we sourced data from the Faculty of Sciences and Techniques meteorological station in Errachidia, Morocco. This meteorological facility, illustrated in Fig. 1, constitutes a crucial component of the PROPRE.MA project, an initiative led by the University of Cadi Ayyad and funded by the Research Institute for Solar Energy and Renewable Energies (IRESEN) [17]. Operating around the clock, seven days a week, this station diligently records various climatic parameters at semi-hourly intervals. These parameters encompass solar radiation, temperature, humidity, atmospheric pressure, rainfall, and various other meteorological characteristics. The overarching objective of the PROPRE.MA project is to construct photovoltaic yield maps specific to Morocco. This endeavor leverages a ground calibration strategy involving 20 identical installations strategically positioned across several urban centers within the country [18, 19].

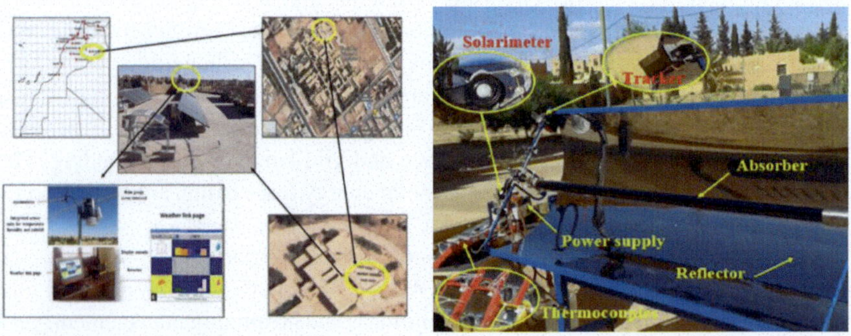

Fig. 1. The different equipment and sensors of the meteorological station of the Faculty of Sciences and Technologies of Errachidia.

For our research, we harnessed data spanning the years 2018 and 2019, constituting a total of 33,675 observations. These observations encompass 28 distinct features. The data include critical variables for solar energy generation, such as temperature, humidity, wind speed, solar radiation, etc. Notably, these variables are recorded hourly and are paramount in predicting solar radiation levels.

4 Methodology

In this section, we detail the comprehensive methodology employed to optimize solar power generation in Errachidia, Morocco. The methodology is systematically divided into distinct phases, each contributing to the overarching goal of enhancing solar energy efficiency in arid regions. The research approach encompasses data collection and preprocessing, feature selection, advanced anomaly detection methods, threshold definition, quality level classification, and model evaluation. Each step is designed to fortify solar energy generation against climatic challenges, ensuring a robust and reliable framework for regions facing similar environmental conditions.

4.1 Data Preprocessing and Feature Selection

The initial phase of our methodology involves the meticulous collection and preprocessing of meteorological data from the Errachidia meteorological station, spanning the years 2018 and 2019. This raw data undergoes rigorous cleaning to address missing values and outliers, ensuring data reliability. Feature engineering techniques are then thoughtfully applied to extract valuable insights from the dataset. To harmonize the dataset and mitigate discrepancies in variable influence, feature scaling is systematically implemented.

In the subsequent phase, feature selection plays a pivotal role in refining the dataset for solar irradiance prediction, utilizing the robustness and well-documented performance of Random Forest. This choice is substantiated by its effectiveness in previous works [20] in the same data location and with the same features. We assess feature importance to identify the most relevant variables for our solar irradiance prediction task. The dataset is then streamlined to retain only those variables that hold significant sway in improving the accuracy of solar power generation forecasts in Errachidia, Morocco.

4.2 Anomaly Detection

Three advanced anomaly detection methods—IF, LOF, and PCA—are introduced in this phase. These methods are selected for their proficiency in handling high-dimensional data and their adaptability to the unique characteristics of solar energy generation patterns. IF efficiently identifies anomalies [21], LOF measures density deviation [22], and PCA uncovers patterns based on reconstruction errors [23].

4.3 Threshold Definition and Quality Level Classification

Threshold definition entails the derivation of climate-specific thresholds for meteorological parameters, including solar radiation, temperature, and humidity, meticulously tailored to Errachidia's unique climate patterns. This process relies on a comprehensive statistical analysis of historical meteorological data, ensuring that the thresholds accurately capture local environmental conditions.

Quality level classification follows, integrating threshold-based classification with machine learning techniques to categorize solar energy quality into distinct levels, such as "Excellent," "Good," "Moderate," or "Poor." This combined approach enhances the accuracy and granularity of quality level predictions. It takes into account the severity and frequency of detected anomalies, allowing us to provide a comprehensive assessment of solar energy quality in Errachidia, which is crucial for effective solar power generation management. The specific criteria used for classifying solar energy quality into these levels are based on a combination of factors, including the magnitude and frequency of anomalies, the impact of anomalies on energy generation, and their historical occurrence patterns in Errachidia's climate. These criteria enable us to make nuanced and data-driven quality level predictions, providing valuable insights for optimizing solar power generation.

Here's the Gaussian distribution function (F(x)) utilized for threshold definition, presented in its mathematical form:

$$F(x) = \frac{1}{\sigma * \sqrt{(2\pi)}} * e^{-\frac{-((x-\mu)^2)}{2\sigma^2}} \tag{1}$$

(1) This statistical function, grounded in F(x), underpins the establishment of climate-specific thresholds, ensuring precise anomaly identification and the effective enhancement of solar power generation in Errachidia, Morocco.

5 Results

Table 1 presents a comprehensive overview of the AUC-PR scores obtained from the evaluation of four distinct anomaly detection methods - Isolation Forest, Local Outlier Factor (LOF), Principal Component Analysis (PCA), and DBSCAN. These scores serve as a quantitative measure of the methods' efficiency in identifying and classifying anomalies in solar energy generation data from Errachidia, Morocco. Isolation Forest, LOF, and PCA stand out with AUC-PR scores of 0.8423, reflecting their prowess in capturing anomalies within varying data densities and high-dimensional spaces. Notably, these three methods exhibited comparable performance, underscoring their adaptability to complex solar energy patterns. In contrast, DBSCAN obtained a considerably lower AUC-PR score of 0.0006, indicating its limitations in aligning with the dataset's distribution of anomalies. Despite this discrepancy, DBSCAN's unique approach may find applicability in specific datasets characterized by distinct anomaly patterns. Together, these AUC-PR scores provide valuable insights into the relative strengths and weaknesses of each anomaly detection method, aiding in the selection of the most suitable approach for optimizing solar power generation in arid regions.

Table 1. AUC-PR scores for anomaly detection methods

Method	AUC-PR score
Isolation forest	0.8423281
Local outlier factor (LOF)	0.8423283
Principal component analysis (PCA)	0.8423284
DBSCAN	0.0005910

The threshold definition phase played a pivotal role in our approach to enhancing solar power generation in Errachidia, Morocco. Through a statistical analysis method tailored to Errachidia's unique climate patterns, we established climate-adaptive thresholds. These thresholds provide a robust foundation for identifying deviations from standard solar energy generation patterns in Errachidia.

The application of an ensemble model significantly enhanced anomaly detection accuracy, yielding an impressive accuracy score of 0.9974. This result underscores the effectiveness of ensemble modeling in improving the identification of anomalies in solar energy generation data.

Before delving into the details of the confusion matrix, it's essential to highlight the class distribution within our dataset. The solar energy quality levels were categorized into four classes: "Excellent," "Good," "Moderate," and "Poor." The distribution is as presented in Fig. 2(a).

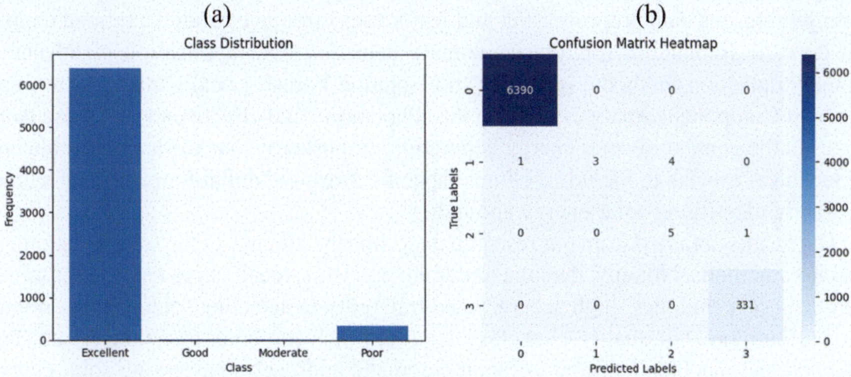

Fig. 2. The results of class distribution and Confusion Matrix for the model

The distribution clearly indicates a significant class imbalance, with the majority of instances falling into the "Excellent" category. This imbalance underscores the challenge of accurately classifying the rarer "Good," "Moderate," and "Poor" quality levels.

Figure 2(b) illustrates the confusion matrix for the ensemble model. This matrix offers valuable insights into the distribution of true positives, true negatives, false positives, and false negatives, facilitating a comprehensive evaluation of model performance.

The confusion matrix, a fundamental tool for assessing classification models, provides insights into the model's ability to correctly classify solar energy quality into different levels. In our case, the confusion matrix showcases an impressive level of accuracy and reliability. With a vast majority of values along the diagonal, indicating true positives and true negatives, our model showcases robust performance.

Specifically, out of 6735 instances, 6390 instances were correctly classified as "Excellent," representing pristine solar energy quality. Additionally, there were no false negatives, signifying that the model effectively identified all "Excellent" solar energy quality instances.

Furthermore, the model demonstrated its precision in identifying solar energy quality levels in other categories. It correctly classified all "Good" and "Moderate" quality levels with three and five instances, respectively. Additionally, in the "Poor" quality level category, the model's precision was exceptionally high, correctly classifying 330 instances.

This high level of precision across different quality levels underscores the methodology's reliability and effectiveness in managing solar power generation in Errachidia's variable climate conditions. Our methodology, combining climate-adaptive threshold definitions and advanced machine learning techniques, holds great promise for regions facing similar challenges in optimizing solar energy generation.

6 Conclusion

In conclusion, this research contributes novel insights and practical solutions to enhance solar power generation in Errachidia, Morocco, amid its challenging climatic conditions. Through rigorous data preprocessing and feature selection, we curated a refined dataset that lays the foundation for robust anomaly detection. Our application of advanced anomaly detection methods, including the Isolation Forest, Local Outlier Factor, and Principal Component Analysis, reflects the adaptability and effectiveness of these techniques in the context of solar energy generation. Importantly, our work extends beyond the technical aspects to include a climate-specific threshold definition, a critical step in accurately identifying solar energy anomalies.

The results obtained from our research underline the efficacy of the selected anomaly detection methods. Notably, the area under the precision-recall curve (AUC-PR) scores affirms these techniques' high accuracy and reliability in detecting solar energy anomalies. Furthermore, our quality level classification, which combines threshold-based classification and machine learning, offers a pragmatic approach to assessing solar energy quality. The delineation of solar quality into distinct levels, such as "Excellent," "Good," "Moderate," or "Poor," provides energy managers and authorities with actionable insights into system performance.

As we look toward the future, several avenues of research beckon. The integration of real-time data acquisition and monitoring, coupled with the development of adaptive anomaly detection models, holds the promise of further improving solar energy generation in Errachidia. Additionally, expanding this work to include predictive maintenance and energy forecasting can enhance solar energy systems' overall resilience and sustainability. Overall, this study lays a foundation for ongoing efforts to harness the potential of solar power generation in challenging climatic regions, contributing to a greener and more sustainable energy future.

References

1. Sobri, S., et al.: Solar photovoltaic generation forecasting methods: a review. Energy Convers. Manage. **156**, 398–411 (2020)
2. Luna, A.S., et al.: Solar forecasting methods applied to the prediction of photovoltaic power production: A review. Renew. Sustain. Energy Rev. **153**, 111660 (2022)
3. Rezende, L.S.M., et al.: Anomaly detection in solar power generation: a systematic literature review. Renew. Sustain. Energy Rev. **151**, 111564 (2021)
4. Zameer, A., et al.: Intelligent and robust prediction of photovoltaic power: a review. IEEE Access **8**, 128356–128371 (2020)
5. Luna, A.S., et al.: Solar power forecasting based on machine learning and ephemeris for blueprints of photovoltaic plants. Electronics **10**(3), 305 (2021)
6. Sobri, S., Koohi-Kamali, S., Rahim, N.A.: Solar photovoltaic generation forecasting methods: a review. Energy Convers. Manage. **156**, 398–411 (2020)
7. Voyant, C., et al.: Machine learning for solar radiation forecasting: a review. Renewable Energy **105**, 569–582 (2017)
8. Antonanzas, J., Osorio, N., Escobar, R., Urraca, R., Martinez-de-Pison, F.J., Antonanzas, F.: Review of photovoltaic power forecasting. Sol. Energy **136**, 78–111 (2016)
9. Voyant, C., Randimbivololona, P., Nivet, M.L., Paoli, C., Muselli, M.: Twenty four hours ahead global irradiation forecasting using multi-model approach: application in Reunion Island. Renewable Energy **118**, 870–880 (2018)
10. Rezende, L.S.M., Lyra, C., Leite, W.N., Batista, G.P., Silva, I.N.: Anomaly detection in solar power generation: A systematic literature review. Renew. Sustain. Energy Rev. **151**, 111564 (2021)
11. Sarmas, E., Spiliotis, E., Stamatopoulos, E., Marinakis, V., Doukas, H.: Short-term photovoltaic power forecasting using meta-learning and numerical weather prediction independent Long Short-Term Memory models. Renewable Energy **216**, 118997 (2023). https://doi.org/10.1016/j.renene.2023.118997
12. Kaur, D., Islam, S.N., Mahmud, M.A., Haque, M.E., Anwar, A.: A VAE-Bayesian deep learning scheme for solar power generation forecasting based on dimensionality reduction. Energy and AI **14**, 100279 (2023). https://doi.org/10.1016/j.egyai.2023.100279
13. Dey, M., Rana, S.P., Simmons, C.V., Dudley, S.: Solar farm voltage anomaly detection using high-resolution μPMU data-driven unsupervised machine learning. Appl. Energy **303**, 117656 (2021). https://doi.org/10.1016/j.apenergy.2021.117656
14. Moulad, L., Belhadaoui, H., Rifi, M.: Implementation of an hierarchical hybrid intrusion detection mechanism in wireless sensor network based on energy management. In: Mizera-Pietraszko, J., Pichappan, P., Mohamed, L. (eds.) Lecture Notes in Real-Time Intelligent Systems. RTIS 2017. Advances in Intelligent Systems and Computing, vol. 756. Springer, Cham (2019). https://doi.org/10.1007/978-3-319-91337-7_33
15. Boutahir, M.K., Farhaoui, Y., Azrour, M.: Towards an effective anomaly detection in solar power plants using the AE-LSTM-GA approach. In: Farhaoui, Y., Rocha, A., Brahmia, Z., Bhushab, B. (eds) Artificial intelligence and smart environment. ICAISE 2022. Lecture Notes in Networks and Systems, vol. 635. Springer, Cham (2023). https://doi.org/10.1007/978-3-031-26254-8_115

16. Belcaid, A., Douimi, M.: A novel online change point detection using an approximate random blanket and the line process energy. Int. J. Artif. Intell. Tools **29**(06), 2050018 (2020). https://doi.org/10.1142/S0218213020500189
17. Halimi, M., Outana, I., El Amrani, A., Diouri, J., Messaoudi, C.: Prediction of captured solar energy for different orientations and tracking modes of a PTC system: technical feasibility study (Case study: South eastern of Morocco). Energy Convers. Manag. **167**, 21e36 (2018)
18. Hessane, A., El Youssefi, A., Farhaoui, Y., Aghoutane, B., Amounas, F.: A machine learning based framework for a stage-wise classification of date palm white scale disease. Big Data Mining Anal. **6**(3), 263–272 (2023). https://doi.org/10.26599/BDMA.2022.9020022
19. Mohamed, K., Abouzid, H., Teidj, S. (2021). Prédiction de Rayonnement Solaire Global (RSG) : Par les Réseaux de Neurones Artificiels Cas d'étude : la ville d'Er-Rachidia, Maroc
20. Boutahir, M.K., Farhaoui, Y., Azrour, M., Zeroual, I., El Allaoui, A.: Effect of feature selection on the prediction of direct normal irradiance. Big Data Mining Anal. **5**(4), 309–317 (2022). https://doi.org/10.26599/BDMA.2022.9020003
21. Liu, F.T., Ting, K.M., Zhou, Z-H.: Isolation forest. In: Proceedings of the 2008 Eighth IEEE International Conference on Data Mining, pp. 413–422 (2008)
22. Breunig, M.M., Kriegel, H.-P., Ng, Raymond, T., Sander, J.: LOF: identifying density-based local outliers. In Proceedings of the 2000 ACM SIGMOD International Conference on Management of Data, pp. 93–104 (2000)
23. Jolliffe, I.T.: Principal component analysis. In: Principal Component Analysis, pp. 1–19. Springer (2014)

Comparative Review: Leadership Styles in the Context of Smart Environments

Mitra Madanchian[1,2(✉)], Hamed Taherdoost[1,3,4], and Nachaat Mohamed[5]

[1] University Canada West, Vancouver, BC, Canada
mitra.madanchian@gmail.com
[2] Fairleigh Dickinson University, Vancouver, BC, Canada
[3] Q Minded, Quark Minded Technology Inc, Vancouver, BC, Canada
[4] Hamta Business Corporation, Vancouver, BC, Canada
[5] Rabdan Academy, HLS, Abu Dhabi, UAE

Abstract. The fusion of leadership and technology has ushered in a new era of possibilities in our quickly developing digital age. The relevance of leadership within these settings has increased as smart surroundings become more interwoven into our daily lives, posing both difficulties and possibilities. The goal of this mini-review is to analyze the leadership philosophies that work best in smart workplaces and to consider how machine learning may be used to improve leadership skills. The distinctive characteristics and advantages of diverse leadership styles will be revealed as the investigation goes further into the complexities of leadership within smart settings. This analysis evaluates leadership styles in smart environments, emphasizing adaptability, sustainability, and emerging trends, while also stressing inclusivity and diversity, providing valuable insights for leaders and organizations in this dynamic landscape. This analysis of leadership styles in smart environments offers practical insights and guidance for leaders, enhancing their adaptability, sustainability, and alignment with emerging trends.

Keywords: Leadership · Smart Environment · Smart Workplace

1 Introduction

Smart environments are physical settings that have been improved using cutting-edge technology and integrated systems. They are also known as intelligent or responsive places. Their main objective is to improve the inhabitants' level of convenience, efficiency, and quality of life in general [1]. Smart environments use sensors and data tools for real-time monitoring of temperature, lighting, occupancy, and user activity. They facilitate device communication through the Internet of Things (IoT), and employ data analytics and machine learning to uncover insights [2].

Smart settings are distinguished by their ability to automate and regulate. Based on real-time data and user preferences, these spaces may independently control a variety of technologies, including security, heating, and cooling, to maximize energy consumption and improve comfort [3]. They have a strong emphasis on user-centric design, providing individualized experiences and flexibility to changing circumstances. From smart homes

Y. Farhaoui et al. (Eds.): ICAISE 2023, LNNS 837, pp. 531–536, 2024.
https://doi.org/10.1007/978-3-031-48465-0_71

[4, 5] and businesses [6] to healthcare facilities [7, 8] and smart cities [9, 10], smart environments find uses in a variety of contexts. They play a crucial role in the continual transformation of urban environments and the incorporation of technology into everyday life, promoting sustainability and well-being.

Leadership in the field of smart environments serves as the driving force for successful installation and operation, with visionary leaders providing the rationale and plan for integrating cutting-edge technology in alignment with the firm's aims and objectives [11, 12]. They manage difficult choices, such as budget allocation and technology selection, while also juggling the significant adjustments that smart system adoption entails, such as changed corporate cultures and changing processes [13, 14]. Leaders prioritize user-centric design, cybersecurity, data privacy, eco-friendly practices, innovation, stakeholder inclusion, and governance frameworks to create smart environments that enhance productivity, foster sustainable community growth, and improve the quality of life while staying aligned with technological trends [15, 16].

The subject of leadership styles within smart environments has been extensively explored in various publications and papers. Cortellazzo et al. [15, 17] conducted a review of research in this area, categorizing it into macro and micro-level analyses. They highlight how leadership is evolving in the digital age. Elkhwesky et al. [18] reviewed leadership trajectories in the modern hospitality industry, summarizing findings from 79 publications over 13 years. They identified theories and research gaps, providing a valuable foundation for future studies, especially in industries like hospitality, which are transforming due to technology integration.

To evaluate and compare leadership styles in intelligent settings, this comparative evaluation was conducted. It offers insights that may be used to improve practices, support sustainability, and influence leadership choices. It eliminates the gap between leadership and technology, promotes ongoing development, and foresees emerging trends, giving leaders and organizations functioning in these dynamic, technologically advanced environments invaluable direction.

2 Leadership Styles

There are many different types of leadership, each with its distinctive traits and adaptability for various circumstances (Fig. 1). For instance, transformational leadership thrives on creativity and innovation, inspiring people to pursue challenging objectives. In contrast, servant leadership places a higher priority on team members' welfare and promotes trust [19]. Transactional leadership focuses on unambiguous expectations and incentives, while distributed leadership promotes cooperation and shared decision-making. While autocratic leadership and laissez-faire leadership reflect more authoritative and hands-off styles, respectively, adaptive leadership excels at negotiating uncertainty. Transactional leadership depends on contingencies and remedial measures, but charismatic leaders utilize their charm and vision to inspire.

A combination of these styles, properly adapted to the requirements of the business and the current circumstance, is often the key to effective leadership. Effective leaders are flexible and able to use this wide range of leadership styles, knowing when to

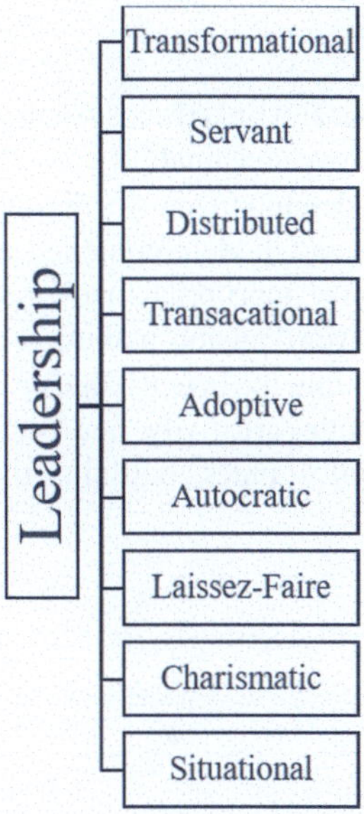

Fig. 1. Key leadership styles in brief

motivate, when to delegate, and when to provide precise instructions [20]. At its foundation, leadership is a dynamic discipline that adapts to the changing problems of the contemporary world rather than being a one-size-fits-all undertaking.

Leaders need to first take into account the culture of the company they lead. Transformational or laissez-faire leadership may prosper in creative contexts, encouraging innovation and independence [21]. In contrast, situations that are more conventional or risk-averse could need transactional or structured leadership [22]. The team's makeup is important as well; motivated and experienced teams may benefit from a hands-off approach, while less experienced teams may need more direct supervision. Crisis circumstances call for adaptive leadership [23], where adaptability and speedy decision-making are crucial, but they often include a combination of leadership philosophies. Leadership is influenced by the nature of the job, whether it is project-based or continuing operations. Coaching or servant leadership may be necessary for employee development and progress [24]. A diverse and inclusive leadership style is necessary. Transformational leadership may be more prevalent in sectors driven by technology [25]. Leadership decisions are also influenced by market circumstances and worldwide operations. Effective leaders are those who can successfully traverse different environments, adapting their leadership philosophies as necessary, and demonstrating the adaptability to succeed in a variety of circumstances.

In smart settings, leadership plays a diverse and crucial role. It involves setting a clear vision, embracing cutting-edge technology, and promoting data-driven decision-making. Leaders must also prioritize cybersecurity, ethical standards, stakeholder inclusion, and adaptability to rapid changes. They should support sustainable development, oversee smart systems, and ensure resilience and disaster recovery measures [26].

3 Comparative Analysis

Smart environments are a dynamic and linked ecosystem where human interactions, data, and technology combine to improve sustainability, accelerate innovation, and optimize operations [27, 28]. Effective leadership in today's digital environments requires a blend of traditional leadership theories and modern context. Leaders must understand various leadership styles, choosing the most appropriate one for each situation, as no single style fits all. Flexibility and context-driven decision-making are key [29]. Leaders should adapt their leadership styles, understanding the benefits and drawbacks, to positively influence company culture, employee morale, creativity, and performance. Table 1 summarizes the core of many leadership philosophies, outlining their distinctive advantages and disadvantages.

Table 1. Summary of leadership styles: strengths and weaknesses

Leadership style	Strengths	Weaknesses
Transformational	• Inspires and motivates • Encourages innovation	• Overdependence on leader charisma • Lack of operational focus
Servant	• Prioritizes follower needs • Promotes collaboration	• Perceived as idealistic • Slower decision-making
Distributed	• Collective decision-making • Adapts to complexity	• Role ambiguity • Coordination challenges
Transactional	• Clarity and accountability • Effective for short-term goals	• Stifles creativity and innovation • Limited long-term effectiveness
Adaptive	• Agility and resilience • Fosters a learning culture	• Resistance to change • Requires high adaptability
Autocratic	• Efficiency and quick decisions • Clear direction	• Low morale • Suppression of creativity
Laissez-Faire	• Empowerment and autonomy • Encourages creativity	• Lack of guidance • Potential for chaos
Charismatic	• Inspires and motivates • Visionary	• Overreliance on charisma • Potential manipulation
Situational	• Flexibility and adaptability • Effective in diverse situations	• Complexity • Time-consuming

4 Conclusion and Future Trends

The analysis of leadership styles in smart environments shows that there's no one-size-fits-all solution for leadership in these cutting-edge settings. Various leadership philosophies, like transformational and servant leadership, are important. Leaders must adapt to changing circumstances, foster creativity, teamwork, and sustainable success. The future of leadership in smart environments will be shaped by integrating AI into leadership practices, focusing on sustainability, dealing with remote leadership, addressing ethical concerns in technology-driven settings, and promoting inclusivity and diversity. Leaders who embrace these trends and emphasize adaptability, sustainability, and collaboration will position their organizations for success in the rapidly evolving world of smart environments.

References

1. Wolter, D., Kirsch, A.: Smart environments: what is it and why should we care? KI-Künstliche Intelligenz **31**, 231–237 (2017)
2. García-Tudela, P.A., Prendes-Espinosa, P., Solano-Fernández, I.M.: Smart learning environments: a basic research towards the definition of a practical model. Smart Learning Environ. **8**(1), 1–21 (2021)
3. Vattano, S.: Smart buildings for a sustainable development. J. Econ. World **2**, 310–324 (2014)
4. Alirezaie, M., et al.: An ontology-based context-aware system for smart homes: E-care@ home. Sensors **17**(7), 1586 (2017)
5. Farhaoui, Y.: Design and implementation of an intrusion prevention system. Int. J. Netw. Secur. **19**(5), 675–683 (2017). https://doi.org/10.6633/IJNS.201709.19(5).04
6. Mannweiler, C., Simoes, J., Moltchanov, B.: Context-aware smart environments enabling new business models and services. In: 2010 ITU-T Kaleidoscope: Beyond the Internet?—Innovations for Future Networks and Services. IEEE (2010)
7. Moore, P., Pham, H.V.: A fog computing model for pervasive connected healthcare in smart environments. Int. J. Grid Util. Comput. **10**(4), 375–391 (2019)
8. Farhaoui, Y., et al.: Big Data Mining Anal. **6**(3), I–II (2023). https://doi.org/10.26599/BDMA.2022.9020045
9. McKenna, H.P.: Human-smart environment interactions in smart cities: exploring dimensionalities of smartness. Future Internet **12**(5), 79 (2020)
10. Farhaoui, Y.: Intrusion prevention system inspired immune systems. Indonesian J. Electr. Eng. Comput. Sci. **2**(1), 168–179 (2016)
11. Iannotta, M., Meret, C., Marchetti, G.: Defining leadership in smart working contexts: a concept synthesis. Front. Psychol. **11**, 556933 (2020)
12. Farhaoui, Y.: Big data analytics applied for control systems. Lect. Notes Netw. Syst. **25**, 408–415 (2018). https://doi.org/10.1007/978-3-319-69137-4_36
13. Nadkarni, S., Prügl, R.: Digital transformation: a review, synthesis and opportunities for future research. Manage. Rev. Quarterly **71**, 233–341 (2021)
14. Alaoui, S.S., Farhaoui, Y.:Data openness for efficient e-governance in the age of big data. Int. J. Cloud Comput. **10**(5–6), 522–532 (2021). https://doi.org/10.1504/IJCC.2021.120391
15. Cortellazzo, L., Bruni, E., Zampieri, R.: The role of leadership in a digitalized world: a review. Front. Psychol. **10**, 1938 (2019)
16. Farhaoui, Y. et al.: Big Data Mining Anal. **5**(4), I–II (2022). https://doi.org/10.26599/BDMA.2022.9020004
17. Alaoui, S.S., Farhaoui, Y.: Hate speech detection using text mining and machine learning. Int. J. Decis. Support System Technol. **14**(1), 80 (2022). https://doi.org/10.4018/IJDSST.286680
18. Elkhwesky, Elbayoumi Salem, Z.I., Ramkissoon, H., Castañeda-García, J.: A systematic and critical review of leadership styles in contemporary hospitality: a roadmap and a call for future research. Int. J. Contemporary Hospitality Manage. **34**(5), 1925–1958 (2022)
19. Dastane, D.O.: Impact of leadership styles on employee performance: a moderating role of gender. Australian J. Bus. Manage. Res. **5**(12), 27–52 (2020)

20. Danquah, V.B.: Analysis of leadership style in organizations: a case study of the movie disclosure. In: Analysis of Leadership Style in Organizations: A Case Study of the Movie Disclosure: Danquah, Victor Birikorang (2021) [Sl]: SSRN
21. Rosing, K.: Transformational leadership and follower creativity: A review of underlying mechanisms and boundary conditions. In: Handbook of Research on Leadership and Creativity, pp. 249–272 (2017)
22. Kark, R., Van Dijk, D., Vashdi, D.R.: Motivated or demotivated to be creative: the role of self-regulatory focus in transformational and transactional leadership processes. Appl. Psychol. **67**(1), 186–224 (2018)
23. Sunderman, H.M., Headrick, J., McCain, K.: Addressing complex issues and crises in higher education with an adaptive leadership framework. Change: Magazine Higher Learn. **52**(6), 22–29
24. Chiniara, M., Bentein, K.: Linking servant leadership to individual performance: Differentiating the mediating role of autonomy, competence and relatedness need satisfaction. Leadersh. Q. **27**(1), 124–141 (2016)
25. Gumusluoğlu, L., Ilsev, A.: Transformational leadership and organizational innovation: the roles of internal and external support for innovation. J. Prod. Innov. Manag. **26**(3), 264–277 (2009)
26. Huang, Z., et al.: The role of leadership in collective creativity and innovation: examining academic research and development environments. Front. Psychol. **13**, 1060412 (2022)
27. Vermesan, O., Friess, P.: Internet of Things: converging technologies for smart environments and integrated ecosystems. River Publishers (2013)
28. Appio, F.P., Lima, M., Paroutis, S.: Understanding smart cities: innovation ecosystems, technological advancements, and societal challenges. Technol. Forecast. Soc. Chang. **142**, 1–14 (2019)
29. Aheleroff, S., et al.: Mass personalisation as a service in industry 4.0: a resilient response case study. Adv. Eng. Inf. **50**, 101438 (2021)

Leading with Intelligence: Harnessing Machine Learning for Effective Leadership

Mitra Madancian[1,2(✉)], Hamed Taherdoost[1], Nachaat Mohamed[3], and Alaeddin Kalantari[4]

[1] University Canada West, Vancouver, BC, Canada
mitra.madanchian@gmail.com
[2] Fairleigh Dickinson University, Vancouver, BC, Canada
[3] Rabdan Academy, HLS, Abu Dhabi, UAE
[4] Hamta Business Corporation, Vancouver, BC, Canada

Abstract. The nexus of leadership and machine learning offers enormous promise for changing how we understand and apply leadership. The purpose of this review is to investigate this complex connection. Leaders may improve their decision-making processes, foresee trends, and optimize tactics with unmatched accuracy by using the power of machine learning. This study highlights the relevance of this convergence and shows how it might enable leaders to handle difficult situations and promote creativity. In the parts that follow, we will go into practical examples, examine machine learning algorithms, solve difficulties, talk about ethical issues, and speculate about the future of leadership in the machine learning era.

Keywords: Intelligence · Leadership · Machine learning

1 Introduction

A fascinating confluence of technology and human decision-making may be seen in the interaction between leadership and machine learning [1]. Fundamentally, this connection represents the use of machine learning methods and data-driven insights to improve leadership practices and promote more sensible decision-making inside enterprises [2].

Leadership has traditionally guided teams and organizations through experience, intuition, and human judgment. However, as data grows increasingly numerous and complicated, machine learning offers leaders a potent tool to filter, examine, and glean insightful information from this data [3, 4]. This may therefore result in more strategic and informed decision-making, which will eventually improve the performance and competitiveness of the business. Machine learning algorithms are capable of processing enormous volumes of data, finding patterns, and making rapid and accurate predictions of future trends or events [5]. This may include better resource allocation, risk management, and forecasting in the context of leadership.

Additionally, machine learning may help executives get a greater knowledge of worker behavior, consumer preferences, and market dynamics [6]. Leaders may modify their plans and tactics to better meet the requirements and expectations of their teams and

Y. Farhaoui et al. (Eds.): ICAISE 2023, LNNS 837, pp. 537–542, 2024.
https://doi.org/10.1007/978-3-031-48465-0_72

stakeholders by using data-driven insights. Additionally, by properly tying goods and services to consumer demands, executives may encourage customer-centric methods as a result of this inquiry. Tackling prejudice and transparency issues in algorithms also encourages moral and inclusive leadership behaviors. Additionally, adopting machine learning demonstrates a commitment to ongoing learning and flexibility, two traits that are crucial in a world that is always changing and driven by technology.

The majority of earlier articles on the topic of leadership and machine learning concentrate on either one particular leadership area or one particular feature of machine learning. For instance, Audibert et al. [7] examined the impact, influence, and leadership of academics throughout the previous decades to give an overview of the development of AI and machine learning across time. Sarker [8] gave an in-depth analysis of several machine learning algorithms and how they may be used in various contexts.

This brief review aims to foster discussion and understanding among leaders and decision-makers in our tech-driven era. It explores the connection between leadership and machine learning, highlighting how machine learning can enhance leadership by enabling data-driven decision-making, fostering innovation, and addressing contemporary challenges. The paper delves into practical applications of machine learning in leadership, anticipates future developments, and delves into ethical considerations.

2 The Role of Machine Learning in Enhancing Leadership Effectiveness

Machine learning is a powerful catalyst for increasing leadership effectiveness in the contemporary world when data is everywhere and corporate settings are becoming ever more complex [9]. The ability of machine learning to alter decision-making procedures plays a crucial role in leadership. Leaders may now make judgments based on data-driven insights, enhancing their ability to make decisions beyond simple intuition or prior knowledge. When CEOs have access to machine learning algorithms, they have a powerful tool for analyzing market trends, comprehending consumer behavior, and streamlining internal processes [10].

Additionally, machine learning gives leadership access to the strength of predictive analytics [11]. Leaders may use predictive models to foresee future trends and outcomes, a skill that is especially useful in industries like banking since it enables proactive risk management and opportunity exploitation. Machine learning also improves efficient resource allocation, which is a sign of strong leadership [12]. These algorithms carefully examine past data to spot trends, assisting decision-makers in taking actions that optimize effectiveness and return on investment.

Further, machine learning encourages customization in leadership, allowing leaders to adapt their style to the distinct requirements and preferences of stakeholders and team members [13]. This individualized leadership style fosters participation and cooperation. Additionally, machine learning encourages ongoing learning and adaptation, reflecting the flexibility needed in businesses with a high rate of change [14]. Leaders who adopt machine learning improve their agility and strategic thinking as well as access to strong data-driven insights, positioning themselves to succeed in an increasingly competitive and data-centric corporate environment.

3 Real-World Examples of Machine Learning Applications in Leadership

Machine learning is quickly changing how leaders operate in a variety of sectors, providing practical solutions that improve decision-making, resource management, and flexibility. These data-driven technologies are being used by leaders more often to help them better manage challenging corporate environments. The revolutionary effect of machine learning in leadership positions and decision-making processes is clearly shown through real-world instances.

Machine learning algorithms enable human resource professionals to rapidly filter through resumes and forecast candidates' potential for certain tasks in the area of talent management [15]. These predictive algorithms assist in creating high-performing teams by locating people who have the necessary skills and cultural fit. Leaders in marketing and product development also use machine learning to have a better understanding of their customers. Leaders may efficiently customize marketing tactics and product offers by analyzing consumer data, including buying patterns and online interactions [16].

To analyze massive datasets, forecast market trends, and optimize portfolios, financial professionals depend on machine learning [17]. This technology helps them make better investment choices and forecast market trends. Machine learning helps healthcare executives by better patient care through enhancing diagnosis, treatment plans, and resource allocation [18, 19]. Manufacturing and logistics processes are streamlined by machine learning, which also gives managers data-driven insights into supply chain management and production optimization [20, 21]. Additionally, machine learning's contributions to cybersecurity, customer service, education, and other areas assist leaders across a range of industries, making it a revolutionary force in contemporary leadership practices [22]. Machine learning is reshaping leadership, offering efficient techniques for success in data-driven business environments. Table 1 showcases applications and benefits across various industries.

4 Ethical Issues with Machine Learning in Leadership

Applying machine learning in leadership roles raises ethical concerns, such as bias in training data, transparency issues, and privacy risks. Leaders must plan for staff reskilling due to automation, address accountability and discrimination issues, protect data security, and assess long-term societal impacts while striking a balance between machine learning and ethical standards.

Leaders need to establish governance frameworks and ethical guidelines for machine learning to address ethical issues and legal requirements. Transparency in AI usage for leadership decisions is crucial to maintain trust. Data security, privacy, and bias mitigation should be top priorities. Leaders should also support reskilling and transition programs to minimize job displacement concerns and ensure widespread benefits. A strategic approach and ethical framework can help effectively integrate machine learning into leadership practices, overcoming technical challenges and ethical dilemmas, as shown in Fig. 1.

Table 1. Leadership applications of machine learning

Industry/Domain	Machine Learning Application	Key Benefits
Talent Acquisition and HR	– Resume Screening	Efficient candidate shortlisting, reduced bias
	– Predictive Hiring	Informed hiring decisions based on data
	– Employee Retention	Proactive retention strategies
Marketing and Customer Insights	– Customer Segmentation	Targeted marketing campaigns
	– Recommendation Systems	Enhanced user engagement, increased revenue
	– Churn Prediction	Reduced customer churn
Finance and Investment	– Algorithmic Trading	Faster and data-driven trading decisions
	– Credit Risk Assessment	Lower default risk
	– Fraud Detection	Improved security, reduced financial losses
Healthcare Leadership	– Disease Diagnosis	Enhanced patient care
	– Drug Discovery	Accelerated drug development
	– Resource Allocation	Efficient resource utilization
Manufacturing and Supply Chain	– Demand Forecasting	Optimized inventory management
Management	– Quality Control	Improved product quality
	– Supply Chain Optimization	Enhanced efficiency
Cybersecurity Leadership	– Anomaly Detection	Real-time threat detection
	– User Authentication	Enhanced security

5 Trends and Transformation

Machine learning is revolutionizing leadership by providing data-driven insights, promoting flexibility in decision-making, and allowing for dynamic adaptation in a rapidly changing world. Human-AI cooperation is reshaping leadership dynamics, enhancing leadership effectiveness through tailored development and predictive analytics. Machine learning also improves team interactions, supports diversity and inclusion efforts, and contributes to sustainability, risk reduction, and crisis management projects.

This change inspires executives to embrace a philosophy of continuous learning and use AI for innovation. Data-driven, flexible, and morally responsible leadership is made possible by the incorporation of machine learning. It enables leaders to make

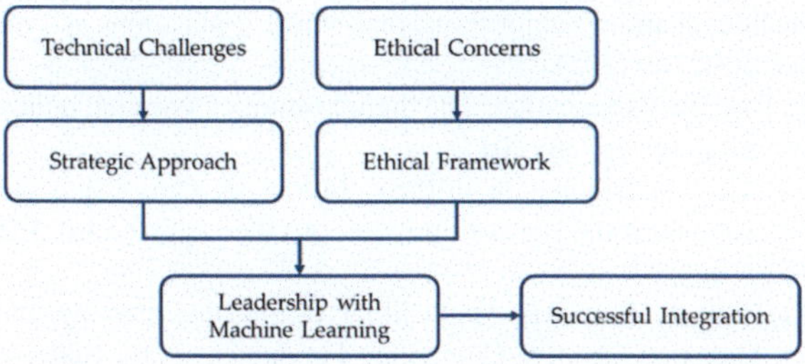

Fig. 1. Integration: challenges and ethical navigation

unbiased judgments, approach problems in a proactive manner, and develop their unique leadership philosophies. Leaders concentrate on strategic thinking and mentoring while AI tackles everyday duties. AI helps risk management and employee engagement, ethical concerns and transparency become essential, and strategy planning benefits from AI-driven insights.

6 Conclusion

This analysis explores the intersection of leadership and machine learning in today's fast-paced technological landscape. It highlights the potential of machine learning to enhance leadership performance and emphasizes the need for careful analysis. Real-world examples demonstrate the competitive advantage of data-driven decision-making enabled by machine learning. The future is expected to bring enhanced leadership, AI-driven decision-making, and evolving leadership philosophies. Overall, the review underscores the promise of intelligence-driven leadership and encourages leaders to leverage machine learning for better outcomes in today's complex environment.

References

1. Sarirete, A., et al.: Artificial intelligence and machine learning research: towards digital transformation at a global scale, pp. 1–3. Springer
2. Teng, Y., Zhang, J., Sun, T.: Data-driven decision-making model based on artificial intelligence in higher education system of colleges and universities. Expert. Syst. **40**(4), e12820 (2023)
3. Barga, R., et al.: Predictive Analytics with Microsoft Azure Machine Learning. Springer (2015)
4. Dean, J.: Big Data, Data Mining, and Machine Learning: Value Creation for Business Leaders and Practitioners. Wiley (2014)
5. Dogan, A., Birant, D.: Machine learning and data mining in manufacturing. Expert Syst. Appl. **166**, 114060 (2021)
6. Vallarino, D.: Buy when? Survival machine learning model comparison for purchase timing (2023). arXiv:2308.14343
7. Audibert, R.B., et al.: In: On the Evolution of AI and Machine Learning: Towards Measuring and Understanding Impact, Influence, and Leadership at Premier AI Conferences (2022). arXiv:2205.13131

8. Sarker, I.H.: Machine learning: algorithms, real-world applications and research directions. SN Comput. Sci. **2**(3), 160 (2021)
9. Williamson, B.: Big data in education: The digital future of learning, policy and practice. In: Big Data in Education, pp. 1–256 (2017)
10. Zhou, M., et al.: Consumer behavior in the online classroom: Using video analytics and machine learning to understand the consumption of video courseware. J. Mark. Res. **58**(6), 1079–1100 (2021)
11. Kashyap, P., Kashyap, P.: Industrial applications of machine learning. In: Machine Learning for Decision Makers: Cognitive Computing Fundamentals for Better Decision Making, pp. 189–233 (2017)
12. Lin, S., et al.: Exploring the relationship between abusive management, self-efficacy and organizational performance in the context of human–machine interaction technology and artificial intelligence with the effect of ergonomics. Sustainability **14**(4), 1949 (2022)
13. Villalba-Diez, J., et al.: Characterization of industry 4.0 lean management problem-solving behavioral patterns using EEG sensors and deep learning. Sensors **19**(13), 2841 (2019)
14. Madni, A.M., Madni, C.C., Lucero, S.D.: Leveraging digital twin technology in model-based systems engineering. Systems **7**(1), 7 (2019)
15. Kulkarni, S.B., Che, X.: Intelligent software tools for recruiting. J. Int. Technol. Inf. Manage. **28**(2), 2–16 (2019)
16. Chen, I.J., Popovich, K.: Understanding customer relationship management (CRM): People, process and technology. Bus. Process. Manag. J. **9**(5), 672–688 (2003)
17. Goodell, J.W., et al.: Artificial intelligence and machine learning in finance: Identifying foundations, themes, and research clusters from bibliometric analysis. J. Behav. Exp. Financ. **32**, 100577 (2021)
18. Dash, S., et al.: Big data in healthcare: management, analysis and future prospects. J. Big Data **6**(1), 1–25 (2019)
19. Farhaoui, Y. et al., Big Data Mining Anal. **6**(3), I–II (2023). https://doi.org/10.26599/BDMA.2022.9020045
20. Johnson, M., et al.: Impact of big data and artificial intelligence on industry: developing a workforce roadmap for a data driven economy. Glob. J. Flex. Syst. Manag. **22**(3), 197–217 (2021)
21. Farhaoui, Y.: Big data analytics applied for control systems. Lecture Notes Netw. Syst. **25**, 408–415 (2018). https://doi.org/10.1007/978-3-319-69137-4_36
22. Awosola, I.: What effect does technology have on a responsive leadership style? In: Exploring Ethical Problems in Today's Technological World, pp. 23–44. IGI Global (2022)

TempoJCM++: An Extension of TempoJCM to Support Schema Change Modeling

Zouhaier Brahmia[1(✉)], Safa Brahmia[1], Fabio Grandi[2], and Rafik Bouaziz[1]

[1] Faculty of Economics & Management, MIRACL Laboratory, University of Sfax, Sfax, Tunisia
zouhaier.brahmia@fsegs.rnu.tn, rafik.bouaziz@usf.tn

[2] Department of Computer Science and Engineering, University of Bologna, Bologna, Italy
fabio.grandi@unibo.it

Abstract. TempoJCM (Temporal JSON Conceptual Model) is a graphical model for conceptual modeling of temporal JSON data. Like the other conceptual data models, TempoJCM allows only to model the current structure of the real world, i.e., the current entities of this world, with their current properties and their current relationships, and the current temporal format of each time-varying model component. It does not support modeling future schema changes, e.g., a property that will be added to an entity after one year, or a temporal format, associated to an entity, which will be changed after some months. Since sometimes the designer, when defining a TempoJCM model, could have information on some schema changes that should be applied at a precise time in the future, it is interesting to also model such future schema changes. In this paper, we extend TempoJCM to support future schema change modeling. The extension consists in allowing the designer to specify, on the conceptual model, (i) the valid-time interval for each model component that will be involved by a future schema change, (ii) a future change to the temporal format of a component, (iii) a future renaming of some component, and (iv) a future replace of a model component with a new simple component. We also provide an application example to motivate and illustrate our proposal.

Keywords: Conceptual Modeling · Graphical conceptual data model · JSON · Temporal JSON · Temporal data model · TempoJCM · Schema change · NoSQL data stores

1 Introduction

Nowadays, temporal JSON-based NoSQL databases [1–3] are being used in several applications fields (e.g., smart grid computing, IoT, cloud computing, online social networks, mobile systems). Therefore, a good attention is being devoted to their conceptual modeling, implementation, exploitation, and maintenance. In a previous work, we have proposed TempoJCM (Temporal JSON Conceptual Model) [4], a graphical conceptual data model for time-varying JSON data. It is the first model of its kind in the temporal JSON-based NoSQL world. It is based on the following JSON model components: a JSON object, a JSON object member, a JSON array, and a JSON array element. The

Y. Farhaoui et al. (Eds.): ICAISE 2023, LNNS 837, pp. 543–549, 2024.
https://doi.org/10.1007/978-3-031-48465-0_73

designer or the NoSQL Database Administrator (NSDBA) could declare any component in a TempoJCM model as of transaction-time [5] format, valid-time [6] format, or bitemporal format, in order to denote that such a component is evolving along the transaction-time dimension, the valid-time dimension, or along both, respectively. Notice that, in the temporal database literature, the transaction time denotes when some fact is current in the database, and the valid time denotes when some fact is valid in the real world.

Similar to other conceptual data models, whether temporal (e.g., TEER [7]) or non-temporal (e.g., EER), TempoJCM does not support modeling schema changes that will occur in the future, e.g., an entity that will be added after two years, a property that will be replaced with another after one year, or a temporal format, associated to a model component, which will be changed after some time period. TempoJCM allows only to model the current structure of the real world, i.e., to model the current entities of this world, with their current properties and their current relationships, and the current temporal format (valid-time, transaction-time, or bitemporal) of each time-varying model component. Since in some cases the NSDBA, while defining a TempoJCM model, could have information on some schema changes that will take place at a precise time in the future, it is necessary, or even mandatory, to also model such changes so that they will be automatically executed when their start times will be reached. With the current version of TempoJCM, the NSDBA proceeds in an *ad hoc* manner to satisfy requirements dealing with modeling of future schema changes.

For this reason, we propose in this paper TempoJCM++, an extension of TempoJCM to support the specification of future schema changes, and provide an example that motivates and illustrates our proposal. In particular, TempoJCM++ allows the NSDBA to specify the following future schema changes:

- the valid-time interval of each model component that will be involved by a future schema change. This interval defines the time during which this component is valid in the real world, which means that only "at the beginning of this interval", such a component will start being considered as belonging to the model, and just "after this interval", such a component will be considered as being dropped from the model;
- a future change to the temporal format of a model component;
- a future renaming of a model component;
- a future replace operation of a model component with a new simple component.

Notice that, to the best of our knowledge, we are the first to propose such an extension to a temporal conceptual JSON data model, in the NoSQL world. Notice also that we were inspired by an old paper of Gregersen [8] dealing with a temporal enhanced entity-relationship model that supports schema changes. Indeed, similarly to [8], TempoJCM++ allows the NSDBA to specify a valid-time interval for any desired JSON model component. However, our model allows also to specify the above mentioned future schema changes, which are not considered in [8].

The rest of the paper is organized as follows. Section 2 briefly presents our TempoJCM data model via an application example. Section 3 proposes TempoJCM++, the extension of TempoJCM to support schema change modeling, and illustrates its use on the same example. Section 4 concludes the paper.

2 Background and Motivation

TempoJCM allows NSDBAs to graphically define conceptual models for time-varying JSON data, in JSON-based NoSQL database applications. In a TempoJCM model, any JSON component (i.e., an object, an object member, an array, or an array element) could be declared as of transaction-time format (TT), as of valid-time format (VT), or as of bitemporal format (BT).

To illustrate the use of TempoJCM, let us consider the example of an application for public library management. We suppose that the NSDBA has defined a TempoJCM model, as shown by Fig. 1, in which the library subscribers are modeled as an array of objects. Each subscriber (defined as an object) has a name (string), an address (string) declared as transaction time, a category (string; e.g., schoolchild, middle schooler, student, …) declared as valid time, and a list of reading interests (an array of strings) declared as valid time.

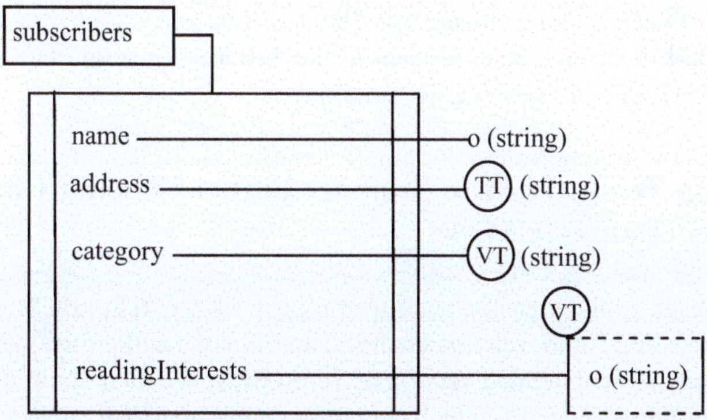

Fig. 1. A TempoJCM model for library subscriber management

However, if the NSDBA wants to specify, on the same model of Fig. 1, some schema changes that will happen in the future, he/she could not do it, as TempoJCM, in its current version, does not provide any support for modeling future schema changes. For example, let us suppose that, when preparing the TempoJCM model of Fig. 1 (on January 10, 2023), the NSDBA has all necessary information on the four following future schema changes:

- *SC1*: to replace the subscriber address with an email (string) of valid-time format, effective from 2025-01-01, since, on that date, the public library will launch its web solution and therefore the email will be mandatory for each subscriber;
- *SC2*: to change the temporal format of the subscriber category as follows: it will be of valid-time format from the beginning until 2024-12-31, and of bitemporal format from 2025-01-01 until a new change occurs;
- *SC3*: to define the subscriber loyalty card (an object characterized by a card identifier (string), and an accumulated loyalty points (integer) of transaction-time format), effective from 2025-01-01, since the library director has decided that from such

a date the subscribers deserve more encouragement (in particular schoolchild and middle schoolers).

- *SC4*: to rename the "readingInterests" member to "mediaInterests", effective from 2026–01-01, since the public library will be transformed into a public media library from that date (so, the library will provide not only paper-based documents but also digital documents which are saved on modern media like CDs, and DVDs).

Unfortunately, the NSDBA could not integrate these four schema changes into the model of Fig. 1 when constructing such a model, since they deal with model changes that will occur in the future and TempoJCM does not support modeling such changes. Therefore, he/she must wait until the valid-time start date of each schema change is reached in order to manually apply it (in our example, he/she must wait until 2025-01-01 to be able to apply SC2, SC2, and SC3, and until 2026-01-01 to apply SC4).

It is clear that, based on what precedes, it will be very helpful for the NSDBA to be able to specify, from the beginning, the above four schema changes on the initial conceptual temporal data model. After that, when the current time reaches the valid-time start date of each schema change, this latter will be executed by the system either automatically and in a transparent manner to the NSDBA, or semi-automatically after notifying the NSDBA and receiving his/her approval.

3 Extending TempoJCM to Support Future Schema Change Modeling: TempoJCM++

This section presents the extensions that we propose for the TempoJCM model so that it will allow modeling future schema changes, and illustrates their use while resuming the above example. The extended version of TempoJCM is called TempoJCM++.

Taking inspiration from [8], which proposes a temporal EER model that supports schema changes, and based on our several previous works [9, 10] dealing with schema changes (at logical level) in an environment that supports schema versioning [11, 12], we have proposed the four following extensions to TempoJCM.

Extension 1: To specify a valid-time interval for a JSON model component. This interval is delimited by two times: (i) a valid-time start, which denotes the time from which the corresponding model component will be considered as effectively belonging to the model and therefore it could be used in both queries and update operations; we use "Beginning" as a special value for the valid-time start if the component will be used from the beginning of the application deployment; (ii) a valid-time end, which denotes the time after which the corresponding component will no longer belong to the model and consequently it could not be queried/updated; we use "UntilSC" as a special value for the valid-time end if the component will be used until some possible schema change occurs. Graphically, the valid-time interval will be linked to the corresponding component, as follows.

[valid-time start, valid-time end]
component_name

Notice that at least one of the boundaries of the time interval must be a precise temporal value. By default, we could say that every model component could have an implicit valid-time interval equal to [Beginning, UntilSC]. However, it is not allowed to explicitly specify such an interval for a component since this will not provide any significant information, and therefore will decrease the readability of the model.

Extension 2: To specify a future change to the temporal format of a model component, by allowing the NSDBA to indicate the valid-time interval of each temporal format that could be assigned to this component. Graphically, the valid-time interval will be linked to the corresponding format of the component, as follows.

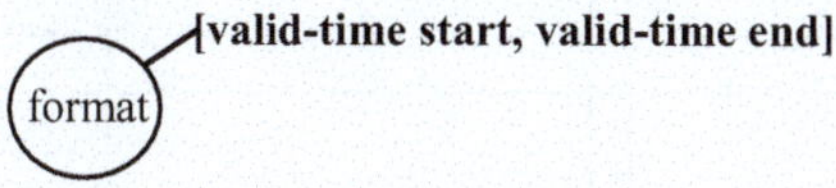

Extension 3: To specify a future renaming of a model component, by allowing the NSDBA to indicate the new name of a component and its valid-time interval. Graphically, the schema change will be specified as a 3-tuple (Ren, new_name, [valid-time start, valid-time end]) and linked to the corresponding component, as follows ("Ren" is used to denote the rename operation):

(Ren, new_name, [valid-time start, valid-time end])
component_name

Extension 4: To specify a future replace operation of a model component with a new simple model component, by allowing the NSDBA to indicate the name, the type (string, number, integer, date, boolean, or null), the valid-time interval, and possibly the temporal format of the replacing component. Graphically, the schema change will be specified as a 5-tuple (Rep, name_new_comp, type_new_comp, [valid-time start, valid-time end], format) and linked to the corresponding component, as follows ("Rep" is used to denote the replace operation):

(Rep, name_new_comp, type_new_comp, [valid-time start, valid-time end], format)
component_name

To illustrate the use of our TempoJCM++ model, let us resume our example of Sect. 2 and suppose that, when preparing the model of Fig. 1, the NSDBA has accurate information on some future schema changes. For that, based on the features proposed by the TempoJCM++ model (as described above), he/she has integrated the four schema changes (SC1, SC2, SC3, and SC4), presented in paragraph 3 of Sect. 2, into the model of Fig. 1. Hence, he/she obtains the TempoJCM++ model shown in Fig. 2.

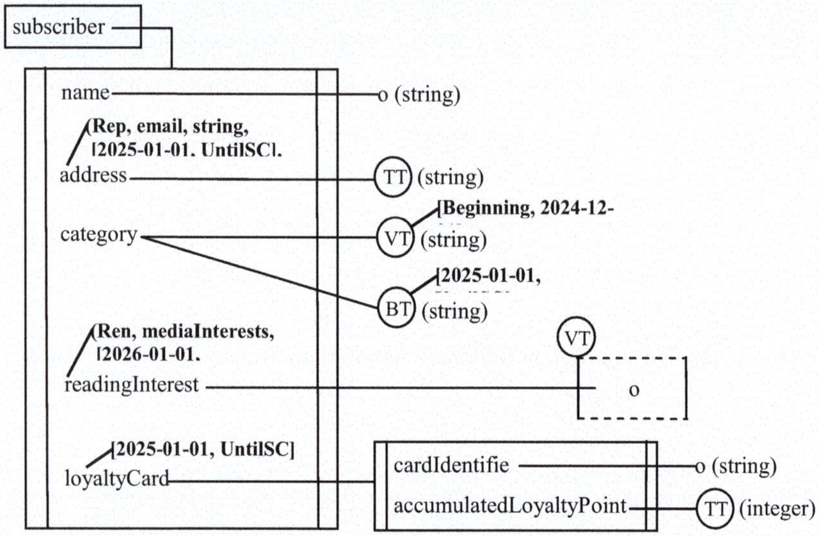

Fig. 2. A TempoJCM++ model for library subscriber management

4 Conclusion

In this paper, we have proposed TempoJCM++, an extension of TempoJCM (our graphical conceptual model for temporal JSON data) in order to make it support modeling of future schema changes. Indeed, TempoJCM++ allows the NSDBA to specify, on the conceptual model and in a user-friendly manner, the following future schema changes: (i) the valid-time interval of any JSON model component which will be concerned by a future schema change, (ii) a future change of the temporal format of a model component, (iii) a future renaming of a model component, and (iv) a future replace of a model component with a new simple component. Notice that each conceptually specified schema change will be physically executed by the system when the current time reaches the beginning of its valid-time interval; the execution could be either in a transparent manner to the NSDBA, or after notifying this latter and receiving his/her approval. We think that our TempoJCM++ model fills a gap in the state of the art of conceptual modeling of temporal JSON-based NoSQL databases.

In the future, we plan to develop a tool that supports TempoJCM++ and shows the feasibility of our proposal. Moreover, we intend to extend TempoJCM++ to support modeling further schema changes and in particular complex ones, like replacing a model component with a complex component (i.e., a JSON object or a JSON array), or merging several model components into one component.

References

1. Brahmia, S., Brahmia, Z., Grandi, F., Bouaziz, R.: τJSchema: a framework for managing temporal JSON-based NoSQL databases. In: Proceedings of the 27th International Conference on Database and Expert Systems Applications (DEXA'2016), Part 2, pp. 167–181 (2016)

2. Mehmood, N.Q., Culmone, R., Mostarda, L.: Modeling temporal aspects of sensor data for MongoDB NoSQL database. J. Big Data **4**(1), Article no. 8 (2017)
3. Goyal, A., Dyreson, C.: Temporal JSON. In: Proceedings of the 5th IEEE International Conference on Collaboration and Internet Computing (CIC 2019), pp. 135–144 (2019)
4. Brahmia, Z., Grandi, F., Brahmia, S., Bouaziz, R.: A Graphical conceptual model for conventional and time-varying JSON data. Procedia Comput. Sci. **184**, 823–828 (2021)
5. Jensen, C.S., Snodgrass, R.T.: Transaction time. In: Liu, L., Özsu, M.T. (eds.) Encyclopedia of Database Systems, 2nd edn., pp. 4200–4201. Springer, USA (2018)
6. Jensen, C.S., Snodgrass, R.T.: Valid time. In: Liu, L., Özsu, M.T. (eds.) Encyclopedia of Database Systems, 2nd edn., pp. 4359–4360. Springer, New York, NY, USA (2018)
7. Gregersen, H., Jensen, C.S.: Temporal entity-relationship models—A survey. IEEE Trans. Knowl. Data Eng. **11**(3), 464–497 (1999)
8. Gregersen, H.: TimeERplus: a temporal EER model supporting schema changes. In: Proceedings of the 22nd British National Conference on Databases (BNCOD 2005), pp. 41–59 (2005)
9. Brahmia, Z., Grandi, F., Oliboni, B., Bouaziz, R.: Schema change operations for full support of schema versioning in the τXSchema framework. Int. J. Inf. Technol. Web Eng. (IJITWE) **9**(2), 20–46 (2014)
10. Brahmia, Z., Brahmia, S., Grandi, F., Bouaziz, R.: Versioning schemas of JSON-based conventional and temporal big data through high-level operations in the τJSchema framework. Int. J. Cloud Comput. **10**(5/6), 442–479 (2021)
11. Brahmia, Z., Grandi, F., Oliboni, B., Bouaziz, R.: Schema versioning in conventional and emerging databases. In: Khosrow-Pour, M. (ed.) Encyclopedia of information science and technology, 4th edn., pp. 2054–2063. IGI Global, Hershey, PA, USA
12. Roddick, J.F.: Schema versioning. In: Liu, L., Özsu, M.T. (eds.) Encyclopedia of database systems, 2nd edn., pp. 3311–3314. Springer, New York, NY, USA (2018)

Analyzing the Shortest Path in a 3D Object Using a Reinforcement Learning Approach

Hakimi Fatima-Zahra(✉) and Yousef Farhaoui

STI Laboratory, T-IDMS Faculty of Sciences and Techniques, Moulay Ismail University of Meknes, Meknes, Morocco
hakimifatimazahra0@gmail.com

Abstract. Finding the shortest path in a 3D mesh is a fundamental problem in various fields, from robotics to fluid simulation. In this article, we compare three popular approaches to solving this problem: the Dijkstra algorithm, the A* algorithm, and reinforcement learning methods (Q-learning and SARSA). We also provide a detailed analysis of the results obtained for each approach.

Keywords: Reinforcement learning · Shortest path · Q-learning · SARSA · Dijkstra · A*

1 Introduction

Finding the shortest path in a 3D mesh involves finding the optimal path between two points in a three-dimensional environment while avoiding obstacles. This problem has important applications in trajectory planning for robots, fluid simulation, autonomous navigation, and many other domains.

In this article, we examine three different approaches to solve this problem:

- Dijkstra: An exhaustive search algorithm that explores all possible paths from the starting point until the destination is reached. We implemented this algorithm in a 3D mesh using a graph structure to represent the mesh and compute the optimal path.
- A*: The A* algorithm improves the efficiency of Dijkstra by using a heuristic function to guide the search towards the solution more efficiently than Dijkstra while ensuring optimality.
- Reinforcement Learning: We set up a reinforcement learning environment to train an agent to learn the optimal policy for pathfinding. We used both Q-learning and SARSA algorithms to train the agent by providing positive rewards when a shorter path is found and negative rewards otherwise.

2 Research Outline

We want to find the shortest distance and path between two faces on a polygon mesh. Polygon mesh is the most basic representation to represent objects in computer graphics. In 3D games, characters and various 3D objects are represented by various polygon

Y. Farhaoui et al. (Eds.): ICAISE 2023, LNNS 837, pp. 550–561, 2024.
https://doi.org/10.1007/978-3-031-48465-0_74

meshes [1–4]. A polygon mesh consists of a number of vertices. An edge connects two vertices and a face is composed of three or more edges. Figures 1 is polygon mesh with 1500 faces and 752 vertices respectively. With a greater number of vertices and faces, we can represent the original shape more accurately but operations or calculations on polygon mesh become more expensive.

Fig. 1. Mesh model with 1500 faces.

For example, in the case of Fig. 2, there are two vertices on the surface of a polygon mesh body. We want to find shortest path on surface of polygon mesh between two points. This paper discusses how to find the shortest distance and shortest path between two vertices S and D on a polygon mesh by using the Dijkstra and A* algorithms[5–7].

2.1 Methodology

This code snippet presents an implementation of Dijkstra's algorithm for finding the shortest path in a graph or mesh, with functions for initialization, traversal, and path reconstruction. It efficiently computes the shortest path between a specified source vertex and destination vertex while maintaining a priority queue to prioritize nodes with lower distances. This versatile algorithm can be applied to various scenarios where finding the optimal route is essential [8–10].

The adaptation of the Dijkstra algorithm to a 3D environment relies on the addition of the vertical dimension (z) in the distance calculations and the representation of nodes

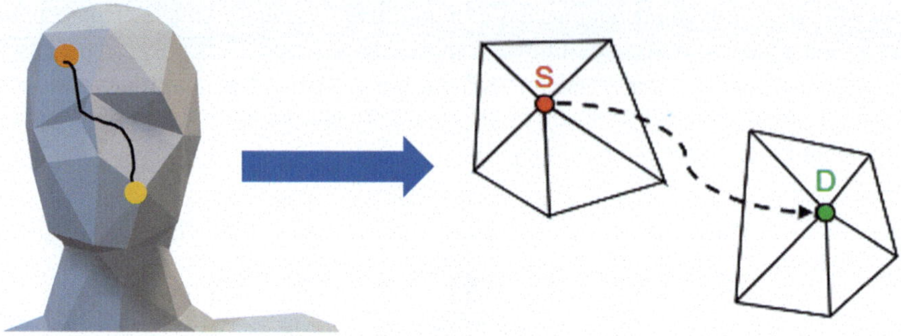

Fig. 2. Two points on polygon mesh.

as three-dimensional coordinates [11].

$$D = \sqrt{(x_2 - x_1)^2 + (y_2 - y_1)^2 + (z_2 - z_1)^2} \tag{1}$$

The A * algorithm is a general graph search algorithm that finds the shortest path consisting of nodes and edges [11]. In our problem, the graph is defined by each face as a node and the edge on the mesh (the line connecting the two faces) as the edge. This expresses the shortest path length (price) through a node n in the A * algorithm.

$$F(n) = G(n) + H(n). \tag{2}$$

G is the length of the calculated shortest path from the starting node to the n, H is the estimate of the shortest distance from the n to the target, F is the sum of G and H.

The A* algorithm commences its journey from the source node and maintains a candidate set, which comprises nodes awaiting selection [12–15]. Initially, the source node is included in the candidate set. During each iteration, a node from the open set with the lowest F value is chosen and moved to the visited set. Subsequently, the neighbors of the selected node are incorporated into the candidate set.

This iterative process persists until the destination node is eventually chosen from the candidate set. The overview of our algorithm is as shown in Fig. 4.[12].

2.2 The Difference Between Dijkstra and A*

Figure 5 shows a simplified representation of the A* algorithm compared with the Dijkstra algorithm. The Dijkstra algorithm proceeds all the way from the starting vertex when searching for the shortest distance, but A * is a more efficient algorithm because it looks toward the target vertex like the black arrows in Figs. 5. [12].

Upon implementing both algorithms within our environment (mesh), we achieved the following outcome.

During our observation, we observed a slight difference in the number of facets and a change in the facets along our trajectory.

This difference is notably evident in terms of the distance traveled and execution time.

```
function Dijkstra(mesh, source):
  Initialize all vertex distances to infinity, except for the source, which has a distance of 0.
  Create an empty priority queue Q.
  Add the source vertex to Q with a priority of 0.
  while Q is not empty:
    Extract the vertex u with the smallest distance from Q.
    Mark u as visited.
    for each neighbor v of u:
      Calculate the distance from the source to v through u.
      If the calculated distance is smaller than the current distance of v:
        Update the distance of v with the calculated distance.
        Set the predecessor vertex of v to u.
        Add v to Q with a priority equal to its distance.
  Return the arrays of distances and predecessor vertices.

function ShortestPath(mesh, source, destination):
  distances, predecessors = Dijkstra(mesh, source)
  path = []
  current = destination
  while current is not None:
    Add current to the beginning of the path.
    Set current as the predecessor vertex of current.
  Return the path
```

Fig. 3. Dijkstra's algorithm pseudocode

In conclusion, the difference between the Dijkstra and A* algorithms are clearly visible in the figure, primarily due to the use of the heuristic function in the A* algorithm. This distinction can result in variations in the obtained paths in terms of distance and execution time.

3 Reinforcement Learning

Reinforcement learning is a machine learning technique in which an agent learns to make decisions by interacting with an environment and receiving rewards or penalties based on its actions [1]. This approach enables the agent to acquire decision-making skills through experience, as it takes actions and receives rewards or penalties that are

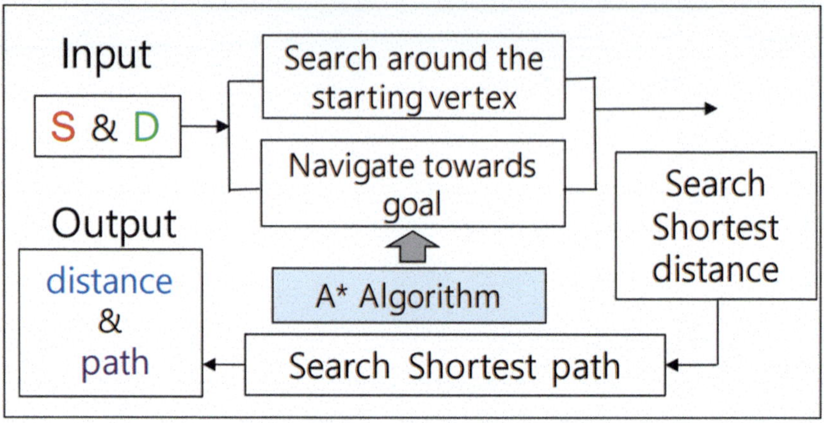

Fig. 4. Research composition diagram

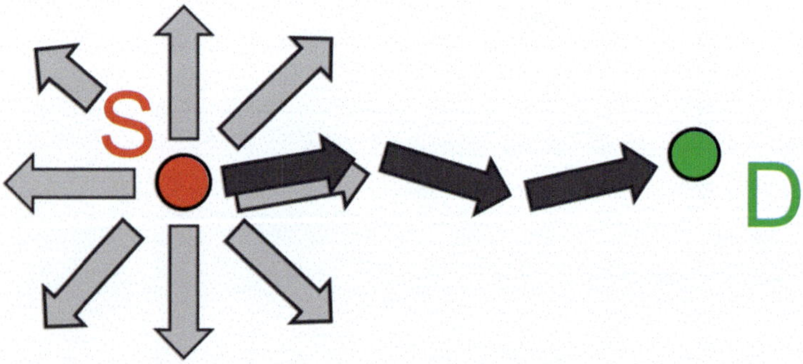

Fig. 5. Comparison of A* and Dijkstra algorithms.

inherently tied to its choices. Reinforcement learning stands out for its ability to solve complex problems through an iterative approach, wherein the agent gradually adjusts to optimize its performance throughout the learning process.

3.1 The Principle of Reinforcement Learning in Our Project

- State space: represents the current state of the agent in the environment. The agent's position is defined by the facets.
- Action space: represents the actions that the agent can take. Actions consist of movements towards accessible neighboring facets.
- The agent must receive rewards or penalties for its actions in order to learn how to maximize its performance. In our case, rewards are based on the distance between facets.

$$Reward = -d,\ d = ||cc - cn||. \tag{3}$$

cc the current state

Table 1. Comparison of A* and Dijkstra algorithms in our mesh.

Dijkstra algorithm	A* algorithm
le plus court chemin entre facette de depart: [356, 316, 360] et facette d arrivee [5, 6, 514] est	le plus court chenin entre tacette de depart: [356, 316, 360] et facette d arrivee [5,6, 514] est: [
[[493 491 492]	[497 502 499]
[497 502 499]	[502 501 499]
[504 502 505]	[601 109 110]
[601 109 110]	[142 611 607]
[608 607 611]	[485 491 482]
[485 491 4S2]	[636 474 477]
[636 474 477]	[7 612 139]
[7 612 139]	[361 337 103]
[628 159 566]	[403 375 384]
[403 375 384]	[566 168 628]
[566 168 628]	[393 380 395]
[393 389 395]	[394 704 241]
[705 704 394]	[391 403 392]
[391 403 392]	[391 375 403]
[391 375 403]	[334 331 335]
[334 331 335]	[366 336 337]
[366 336 337]	[343 339 338]
[343 339 338]	[349 370 369]
[369 348 349]	[349 371 370]
[351 353 363]	[371 349 350]
[52 547 49]	[55 46 1]
[48 538 546]	[64 549 51]
[538 547 546]	[51 2 64]
[46 55 554]	[548 549 557]]
[548 549 557]]	le nodibre des facettes entre le depart et 1 arrivee: 24
le nombre des facettes entre le depart et 1 arrivee; 25	

cn the next state.

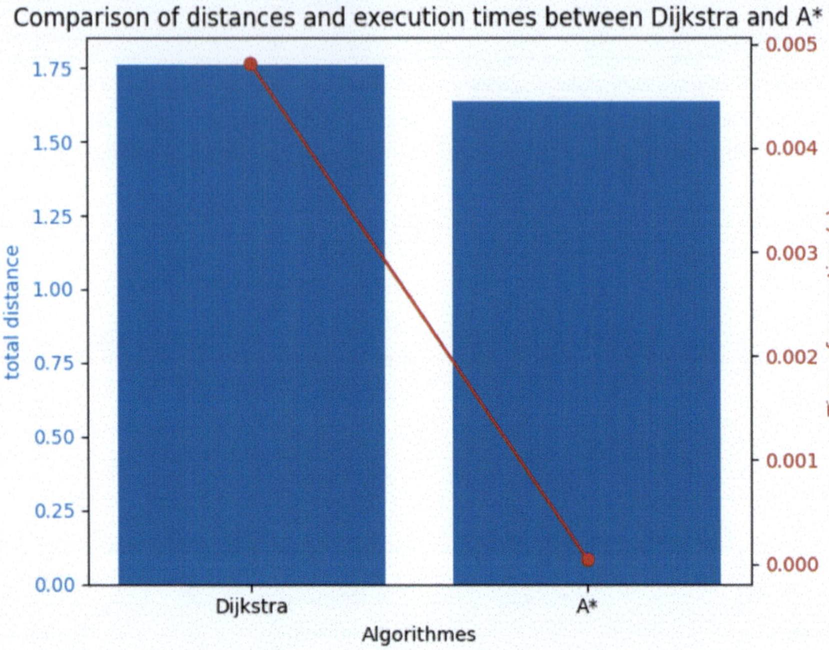

Fig. 6. Comparison of distances and execution times between Dijkstra and A*.

4 Learning Algorithms

4.1 Q-learning

Q-learning is a popular reinforcement learning algorithm used to solve Markov Decision Processes (MDP). It is a model-free approach, which means it does not require prior knowledge or understanding of the environment's dynamics [16].

Here is the pseudo-code for Q-learning:

We applied this pseudo-code to our mesh, resulting in the following outcomes.

Hence, this table contains reward values for each pair (state, action). Some values equal to 0.000000 indicate a zero reward, while values equal to 0.0 signify that no change is associated with that specific state-action pair.

By utilizing this table, we can determine the shortest path by leveraging the values it contains. The result is presented in the form of a list that includes the facets comprising our trajectory.

4.2 SARSA

The SARSA (State-Action-Reward-State-Action) algorithm is a reinforcement learning algorithm used to solve sequential decision-making problems in an environment based on an agent's actions. It is an On-Policy algorithm, meaning it uses the policy currently being learned to update its internal learned values [17].

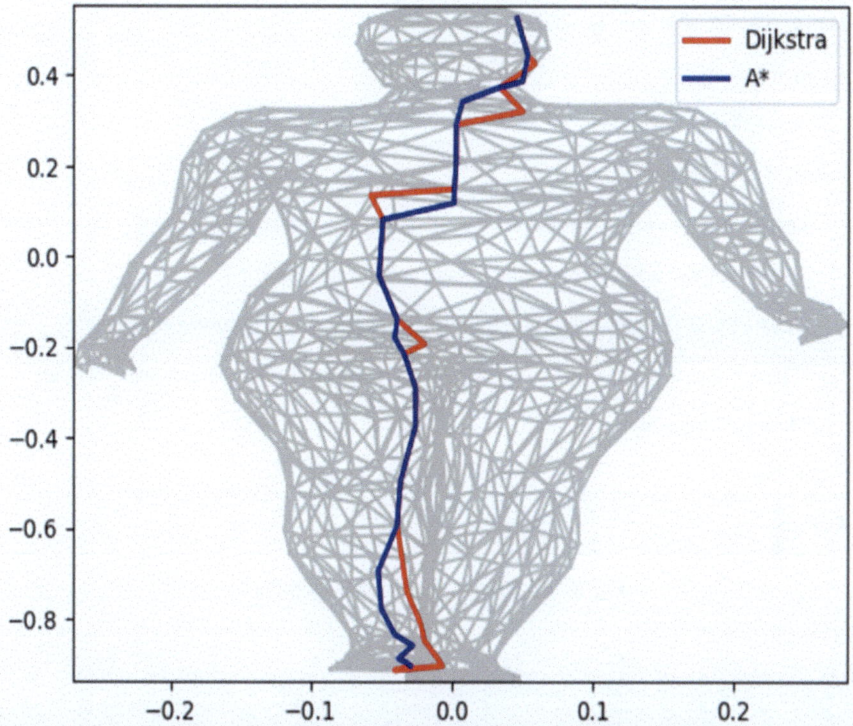

Fig. 7. The shortest path of the Dijkstra and A* algorithms.

The main difference between SARSA and Q-learning lies in how the utility values (Q-values) are updated during the learning process.

The SARSA update formula is as follows:

$$Q[s, a] = (1 - \alpha)Q[s, a] + \alpha(r + \gamma Q[s', a']).$$

s′ is the new state, s is the previous state.

a is the chosen action.

r is the reward received by the agent.

α is a number between 0 and 1, called the learning rate.

γ is the discount factor.

However, in Q-learning, the updates to Q-values are made using the current state-action pair (S, A) and the best possible action in the next state (S′). This means that Q-learning is based on estimating the maximum utility value of the next action, regardless of the agent's current policy.

The Q-learning update formula is as follows:

$$Q[s, a] = (1 - \alpha)Q[s, a] + \alpha(r + \gamma max\, Q[s', a']). \tag{5}$$

```
Initialize Q-table with zeros for all state-action pairs
For each episode:
  Initialize the starting state
  Repeat until the episode is finished:
    Choose an action using an exploration strategy (e.g., epsilon-greedy)
    Take the chosen action, observe the reward, and the new state
    Update the Q-value for the current state-action pair using the Q-learning equation:
    Q (s, a) = Q (s, a) + alpha * [reward + gamma * max (Q (new_state, all_actions)) – Q (s, a)]
    Move to the new state
  End Repeat
End For
```

Fig. 8. The pseudo-code for Q-learning.

```
             0         1         2         3         4         5         6  \
0    -0.127764 -0.132363 -0.050865 -0.131719 -0.128047 -0.132363 -0.131061
1     0.000000  0.000000  0.000000  0.000000  0.000000  0.000000  0.000000
2    -0.125840 -0.129998 -0.054102 -0.129817 -0.126154 -0.129998 -0.129680
3    -0.129711 -0.133807 -0.056883 -0.133595 -0.129952 -0.133807 -0.133579
4    -0.125849 -0.130028 -0.051900 -0.129671 -0.126033 -0.130028 -0.132987
...        ...       ...       ...       ...       ...       ...       ...
1495  0.000000  0.000000  0.000000  0.000000  0.000000  0.000000  0.000000
1496  0.000000  0.000000  0.000000  0.000000  0.000000  0.000000  0.000000
1497  0.000000  0.000000  0.000000  0.000000  0.000000  0.000000  0.000000
1498  0.000000  0.000000  0.000000  0.000000  0.000000  0.000000  0.000000
1499  0.000000  0.000000  0.000000  0.000000  0.000000  0.000000  0.000000

             7         8         9  ...  2240      2241  2242  2243  2244  \
0    -0.063997 -0.128047 -0.131000  ...   0.0  0.000000   0.0   0.0   0.0
1     0.000000  0.000000  0.000000  ...   0.0  0.000000   0.0   0.0   0.0
2    -0.067899 -0.126154 -0.128753  ...   0.0  0.000000   0.0   0.0   0.0
3    -0.070103 -0.129952 -0.132597  ...   0.0 -0.016327   0.0   0.0   0.0
4    -0.065115 -0.126033 -0.128807  ...   0.0 -0.015578   0.0   0.0   0.0
...        ...       ...       ...  ...   ...       ...   ...   ...   ...
1495  0.000000  0.000000  0.000000  ...   0.0  0.000000   0.0   0.0   0.0
1496  0.000000  0.000000  0.000000  ...   0.0  0.000000   0.0   0.0   0.0
1497  0.000000  0.000000  0.000000  ...   0.0  0.000000   0.0   0.0   0.0
1498  0.000000  0.000000  0.000000  ...   0.0  0.000000   0.0   0.0   0.0
1499  0.000000  0.000000  0.000000  ...   0.0  0.000000   0.0   0.0   0.0

      2245  2246  2247  2248  2249
0      0.0   0.0   0.0   0.0   0.0
1      0.0   0.0   0.0   0.0   0.0
2      0.0   0.0   0.0   0.0   0.0
3      0.0   0.0   0.0   0.0   0.0
4      0.0   0.0   0.0   0.0   0.0
...    ...   ...   ...   ...   ...
1495   0.0   0.0   0.0   0.0   0.0
1496   0.0   0.0   0.0   0.0   0.0
1497   0.0   0.0   0.0   0.0   0.0
1498   0.0   0.0   0.0   0.0   0.0
1499   0.0   0.0   0.0   0.0   0.0

[1500 rows x 2250 columns]
```

Fig. 9. A table of rewards for a pair (state, action).

```
print("Le plus court chemin trouvé :", shortest_path)
```

```
245
42
179
64
274
126
120
152
93
155
104
82
151
111
122
94
148
Le plus court chemin trouvé : [0, 245, 42, 179, 64, 274, 126, 120, 152, 93, 155, 104, 82, 151, 111, 122, 94, 148]
```

Fig. 10. The shortest path using Q-learning.

4.3 Comparison Between Q-learning and SARSA in Our Mesh.

Similar to the Q-learning algorithm, we have also implemented the SARSA algorithm. The difference between these two algorithms lies in the update formula used. This distinction has an impact on the rewards obtained for each algorithm.

Table 2. Comparison of Q-learning and SARSA in terms of rewards

Q-learning algorithm	SARSA algorithm
−1.3343598603801008	−0.3243598603801008
−1.4359860380100899	−1.243598603801008
−1.7483598603801008	−1.2483598603801008
−1.3243598603801008	−1.3243598603801008
−0.6143598603801987	−1.3143598603801987
−0.6459860380100876	−1.3359860380100876
−1.0243598604801008	−1.3243598603801008
−0.0435766603801998	−1.3435766603801008
−0.3243598603801008	−1.3243598603801008
−0.3243598603456789	−1.3243598603801008

This means that Q-learning is more focused on exploiting acquired knowledge, while SARSA is more inclined towards exploring new actions based on the current policy.

5 Conclusion and Discussion

Finding the shortest path in 3D mesh environments is a complex challenge, but reinforcement learning (RL) methods have proven to be an effective approach to successfully tackle this challenge.

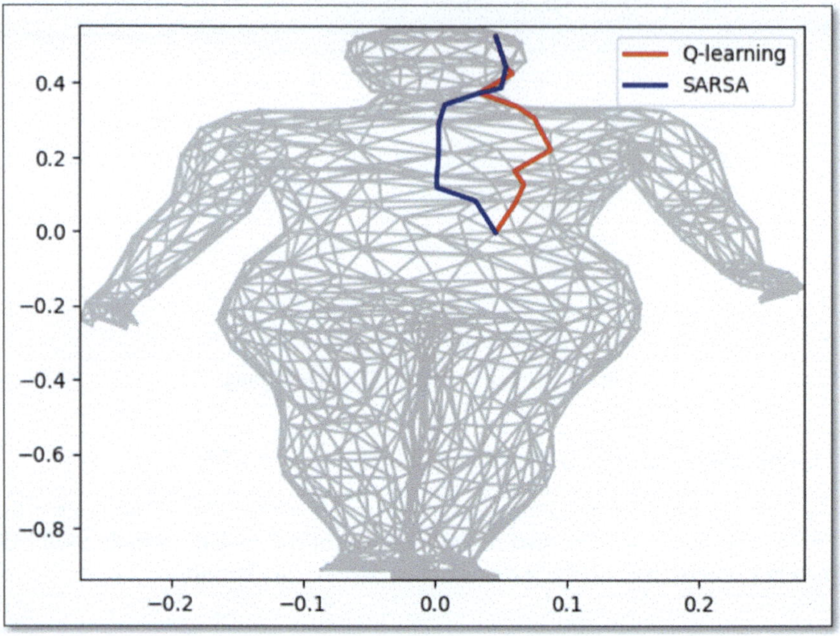

Fig. 11. The shortest path using Q-learning and SARSA algorithms.

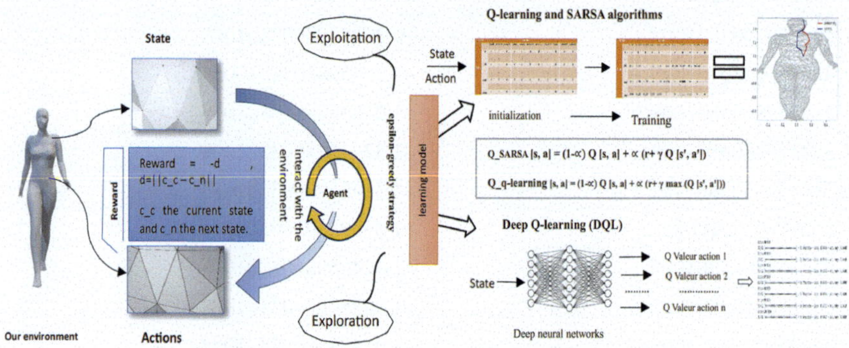

Fig. 12. The principle of reinforcement learning

With RL, our agent can autonomously learn and adjust its decision-making policy based on rewards and interactions with the environment. This approach enables us to solve complex navigation problems and opens up promising application possibilities in fields such as robotics, trajectory planning, and virtual reality.

The shortest path algorithms, whether it's Dijkstra's or A*, are used when the environment is known in advance. In contrast, reinforcement learning can solve the problem even when the environment is not known in advance, using the trial-and-error principle.

References

1. Kanai, T., Suzuki, H.: Approximate shortest path on a polyhedral surface based on selective refinement of the discrete graph and its applications
2. Farhaoui, Y.: Design and implementation of an intrusion prevention system. Int. J. Netw. Secur. **19**(5), 675–683 (2017). https://doi.org/10.6633/IJNS.201709.19(5).04
3. Farhaoui, Y. et al., Big Data Mining Anal. **6**(3), I–II (2023). https://doi.org/10.26599/BDMA.2022.9020045
4. Farhaoui, Y.: Intrusion prevention system inspired immune systems. Indonesian J. Electr. Eng. Comput. Sci. **2**(1), 168–179 (2016)
5. Farhaoui, Y.: Big data analytics applied for control systems. Lecture Notes Networks Syst. **25**, 408–415 (2018). https://doi.org/10.1007/978-3-319-69137-4_36
6. Farhaoui, Y. et al., Big Data Mining Anal. **5**(4), I–II (2022). https://doi.org/10.26599/BDMA.2022.9020004
7. Alaoui, S.S., Farhaoui, Y.: Hate speech detection using text mining and machine learning. Int. J. Decis. Support Syst. Technol. **14**(1), 80 (2022). https://doi.org/10.4018/IJDSST.286680
8. Alaoui, S.S., Farhaoui, Y.: Data openness for efficient e-governance in the age of big data. Int. J. Cloud Comput. **10**(5–6), 522–532 (2021). https://doi.org/10.1504/IJCC.2021.120391
9. El Mouatasim, A., Farhaoui, Y.: Nesterov step reduced gradient algorithm for convex programming problems. Lecture Notes Netw. Syst. **81**, 140–148 (2020). https://doi.org/10.1007/978-3-030-23672-4_11
10. Tarik, A., Farhaoui, Y.: Recommender system for orientation student. Lecture Notes Netw. Syst. **81**, 367–370 (2020). https://doi.org/10.1007/978-3-030-23672-4_27
11. Mitchell, J.S.B., Mount, D.M., Paradimitriou, C.H.: The Discrete Geodesic Problem
12. Choi, J., Zhang, B., Oh, K.: The shortest path from shortest distance on a polygon mesh
13. Sossi Alaoui, S., Farhaoui, Y.: A comparative study of the four well-known classification algorithms in data mining. Lecture Notes Netw. Syst. **25**, 362–373 (2018). https://doi.org/10.1007/978-3-319-69137-4_32
14. Farhaoui, Y.: Teaching computer sciences in morocco: an overview. IT Professional **19**(4), 12–15, 8012307 (2017). https://doi.org/10.1109/MITP.2017.3051325
15. Farhaoui, Y.: Securing a local area network by IDPS open source. Procedia Computer Science **110**, 416–421 (2017). https://doi.org/10.1016/j.procs.2017.06.106
16. Mnih, V., Kavukcuoglu, K., Silver, D.: Rusu, A.A., Veness, J., Bellemare, M.G., Graves, A., Riedmiller, M., Fidjeland, A.K., Ostrovski, G., et al.: Human-level control through deep reinforcement learning
17. Chaplot, D.S., Lample, G., Sathyendra, K.M., Salakhutdinov, R.: Transfer deep reinforcement learning in 3d environments: An empirical study. In: NIPS Deep Reinforcemente Leaning Workshop (2016)
18. Kirsanov, D., Gortler, S.J., Hoppe, H.: Fast exact and approximate geodesic paths on meshes
19. Kimmel, R., Sethian, J.A.: Computing geodesic paths on manifolds
20. Chen, J., Han, Y.: Shortest paths on a polyhedron, Part I
21. Zhu, Y., Mottaghi, R., Kolve, E., Lim, J.J., Gupta, A., Fei-Fei, L., Farhadi, A.: Target-driven visual navigation in indoor scenes using deep reinforcement learning

A New Fast Algorithm for Computation of $log_{\omega}(2)$ on Finite Fermat Fields

Mohamed Bamarouf(✉), Ahmed Asimi, and Lahdoud Mbarek

Department of Mathematics, Faculty of Sciences, Laboratory: Computer Systems & Vision Laboratory (LabSiv), Team: Security, Cryptology, Access Control and Modelling (SCCAM), University Ibnou Zohr-Agadir, Agadir, Morocco
bamaroufmohamed@gmail.com, asimiahmed@uiz.ac.ma, mbarek.lahdoud@telecom-paristech.or

Abstract. The quantity of data generated by the Internet of Things creates new challenges for the pillars of security (confidentiality, integrity and availability). These challenges are due to the diversity of characteristics of connected objects: size, computing power, storage capacity and autonomy. Based on this observation, the criticality of this diversity requires us to define new security algorithms capable of processing all objects in all circumstances. To overcome the majority of these new challenges, the use of cryptographic elliptic curves (ECC) is essential for any development in the field. The difficulty of research in this area is linked to the complexity of arithmetic in a finite field. The latter is a central axis to be optimized in the coming years to make ECCs more adapted to all the constraints of connected objects. The aim of this work is to propose a fast algorithm allowing to calculate the discrete logarithm of number two in a finite Fermat field $\mathbb{F}_p$. Therefore, this problem consists to find an integer $k \in \{1,\ldots, p-1\}$ such that $\omega^k = 2$ to gather ω is a primitive element of a Fermat field $\mathbb{F}_p$.

Keywords: Cryptography · Finite Fields · Elliptic Curves · Discrete Logarithms

1 Introduction and Notations

The rapid advancements in computer processing power and the widespread adoption of miniature connected devices have presented significant challenges for security algorithm designers. This article addresses two critical observations in this context. Firstly, the increasing need for larger cryptographic keys has resulted in higher computational requirements, longer latency, and increased energy consumption. Secondly, there is a growing demand to reduce the computational power required, particularly to accommodate the limitations of miniature connected devices.

Modern cryptosystems (RSA [1], Diffie-Helleman [2], El Gamal [3, 4], Rabin, [5] Merkle-Hellman, [6] Paillier, ECC [7, 8]) are based on mathematical problems namely: Prime number factorization [7], the discrete logarithm in a finite cyclic group [4], the problem of the sum of the subsets in a finite field [9], the discrete Logarithm in the group of points of an elliptic curve.

Y. Farhaoui et al. (Eds.): ICAISE 2023, LNNS 837, pp. 562–567, 2024.
https://doi.org/10.1007/978-3-031-48465-0_75

The different calculations established in these cryptosystems are carried out in a finite field, which reinforces our interest in optimizing arithmetic in these fields.

Finite field arithmetic plays a crucial role in elliptic curve cryptography (ECC). The performance of arithmetic operations, such as addition, subtraction, and multiplication, in finite fields has a significant impact on the efficiency and security of elliptic curve cryptography operations.

Most finite fields used in cryptography are of type $\mathbb{F}_p = (\mathbb{Z}/p\mathbb{Z}, +,.)$ with p is a prime number, and the fourth arithmetic operations used are: Modular addition, subtraction, multiplication and inversion module p. To describe the elements of $\mathbb{F}_p$ we have two writings:

(1) Additive writing: $\mathbb{F}_p = \{0, 1, 2,\ldots, p-1\}$
(2) Multiplicative writing: $\mathbb{F}_p = \left\{0, \omega, \omega^2, \ldots, \omega^{p-1}\right\}$ With ω is a primitive root of $(\mathbb{F}_p{}^*, .)$.

Each script has advantages and disadvantages. For additive (multiplicative) writing, addition and subtraction (multiplication and modular inversion) modulo p are simpler compared to the other two writings which are more complex. However, in terms of computing power and energy consumed, multiplication and modular inversion are the most expensive operations, for this we will choose multiplicative writing [10–15].

The aim of this work is to propose a fast algorithm allowing to calculate the discrete logarithm of number two in a finite Fermat field $\mathbb{F}_p$. Therefore, this problem consists to find an integer $k \in \{1,\ldots, p-1\}$ such that $\omega^k = 2$ to gather ω is a primitive element of a Fermat field $\mathbb{F}_p$.

Definition 1.1. A finite Fermat field is a prime finite field $\mathbb{F}_p$ with p is a prime Fermat number.

2 Computation of $\log_w(2)$

Propriety 2.1. In a finite Fermat field $\mathbb{F}_p$ with p is a prime number, we get: $\mathrm{ord}(2) = 2^{n+1}$.

Proof. Let $\mathbb{F}_p$ be a finite Fermat field. We get:

$$\begin{aligned} 2^{2^{n+1}} &= \left(2^{2n}\right)^2 \mod p \\ &= (-1)^2 \mod p \\ &= 1 \mod p \end{aligned}$$

Therefore θ(2) divides 2^{n+1} Thus, it exists $s \in \{1,\ldots, n+1\}$ such that $\theta(2) = 2^s$. So p divides $2^s - 1$ which is absurd with $p > 2^s$ -1 if $s < n+1$. What proves $s = n+1$, thus $\theta(2) = 2^{n+1}$.

Propriety 2.2. Let (G,.) be a cyclic group of order $N = 2^{2^n}$ with generator ω and an element $h \in \mathbf{G}$ with order $\theta(h) = 2^{n_h}$. Then there exit the only $k \in \{0,\ldots, 2^{n_h-1} - 1\}$ such that $\log_\omega(h) = (2k+1)2^{2^n - n_h}$.

Proof. Since $\theta(h) = 2^{n_h}$, then the only sub-goup of (**G**,.) of order $\theta(h) = 2^{n_h}$ is $H = \langle \omega^{2^{2^n - n_h}} \rangle$.

Therefore h ∈ H, so there exit k ∈ $\{0,\ldots, 2^{n_h-1} - 1\}$ such that $h = \omega^{(2^{2^n-n_h})(2k+1)}$. We deduce $\log_\omega(h) = (2k+1)2^{2^n-n_h}$ because $2^{2n-n_h} \leq (2k+1)\,2^{2^n-n_h} < 2^{2n}$.

Propriety 2.3. In a finite Fermat field, we get $\log_\omega(2) = 2^{2^n-n-1}t_w$ with t_w is odd strictly less than 2^{n+1}.

Proof. The proof of this theorem relies on Propriety 2.2.

Theorem 2.1. Let $\mathbb{F}_p$be a Fermat field with $p = 1 + 2^{2^n}$ a prime number, $h_k = (2\omega^{-x_k})^{2^{2^n-1-k}}$, such that $h_k = (-1)^{d_k}$ and $x_{k+1} = x_k + 2^k d_k$ for all k ∈ $\{0,\ldots, 2^n - 1\}$. We then get.

$\log_\omega(2) = 2^{2^n-n-1}\left(1 + \sum_{s=1}^{s=n} 2^s d_{2^n-n-1+s}\right)$ such that h_k and d_k satisfy the following conditions:

$$\begin{cases} x_k = d_k = 0 & \forall k \in \left\{0, ..., 2^n - n - 2\right\} \\ h_k = 1 & \forall k \in \left\{0, ..., 2^n - n - 2\right\} \\ x_{2^n-n-1} = 2^{2^n-n-1} & \\ h_{2^n-n-1} = -1 & \\ d_{2^n-n-1} = 1 & \\ d_{2^n-n-1+s} \in \{0, 1\} & \forall s \in \{1, ..., n\} \\ h_{2^n-n-1+s} \in \{-1, 1\} & \forall s \in \{1, ..., n\} \\ d_{2^n-n-1+s} = 1 & \textit{If and only if} \;\; h_{2^n-n-1+s} = -1 \\ h_{2^n-n-1+s} = 0 & \textit{If and only if} \;\; h_{2^n-n-1+s} = 1 \end{cases}$$

Proof. The proof **of** this theorem relies on propriety 2.3.

Theorem 2.2. Let $\mathbb{F}_p$ be a Fermat field with $p = 1 + 2^{2^n}$ a prime number, $h_k = (2\omega^{-x_k})^{2^{2^n-1-k}}$, such that $h_k = (-1)^{d_k}$ and $(S_s) : \left(\omega^{2^{n-s}}\right)^x = 2^{2^{n-s}}$. We then get:

For each s ∈ {1, · · ·, n}, if there exist $x \in \left\{2^{2^n-n-1}(2i+1);\; i \in \{0, ..., 2^n\}\right\}$ such that $\left(\omega^{2^{n-1}}\right)^x = 2^{2^{n-s}}$, then $d_{2^n-n-1+s} = 0$ otherwise $d_{2^n-n-1+s} = 1$.

Proof. The proof of this theorem relies on $\theta(2) = 2^{n+1}$.

Corollary 2.1. Let $\mathbb{F}_p$ be a Fermat field with $p = 1 + 2^{2^n}$ a prime number with generator ω, an element $h \in \mathbb{F}_p$ with order $\theta(h) = 2^{n_h}$ and $h_k = (2\omega^{-n_k})^{2^{2^n-1-k}}$, such that $h_k = (-1)^{d_k}$.

We then get:

For each s ∈ {1, · · ·, n}, if there exit i ∈ $\{0,\ldots, 2^n\}$ such that $\omega^{2^{2^n-s-1}(2i-1)}$ then $d_{2^n-n-1+s} = 0$ otherwise $d_{2^n-n-1+s} = 1$.

Proof. The proof of this corollary relies on $\theta(2) = 2^{n+1}$.

Algorithm 1: Computation of $\log_{\omega}(2)$ algorithm

Inputs: $p = 1 + 2^{2^n}$ prime number Fermat, ω primitive element of $\mathbb{F}_p$
Output: The unique integer $h \in \{1,\ldots, p-1\}$ such that $\omega^h = 2$
Begin
for $s \in \{1,\ldots, n\}$ do
  i: = 0
  Compute $\omega^{2^{2^n-s-1}(2i-1)}$ mod p
  Compute $2^{2^{n-s}}$ mod p
  while $\omega^{2^{2^n-s-1}(2i-1)} != 2^{2^{n-s}}$ mod p and $i \leq 2^n$ **do**
    i: = i + 1
  end while
  if $i == 2^n + 1$ then
    $d_{2^n-n-1+s} = 0$
  else
    $d_{2^n-n-1+s} = 1$
  end if
end for
Set h: $= 2^{2^n-n-1}\left(1+\sum_{s=1}^{s=n} 2^s d_{2^n-n-1+s}\right)$
Return h
End

---

3 Implementation

By using the logiciel PARIgp under windows 10, we calculate the following results.

Let $\mathbb{F}_p$ be a Fermat field with $p = 1+2^{2^n}$ a prime number and ω its primitive element.

(1) Assume n = 1, p = 5 and ω = 2. Then we get $\log_2(2) = 1 = 2^{2^n-n-1}(1+2d_{2^n-n-1+1})$, $d_{2^n-n-1+s} = 0$ and $d_{2^n-n-1+s} = 1$.
(2) Assume n = 2, p = 17 and ω = 3. Then we get for $t \in \{3, 5, 7\}$:
(2.1) $\log_3(2) = 14 = 2.7 = 2(1+2+4) = 2^{2^n-n-1}\left(1+\sum_{s=1}^{s=n} 2^s d_{2^n-n-1+s}\right)$, and $d_{2^n-n-1+1} = 1$ and $h_{2^n-n-1+1} = -1, d_{2^n-n-1+2} = 1$, $h_{2^n-n-1+2} = -1$
(2.2) $\log_{3^3}(2) = 3^{15}14\%16 = 10 = 2,5 = 2(1+4) = 2^{2^n-n-1}\left(1+\sum_{s=1}^{s=2} 2^s d_{2^n-n-1+s}\right)$ $d_{2^n-n-1+1} = 0$ *and* $h_{2^n-n-1+1} = 1$, $d_{2^n-n-1+2} = 1$ *and* $h_{2^n-n-1+2} = -1$.
(2.3) $\log_{3^5}(2) = 5^{15}14\%16 = 6 = 2.3 = 2(1+2) = 2^{2^n-n-1}\left(1+\sum_{s=1}^{s=2} 2^s d_{2^n-n-1+s}\right) d_{2^n-n-1+1} = 1$ *and* $h_{2^n-n-1+1} = -1$, $d_{2^n-n-1+2} = 0$ *and* $h_{2^n-n-1+2} = 1$.

(3) Assume **n = 3, p = 257 and** ω **= 3. Then we get** $t \in \{3, 5, 7\}$**:**

(3.1) $\log_3(2) = 48 = 2^4.3 = 2^4(1+2) = 2^{2^n-n-1}\left(1+\sum_{s=1}^{s=3} 2^s d_{2^n-n-1+s}\right)$

$$d_{2^n-n-1+1} = 1 \ \ and \ \ h_{2^n-n-1+1} = -1,$$
$$d_{2^n-n-1+2} = 0 \ \ and \ \ h_{2^n-n-1+2} = 1, \ .$$
$$d_{2^n-n-1+3} = 0 \ \ and \ \ h_{2^n-n-1+3} = 1$$

(3.2) $\log_{3^3}(2) = 3^{255} = 48\%256 = 2^4(1) = 2^{2^n-n-1}\left(1+\sum_{s=1}^{s=3} 2^s d_{2^n-n-1+s}\right)$

$$d_{2^n-n-1+1} = 0 \ \ and \ \ h_{2^n-n-1+1} = 1,$$
$$d_{2^n-n-1+2} = 0 \ \ and \ \ h_{2^n-n-1+2} = 1,$$
$$d_{2^n-n-1+3} = 0 \ \ and \ \ h_{2^n-n-1+3} = 1$$

(3.3) $\log_{3^5}(2) = 5^{255} = 48\%256 = 2^4.7 = 2^4(1+2+4) = 2^{2^n-n-1}\left(1+\sum_{s=1}^{s=3} 2^s d_{2^n-n-1+s}\right)$

$$d_{2^n-n-1+1} = 1 \ \ and \ \ h_{2^n-n-1+1} = -1,$$
$$d_{2^n-n-1+2} = 1 \ \ and \ \ h_{2^n-n-1+2} = -1,$$
$$d_{2^n-n-1+3} = 0 \ \ and \ \ h_{2^n-n-1+3} = 1$$

(3.4) $\log_{3^7}(2) = 7^{255} = 48\%256 = 2^4.5 = 2^4(1+4) = 2^{2^n-n-1}\left(1+\sum_{s=1}^{s=3} 2^s d_{2^n-n-1+s}\right)$

$$d_{2^n-n-1+1} = 0 \ and \ \ h_{2^n-n-1+1} = 1,$$
$$d_{2^n-n-1+2} = 1 \ \ and \ \ h_{2^n-n-1+2} = -1,$$
$$d_{2^n-n-1+3} = 0 \ \ and \ \ h_{2^n-n-1+3} = 1$$

(4) Assume **n = 4, p = 65537 and** ω **= 3. Then we get** $t \in \{3, 5, 7\}$**:**

(4.1) $\log_3(2) = 55296 = 2^{11}.3^3 = 2^{11}.(1+2+8+16) = 2^{2^n-n-1}\left(1+\sum_{s=1}^{s=4} 2^s d_{2^n-n-1+s}\right)$

$$d_{2^n-n-1+1} = 1 \ and \ \ h_{2^n-n-1+1} = -1,$$
$$d_{2^n-n-1+2} = 0 \ \ and \ \ h_{2^n-n-1+2} = 1,$$
$$d_{2^n-n-1+3} = 1 \ \ and \ \ h_{2^n-n-1+3} = -1,$$
$$d_{2^n-n-1+4} = 1 \ \ and \ \ h_{2^n-n-1+4} = -1$$

(4.2) $\log_{3^3}(2) = 3^{65535}.2^{11}.3^3\%65536 = 2^{11}.9 = 2^{11}.(1+8) = 2^{2^n-n-1}\left(1+\sum_{s=1}^{s=4} 2^s d_{2^n-n-1+s}\right)$

$$d_{2^n-n-1+1} = 0 \ and \ \ h_{2^n-n-1+1} = 1,$$
$$d_{2^n-n-1+2} = 0 \ \ and \ \ h_{2^n-n-1+2} = 1,$$
$$d_{2^n-n-1+3} = 1 \ \ and \ \ h_{2^n-n-1+3} = -1,$$
$$d_{2^n-n-1+4} = 0 \ \ and \ \ h_{2^n-n-1+4} = 1$$

$$\log_{3^5}(2) = 5^{65535}.2^{11}.3^3\%65536 = 2^{11}.31 = 2^{11}.(1+2+4+8+16)$$

(4.3)
$$= 2^{2^n-n-1}\left(1+\sum_{s=1}^{s=4} 2^s d_{2^n-n-1+s}\right)$$

$$d_{2^n-n-1+1} = 1 \; and \quad h_{2^n-n-1+1} = -1,$$
$$d_{2^n-n-1+2} = 1 \quad and \;\; h_{2^n-n-1+2} = -1,$$
$$d_{2^n-n-1+3} = 1 \quad and \;\; h_{2^n-n-1+3} = -1,$$
$$d_{2^n-n-1+4} = 1 \quad and \;\; h_{2^n-n-1+4} = -1$$

(4.4) $\log_{3^7}(2) = 7^{65535}.2^{11}.3^3\%65536 = 2^{11}.13 = 2^{11}.(1+4+8) = 2^{2^n-n-1}\left(1+\sum_{s=1}^{s=4} 2^s d_{2^n-n-1+s}\right)$

References

1. Diem, C.: On the discrete logarithm problem in elliptic curves. Compositio Mathematica **147**, 75–104 (2011)
2. Diffie, W., et Hellman, M.: New directions in cryptography. IEEE Trans. Inf. Theory **22**, 644–654 (1976)
3. ElGamal, T.: A public key cryptosystem and a signature scheme based on discrete logarithms. IEEE Trans. Inf. Theory **31**, 469–472 (1985)
4. Koblitz, N.: Elliptic curve cryptosystems. Math. Comput. **48**, 203–209 (1987)
5. Li J., et Wan, D.: On the subset sum problem over finite fields. Finite Fields Appl. **14**, 911–929 (2008)
6. McCurley K.S.: «The discrete logarithm problem. Chez Proceedings of Symposium in Applied Mathematics (1990)
7. Merkle R., et Hellman, M.: Hiding information and signatures in trapdoor knapsacks. IEEE Trans. Inf. Theory **24**, 525–530 (1978)
8. Miller, V.S.: Use of elliptic curves in cryptography. In: Chez Conference on the Theory and Application of Cryptographic Techniques (1985)
9. Rabin, M.O.: Digitalized signatures and public-key functions as intractable as factorization (1979)
10. Farhaoui, Y.: Securing a local area network by IDPS open source. Procedia Computer Science **110**, 416–421 (2017). https://doi.org/10.1016/j.procs.2017.06.106
11. Farhaoui, Y. et al., Big Data Mining Anal. **5**(4) (2022), I–II. https://doi.org/10.26599/BDMA.2022.9020004
12. Farhaoui, Y.: Big data analytics applied for control systems. Lecture Notes Netw. Syst. **25**, 408–415 (2018). https://doi.org/10.1007/978-3-319-69137-4_36
13. Farhaoui, Y.: Intrusion prevention system inspired immune systems. Indonesian. J. Electr. Eng. Comput. Sci. **2**(1), 168–179 (2016)
14. Farhaoui, Y., et al.: Big Data Mining Anal. **6**(3), I–II (2023). https://doi.org/10.26599/BDMA.2022.9020045
15. Farhaoui, Y.: Design and implementation of an intrusion prevention system. Int. J. Netw. Secur. **19**(5), 675–683 (2017). https://doi.org/10.6633/IJNS.201709.19(5).04
16. Paillier, P.: Public-key cryptosystems based on composite degree residuosity classes. In: Chez International Conference on the Theory and Applications of Cryptographic Techniques (1999)
17. Montgomery, P.L.: A survey of modern integer factorization algorithms. CWI Quarterly **7**, 337–366 (1994)

Author Index

Y. Farhaoui et al. (Eds.): ICAISE 2023, LNNS 837, pp. 569–571, 2024.
https://doi.org/10.1007/978-3-031-48465-0

The manufacturer's authorised representative in the EU is Springer Nature Customer Service Centre GmbH, Europaplatz 3, 69115 Heidelberg, Germany. If you have any concerns regarding our products, please contact ProductSafety@springernature.com

Printed and bound by CPI Group (UK) Ltd, Croydon, CR0 4YY
28/09/2026
02224066-0002